FOOD TOXICOLOGY

Current Advances and Future Challenges

FOOD TOXICOLOGY

Current Advances and Future Challenges

Edited by
Ashish Sachan, DVM, MVSc, PhD
Suzanne Hendrich, PhD

APPLE ACADEMIC PRESS

Apple Academic Press Inc.
3333 Mistwell Crescent
Oakville, ON L6L 0A2 Canada

Apple Academic Press Inc.
9 Spinnaker Way
Waretown, NJ 08758 USA

© 2018 by Apple Academic Press, Inc.
First issued in paperback 2021
No claim to original U.S. Government works
ISBN-13: 978-1-77463-055-6 (pbk)
ISBN-13: 978-1-77188-617-8 (hbk)

Library and Archives Canada Cataloguing in Publication

Food toxicology : current advances and future challenges / edited by Ashish Sachan, DVM, MS, PhD, Suzanne Hendrich, PhD.

Includes bibliographical references and index.
Issued in print and electronic formats.
ISBN 978-1-77188-617-8 (hardcover).--ISBN 978-1-315-16107-5 (PDF)

1. Food--Toxicology. I. Sachan, Ashish, editor II. Hendrich, Suzanne, editor

RA1258.F6615 2017   615.9'54   C2017-906617-X   C2017-906618-8

Library of Congress Cataloging-in-Publication Data

Names: Sachan, Ashish, editor. | Hendrich, Suzanne, editor.
Title: Food toxicology : current advances and future challenges / editors, Ashish Sachan, Suzanne Hendrich.
Other titles: Food toxicology (Sachan)
Description: Toronto ; New Jersey : Apple Academic Press, 2018. | Includes bibliographical references and index.
Identifiers: LCCN 2017046191 (print) | LCCN 2017047026 (ebook) | ISBN 9781315161075 (ebook) | ISBN 9781771886178 (hardcover : alk. paper)
Subjects: | MESH: Food--toxicity | Food Contamination
Classification: LCC RA601 (ebook) | LCC RA601 (print) | NLM WA 701 | DDC 363.19/2--dc23
LC record available at https://lccn.loc.gov/2017046191

Apple Academic Press also publishes its books in a variety of electronic formats. Some content that appears in print may not be available in electronic format. For information about Apple Academic Press products, visit our website at **www.appleacademicpress.com** and the CRC Press website at **www.crcpress.com**

ABOUT THE EDITORS

Ashish Sachan, DVM, MVSc, PhD

Ashish Sachan, DVM, MVSc, PhD, is a veterinarian licensed in toxicology by the College of Veterinarians of Ontario (CVO), Canada. Dr. Sachan has been involved with toxicological sciences for more than twenty years in both the university and industrial settings. Dr. Sachan's publications and books have widely covered advancements in the field of pharmacology and toxicology, including research topics related to ethnopharmacology, pesticide toxicology, and nanosensor technologies. He received his PhD in toxicology from the Department of Biochemistry, Biophysics and Molecular Biology, Iowa State University, USA. His PhD involved the development of aptasensor technologies to detect toxic chemical species of forensic significance. Dr. Sachan has been inducted into the Iowa State University chapter of the Honors Society of Agriculture–Gamma Sigma Delta. Currently he is the Director of Toxam, Inc. and also serves on the board of directors for the Society of Toxicology of Canada (STC). His current professional interests include the regulatory affairs and the scientific and business development of agricultural and veterinary products.

Suzanne Hendrich, PhD

Suzanne Hendrich, PhD, is a University Professor and the Lura M. Lovell Fellow at the Department of Food Science and Human Nutrition at Iowa State University, where she has been on the faculty since 1987. Dr. Hendrich has authored more than 150 research papers and abstracts, mainly on the bioavailability of dietary constituents, such as soybean isoflavones, which may prevent chronic diseases. Her mycotoxin research has focused mainly on fumonisin and deoxynivalenol metabolism, bioavailability, and detoxification. Her current food toxicology interests include safety of dietary ingredients such as digestion-resistant starches, interactions between gut bacteria and foodborne toxicants, and safety of dietary supplements. Dr. Hendrich compiles an annual report based on data from the American Association of Poison Control Center comparing foodborne toxicants, including dietary botanical, vitamin and mineral supplements, seafood toxins, and other food-borne toxicants for their adverse effects. She writes a "foodtox" blog (https://foodtoxicologyprof.wordpress.com/) for which she critiques research on food chemical and dietary supplement safety. Dr. Hendrich earned her BA in biology from UCLA and her PhD in nutrition, UC Berkeley.

CONTENTS

LIST OF CONTRIBUTORS

Mohamed Abdelrazek Abdelaleem
Nuclear Research Center, Atomic Energy Authority, Cairo 13759, Egypt. E-mail: Abdelrazek_MD@ yahoo.com

J. E. Aguilar-Toalá
Centro de Investigación en Alimentación y Desarrollo A. C., carretera a la Victoria Km. 0.6, Hermosillo, Sonora 83304, México

M. Carmen Rubio Armendáriz
Área de Toxicología, Universidad de La Laguna, 38071 La Laguna, Tenerife, Islas Canarias, España. E-mail: crubiotox@gmail.com

Juan Antonio Gimenez Bastida
Division of Clinical Pharmacology, Department of Pharmacology, Vanderbilt University Medical Center, Nashville, TN 37212, United States

Alison Behling
Ecolab, St. Paul, MN, United States

Luciano Zanetti Pessôa Candiotto
Group of Advanced Studies in Health Sciences, State University of West Paraná, UNIOESTE, Campus Francisco Beltrão, Paraná, Brazil

Yi Cao, PhD
Key Laboratory of Environment-Friendly Chemistry and Applications of Ministry of Education, Lab of Biochemistry, College of Chemistry, Xiangtan University, Hunan 411105, PR China. E-mail: caoyi39@ xtu.edu.cn

Arturo Hardisson de la Torre
Área de Toxicología, Universidad de La Laguna, 38071 La Laguna, Tenerife, Islas Canarias, España

Luneia Catiane de Souza
Group of Advanced Studies in Health Sciences, State University of West Paraná, UNIOESTE, Campus Francisco Beltrão, Paraná, Brazil

Ángel J. Gutiérrez Fernández
Área de Toxicología, Universidad de La Laguna, 38071 La Laguna, Tenerife, Islas Canarias, España

H.S. García, PhD
Unidad de Investigación y Desarrollo de Alimentos, Instituto Tecnológico de Veracruz, M.A. de Quevedo 2779, Col. Formando Hogar, Veracruz, Veracruz 91897, México

R. García-Varela, PhD
CIATEJ, Autopista Monterrey-Aeropuerto, Km 10. Parque PIIT. Via de Innovacion 404, Apodaca, NL 66629, México

Maeva Giraudo, PhD
Environment and Climate Change Canada, Aquatic Contaminants Research Division, 105 McGill Street, Montreal, QC, Canada H2Y 2E7. E-mail: maeva.giraudo@gmail.com. Tel.: +1 514 283 5036. E-mail: ludomailinra@gmail.com

Consuelo Revert Gironés
Área de Toxicología, Universidad de La Laguna, 38071 La Laguna, Tenerife, Islas Canarias, España

A.F. González-Córdova, PhD
Centro de Investigación en Alimentación y Desarrollo A. C., carretera a la Victoria Km. 0.6, Hermosillo, Sonora 83304, México

Suzanne Hendrich
University Professor and Lura M. Lovell Faculty Fellow, Food Science & Human Nutrition-H Science, Iowa State University, Ames, IA, USA. E-mail: shendric@iastate.edu

A. Hernández-Mendoza, PhD
Centro de Investigación en Alimentación y Desarrollo A. C., carretera a la Victoria Km. 0.6, Hermosillo, Sonora 83304, México. E-mail: ahernandez@ciad.mx

Martin Hoagland
Nalco Water, An Ecolab Company, Naperville, IL, United States

Shruti V. Kabadi, PhD
FDA/CFSAN/OFAS/DFCN, 5100 Paint Branch Pkwy, HFS 275, College Park, MD 20740, United States

José Moisés Laparra Llopis
Immunonutrition and Health Group, Valencian International University, C/Gorgos, 5-7, 46021 Valencia, Spain. E-mail: j.moises.laparra@uv.es

Alberto Mantovani
Noodles Onlus, Nutrition & food safety and wholesomeness, Rome, Italy; Istituto Superiore di Sanità, Viale Regina Elena 299, 00161 Rome, Italy

José M. Caballero Mesa
Área de Toxicología, Universidad de La Laguna, 38071 La Laguna, Tenerife, Islas Canarias, España

Yuseok Moon, PhD ATS
Laboratory of Mucosal Exposome and Biomodulation, Department of Biomedical Sciences, Pusan National University School of Medicine, Yangsan, South Korea; Immunoregulatory Therapeutics Group in Brain Busan 21 Project, Busan, South Korea. E-mail: moon@pnu.edu

April Neal-Kluever, PhD, DABT
FDA/CFSAN/OFAS/DFCN, 5100 Paint Branch Pkwy, HFS 275, College Park, MD 20740, United States. E-mail: April.kluever@fda.hhs.gov

Carolina Panis
Group of Advanced Studies in Health Sciences, State University of West Paraná, UNIOESTE, Campus Francisco Beltrão, Paraná, Brazil; Laboratory of Inflammatory Mediators, State University of West Paraná, UNIOESTE, Campus Francisco Beltrão, Paraná, Brazil. E-mail: carolpanis@sercomtel.com.br

Nathan Pechacek
Ecolab, Eagan, MN, United States. E-mail: Nathan.Pechacek@ecolab.com

Ludovic Peyre, PhD
UMR 1331 TOXALIM (Research Center in Food Toxicology), Institut National de la Recherche Agronomique (INRA), Laboratory of Xenobiotic's Cellular and Molecular Toxicology, 400 route des Chappes, BP 167, 06903 Sophia-Antipolis Cedex, France. E-mail: ludovic.peyre-teisseire@hotmail.fr

Ilaria Proietti
European Commission, Joint Research Centre (JRC), Economics of Agriculture—Sustainable Resources, Edificio Expo. C/Inca Garcilaso 3, 41092 Seville, Spain; Noodles Onlus, Nutrition & food safety and wholesomeness, Rome, Italy

S. Raisuddin
Department of Medical Elementology & Toxicology, Jamia Hamdard (Hamdard University), New Delhi 110062, India. E-mail: sraisuddin@jamiahamdard.ac.in

Ashish Sachan
Toxam Inc., Guelph, ON, Canada. E-mail: asachan@toxam.ca

Kathryn Sande
Ecolab, Eagan, MN, United States

Shikha Sharma
Department of Medical Elementology & Toxicology, Jamia Hamdard (Hamdard University), New Delhi 110062, India

B. Vallejo-cordoba, PhD
Centro de Investigación en Alimentación y Desarrollo A. C., carretera a la Victoria Km. 0.6, Hermosillo, Sonora 83304, México

Vanessa Jacob Victorino
Group of Advanced Studies in Health Sciences, State University of West Paraná, UNIOESTE, Campus Francisco Beltrão, Paraná, Brazil

Dailos González Weller
Área de Toxicología, Universidad de La Laguna, 38071 La Laguna, Tenerife, Islas Canarias, España; Servicio de Inspección Sanitaria y Laboratorio, Área de Salud de Tenerife, Servicio Canario de Salud, Rambla de Santa Cruz, 38006 Santa Cruz de Tenerife, España

Yen-Ching Wu, PhD
FDA/CFSAN/OFAS/DFCN, 5100 Paint Branch Pkwy, HFS 275, College Park, MD 20740, United States

LIST OF ABBREVIATIONS

AD	antimicrobial division
ADHD	attention deficit hyperactivity disorder
ADI	acceptable dietary intake
AECOSAN	Spanish Consumer Agency for Food Safety and Nutrition
AhRs	aryl hydrocarbon receptors
AJs	adherent junctions
AMU	atomic mass units
AOP	adverse outcome pathway
APEs	alkylphenol ethoxylates
APs	alkylphenols
ASD	autism spectrum disorder
ASR	auditory startle reflex
ATSDR	Agency for Toxic Substances and Disease Registry United States
AuNPs	gold nanoparticles
BBB	blood–brain barrier
BCB	blood–cerebral spinal fluid barrier
BHA	butylated hydroxyanisole
BHT	butylated hydroxytoluene
BMD	benchmark dose
BMP	bone-morphogenetic protein
BMR	benchmark response
BPA	bisphenol A
CCK	cholecystokinin
cEDI	cumulative estimated daily intake
CFR	Code of Federal Regulations
CFs	consumption factors
CFSAN	Center for Food Safety and Applied Nutrition
CI	confidence interval
CLA	conjugated linoleic acid
CNS	central nervous system
CS-UCNPs	core/shell upconversion nanoparticles
CTC	circulating tumor cells
DBP	di-n-butyl phthalate

DC	dietary concentration
DDD	daily dietary dose
DEHP	di-(2-ethylhexyl) phthalate
DiDP	di-isobutylpthalate
DILs	Derived Intervention Levels
DiNP	di-isononylpthalate
DL-PCB	dioxin-like polychlorinated biphenyls
DLPFC	dorsolateral prefrontal cortex
DnBP	di-*n*-butylpthalate
DNT	developmental neurotoxicity testing
DON	deoxynivalenol
DRCs	dioxin-related compounds
ECM	extracellular matrix
EDCs	endocrine-disrupting chemicals
EDI	estimated daily intake
EDs	endocrine disruptors
EECs	enteroendocrine cells
EFSA	European Food Safety Authority
EGF	epidermal growth factor
ELONA	enzyme-linked oligonucleotide assay
EMT	epithelial-to-mesenchymal transition
EPA	US Environmental Protection Agency
EPS	exopolysaccharides
ERs	estrogen receptors
EU	European Union
FAO/WHO	United Nations for Food and Agriculture/World Health Organization
FAP	Food Additive Petition
FCN	Food Contact Notification
FCSs	food contact substances
FCSS	food contact surface sanitizers
FDA	US Food and Drug Administration
FFQ	food frequency
FGF	fibroblast growth factor
FIFRA	Fungicide and Rodenticide Act
FOB	functional observation battery
FQPA	Food Quality Protection Act
FSIS	Food Safety and Inspection Service
GBS	Guillain–Barre syndrome
GIT	gastrointestinal tract

GJs	gap junctions
GL	Guideline Levels
GnRH	gonadotropin-releasing hormone
GPCRs	G-protein-coupled receptors
GR	glutathione reductase
GRAS	generally recognized as safe
HDAC	histone deacetylase
HEDP	1-hydroxyethane 1,1-diphosphonic acid
HGF	hepatocyte-growth factor
HPA	hypothalamic–pituitary–adrenal
HPG	hypothalamic–pituitary–gonadal
HRP	horseradish peroxidase
HTS	high-throughput screening
IARC	International Agency for Research on Cancer
IBD	inflammatory bowel disease
IGF	insulin-like growth factor
IQ	intelligence quotient
JAK	janus kinase
JECFA	Joint FAO/WHO Expert Committee on Food Additives
LCR	lifetime cancer risk
LOD	limit of detection
LRET	luminescence resonance energy transfer
LSPR	localized surface plasmon resonance
MAPKs	mitogen-activated protein kinases
MEHP	mono(2-ethyl-hexyl) phthalate
MEP	monoethyl phthalate
MET	mesenchymal-to-epithelial transition
MiBP	mono-isobutyl phthalate
MnBP	mono-n-butyl phthalate
MNPs	magnetic nanoparticles
MOG	Modified One Generation
MPCs	maximum permissible concentrations
MWM	Morris Water Maze
NCGC	NIH Chemical Genomics Center
NM	nanomaterial
NOAEL	no observed adverse effect level
NP	nonylphenol
NPEs	nonylphenol ethoxylates
NTP	National Toxicology Program
OBs	osteoblasts

OCPs	organochlorine pesticides
OECD	Organization for Economic Cooperation and Development
OP	octylphenol
OPP	Office of Pesticide Programs
OPPTS	US EPA's Office of Prevention, Pesticides, and Toxic Substances
OSHA	US Occupational Safety and Health Administration
PAGs	Protective Action Guides
PAHs	polycyclic aromatic hydrocarbons
PAs	protective actions
PBBs	polybrominated biphenyls
PBDEs	polybrominated diphenyl ethers
PC	phosphatidylcholine
PCBs	polychlorinated biphenyls
PCDD	polychlorinated dibenzo-p-dioxins
PCR	polymerase chain reaction
PFC	prefrontal cortex
PFCs	perfluorinated compounds
PFOA	perfluorooctanoic acid
PFOS	perfluorooctane sulfonate
PKC	protein kinase C
PMTDI	provisional maximum tolerable daily intake
PNDs	postnatal days
POD	point of departure
POPs	persistent organic pollutants
PPARs	peroxisome proliferator-activated receptors
ppb	parts per billion
PPI	prepulse inhibition
PTSs	persistent toxic substances
PTWI	provisional tolerable weekly intake
PVC	poly(vinyl chloride)
QD	quantum dots
Ras/ERK	Ras-extracellular signal-regulated kinase
RBE	relative biological efficiency
RNS	reactive nitrogen species
ROS	reactive oxygen species
RTK	receptor tyrosine kinase
RT-PCR	reverse-transcription polymerase chain reaction
RXRs	retinoid X receptors
SCFAs	short-chain fatty acids

SFB	segmented filamentous bacteria
SOD	superoxide dismutase
ssDNA	single-stranded DNA
TAS	total antioxidant status
TBT	tributyltin
TCDD	2,3,7,8-tetrachlorodibenzodioxin
TDI	tolerable daily intake
TEQs	TCDD equivalents
TGF-β	transforming growth factor beta
TH	thyroid hormone
TJs	tight junctions
TK	toxicokinetic
TMA	trimethylamine
TMAO	TMA N-oxide
TOR	Threshold of regulation
TPT	triphenyltin
TRs	thyroid receptors
TSCA	Toxic Substances Control Act
TWI	tolerable weekly intake
UCNPs	upconversion nanoparticles
URF	unit risk factor
US	unauthorized substances
USDA	US Department of Agriculture
WHO	World Health Organization
ZEN	zearalenone

PREFACE

The objective of this book is to offer academia, government, regulatory agencies, consumers, and those in the food industry knowledge of the breadth and depth of the multifaceted field of food toxicology. There are more than seven billion people and counting in the world today, and advances in food toxicology have a direct bearing on food safety issues that are of concern to all of humanity for the foreseeable future. Massive globalization, industrialization, and commercialization have affected every aspect of food production, the food supply chain, and food consumption. The challenge of sustainable and safe food for everyone needs a multidisciplinary and multi-sectorial approach from related industries and the government sector alike.

This volume, Food Toxicology: Current Advances and Future Challenges, offers important global perspectives of scientists and experts in areas related to biomarkers and nanosensors in food toxicology, toxicology of nanomaterials, chemicals in sanitation and packaging, food additives, mycotoxins, endocrine disruptors, radionuclides, toxic metals, and waste-burning residues in food. The fourteen chapters in this the book effectively cover this broad range of topics. The discussion on biomarkers in food toxicologyprovides interesting information surrounding the biomarkers of exposure, effect, carcinogenesis, and susceptibility. The chapter on the mucosal exposome delivers an interesting perspective on the effect of dietary components on the endogenous mucosal microenvironment affecting host physiology. This chapter also discusses biomarkers involving responses ofthe mucosal exposome to toxicants.

The toxicity of engineered nanomaterials used in food additives and food packaging is discussed in the chapter on the toxicity of ingested nanomaterials. Food incorporates numerous potentially toxic chemicals in its path from raw material to ingestion. Safety evaluation for such chemicals, especially during the processing and packaging of food, is detailed in the chapter on safety evaluation of chemistries used in the food and beverage processing and packaging industries. The chapter also puts in detailed context the direct and indirect dietary exposure of food contact substances as regulated primarily by the United States Food and Drug Administration (FDA) and the United States Environmental Protection Agency (EPA).

The succeeding chapter on developmental neurotoxicity considerations for food additive safetyprovides adetailed overview of the developmental neurotoxicity (DNT) caused by food additives and theevolution of regulatory approaches for risk assessment through DNT testing methodologies. In context the chapter also sheds light on the structural neural development and the susceptibility of the developing brain to neurotoxicity.

The chapter on the protective effect of food-grade lactic acid bacteria highlights the importance of exogenous antioxidants in our daily diet against oxidative stress due to reactive oxygen species (ROS) and reactive nitrogen species (RNS). In this volume the information on mycotoxins focuses on the important aspect of the fungal contamination of food grains and discusses risk assessment procedures and current and future mycotoxin mitigation strategies. The information in the book on endocrine disrupting chemicals lays out the common endocrine disruption chemicals in food harmful to human health along with an example-based exposure assessment of such chemicals. Further, the chapter on radionuclides in food provides an interesting and detailed perspective surrounding the pathways of human exposureand related health risks due to radionuclides through food sources.

The information in the book on metal toxicity delineates recent advances in assessment of heavy metals in food affecting human health. The chapter on waste-burning residues in foods focuses on toxicological risks due to improper waste management in low-income countries. The chapter focuses on the routes of exposure and adverse health effects due the toxic residues found in food because of open-air solid waste burning. The chapter on epithelial to mesenchymal transition (EMT) provides an excellent overview on the biological process of EMT and explains its developing role as a valuable marker in ecotoxicology and food toxicology. Subsequently, the chapter on pesticide residues provides a great overview of the regulatory challenges, residue monitoring in food and water, and the detrimental impact of such pesticide residues to human and animal health. Finally the book also provides introduction to aptamers and focuses on the use of aptamers as powerful nanosensing tools in food toxicology.

The book has involved the diligent efforts of 36 scientists and experts from 11 countries around the globe. It provides informative research for food scientists and researchers and others involved with food safety and toxicology.

—**Ashish Sachan,** Canada
—**Suzanne Hendrich,** USA

CHAPTER 1

BIOMARKERS IN FOOD TOXICOLOGY

JUAN ANTONIO GIMENEZ BASTIDA[1] and
JOSÉ MOISÉS LAPARRA LLOPIS[2*]

[1]*Division of Clinical Pharmacology, Department of Pharmacology, Vanderbilt University Medical Center, Nashville, TN 37212, United States*

[2]*Immunonutrition and Health Group, Valencian International University, C/Gorgos, 5-7, 46021 Valencia, Spain*

[*]*Corresponding author. E-mail: j.moises.laparra@uv.es*

CONTENTS

ABSTRACT

Major food toxicological interests arise from the need to estimate exposure, early effects, and individual variation in sensitivity reflecting the interaction between exogenous and endogenous factors in the human body. To this end, "biomarkers" represent tools to measure response(s) (functional, physiological, and biochemical) at the cellular or molecular level reflecting interactions between a biological system and a potential hazard of a chemical, biological, and physical nature. Overall, three classes of biomarkers are identified: *biomarkers of exposure*, *biomarkers of effect*, and *biomarkers of susceptibility*. Selection of appropriate "priority" biomarkers is of critical importance to assess the risk and possible health-related outcomes of exposure to individuals and population subgroups. The usefulness and characteristics of pertinent analytical methods and technical considerations of each are discussed. Additionally, dietary toxicants which target gut microbiota can significantly impact different organs and tissues. Any adverse effect on the gut microbiota, which plays an important role in metabolism, can potentiate some of these toxicants. Importantly, developmental processes during perinatal period in the nervous system are especially vulnerable by some food toxicants at doses that may not be toxic to mature systems. This chapter summarizes the current knowledge about biomarkers suitable for application to the hepatic, renal, hematological, immune, pulmonary, reproductive, developmental, nervous system, and cognitive functions and those associated with carcinogenic mechanisms, as well as the assessment of risk to human health derived from food toxicants.

1.1   INTRODUCTION

In recent years, the nutritional richness of some foods as well as the innovative marketing of several other fortified foods with potentially beneficial defined ingredients for health constituted an important driving force favoring their worldwide commercialization. For example, there is an increasing interest in some kinds of seaweed as a source of fiber and minerals as well as foods fortified with phytochemicals because of their hypocholesterolemic and antioxidant effects. However, these foods can be a source of naturally occurring toxicants such as heavy metals (i.e., arsenic, cadmium, lead) and/or oxidized products produced during food processing and storage (i.e., phytosterol oxides). Moreover, the consumption of some staple foods also constitutes a significant source of some toxicants, for example, rice as source of inorganic arsenic, and potentially immunogenic components, including gluten-containing cereals as innate modulators for celiac and overweight/ obese patients.

The toxicity and resulting threat to human health of food toxicant(s) is, of course, related to its concentration in the food. Increasing research efforts have been made to better understand and estimate exposure levels of naturally occurring food toxicants or those produced during food processing and storage.[1,2] Several scientific studies reveal that traditionally held viewpoints of food toxicology in evaluating the concentration, physicochemical species, and evolution of the toxicant(s) as well as questionnaires about food frequency (FFQ) consumption usually overestimate toxicant(s) intake meanwhile food records often underestimate their dietary intake. Thus, FFQ need to be corrected according to physiological factors (i.e., biotransformation or factors that can affect the absorption of food/toxicant) to better compile toxicological information and predict health-risk assessments.

Food toxicological risk assessment implies major challenges because of the low doses to which humans are frequently exposed, so limiting the usefulness of experimental in vivo and in vitro models. Biomarkers constitute biological parameters reflecting interactions between a biological system and a hazard (physical, chemical, and biological or their metabolites) at cellular and molecular level.[3,4] These biomarkers are, among others, tools to evaluate the risk and possible health-related outcomes to individuals and population subgroups. Because of their increasing availability and sensitivity, biomarkers are increasingly being used to explore the links between exposure and effect in order to establish causality.[5] The contrast between individual variability and common characteristics becomes important to understand how behavioral phenotypes, constructed as an environmentally driven diet physiological adaptation, influence susceptibility to food toxicants.

Analysis of tissues and body fluids for defined chemicals or derived metabolites as well as enzymes and other biochemical substances to identify the interaction of toxicant(s) with biological systems has been used.[6] Measurement of biomarkers has been recognized as an advantageous tool directly linking exposure with an internal concentration and potential outcomes as relevant to the health-risk assessment. In this scenario, data from experimental in vivo and in vitro models can be compared to the biomarkers response(s) to establish and refine biological parameters that may be used in hazard identification and exposure potentially approaching biological response(s) in humans. The reaction and health outcome to toxicant(s) exposure through food consumption highly depends on inherited or behavioral patterns, life-style, chemical form(s) of the toxic, and how it is affected by household processes. Moreover, the critical dose–response relationship is difficult to estimate because of the physiological accumulation and biotransformation into tissues and organs as well as toxicant(s)-induced phenotype adaptations of these that will condition health outcomes to future exposure. Beyond risk factors, recent data suggest that food composition and its influence—on and interaction—with other dietary components as well as gut microbiota, and finally their crosstalk with the host's intestinal immune system are important determinants of food toxicant(s).[7,8]

Importantly, there is a growing body of evidence that the intrauterine or early childhood exposure to food toxicants induce phenotype changes and important health outcomes later in life. Currently, this information is foremost limited to heavy metals exposure through rice and water consumption (i.e., arsenic),[9–11] but it is largely inferential after long-term consumption of some oxidized derived phytochemicals and immunogenic components such as gluten. Thus, it points out the enormous importance to establish preventive nutritional intervention strategies stressing the need to limit the dietary intake of such components and the use of such foods as part of a balanced diet.

In this chapter, we summarized current knowledge about biomarkers suitable for application to several different physiological compartments and relevant systems. Additionally, identification of practicable biomarkers associated with different toxic end points or outcomes and usefulness and characteristics of pertinent analytical methods and technical considerations of each are discussed.

1.2   TAXONOMY OF BIOMARKERS

The term biomarker was first used in relation to a search for several classes of porphyrins in an attempt to establish the antiquity of terrestrial life.[12] This term receives different meanings depending upon the focus of the study. Currently, the most relevant biomarkers refer to exposure, effect, carcinogenesis, and susceptibility to other disease endpoints or adverse conditions and will be explained in further sections in this chapter. The National Institutes of Health Biomarkers Definitions Working Group defined the term biomarker as "characteristic that is objectively measured and evaluated as an indicator of normal biological processes, pathogenic processes, or pharmacologic responses to a therapeutic intervention."[13] According to this definition, biomarkers are not only used as indicators of the state/progress/impact of a toxic process but is also as an important instrument to investigate the relevance of the underlying toxicological mechanisms (Fig. 1.1).

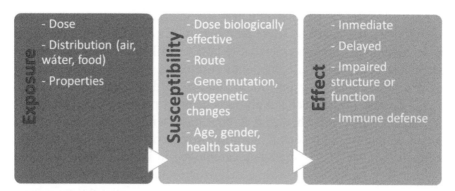

FIGURE 1.1 Schematic temporally relationship between biomarkers of exposure, susceptibility (carcinogenesis), and effect.

Biomarkers have been widely used to detect and monitor chronic diseases. Although important attempts have been made to identify "predictive" biomarkers, there are still significant uncertainties about them, thus resulting in an incessant growing research area. Nowadays, there is a considerable interest in the use and application of biomarkers to identify the nature and amounts of toxicological exposures that take place through food intake. Monitoring of circulating microRNAs (miRNAs) and identification of the excreted urinary metabolites represent important and complementary indicators of early exposures.[14] Thus, human cellular-based in vitro test systems rather than animal studies can be important elements for the prediction of various hazardous properties. However, important aspects such as how accurate and predictive, in terms of false negatives and false positives, or how they must be configured as replacement tests must be discussed. The need to continue identifying biomarkers that are specific to individual foods has arisen, for example, from successes in the biomarker-directed subset of nutrition status.[15,16]

Biomarkers of exposure could be predictive of "risk," as long as appropriate data are available to quantify the dietary intake–physiological effects relationship. Biomarkers of "effect" include a broad umbrella of measurements indicative of exposure to a particular food, although physicochemical interactions and microbiota-driven biotransformation processes in the gastrointestinal tract can decrease the specificity of those. Biomarkers of susceptibility recognize interindividual/population variability that most notably refers to genotype and can be influenced by ethnic and/or sociocultural aspects.

Several different factors influence the interaction between host and food-derived components. These factors may be considered in the context of a food/diet-compound/metabolite-host. Availability and release of potential hazardous/toxic compounds from food matrix is a prerequisite to be biotransformed by gut microbiota or be directly absorbed. Thus, it is important to consider the physicochemical properties and composition of the food, and whether it is consumed within a diet determining the bioavailability of potential hazardous/toxic compounds. For example, inorganic arsenic is classified as a carcinogen in humans (group I)[17] and foods and drinking water are the main routes for human exposure (i.e., seaweed, rice, etc.). In Western countries, seaweeds have not only been used mainly as a source of thickening and/or gelling agents with a wide range and various applications in the food industry, but also the general population started to increase their consumption. Because of the seaweed's advantageous nutritional features (i.e., fibrous food with high concentration of mineral/vitamins and antioxidants), they received high attention. Moreover, their consumption has been continuously increased worldwide and constitutes the causal relationship between popularity of seaweed and collateral human exposure to toxicants. In Asian countries, however, seaweeds have been consumed throughout the history. For example, the Japanese population had an average intake of 1.6 kg/year including species such as *Hizikia fusiforme* with an inorganic arsenic content greater than 50 μg/g dry weight.[18] However, there is a relative low incidence of arsenic-related cancer in this population. The latter can be explained by significant differences between the exposure dose and the effective dose reaching the target site. This points out the importance of factors that can affect the absorption of food/toxicant or the host's capacity for biotransformation of the toxicant.

Finally, information about host characteristics such as age, race, gender, health status, genetic susceptibility, and previous exposure to the same or other food-derived toxicants can provide clues as to the types of biomarkers that may be used to assess exposure, effect, and susceptibility. Biomarkers of carcinogenesis should differentiate interactions with DNA (genotoxic) and those that regulate gene expression (epigenetic or nongenotoxic).

1.3 "OMIC" BIOMARKERS

In the past few years, "omic" technologies have acquired an increasing interest, since they allow identifying new potential candidate biomarkers much faster than ever, leading to sensitive, specific, reproducible, rapid, and relatively cheap assays in comparison with traditional methods.[19] The main categories to classify these technologies are genomics, proteomics, and metabolomics/metabonomics.[20] DNA array technologies allow the analysis of thousands of genes at the same time providing information of the transcriptional profile of a sample after exposition to a compound.[21] Nowadays, an important number of high-throughput sequencing technologies are available to investigate the effects of chemical pollutants, dietary components (i.e., acrolein, dioxins, lead), and other external stresses on the structure and function of genomes determining mRNA and miRNA expression profile, as well as epigenetic modifications.[22,23] Proteomics encompasses as range of techniques intended to measure simultaneously a wide range of proteins associated with a biological process. Proteomic approaches can be used for (1) proteome profiling, (2) comparison of two different samples, (3) posttranslational modifications, and (4) interactions between proteins. The most common proteomic techniques are two-dimensional polyacrylamide gel electrophoresis, mass spectrometry, and protein microarray technology.[24] Additionally, recent direct proteome analyses using nanoliquid chromatography coupled to tandem mass spectrometry have proven accurate enough in the identification and quantification (Exponentially Modified Protein Abundance Index values) of complex mixtures of proteins.[2,25] Metabolomics is the study of all metabolites or low molecular weight organic or inorganic chemicals that are products or substrates of processes mediated by enzymes. Technically, nuclear magnetic resonance spectroscopy and MS analysis are the most commonly used techniques.[26] In addition to these omic techniques, in a recent revision,[27] the authors highlighted the development in lipidomic measurement technologies as well as high-content screening as new omic technologies to be applied in toxicological studies. The UK Committee on Toxicity of Chemicals in Food lists seven roles for -omics in chemical risk assessment in relation to food and feed[28]: (1) evaluation of toxicological mode of action, (2) determination of interspecies variability and extrapolations, (3) description of changes in molecular events at low doses, (4) identification of biomarkers that can be measured more easily, (5) interpretation or extrapolation between similar chemical structures (read across), (6) candidate molecule comparison and selection, and (7) development of reliable in vitro models to reduce the use of animal models.

It is important for omic biomarkers to differentiate those generated by environmental exposures that lead to disease from those that are consequences of a disease. For example, a change in the environment (i.e., chemical pollutants, temperature) can reveal the role of a single nucleotide polymorphism in the risk of developing, among other, alterations in lung and skin diseases.[29] Otherwise, genetic factors have been directly associated to cancer in about only 10% as shown by studies on twins.[30]

The biomarker-identification process involves two main stages: discovery and validation.[31] The main objective in the discovery phase is the determination of common consistent features in the biomarkers in different studies (ideally, in a prospective cohort study). In this phase, the data are analyzed using potential analytical methods (statistical analysis, database, literature-mining or bioinformatic techniques) for the identification of genes, proteins, or metabolites in normal subjects compared to patients. After the identification of candidate biomarker(s), the validation step begins in a long and complex process. In this phase, researchers try to determine causality, prevention, diagnosis, prognosis, and treatment of diseases of candidate biomarkers.[27,32] Numerous studies identifying biomarkers related to chronic human diseases, including cancer,[33,34] diabetes,[35] or cardiovascular diseases[36] can be found in the literature. However, despite the numerous studies identifying biomarkers associated with human diseases (a PubMed search for "biomarkers identification and human diseases" will give more than about 12,000 hits), the lack of universal validation guidelines makes the validation of biomarkers difficult. There have been described some of the potential pitfalls in biomarker development[32]: (1) variability; (2) limited availability of samples; (3) lack of standard methods, quality assurance, and quality control; (4) intellectual property protection—lack of collaboration; (5) inadequate sharing of data between academia and industry partners; (6) lack of clear regulatory guidance; (7) insufficient long-term data; (8) insufficient evidence of safety and efficacy. Therefore, much more studies need to be conducted in order to drive a search for improvements in the identification of omic biomarkers.

Omic technologies play a key role in a process to discover mechanisms that connect biomarkers of exposure with biomarkers of effect in studies identifying chemical hazards that can reach the organism through feeding.

The transcriptomic analysis of animal tissues and cells exposed to chemical pollutants/contaminants as well as dietary components is a potent tool in the identification of biomarkers of effect. For example, consumption of a vitamin A-enriched diet modified the expression of multiple genes such as CYP26A1, CYP26B1, or CYP2C22[37] exposed breast cancer cells to lycopene (10 μM; 48 h) and investigated the gene expression changes. Lycopene exhibited ability to modulate 391 genes, and the systematic pathway analysis revealed that these genes were involved in mechanisms of apoptosis, cell communication, MAPK and cell cycle as well as xenobiotic metabolism. Although these studies help in the identification of biomarkers of effect associated with xenobiotics exposure, it should be noted that microarray is a technique that can show variability and the results thus obtained need further confirmation. Therefore, quantitative real-time PCR is a method to validate the changes in gene expression observed.[38] Jobgen et al. documented differential expression in an important number of genes in adipose tissue of rats fed high-fat diet enriched with L-arginine.[39] The analysis of a group of the affected genes (fatty-acid synthase, PPAR-γ, or hormone-sensitive lipase, among others) by reverse-transcription polymerase chain reaction (RT-PCR) validated the results of the microarray, thus helping to understand the molecular mechanism whereby this compound exerts its effects. Gene expression changes (microarray and RT-PCR) observed in human intestinal cells exposed have been linked to orange extract rich in phenolic compounds together with the changes observed in protein levels (Western blot). Thus, the extract modulated the expression of PAI-1 and MMP12 (gene and protein level), two molecules related to the inflammatory response at the intestinal level.[40]

The studies described above have very little concerning hazards associated with natural compounds. However, approximately 30,000 existing chemical compounds require further toxicological evaluation.[28] Caffeine is a compound that due to its possible adverse effects is under investigation. Transcriptomic analyses revealed that caffeine induced changes in the expression of genes associated with lipid metabolism in rat postimplantation whole-embryo cultures.[41] Food additives, including butylated hydroxytoluene (BHT) and butylated hydroxyanisole (BHA), are chemical compounds that confer a desirable property to a food while also may have side effects on the consumer.[42] Transcriptomic analysis of liver samples from animals fed with BHT-enriched diet revealed a change in the expression of 10 genes associated with dietary exposure levels (28 days). Five of these genes were related to drug metabolism, including CYP2B1/2, CYP3A9, CYP2C6, glutathione S-transferase m type 2, and carboxylesterase 10 precursor. These authors confirmed these results by RT-PCR and protein expression levels by Western blotting.[43] In 2006, Nair et al. investigated the gene expression profile in small intestine and liver of animals exposed to BHA (200 mg/ kg). They described an Nrf2-dependent modulation of 2580 genes in the small intestine and 1317 in the liver. However, whether these changes in gene expression were also observed at the protein level was not studied.[44]

Besides, the essential role of Nrf2 in the regulation of the response against BHA by modulating a defined set of proteins (intermediate marker of effect) has been described using proteomic (2D-MALDI-Ms) techniques.[45]

Perfluorinated compounds (PFCs) have been extensively used in consumer products, being perfluorooctane sulfonate (PFOS) one of the best known PFCs causing immunotoxicity[46] as well as perfluorooctanoic acid (PFOA), a carcinogen, a liver and developmental toxicant that also alters thyroid hormonal effects.[47] Using transcriptome techniques, there have been reported significant changes (up/down-regulation) in the expression of several different genes associated, among other, with hepatic peroxisomal proliferation, fatty acid activation, transport, and oxidation indicates that PFOS may exert its influence through PPAR-α regulatory pathway.[48] The aforementioned studies show the potential of genomic and proteomic tools in food toxicology research. However, the dose–response relationship still remains elusive and these potential biomarkers are not conclusive about the mechanism of action. As shown, changes in the transcriptional profile after exposure do not always imply a variation at protein level. Therefore, the genomic studies should be accompanied by protein expression analysis for validation.

1.4   BIOMARKERS IN FOOD TOXICOLOGY

Ideally, biomarkers must reflect the absorbed amount or proportion of the toxicant or nutrient or potential metabolites derived from them that reach a targeted tissue affecting physiological functions. Thus, biomarkers should be able to identify a defined hazard and estimate the health-risk assessment defining cause/dose–effect relationship. This assumption attributes to the biomarkers an integrated measure of the internal dose (*biomarkers of exposure*) as well as the potential alterations of altered structures and/or their function to gain insights about its clinical significance (*biomarkers of effect*). Although human carcinogenesis results from several and different complex interactions between exogenous factors and host's genotype, biomarkers have also been used to prevent and determine cancer development (*biomarkers of carcinogenesis*). Understanding and validation of biomarkers in cancer development is essential, since the prognosis of cancers greatly improves with the earliest diagnosis. In this context, individual variability constitutes an essential feature to help elucidating the inherent ability and degree of response to the exposure to a defined food ingredient (*biomarkers of susceptibility*).

1.4.1   BIOMARKERS OF EXPOSURE

The term biomarker of exposure can refer to a food-derived substance or its metabolites as well as the product resulting from its/their interaction between those and target molecules that is measured in the host's physiological compartments. These assumptions are reflected, among other, in the rotenone-induced toxicity associated to the development of Parkinson's disease.[49] Rotenone intake takes place through contaminated food due to its use as insecticide. Using metabolomic studies on human peripheral blood lymphocytes, it has been shown that rotenone-induced mitochondrial dysfunction is associated with cardiolipin-derived oxidation/hydrolysis products formed early in apoptosis. The different chemical species that result further oxidized as well as the appearing *sn-1* and *sn-2* structures could represent new biomarkers of rotenone exposure.

Measurements can be estimated from substances/metabolites concentrations in food as well as measures of the level currently experienced by the individual or population through FFQs. These provide useful information about representative of habitual dietary intake allowing ranking individuals or population. Besides, FFQs can be insensitive to absolute intake for specific nutrients and exclude foods due to ethnicity. This could be of critical importance causing over/underestimate of toxicant(s) intake because of the need to consider individual physiological factors to better compile toxicological information and predict health-risk assessments.

Biomarkers of exposure confirm and quantify the link between ingested food and resultant internal concentrations. Their linkage to biomarkers of effect defines dose–response relationships, for which purpose, it results critical to obtain information about frequency and duration of the exposure. Moreover, there can be differentiated "intake," defined as the amount of food toxicant(s) provided with the diet, and the "exposure dose," which is the total amount absorbed according to the frequency and duration of the exposure. Measurements of substances/metabolites concentrations in cells, tissue, body fluids (blood, cerebrospinal fluid) or urine and feces reflect their distribution throughout the organism. Specific biological (geno/phenotype) responses of "exposure dose" quantify covalent adducts formed between food toxicant(s) and cellular macromolecules (proteins, DNA), or their excretion products and will be indicative of the "amount" delivered to target sites. Importantly, it has to be considered the "half-life" of the biomarker that will provide valuable information about the period of exposure. Metabolites poorly distributed into the organism as well as those easily removable from the organism through exhaled breath[50,51] or urine[52] may show significant variation of their concentrations during or immediately following an exposure, but not detectable at later times. Specific enzymes and reactive compounds generated during inflammation can also exhibit important variations if the altered enzyme activity can be both beneficial and harmful, or the products of primary reactions can be further modified; for example, myeloperoxidase and the generation of hypochlorous acid aid host defense against the invasion of microorganisms but may also be increased in disease states such as colitis.[53]

Thus, the analysis of several biomarkers (i.e., hemoglobin adducts in the blood, urine metabolites, parent substances/compounds) not only at a single time point but also measuring biomarkers over time is crucial. Importantly, if those possess different biological significances, they will provide a more complete picture about the exposure.

1.4.2 BIOMARKERS OF EFFECT

The term biomarkers of effect refer to alterations that take place within an organism and can be associated with a health disease. These markers show biochemical, metabolic, or physiological changes in an organism following exposure at level[19] and document either preclinical alterations or adverse health effects elicited by the absorption of food toxicants and/or their metabolites. Thus, biomarkers of effect may be used directly in hazard identification and dose–response measurements. For preventive purposes, an ideal biomarker of effect is one that measures alteration that is still reversible, but nevertheless, those that measure nonreversible effects are still important allowing early intervention. Notably, most accessible tissues and fluids are commonly used to gain valuable information about alterations taking place in target compartments.

Hematotoxicity: Alterations in enzyme synthesis and/or activity as well as routine leukocyte and erythrocyte counts, or their metabolic rates in relation to essential micronutrients have been used to monitor toxic effects derived from food substances/compounds.[54,55] Importantly, the interference caused by dietary deficiencies in nutritional minerals (i.e., iron) can cause lower numbers of red blood cells that can be misinterpreted.

Nephrotoxicity: Exposure to drugs often results in toxicity in kidney which represents the major control system maintaining homeostasis of body and thus is especially susceptible to xenobiotics. In this line, thiazoles (i.e., thiabendazole used as fungicide and food preservative) seem to exert its nephrotoxicity due to a ring cleavage metabolite, thus, causing tubular necrosis, increases in relative kidney weight and serum urea nitrogen concentration, and a decrease in renal GSH concentration.

Several different parameters such as functional (i.e., serum creatinine and β2-microglobulin), biochemical (i.e., eicosanoids, fibronectin, kallikrein activity, sialic acid and glycosaminoglycans in urine, and red blood cell negative charges), cytotoxicity (i.e., tubular antigens) markers as well as urinary proteins (i.e., albumin, transferrin, retinol-binding globulin), and enzymes (i.e., N-acetylglucosaminidase, β-galactosidase) have been used as markers of renal damage.

Hepatotoxicity: The liver is the central organ for metabolic processes where nutrients and other metabolites as well as endocrine and immune signals are transported allowing the liver to play its central role in mammalian physiology. Traditionally, the activities of specific isoenzymes (i.e., ALT1 and ALT2 to discriminate liver toxicity and adrenal and mitochondrial toxicity) and the concentrations of serum proteins (i.e., gamma-glutamyl-transferase and albumin) and bile acids synthetized in the liver and mitochondrial and nuclear DNA fragments have been used to estimate liver damage. It is important to take into consideration that some of these proteins can also be elevated in severe liver diseases. Differences in the microbiome functionality could also have important consequences in the alteration of serum bile acids profile.[56] Currently, hepatotoxicity biomarkers sometimes fail to identify a toxic compound. This clear need to find specific biomarkers motivated an intensive research on omic methods.

Immunotoxicity: The gut not only supplies nutrients and metabolites but also generates signals that stem from the intestinal mucosa and immune system. For a long time, cereal proteins such as those in the gluten fraction received clinical interest because their potential to promote allergic processes and intestinal disorders. Recently, different nongluten components of the α-amylase/trypsin inhibitors family (CM3 and 0.19) have revealed their immunotoxic potential participating as important novel contributors to the severity of intestinal innate immune responses.[57] These physiological response(s) are associated to changes in the pathogen-associated molecular patterns (i.e., Toll-like receptor 4 [TLR4]). This interaction enhances the basal threshold of innate immunity signaling resulting in a more severe immune response to subsequent gluten exposure. In humans, there have also been described adverse immunosuppressive effects after heavy metals (i.e., inorganic arsenic, cadmium, lead) intake in chronic and acute studies.[58] In this line, immune deficiencies at early stages of life due to gestational exposure to arsenic have also been associated to increased fetal loss and infant mortality.[10,59]

Pulmonary toxicity: From a food safety point of view, lung injury is derived from the activation (i.e., lipopolysaccharide from Gram-negative bacteria) of innate immune receptors that takes place by increased plasma nonesterified fatty acids in response to severe systemic inflammatory response syndrome secondary to sepsis. Measurements of gross effects on lung function (i.e., expiratory flow, forced expiratory volume) are commonly used as well as analysis of bronchoalveolar lavage fluid to detect lung injury. Otherwise, the production of cytokines has been mainly used to answer research questions.[60]

Reproductive and developmental toxicity: Physiological indicators of testicular and ovary function and measurements of circadian production of hormones, taking into account their large inter/intraindividual variations, can be used as biomarkers. Recently, the significant improvement of molecular miRNA techniques has led to propose them as potential biomarkers for reproductive toxicity.[61] Notably, environmental chemicals such as bisphenol A and phthalates have received intense attention because of the established epigenetic modifications caused by in utero exposure to these compounds. Thus, bisphenol A and phthalates induce alterations in gene expression that may persist throughout life.[62]

There is evidence that reproductive toxicity affects development or may result in an offspring unhealthy phenotype. This process is being increasingly thought to be attributed to epigenetic changes as integral underlying mechanism. Despite many different biomarkers (i.e., cytogenetic studies, hormones, and DNA probes) have been previously used to monitor the development during pregnancy and the well-being of the fetus, their association with epigenetic changes (i.e., methylation of DNA, modification of histones, including acetylation, RNA silencing or modification, or a combination of those) still represents novel aspects.

Neurotoxicity: The functions of the nervous system are complex and biomarkers may focus on molecular processes, cellular effects, and behavioral measurements. Communication between the gut and brain occurs constantly and several different environmental factors at the intestinal level appear to be largely responsible for the development, maturation, and function of the enteric nervous system as well as many aspects of the central nervous system structure and function. To date, there is good evidence of the close relationship and dependence of gut microbiota composition on dietary habits and its influence on brain development,[63] stress response(s), anxiety, and memory.[64] Gut microbiota has been shown to condition the profile of secondary bile acids, increase hydroxylations of several different steroid substrates as well as biotransformation of heavy metals either improving or worsening physiological functions or disease development.

1.4.3 BIOMARKERS OF CARCINOGENESIS

The World Health Organization has already warned about the enormous appearance and development of new cases of cancer during the next 20 years estimated to be up to a 70% increase globally.[65] Notably, these cancers are located in different physiological areas as a function of gender—(1) the top five most-diagnosed cancers in men: lung, prostate, colorectal, stomach, and liver, and (2) the top five most-diagnosed cancers in women: breast, colorectal, lung, cervix, and stomach.

Conclusions on dietary factors and cancer reveal that about one-third of cancers are related to dietary factors. Despite many clinical and molecular markers for predicting outcomes in several different diet-related cancers have been reported (Table 1.1), in most cases, their application in clinical practice remains still unclear. Most food-derived substances/compounds undergo enzyme-mediated reactions in the intestinal epithelia and are further metabolized in the liver, where reactive compounds can be generated, thus resulting in the formation of DNA adducts leading to initiating agents of cancer or receptor occupancy to promote tumors. Some DNA adducts can exert mutagenic effects, where replication of the damaged DNA may result in altered gene expression. Overweight/obesity convincingly increases the risks of several common cancers (i.e., breast, endometrial, colorectal cancer).[66,67] In this context, the gut microbiome structure and function has emerged as a novel modifiable factor and which may therefore be considered as a biomarker itself. Additionally, it is worth to point out the common practice extended with the incessant globalization promoting the use of additives (i.e., certain whiteners such as TiO_2) that can be associated to an exacerbated tumor formation in colitis associated cancer models.[68]

Popular foods such as snacks and popcorn are quite common today in our society. These products are subjected to a baking or microwave cooking at a fairly high temperature that generate cytotoxic compounds. These cooking procedures usually generate glycidamide that yields an epoxide derivative with a strong alkylating capacity causing a reasonable concern about possible health risks. In order to gain more information to better estimate the assessment of the human carcinogenic risk, there can be determined its metabolite *N*-(*R*,*S*)-acetyl-*S*-(2-carbamoyl-2-hydroxyethyl)-L-cysteine in human urine.[69] The proportion of this metabolite is important in this context because epoxide glycidamide is thought to be the ultimate carcino-genic metabolite. Besides, measurements of hemoglobin adducts of epoxide glycidamide can provide useful information about glycidamide reactivity and exposure. PFOA and PFOS, the two most known perfluoroalkil acids, can also be found in cooked snacks and popcorn and have been identified as carcinogens, liver and developmental toxicants, and immune system toxi-cants. Extensive amount of data have recently become available describing concentrations of perfluoroalkil acids in environmental media, wildlife, and human tissues. These compounds have been detected in blood (whole blood, plasma, and serum) as well as breast milk, liver, seminal plasma, and umbil-ical cord blood.[47] Thus, urinary and/or blood concentrations of perfluoroalkil acids may serve as early biomarkers to better estimate carcinogenic risk.

TABLE 1.1 Biomarkers Identified in Food-Associated Cancers (Colorectal, Pancreatic, and Bladder Cancer).

Cancer		Biomarkers	References
Gastric	Systemic	Angiogenic factors, cell surface receptors, adhesion molecules, proteins or peptides, downstream-signaling molecules, endothelial progenitor cells, parameters of energy metabolism, miRNAs	[70–73]
	Local	Microvessel density, CD31	[74,75]
Pancreatic	Systemic	Tumor cells and DNA, glypican-1 gene, gene mutations (i.e., KRAS, cyclin-dependent kinase inhibitor 2A, and p21WAF/CIP1)	[76,77]
	Local	Intraepithelial neoplasias	[78,79]
Bladder	Systemic	Circulating tumor cells, exosomes, and circulating miRNAs	[80]
	Local	TP53, RB1, ERBB2, or PTEN genes and miRNAs	[81,82]
Colorectal	Systemic	Circulating DNA and tumor tissue for the KRAS and PIK3CA mutations	[83]
	Local	Microsatellite instability, CpG island methylator phenotype, and DNA global hypomethylation	[84]

It cannot be ruled out the relatively high contribution of commonly considered "healthy" food (i.e., seaweed, rice) to the dietary inorganic arsenic intake. Long-term ingestion of inorganic arsenic through food and drinking water has been associated to several different malignancies such as skin, bladder, lung, and liver cancers. The most common biomarker used to estimate exposure for inorganic arsenic is the measurement of total urinary arsenic. However, the host biotransformation process of the toxicant includes methylation changes and is thought to play a key role increasing its genotoxicity. Currently, the molecular mechanisms involved are not fully understood and several of those (i.e., oxidative stress, gene expression, DNA reactivity) have been proposed for arsenic-induced carcinogenicity, but a scientific consensus has not been achieved.

1.4.4 BIOMARKERS OF SUSCEPTIBILITY

These represent indicators of the ability and degree of an organism to respond to food toxicants and their metabolites. There may be intrinsic factors, genetic characteristic, or preexisting disease states that may causes an increase in the absorbed proportion of the toxicants, or a decrease in the biological effective dose or altered cellular response. Previous efforts on functional polymorphism(s) revealed the association of single nuclear polymorphisms (i.e., rs11200014, rs2981579, and rs2981578 in the promoter of FGFR2 gene) to breast cancer susceptibility.[85] Also, combining earlier results from global differential allele-specific expression analysis and genome-wide association studies can help to identify single nuclear polymorphisms exhibiting expression imbalances favoring breast cancer susceptibility chromosome regions.[86] In this context, the implications for susceptibility of several different factors, such as diet, age, and physiological status and sensitization as well as enzyme induction or inhibition of previous exposure on subsequent susceptibility are of importance. For example, overweight/obesity has been identified as an important risk factor for cancer development, but only defined pro-inflammatory cytokines (i.e., IL-1β) have been significantly associated to breast cancer, whereas other biomarkers such as IL-6 and ox-LDL involved in different processes and different stages did not show association to breast cancer.[87] AMP-activated protein kinase is a fundamental regulator of energy metabolism, stress resistance, and cellular proteostasis. Additionally, together with the hypoxia-inducible factor (HIF)-signaling pathways are evolutionarily conserved survival mechanisms that regulate both longevity and cancer growth.[88]

Phenotypic plasticity as determined by genetic factors influences the response to exposure. Polymorphism(s) for some metabolic activation/deactivation enzymes (i.e., cytochrome P-450 isozymes, glutathione/methyl transferases), and different rates of enzyme activity controlling biotransformation of food-derived substances/compounds with potential toxic effects lead to differences in susceptibility by increasing or decreasing the biologically effective dose able to reach the target physiological compartment.

These indicate which factors may increase or decrease an individual's risk of suffering negative physiological consequences following exposure to a food toxicant(s) or components because of changes in the distribution and half-life of metabolites clearance from the main target physiological compartment. The effect may vary between ethnic, social, and demographic groups. For example, polymorphism encoding for the paraoxonase/arylesterase 1 have been directly associated to the level of lipoperoxides after cholesterol intake.[89] Additionally, there have been established significant associations between the rs11200014 polymorphism and breast cancer risk in Caucasians, not in Asians and Africans. Besides, the rs2981579 polymorphism was associated to an increased risk of breast cancer risk in all ethnicities.[89]

An additional form of susceptibility that receives increasing attention is that coming from the immune system. Prior exposure to a food toxicant or other intrinsic food components may promote alterations in the basal threshold triggering immune response(s) that sensitize or predispose to subsequent exposures. An example is the activation of intestinal innate immune cells via TLR4 and other pro-inflammatory receptors by molecular signals from certain nutrients[57] or from the microbiota. Thus, modulation of bacterial composition and/or function in the gut can greatly impact, worsening or improving, weight gain and insulin resistance,[90] possibly by interaction nucleotide oligomerization domain-1 and -2 proteins and other yet undefined pathways.[91]

1.5 SUMMARY AND RECOMMENDATIONS

There can be used biomarkers in different categories, while recognizing the difficulty to establish a clear distinction between them. Biomarkers can be applied not only to monitor diseases progression but also to identify those groups of populations along with dietary habits that aggravate or reduce the risk of developing toxic response(s). Importantly, biomarkers must be validated establishing the relationship between the exposure and health outcomes derived from food toxicants and/or their metabolites. The use and validation of biomarkers implies research, ethical, sociocultural, and legal aspects, which may vary according to defined features of different countries and heritable customs concerning food selection and dietary habits. In this context, it cannot be ruled out that FFQs potential over/underestimation of toxicant(s) intake, thus, making necessary to consider individual physiological factors to compile toxicological information and predict health-risk assessments. These particular features may hamper research and the use of certain biomarkers. The fast and continuous development of novel foods together with the incessant worldwide globalization provides additional reasons for the quest of reliable biomarkers.

KEYWORDS

- **biomarkers**
- **food**
- **toxicants**
- **health risk**

REFERENCES

1. Laparra, J. M.; Vélez, D.; Montoro, R.; Barberá, R.; Farré, R. Bioavailability of Inorganic Arsenic in Cooked Rice: Practical Aspects for Human Health Risk Assessment. *J. Agric. Food Chem.* **2005**, *53*, 8829–8833.
2. Laparra, J. M.; Alfonso-García, A.; Alegría, A.; Barberá, R.; Cilla, A. Keto-Stigmasterol and 7-Keto-Cholesterol Induce Differential Proteome Changes to Intestinal Epitelial (Caco-2) Cells. *Food Chem. Toxicol.* **2015**, *84*, 29–36.
3. World Health Organization. Environmental Health Criteria 237. Principles for Evaluating Health Risks in Children Associated with Exposure to Chemicals. *WHO Library Cataloguing-in-Publication Data*, 2006, ISBN: 92 4 157237 X.
4. US NRC (US National Research Council). *Biologic Markers in Reproductive Toxicology*. National Academy Press: Washington, DC, 1989; p. 395.
5. Benford, D. J.; Hanley, A. B.; Bottrill, K.; Oehlschlager, S.; Balls, M.; Branca, F.; Castegnaro, J. J.; Descotes, J.; Hemminiki, K.; Lindsay, D.; Schilter, B. Biomarkers as Predictive Tools in Toxicity Testing. *ATLA* **2000**, *28*, 119–131.
6. Sauer, J. M.; Hartung, T.; Leist, M.; Knudsen, T. B.; Hoeng, J.; Hayes, A. W. Systems Toxicology: The Future of Risk Assessment. *Int. J. Toxicol.* **2015**, *34* (4), 346–388.
7. Dangleben, N. L.; Skibola, C. F.; Smith, M. T. Arsenic Immunotoxicity: A Review. *Environ. Health* **2013**, *12*, 73.
8. Rey, F. E.; Gonzalez, M. D.; Cheng, J.; Wu, M.; Ahern, P. P.; Gordon, J. I. Metabolic Niche of a Prominent Sulfate-Reducing Human Gut Bacterium. *PNAS* **2013**, *110* (33), 13582–13587.
9. Vahter, M. Health Effects of Early Life Exposure to Arsenic. *Basic Clin. Pharmacol. Toxicol.* **2008**, *102*, 204–211.
10. Rahman, A.; Persson, L. Å.; Nermell, B.; El Arifeen, S.; Ekström, E. C.; Smith, A. H.; Vahter, M. Arsenic Exposure and Risk of Spontaneous Abortion, Stillbirth, and Infant Mortality. *Epidemiology* **2010**, *21* (6), 797–804.

11. Grandjean, P.; Landrigan, P. J. Neurobehavioural Effects of Developmental Toxicity. *Lancet Neurol.* **2014**, *13*, 330–338.

12. Rho, J. H.; Bauman, A. J.; Boettger, H. G.; Yen, T. F. A Search for Porphyrin Biomarkers in Nonesuch Shale and Extraterrestrial Samples. *Space Life Sci.* **1973**, *4*, 69–77.

13. Strimbu, K.; Tavel, J. A. What Are Biomarkers? *Curr. Opin. HIV AIDS*, **2010**, *5* (6), 463–466.

14. Oclon, E. A.; Latacz, A.; Zubel-Łojek, J.; Pierzchała-Koziec, K. Hyperglycemia-Induced Changes in miRNA Expression Patterns in Epicardial Adipose Tissue of Piglets. *J. Endocrinol.* **2016**, pii: JOE-15-0495.

15. Beckett, E. L.; Martin, C.; Choi, J. H.; King, K.; Niblett, S.; Boyd, L.; Duesing, K.; Yates, Z.; Veysey, M.; Lucock, M. Folate Status, Folate-Related Genes and Serum miR-21 Expression: Implications for miR-21 as a Biomarker. *BBA Clin.* **2015**, *4*, 45–51.

16. King, J. C.; Brown, K. H.; Gibson, R. S.; Krebs, N. F.; Lowe, N. M.; Siekmann, J. H.; Raiten, D. J. Biomarkers of Nutrition for Development (BOND)—Zinc Review. *J. Nutr.* **2016**, pii: jn220079.

17. IARC (International Agency for Cancer Research). IARC Monographs on the EValuation of Carcinogenic Risks to Humans. *OVerall EValuations of Carcinogenicity: An Updating of IARC Monographs*; International Agency for Cancer Research: Lyon, 1987; Vol. 1–42, Suppl. 7.

18. Laparra, J. M.; Vélez, D.; Montoro, R.; Barberá, R.; Farré, R. Bioaccessibility of Inorganic Arsenic Species in Raw and Cooked *Hizikia fusiforme* Seaweed. *Appl. Organomet. Chem.* **2004**, *18*, 662–669.

19. Gundert-Remy, U.; Dahl, S. G.; Boobis, A.; Kremers, P.; Kopp-Schneider, A.; Oberemm, A.; Renwick, A.; Pelkonen, O. Molecular Approaches to the Identification of Biomarkers of Exposure and Effect—Report of an Expert Meeting Organized by COST Action B15. *Toxicol. Lett.* **2005**, *156*, 227–240.

20. Kleno, T. G.; Kiehr, B.; Baunsgaard, D.; and Sidelmann, U. G. Combination of "Omics" Data to Investigate the Mechanism(s) of Hydrazine-Induced Hepatotoxicity in Rats and to Identify Potential Biomarkers. *Biomarkers* **2004**, *9*, 116–138.

21. Tarca, A. L.; Romero, R.; Draghici, S. Analysis of Microarray Experiments of Gene Expression Profiling. *Am. J. Obstet. Gynecol.* **2006**, *195*, 373–388.

22. Paoloni-Giacobino, A. Post Genomic Decade—The Epigenome and Exposome Challenges. *Swiss Med. Wkly.* **2011**, *141*, w13321.

23. Reuter, J. A.; Spacek, D. V.; Snyder, M. P. High-Throughput Sequencing Technologies. *Mol. Cell* **2015**, *58*, 586–597.

24. Chandramouli, K.; Qian, P. Y. Proteomics: Challenges, Techniques and Possibilities to Overcome Biological Sample Complexity. *Hum. Genet. Proteomics* **2009**, *2009*, 239204.

25. Ishihama, Y.; Oda, Y.; Tabata, T.; Sato, T.; Nagasu, T.; Rappsilber, J.; Mann, M. Exponentially Modified Protein Abundance Index (emPAI) for Estimation of Absolute Protein Amount in Proteomics by the Number of Sequenced Peptides Per Protein. *Mol. Cell. Proteomics* **2005**, *4*, 1265–1272.

26. Brennan, L. Metabolomics in Nutrition Research: Current Status and Perspectives. *Biochem. Soc. Trans.* **2013**, *41*, 670–673.

27. Sturla, S. J.; Boobis, A. R.; FitzGerald, R. E.; Hoeng, J.; Kavlock, R. J.; Schirmer, K.; Whelan, M.; Wilks, M. F.; Peitsch, M. C. Systems Toxicology: From Basic Research to Risk Assessment. *Chem. Res. Toxicol.* **2014**, *27*, 314–329.

28. European Food Safety Authority (EFSA). A Foresight Study on Emerging Technologies: State of the Art of Omics Technologies and Potential Applications in Food and Feed Safety. *EFSA Supporting Publication 2013, EN-495*, 2013.

29. Feil, R.; Fraga, M. F. Epigenetics and the Environment: Emerging Patterns and Implications. *Nat. Rev. Genet.* **2012**, *13*, 97–109.

30. Lichtenstein, P; Holm, N. V.; Verkasalo, P. K.; Iliadou, A.; Kaprio, J.; Koskenvuo, M.; Pukkala, E.; Skytthe, A.; Hemminki, K. Environmental and Heritable Factors in the Causation of Cancer—Analyses of Cohorts of Twins from Sweden, Denmark, and Finland. *N. Engl. J. Med.* **2000**, *343* (2), 78–85.

31. Sahab, Z. J.; Semaan, S. M.; Sang, Q. X. Methodology and Applications of Disease Biomarker Identification in Human Serum. *Biomark. Insights* **2007**, *2*, 21–43.

32. Lin, D.; Hollander, Z.; Meredith, A.; McManus, B. M. Searching for 'Omic' Biomarkers. *Can. J. Cardiol.* **2009**, *25* (Suppl A), 9A–14A.

33. Duscharla, D.; Bhumireddy, S. R.; Lakshetti, S.; Pospisil, H.; Murthy, P. V.; Walther, R.; Sripadi, P.; Ummanni, R. Prostate Cancer Associated Lipid Signatures in Serum Studied by ESI-Tandem Mass Spectrometry as Potential New Biomarkers. *PLoS One* **2016**, *11*, e0150253.

34. Guo, J.; Xie, K.; Zheng, S. Molecular Biomarkers of Pancreatic Intraepithelial Neoplasia and Their Implications in Early Diagnosis and Therapeutic Intervention of Pancreatic Cancer. *Int. J. Biol. Sci.* **2016**, *12* (3), 292–301.

35. Zhao, Y.; Xue, Q.; Su, X.; Xie, L.; Yan, Y.; Wang, L.; Steinman, A. D. First Identification of the Toxicity of Microcystins on Pancreatic Islet Function in Humans and the Involved Potential Biomarkers. *Environ. Sci. Technol.* **2016**, *50*, 3137–3144.

36. Gil-Izquierdo, A.; Penalvo, J. L.; Gil, J. I.; Medina, S.; Horcajada, M. N.; Lafay, S.; Silberberg, M.; Llorach, R.; Zafrilla, P.; Garcia-Mora, P.; Ferreres, F. Soy Isoflavones and Cardiovascular Disease Epidemiological, Clinical and -Omics Perspectives. *Curr. Pharm. Biotechnol.* **2012**, *13*, 624–631.

37. Chalabi, N.; Satih, S.; Delort, L.; Bignon, Y. J.; Bernard-Gallon, D. J. Expression Profiling by Whole-Genome Microarray Hybridization Reveals Differential Gene Expression in Breast Cancer Cell Lines after Lycopene Exposure. *Biochim. Biophys. Acta* **2007**, *1769*, 124–130.

38. Morey, J. S.; Ryan, J. C.; Van Dolah, F. M. Microarray Validation: Factors Influencing Correlation between Oligonucleotide Microarrays and Real-Time PCR. *Biol. Proc. Online* **2006**, *8*, 175–193.

39. Jobgen, W.; Fu, W. J.; Gao, H.; Li, P.; Meininger, C. J.; Smith, S. B.; Spencer, T. E.; Wu, G. High Fat Feeding and Dietary L-Arginine Supplementation Differentially Regulate Gene Expression in Rat White Adipose Tissue. *Amino Acids* **2009**, *37*, 187–198.

40. Gimenez-Bastida, J. A.; Martinez-Florensa, M.; Espin, J. C.; Tomas-Barberan, F. A.; Garcia-Conesa, M. T. A Citrus Extract Containing Flavanones Represses Plasminogen Activator Inhibitor-1 (PAI-1) Expression and Regulates Multiple Inflammatory, Tissue Repair, and Fibrosis Genes in Human Colon Fibroblasts. *J. Agric. Food Chem.* **2009**, *57*, 9305–9315.

41. Robinson, J. F.; van Beelen, V. A.; Verhoef, A.; Renkens, M. F.; Luijten, M.; van Herwijnen, M. H.; Westerman, A.; Pennings, J. L.; Piersma, A. H. Embryotoxicant-Specific Transcriptomic Responses in Rat Postimplantation Whole-Embryo Culture. *Toxicol. Sci.* **2010**, *118*, 675–685.

42. Hocman, G. Chemoprevention of Cancer: Phenolic Antioxidants (BHT, BHA). *Int. J. Biochem.* **1988**, *20*, 639–651.

43. Stierum, R.; Conesa, A.; Heijne, W.; van Ommen, B.; Junker, K.; Scott, M. P.; Price, R. J.; Meredith, C.; Lake, B. G.; Groten, J. Transcriptome Analysis Provides New Insights into Liver Changes Induced in the Rat upon Dietary Administration of the Food Additives Butylated Hydroxytoluene, Curcumin, Propyl Gallate and Thiabendazole. *Food Chem. Toxicol.* **2008,** *46,* 2616–2628.

44. Nair, S.; Xu, C.; Shen, G.; Hebbar, V.; Gopalakrishnan, A.; Hu, R.; Jain, M. R.; Lin, W.; Keum, Y. S.; Liew, C.; Chan, J. Y.; Kong, A. N. Pharmacogenomics of Phenolic Antioxidant Butylated Hydroxyanisole (BHA) in the Small Intestine and Liver of Nrf2 Knockout and C57BL/6J Mice. *Pharm. Res.* **2006,** *23,* 2621–2637.

45. Abdullah, A.; Kitteringham, N. R.; Jenkins, R. E.; Goldring, C.; Higgins, L.; Yamamoto, M.; Hayes, J.; Park, B. K. Analysis of the Role of Nrf2 in the Expression of Liver Proteins in Mice Using Two-dimensional Gel-Based Proteomics. *Pharmacol. Rep.* **2012,** *64,* 680–697.

46. Domingo, J. L. Health Risks of Dietary Exposure to Perfluorinated Compounds. *Environ. Int.* **2012,** *40,* 187–195.

47. Lau, C.; Anitole, K.; Hodes, C.; Lai, D.; Pfahles-Hutchens, A.; Seed, J. Perfluoroalkyl Acids: A Review of Monitoring and Toxicological Findings. *Toxicol. Sci.* **2007,** *99* (2), 366–394.

48. Bjork, J. A.; Lau, C.; Chang, S. C.; Butenhoff, J. L.; Wallace, K. B. Perfluorooctane Sulfonate-Induced Changes in Fetal Rat Liver Gene Expression. *Toxicology* **2008,** *251,* 8–20.

49. Tyurina, Y. Y.; Winnica, D. E.; Kapralova, V. I.; Kapralov, A. A.; Tyurin, V. A.; Kagan, V. E. LC/MS Characterization of Rotenone Induced Cardiolipin Oxidation in Human Lymphocytes: Implications for Mitochondrial Dysfunction Associated with Parkinson's Disease. *Mol. Nutr. Food Res.* **2013,** *57* (8), 1410–1422.

50. van Horck, M.; Alonso, A.; Wesseling, G.; de Winter-de Groot, K.; van Aalderen, W.; Hendriks, H.; Winkens, B.; Rijkers, G.; Jöbsis, Q.; Dompeling, E. Biomarkers in Exhaled Breath Condensate Are Not Predictive for Pulmonary Exacerbations in Children with Cystic Fibrosis: Results of a One-Year Observational Study. *PLoS One* **2016,** *11* (4), e0152156.

51. Tahan, F.; Eke, G. H.; Bicici, E.; Saraymen, B.; Akar, H. H. Increased Postexercise Lipoxin A4 Levels in Exhaled Breath Condensate in Asthmatic Children with Exercise-Induced Bronchoconstriction. *J. Investig. Allergol. Clin. Immunol.* **2016,** *26* (1), 19–24.

52. De Craemer, S.; Croes, K.; van Larebeke, N.; Sioen, I.; Schoeters, G.; Loots, I.; Nawrot, T.; Nelen, V.; Campo, L.; Fustinoni, S.; Baeyens, W. Investigating Unmetabolized Polycyclic Aromatic Hydrocarbons in Adolescents' Urine as Biomarkers of Environmental Exposure. *Chemosphere* **2016,** *155,* 48–56.

53. Kato, Y. Neutrophil Myeloperoxidase and Its Substrates: Formation of Specific Markers and Reactive Compounds during Inflammation. *J. Clin. Biochem. Nutr.* **2016,** *58* (2), 99–104.

54. Belury, M. A.; Cole, R. M.; Bailey, B. E.; Ke, J. Y.; Andridge, R. R.; Kiecolt-Glaser, J. K. Erythrocyte Linoleic Acid, But Not Oleic Acid, Is Associated with Improvements in Body Composition in Men and Women. *Mol. Nutr. Food Res.* **2016,** doi: 10.1002/mnfr.201500744.

55. Barden, A.; O'Callaghan, N.; Burke, V.; Mas, E.; Beilin, L. J.; Fenech, M.; Irish, A. B.; Watts, G. F.; Puddey, I. B.; Huang, R. C.; Mori, T. A. n-3 Fatty Acid Supplementation and Leukocyte Telomere Length in Patients with Chronic Kidney Disease. *Nutrients* **2016,** *8* (3), pii: E175.

56. Ridlon, J. M.; Harris, S. C.; Bhowmik, S.; Kang, D. J.; Hylemon, P. B. Consequences of Bile Salt Biotransformations by Intestinal Bacteria. *Gut Microbes* **2016**, *7* (1), 22–39.

57. Junker, Y.; Zeissig, S.; Kim, S.; Barisani, D.; Wieser, H.; Leffler, D. A.; Zevallos, V.; Libermann, T. A.; Dillon, S.; Freitag, T. L.; Kelly, C. P.; Schuppan, D. Wheat Amylase Trypsin Inhibitors Drive Intestinal Inflammation via Activation of Toll-Like Receptor 4. *J. Exp. Med.* **2012**, *209*, 2395–2408.

58. Portales, P. D.; Baranda, L.; Díaz-Barriga, F.; Saavedra-Alanis, V.; Layseca, E.; Gonzalez-Amaro, R. Effect of Arsenic, Cadmium and Lead on the Induction of Apoptosis of Normal Human Mononuclear Cells. *Clin. Exp. Immunol.* **2002,** *129* (1), 69–77.

59. Rahman, A.; Vahter, M.; Ekström, E. C.; Rahman, M.; Golam Mustafa, A. H.; Wahed, M. A.; Yunus, M.; Persson, LA. Association of Arsenic Exposure during Pregnancy with Fetal Loss and Infant Death: A Cohort Study in Bangladesh. *Am. J. Epidemiol.* **2007,** *165* (12), 1389–96.

60. Gonçalves-de-Albuquerque, C. F.; Medeiros-de-Moraes, I. M.; Oliveira, F. M.; Burth, P.; Bozza, P. T.; Castro Faria, M. V.; Silva, A. R.; Castro-Faria-Neto, H. C. Omega-9 Oleic Acid Induces Fatty Acid Oxidation and Decreases Organ Dysfunction and Mortality in Experimental Sepsis. *PLoS One* **2016**, *11* (4), e0153607.

61. Sun, Z.; Zhang, W.; Li, S.; Xue, X.; Niu, R.; Shi, L.; Li, B.; Wang, X.; Wang, J. Altered miRNAs Expression Profiling in Sperm of Mice Induced by Fluoride. *Chemosphere* **2016**, *155*, 109–114.

62. Singh, S.; Li, S. S. Epigenetic Effects of Environmental Chemicals Bisphenol A and Phthalates. *Int. J. Mol. Sci.* **2012,** *13* (8), 10143–10153.

63. Heijtz, R.; Wang, S.; Anuar, F.; Qian, Y.; Björkholm, B.; Samuelsson, A.; Hibberd, M. L.; Forssberg, H.; Pettersson, S. Normal Gut Microbiota Modulates Brain Development and Behavior. *Proc. Nat. Acad. Sci. USA* **2011**, *108*, 3047–3052.

64. Gareau, M. G.; Wine, E.; Rodrigues, D. M.; Cho, J. H.; Whary, M. T.; Philpott, D. J.; Macqueen, G.; Sherman, P. M. Bacterial Infection Causes Stress-Induced Memory Dysfunction in Mice. *Gut* **2011**, *60*, 307–317.

65. World Health Organization (WHO). www.who.int. Fact sheet No. 297, 2015.

66. Agresti, R.; Meneghini, E.; Baili, P.; Minicozzi, P.; Turco, A.; Cavallo, I.; Funaro, F.; Amash, H.; Berrino, F.; Tagliabue, E.; Sant, M. Association of Adiposity, Dysmetabolisms, and Inflammation with Aggressive Breast Cancer Subtypes: A Cross-Sectional Study. *Breast Cancer Res. Treat.* **2016**, *157* (1), 179–189.

67. Schmitz, J.; Evers, N.; Awazawa, M.; Nicholls, H. T.; Brönneke, H. S.; Dietrich, A.; Mauer, J.; Blüher, M.; Brüning, J. C. Obesogenic Memory Can Confer Long-Term Increases in Adipose Tissue But Not Liver Inflammation and Insulin Resistance after Weight Loss. *Mol. Metab.* **2016,** *5* (5), 328–339.

68. Urrutia-Ortega, I. M.; Garduño-Balderas, L. G.; Delgado-Buenrostro, N. L.; Freyre-Fonseca, V.; Flores-Flores, J. O.; González-Robles, A.; Pedraza-Chaverri, J.; Hernandez-Pando, R.; Rodríguez-Sosa, M.; León-Cabrera, S.; Terrazas, L. I.; van Loveren, H.; Chirino, Y. I. Food-Grade Titanium Dioxide Exposure Exacerbates Tumor Formation in Colitis Associated Cancer Model. *Food Chem. Toxicol.* **2016**, *93*, 20–31. pii: S0278-6915(16)30120-X.

69. Scherer, G.; Urban, M.; Hagedorn, H. W.; Serafin, R.; Feng, S.; Kapur, S.; Muhammad, R.; Jin, Y.; Sarkar, M.; Roethig, H. J. Determination of Methyl-, 2-Hydroxyethyl- and 2-Cyanoethyl Mercapturic Acids as Biomarkers of Exposure to Alkylating Agents in Cigarette Smoke. *J. Chromatogr. B:: Anal. Technol. Biomed. Life Sci.* **2010**, *878* (27), 2520–2528.

70. Duda, D. G. Molecular Biomarkers of Response to Antiangiogenic Therapy for Cancer. *ISRN Cell Biol.* **2012**, *2012*, 1–11.

71. de Haas, S.; Delmar, P.; Bansal, A. T.; Moisse, M.; Miles, D. W.; Leighl, N.; Escudier, B.; Van Cutsem, E.; Carmeliet, P.; Scherer, S. J.; Pallaud, C.; Lambrechts, D. Genetic Variability of VEGF Pathway Genes in Six Randomized Phase III trials Assessing the Addition of Bevacizumab to Standard Therapy. *Angiogenesis* **2014**, *17*, 909–920.

72. Konishi, H.; Ichikawa, D.; Komatsu, S.; Shiozaki, A.; Tsujiura, M.; Takeshita, H.; Morimura, R.; Nagata, H.; Arita, T.; Kawaguchi, T.; Hirashima, S.; Fujiwara, H.; Okamoto, K.; Otsuji, E. Detection of Gastric Cancer-Associated MicroRNAs on MicroRNA Microarray Comparing Pre- and Post-operative Plasma. *Br. J. Cancer* **2012**, *106*, 740–747.

73. Song, M. Y.; Pan, K. F.; Su, H. J.; Zhang, L.; Ma, J. L.; Li, J. Y.; Yuasa, Y.; Kang, D.; Kim, Y. S.; You, W. C. Identification of Serum MicroRNAs as Novel Non-invasive Biomarkers for Early Detection of Gastric Cancer. *PLoS One* **2012**, *7*, e33608.

74. Badescu, A.; Georgescu, C. V.; Vere, C. C.; Craitoiu, S.; Grigore, D. Correlations between Her2 Oncoprotein, VEGF Expression, MVD and Clinicopathological Parameters in Gastric Cancer. *Rom. J. Morphol. Embryol.* **2012**, *53*, 997–1005.

75. Carboni, F.; Valle, M.; Camperchioli, L.; Sandri, G. B.; Sentinelli, S.; Garofalo, A. Mesothelial Cyst of the Round Ligament of the Liver. *J. Minim. Access. Surg.* **2016**, *12* (1), 83–85.

76. Sausen, M.; Phallen, J.; Adleff, V.; Jones, S.; Leary, R. J.; Barrett, M. T.; Anagnostou, V.; Parpart-Li, S.; Murphy, D.; Kay Li, Q.; Hruban, C. A.; Scharpf, R.; White, J. R.; O'Dwyer, P. J.; Allen, P. J.; Eshleman, J. R.; Thompson, C. B.; Klimstra, D. S.; Linehan, D. C.; Maitra, A.; Hruban, R. H.; Diaz, L. A., Jr.; Von Hoff, D. D.; Johansen, J. S.; Drebin, J. A.; Velculescu, V. E. Clinical Implications of Genomic Alterations in the Tumour and Circulation of Pancreatic Cancer Patients. *Nat. Commun.* **2015**, *7*, 7686.

77. Melo, S. A.; Luecke, L. B.; Kahlert, C.; Fernandez, A. F.; Gammon, S. T.; Kaye, J.; LeBleu, V. S.; Mittendorf, E. A.; Weitz, J.; Rahbari, N.; Reissfelder, C.; Pilarsky, C.; Fraga, M. F.; Piwnica-Worms, D.; Kalluri, R. Glypican-1 Identifies Cancer Exosomes and Detects Early Pancreatic Cancer. *Nature* **2015**, *523* (7559), 177–182.

78. Koorstra, J. B.; Feldmann, G.; Habbe, N.; Maitra, A. Morphogenesis of Pancreatic Cancer: Role of Pancreatic Intraepithelial Neoplasia (PanINs), *Langenbecks Arch. Surg.* **2008**, *393*, 561–570.

79. Cooper, C. L.; O'Toole, S. A.; Kench, J. G. Classification, Morphology and Molecular Pathology of Premalignant Lesions of the Pancreas. *Pathology* **2013**, *45*, 286–304.

80. Nagata, M.; Muto, S.; Horie, S. Molecular Biomarkers in Bladder Cancer: Novel Potential Indicators of Prognosis and Treatment Outcomes. *Dis. Mark.* **2016**, *2016*, 8205836.

81. Knowles, M. A.; Hurst, C. D. Molecular Biology of Bladder Cancer: New Insights into Pathogenesis and Clinical Diversity. *Nat. Rev. Cancer* **2014**, *15* (1), 25–41.

82. Yoshino, H.; Seki, N.; Itesako, T.; Chiyomaru, T.; Nakagawa, M.; Enokida, H. Aberrant Expression of MicroRNAs in Bladder Cancer. *Nat. Rev. Urol.* 2013, *10* (7), 396–404.

83. Thierry, A. R.; Mouliere, F.; El Messaoudi, S.; Mollevi, C.; Lopez-Crapez, E.; Rolet, F.; Gillet, B.; Gongora, C.; Dechelotte, P.; Robert, B.; Del Rio, M.; Lamy, P. J.; Bibeau, F.; Nouaille, M.; Loriot, V.; Jarrousse, A. S.; Molina, F.; Mathonnet, M.; Pezet, D.; Ychou, M. Clinical Validation of the Detection of KRAS and BRAF Mutations from Circulating Tumor DNA. *Nat. Med.* **2014**, *20* (4), 430–435.

84. Grady, W. M.; Pritchard, C. C. Molecular Alterations and Biomarkers in Colorectal Cancer. *Toxicol. Pathol.* **2014**, *42*, 124–139.

85. Zhou, L.; Yao, F.; Luan, H.; Wang, Y.; Dong, X.; Zhou, W.; Wang, Q. Three Novel Functional Polymorphisms in the Promoter of FGFR2 Gene and Breast Cancer Risk: A HuGE Review and Meta-analysis. *Breast Cancer Res. Treat.* **2012,** *136* (3), 885–897.

86. Lin, W.; Lin, H. D.; Guo, X. Y.; Lin, Y.; Su, F. X.; Jia, W. H.; Tang, L. Y.; Zheng, W.; Long, J. R.; Ren, Z. F. Allelic Expression Imbalance Polymorphisms in Susceptibility Chromosome Regions and the Risk and Survival of Breast Cancer. *Mol. Carcinog.* **2016,** doi:10.1002/mc.22493.

87. Dias, J. A.; Fredrikson, G. N.; Ericson, U.; Gullberg, B.; Hedblad, B.; Engström, G.; Borgquist, S.; Nilsson, J.; Wirfält, E. Low-Grade Inflammation, Oxidative Stress and Risk of Invasive Post-Menopausal Breast Cancer—A Nested Case-Control Study from the Malmö Diet and Cancer Cohort. *PLoS One* **2016,** *11* (7), e0158959.

88. Salminen, A.; Kaarniranta, K.; Kauppinen, A. AMPK and HIF Signaling Pathways Regulate both Longevity and Cancer Growth: The Good News and the Bad News about Survival Mechanisms. *Biogerontology* **2016,** *17* (4), 655–680.

89. Bonafè, M.; Marchegiani, F.; Cardelli, M.; Olivieri, F.; Cavallone, L.; Giovagnetti, S.; Pieri, C.; Marra, M.; Antonicelli, R.; Troiano, L.; Gueresi, P.; Passeri, G.; Berardelli, M.; Paolisso, G.; Barbieri, M.; Tesei, S.; Lisa, R.; De Benedictis, G.; Franceschi, C. Genetic Analysis of Paraoxonase (PON1) Locus Reveals an Increased Frequency of Arg192 Allele in Centenarians. *Eur. J. Hum. Genet.* **2002,** *10* (5), 292–296.

90. Abu-Shanab, A.; Quigley, E. M. The Role of the Gut Microbiota in Nonalcoholic Fatty Liver Disease. *Nat. Rev. Gastroenterol. Hepatol.* **2010,** *7* (12), 691–701.

91. Parnell, J. A.; Raman, M.; Rioux, K. P.; Reimer, R. A. The Potential Role of Prebiotic Fibre for Treatment and Management of Non-alcoholic Fatty Liver Disease and Associated Obesity and Insulin Resistance. *Liver Int.* **2011,** 1–11.

CHAPTER 2

MUCOSAL EXPOSOME AND FOOD TOXICITY BIOMARKERS

YUSEOK MOON[1,2*]

[1]Laboratory of Mucosal Exposome and Biomodulation, Department of Biomedical Sciences, Pusan National University, Yangsan, South Korea

[2]Immunoregulatory Therapeutics Group in Brain Busan 21 Project, Busan, South Korea

*E-mail: moon@pnu.edu

CONTENTS

ABSTRACT

The exposome represents the sum of all elements that exist in the exogenous macroenvironment and the endogenous microenvironment of the host that may influence disease emergence and clinical outcome. In particular, the mucosal exposome is the internal niche-containing food components and microbiome, all of which crosstalk with the host physiological components including the immune system. This integrated network may determine homeostasis, and acute toxicity/disease or chronic disease progression in the host. In this chapter, a mucosal exposome-based understanding of food toxicity and related biomarkers will be described. In the luminal parts of mucosa, foodborne toxicants are mixed with other dietary components and gut microbiota, all of which interact together, leading to complex events in the body. Moreover, the diversity and presence of the toxicant-metabolizing microbes in the gut are another important determinant of the susceptibility to toxicity and may be useful as susceptibility biomarkers. In terms of the mucosal biological responses to foodborne xenobiotics, mucosal effector biomarkers are important in estimating the gut-related diseases including gastroenteritis and systemic disorders. For instance, behavioral responses such as anorexia at the gut–brain axis are mechanistically linked to alteration of eating disorder-linked hormones from the intestinal neuroendocrine cells, which can be useful biomarkers for mucosal exposome-related complications. This chapter will cover issues in terms of food-related diseases including food hypersensitivity, the chronic inflammatory disorders, and the malignant tumors.

2.1 INTRODUCTION TO THE MUCOSAL EXPOSOME IN RELATION TO FOOD COMPONENTS

In response to foodborne insults including pathogens and toxicants, the endogenous mucosal microenvironment of the host is altered, leading to systemic biological outcomes. Alteration of the mucosal niche may determine whether the host can keep homeostasis or enter a disease state. The mucosal exposome represents the sum of all elements that exist in the endogenous microenvironment that may influence host physiology and pathogenic states during the whole of lifespan.[1-3] The most influential external factor toward the mucosal exposome is dietary components with daily exposure. These dietary components include macronutrients, regulatory nutrients, beneficial food microbes, and other adverse xenobiotics including foodborne pathogens and toxic chemicals. Food-derived human mucosal disorders are the result of disrupted homeostasis of the dynamic mucosal interaction of food components, the gut microbiota, and the host responses via immune and neuroendocrine systems.

Nutrients can have acute and chronic effects in shaping the composition and diversity of human microbiota. Moreover, nutrients and gut microbiota are important modulators of signaling systems such as immune responses and neuroendocrine circuits in the intestine. Diet determines its dynamics and composition, and the network of diet–microbiota–host sentinels. Gut microorganisms or their products in response to food-derived factors are responsible for triggering the mucosal immune system and release of the neuroendocrine stimulators by the microbial factors or the immune-related cells.[4] Moreover, there is a lot of evidence for the close crosstalk between the human mucosal immune system and the neuroendocrine circuit, which contributes to the diet-related clinical outcomes. Recent increases in the incidence of chronic disorders including cardiovascular diseases (athero-sclerosis), metabolic disorders (insulin resistance, type 2 diabetes), inflam-matory bowel disease (IBD), asthma, and various epithelial cancers in developed countries have been linked to changes in diet, microbiota, and host-signaling systems including gastrointestinal immune and neuroendo-crine systems (Fig. 2.1).

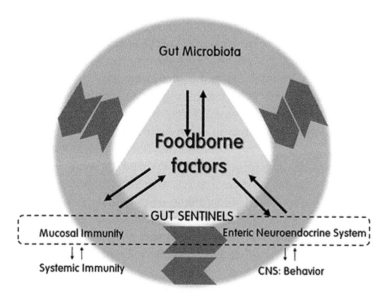

FIGURE 2.1   Food network in the mucosal exposome.

In the following sections, different aspects of the crosstalk in the mucosal exposome and effector biomarkers in the crosstalk are presented to address:

- Monitoring of critical control points in exposomal crosstalk.
- Clinical outcomes from exposomal crosstalk.
- Intervention and prevention of mucosal toxicity and disorders.

2.2   CROSSTALK BETWEEN FOOD COMPONENTS AND GUT MICROBIOTA

Digested food ingredients are the important nutritional source for gut microbes. Depending on the composition of the ingredients, the ecological dominancy and diversity of gut microbiota are modulated.[5] As fatty acids and simple carbohydrates in the diet are absorbed and depleted during transit through the small intestine, the dominant microorganisms are in competitive relationship with the host for simple sugars in the small intestine. Moreover, the intraluminal pH gradually increases in the small intestine from pH 6 to about pH 7.4 in the terminal ileum, but drops to 5.7 in the caecum. However, it again gradually increases, reaching pH 6.7 in the rectum.[6,7] Since the small intestine is more alkaline, and has higher levels of oxygen, bile acids, and antimicrobial peptides than the colon, the microbial community of the small intestine is dominated by fast -growing facultative anaerobes such as phylum Firmicutes (the segmented filamentous bacteria (SFB), Lactobacillaceae) and phylum Proteobacteria (Enterobacteriaceae, and *Helicobacter* spp.) that tolerate these adverse mucosal environment. As most simple sugars are depleted in the small intestine, the colonic bacteria need to ferment the remaining complex polysaccharides as the principal carbon sources that reach the colon. Moreover, lower levels of antimicrobial peptides and slower transit time in colon facilitate the growth of fermentative polysaccharide-degrading anaerobes, notably Bacteroidetes (Bacteroidaceae), Actinobacteria (Bifidobacteriaceae), and Firmicutes (Clostridia). According to a preclinical investigation, a high-fat diet (HFD) alters the composition of the intestinal microbiome of mice independent of obesity. HFD reduces the proportions of phylum Bacteroidetes and increased the proportions of phylum Proteobacteria,[8] whereas high-fiber containing foods such as important nondigestible plant carbohydrate increases proportions of Firmicutes bacteria.[9] In addition to nutritional factors, host-derived mucosal components can select characteristic microbiome profiles. For instance, gut bacteria can utilize host mucin and the bacterial preference to utilize complex glycans of the mucin can provide some gut microbes with a growth advantage and access to spatially privileged niche in the gut mucus layer.[5] Most gut microbes are located in the upper gel-like layer, but the lower dehydrated layer of mucus is not easy for microbes to penetrate.

While nutritional components determine the microbial community in the gut, one of most influential actions of gut microbiota on the diet is the metabolic modification of the food components, leading to changes in their bioavailability in the body after absorption.[10] Through the microbial metabolism, modified food components can be either beneficial or detrimental to human health. For instance, dietary phospholipid phosphatidylcholine (PC) metabolites such as choline, trimethylamine (TMA) N-oxide, and betaine have been positively correlated with the risk of many cardiovascular diseases.[11,12] The dietary phospholipid PC (lecithin) is enriched in foods such as red meat, eggs, milk, and certain fish. Moreover, these food stuffs are also rich in l-carnitine which can be converted into TMA by gut microbiota like PC.[13] Furthermore, TMA is metabolized further by hepatic flavin mono-oxygenase or gut microbial enzymes to TMA N-oxide (TMAO) which is definitely a harmful metabolite in cardiovascular inflammatory diseases.[10,11] TMAO increases cholesterol deposition in the intestines, the liver, and artery wall while reducing the cholesterol clearance from peripheral cells such as those in the artery wall. Although there is no definite mechanism, vegans and vegetarians produce less TMAO from l-carnitine than omnivores, suggesting differential profiles of TMA-metabolizing gut microbiota in the host dependent on dietary fibers and/or fiber-associated components.[13]

In contrast with disease-promoting microbial food metabolites, some food materials such as dietary fibers can be converted into more health-promoting components by virtue of gut microbial metabolism. Soluble parts of the dietary fibers such as pectin, gum, and mucilages can be degraded to bioactive short-chain fatty acids (SCFAs, saturated aliphatic organic acids consisting of one to six carbons). Acetate (C2), propionate (C3), and butyrate (C4) are mostly present in colon at % ratio of 60:20:20.[14–16] In the cecum and large intestine, 95% of the produced SCFAs are rapidly absorbed by the colonocytes, whereas the remaining 5% are excreted in the feces.[14,17] The proximal part of the colon is the principal site of fermentation of indigestible fibers by saccharolytic fermenting phylum Bacteroidetes.[17] Fermentation of bacterial proteins and amino acids by the primary fermenters, like Bacteroidetes, also results in production of SCFA. When fermentable dietary fibers become limiting in the more distal parts of the large intestine with increase in the luminal pH, populations of the butyrate-producing

Bacteroides bacteria decrease and the acetate- and propionate-producing Bacteroides-related bacteria become dominant.[18] However, fermentation of bacterial proteins and amino acids occurs in the more distal part of the colon by secondary proteolytic fermenters, leading to productions of potentially toxic metabolites such as amines, phenolic compounds, and volatile sulfur compounds in the distal colon.[18]

2.3   CROSSTALK-BASED MODULATION OF HUMAN GUT HEATH

Dietary components or toxic contaminants can influence human gut health via direct regulation of mucosa-associated tissues with immunological or neuroendocrine functions. Moreover, some of dietary xenobiotics can be transformed to mucoactive metabolites by gut microbial enzymes. In addition to the microbial xenobiotic metabolism, gut-specific metabolic enzymes of host cells including enterocytes and immune-related cells can reduce or increase the biological activity of the foodborne xenobiotics. Moreover, these gut-derived metabolic activities may influence the efficacy of therapeutics and clinical course in the patients. In the present section, functions and mechanisms of the gut metabolites will be addressed particularly in terms of inflammatory and oncogenic disorders in the gut exposome.

2.3.1   HOST-CELL-DERIVED METABOLISM OF FOODBORNE XENOBIOTICS AND FOOD-MEDIATED ALTERATION OF DRUG METABOLISM

The cytochromes P450 (CYP) are the principal enzymes involved in the biotransformation of drugs and other foreign compounds. These CYPs are responsible for the majority of phase I drug metabolism reactions and the most common P450 cytochrome subfamily expressed in the mucosa of the human small intestine is CYP3A, which represents an average content of approximately 80% of spectrally determined P450 content, followed by CYP2C9 (15%). The CYP1A cytochrome is expressed in the duodenum, together with less abundant levels of CYP2C8–10 and CYP2D6.[19] Many food toxicants can be metabolized by the CYP superfamily such as CYP3A4 (aflatoxin B1, benzo[a]pyrene-7,8-dihydrodil) and CYP1B1 (heterocyclic amines).[19] Disruption of the intestinal barrier allows the translocation of detrimental bile acids including deoxycholic acid and lithocholic acid. These secondary bile acids are metabolized by the gut microbiota and are potentially toxic end products of cholesterol metabolism and their concentrations must be tightly regulated. However, bile-acid-induced CYP3A4 can be an important potential feed-forward protection of mucosal integrity against these hepatoxic or carcinogenic bile acids.[20] Moreover, intestinal CYP enzymes can also be protective against food toxicant-induced pro-inflammatory insults by mediating production of anti-inflammatory colonic corticosterone or arachidonic acid metabolites such as 11,12-DiHETE (dihydroxy-5Z,8Z,11Z,17Z-eicosatetraenoic acid) and 14,15-DiHETE.[21,22]

In addition to the metabolism of food toxicants, dietary components can alter normal drug metabolism and thus influence patient outcome. However, some of these nutrient–drug interactions may have little or no impact on the patient, whereas others may be fatal, depending on various host factors such as age, gender, and presence of underlying diseases.[23] Therefore, the interactions in the risky population need to be recognized, understood, predicted, and then managed to control the food–drug interactions. The clinical effects of grapefruit juice include a significant reduction in the intestinal conversion of drugs metabolized by the CYP3A4 enzyme, resulting in an increase in systemic exposure, leading to adverse drug reactions and toxicity.[24,25] In addition, flavonoids found in grapefruit juice such as naringin and hesperidin also inhibit the influx transporter protein family, namely organic anion transporting peptide and the overall effects include reduced bioavailability and subsequent decreased systemic and tissue concentrations of the exposed drugs. As inhibition of CYP3A4 can last for longer than 3 days after ingestion of grape fruit juice, CYP3A4-dependent medications need to be applied when the new CYP3A4 enzyme can be synthesized in the gut wall after the clearance of the grape fruit components. Moreover, herbal drug components such as andrographolide, curcumin, and lycopene produce no significant induction effects on CYP1A2, CYP2D6, and CYP3A4, whereas bergamottin and resveratrol are significant inducers of enterocyte CYP1A2 in vitro when assessed at the gastrointestinal concentration.[26] Clinical investigations on the herbal dietary supplements showed regulation of CYP metabolism in human such as *Gingko biloba* extract for regulation of CYP3A4 and CYP2C19, milk thistle/silymarin for CYP2C9 inhibition, goldenseal/berberine for CYP3A4 and CYP2D6 inhibition, and Echinacea for inhibition of CYP3A4, CYP1A2, and CYP2C9.[27] Therefore, considering crosstalk between diet and drugs, careful and evidence-based guidelines should be set up for the safe and efficacious medication and appropriate intake of dietary components.

2.3.2   GUT MICROBIOTA-DERIVED METABOLISM OF FOOD COMPONENTS AND CONTAMINANTS AND FUNCTIONS OF METABOLITES IN INTEGRITY OF GUT-RELATED HEALTH

The high metabolic capacity of the intestinal flora is mostly owing to its enormous pool of enzymes that catalyze reactions in phase I and phase II of xenobiotic metabolism of dietary components and foodborne contaminants. This high CYP enzymatic activity of the major bacterial strains come from the human fecal flora.[19] Intestinal *Eubacterium aerofaciens*, *Desulfomonas pigra*, and S*treptomyces coelicolor* A3 isolated from human feces were found to have a CYP -like genes.[28,29] Gut microbiota-mediated xenobiotic metabolism may result in different outcome of bioactive food components, which could be either beneficial or harmful in host integrity. As commented in the previous section, PC and l-carnitine rich in the red meat are converted to TMA and TMAO which trigger macrophage activation and make endothelial injuries, leading to cardiovascular lesions. Moreover, endogenous factors such as the primary bile acids are also converted to the hepatoxic or carcinogenic bile acids such as deoxycholic acid and lithocholic acid by gut microbiota. High physiologic concentrations of the secondary bile acids can cause oxidative and nitrosative stress, DNA damage, and apoptosis and contribute to the flat adenoma lesions, hyperplasia of Peyer's patches and chronic colitis. Frequently repeated and prolonged exposure to the secondary bile acids can lead to the generation of genomic instability, development of apoptosis resistance, and, ultimately, cancer.[30,31]

In contrast to the harmful effects of gut microbiota-mediated metabolism, flavonoids and indoles from foods such as cruciferous vegetables and gut microbial metabolites can be effective in activating aryl hydrocarbon receptor (AhR), an important nuclear receptor for regulating many genes that control immunity and inflammation, whereas some of food components such as resveratrol have antagonistic effects on AhR-linked signals.[10] The essential amino acid tryptophan is another nutrient that might have anti-inflammatory activities. As found in various vegetables and fish, tryptophan is metabolized by the microbiota (e.g., *Lactobacilli*) in stomachs and the small intestine to indole-3-aldehyde, another AhR agonist. Tryptophan can be converted to kynurenine, another AhR ligand, by intestinal indoleamine 2,3-dioxygenase.[32] Food–microbiota-derived AhR activation may contribute to the mucosal immune tolerance and shape the composition of the intestinal microbial community by promoting regulatory T cells and induction of interleukin 22, which affect the microbiota, providing resistance to colonization by pathogens such as *Candida albicans* and protecting the mucosa against inflammation.[33] This "Food–Microbiota–AhR" triad might represent an important strategy of human exposomal co-evolutionary commensalism for fine tuning of host mucosal defense. However, persistent activation and induction of AhR has been extensively associated with poor prognosis of human epithelial cancers, particularly the late stage of tumorigenesis in most human tumors due to its selective advantage for tumor cells and immune suppression as previously commented.[34,35] For instance, tumor cells secrete kynurenine which is produced via tryptophan-2,3-dioxygenase and acts as another tryptophan-derived ligand of the AhR in an autocrine/paracrine fashion, leading to suppression of antitumor immune responses and promotions of tumor-cell survival and motility.[36]

Tryptophan metabolites such as kynurenic acid and niacin also interact with certain G-protein-coupled receptors (GPCRs).[37] In addition to the some tryptophan metabolites, n-3 polyunsaturated fatty acids and gut microbial SCFA can trigger epithelial GPCR-linked sensing of the bioavailability of nutrients while SCFA can inhibit histone deacetylase (HDAC), which promotes the immunological tolerance and anti-inflammatory responses. However, metabolite-sensing GPCRs also have features of regulatory function during gastrointestinal inflammatory responses and metabolic disorders by altering leptin production, adiposity, and insulin secretion.[37] Manipulation of gut immunity and metabolism through metabolite-sensing GPCRs, HDACs, or transcription factors, such as AhR would provide new insights into potential disease-mitigating actions of the food–microbiota–immune axis against many mucosal inflammatory and oncogenic diseases in human populations.

2.4 NEUROENDOCRINE REGULATION IN MUCOSAL EXPOSOME

Scattered throughout the intestinal epithelial layer, the enteroendocrine cells (EECs) which, despite only making up 1% of the population of the epithelial cells, collectively form the largest endocrine system in human body. In addition, EECs are embedded in a majority of nonendocrine cells, including absorptive enterocytes, goblet cells, and Paneth cells and respond to luminal antigens including nutrients, taste molecules, and bacterial metabolites by secreting more than 20 peptide hormones including cholecystokinin (CCK), glucagon-like peptide 1 and 2 (GLP-1, GLP-2), glucose-dependent insulinotropic peptide, peptide YY (PYY), somatostatin and ghrelin, and bioactive amines such as serotonin (5-hydroxy-trytamine [5-HT]).[38] EECs act as chemical sensors of the luminal contents which transduce transepithelial signals to trigger exocytosis of neuroendocrine hormones into the circulation or the neighboring enterocytes, EECs, or afferent neuronal cells. Sometimes, gut peptides can diffuse into the systemic circulation or lymphatics to eventually reach the brain and act on central receptors in an endocrine-cascade fashion. Food components in the intestinal lumen may signal to the afferent nerve terminals two different routes. First, luminal stimulants may be ferried across the epithelial cells by specific transport proteins where they then interact with specialized receptors on the afferent nerve terminals. Second, luminal antigen-recognizing EECs transduce signals via activated receptors and secrete hormones which bind to and activate the afferent nerve cells. Among the EECs, K and I cells are primarily localized in the proximal intestine, whereas L cells are predominant in the distal intestine and colon.[39] Gut peptides, including ghrelin, leptin, CCK, GLP-1, and PYY, whose receptor activation leads to neuronal firing in the vagal and spinal afferents indirectly, via activation of neurons of the enteric nervous system, which have also been shown to express gut peptide receptors. In addition to activation of the afferent nerve terminals and enteric nerve cells, neuroendocrine hormones from some EECs such as the duodenal I cells can modulate the metabolic homeostasis by sensing long-chain fatty acids via activation of GPCRs. The I cells undergo calcium flux and membrane depolarization, culminating in secretion of the hormone CCK which acts through the CCK receptor to cause gall bladder contraction and pancreatic enzyme secretion, allowing efficient assimilation of the dietary long-chain fatty acids from the lumen.[40,41]

2.4.1 EEC-MODULATED IMMUNE REGULATION IN RESPONSE TO DIET AND MICROBIOTA

Beyond their classical role of detecting luminal nutrients and taste molecules, EECs also detect and respond to pathogens via the expression of pattern recognition receptors and to the intestinal microbiome via the expression of GPCRs for the metabolites such as SCFAs or the endocannabinoids (N-oleoylethanolamide, and 2-oleoylglycerol) produced by the commensal microbes.[42,43] In response to pathogens and microbial metabolites, EECs secrete peptide hormones and classical cytokines to the surrounding mucosal leucocytes. In addition to classical cytokine receptors and neuroendocrine hormone receptors, immune cells express various types of neuroendocrine peptide hormones such as adrenocorticotropic hormone, endorphins, thyrotropin, chorionic gonadotropin, and growth hormone which have direct immunomodulatory effects.[44] These neuroendocrine peptide hormones also signal to vagal afferents triggering an anti-inflammatory vagal reflex. The resulting acetylcholine released from vagal efferent nerve downregulates the inflammatory responses in the surrounding mucosal immune cells. This anti-inflammatory pathway was first addressed in the hemorrhagic shock.[45] Nutritional stimulation of CCK via a HFD has protective functions against inflammation via a vagal reflex which results in the release of acetylcholine which inhibits the pro-inflammatory cytokine secretion from macrophage. Ingestion of high amounts of fat induces release of CCK that triggers the peripheral vagal afferent neurons and activation of CCK receptors then activates vagal efferent nerves secreting parasympathetic acetylcholine which downregulates excessive inflammatory responses to luminal antigens including commensal bacteria and dietary components.[45] Therefore, the protective effects of HFD on mucosal inflammation and intestinal permeability are abrogated by vagotomy and administration of antagonists against CCK receptors or nicotinic receptors. The EEC-modulated vagal afferent signals also modulate classical feeding pathways resulting in altered fat deposition. This in turn promotes the adipocytes to secrete adipokines such as leptin which ultimately affect the immune cell functions. Moreover, homing T lymphocytes directly influence the functions of peptide hormones or the population of EECs by regulating self-renewal or differentiation from a large reservoir of stem cells.[42]

2.4.2 DIET/GUT MICROBIOTA-MODULATED NEURONAL REGULATION, NUTRIENT SENSING, AND ENERGY BALANCE

Pre-absorptive nutrients trigger a gut–brain axis by stimulating the release of gut peptides, which activate important multiple neural or humoral pathways in the rhombencephalon and hypothalamus to regulate energy balance by altering both energy intake and energy expenditure.[46] These gut–brain signals are critical to the controls of food intake, especially in termination of a meal (satiation) and length of the intermeal interval (satiety).[38] Gut microbiota can influence the central nervous system and alter behavior in the microbiota–gut–brain axis. Changes in mood and behavior are present during acute inflammation in experimental IBD which is associated with dysbiosis and administration of probiotics could prevent the behavioral defects in host with IBD,[47] which supports the clinical observation that cortical changes and subsequent psychiatric disorders are common in about 50% patients with IBD. Irritable bowel syndrome (IBS) is functional gastrointestinal disorder and up to 50–90% of patients with IBS show psychiatric comorbidity. In particular, probiotic treatment using *Bifidobacterium longum* str. NCC3001 normalizes the anxiety-like behavior in mice with moderate colitis, possibly by a vagal nerve-dependent mechanism to regulate brain-derived neurotropic factor.[48] In line with the psychiatric impact of the gut microbiota, the micro-biota–gut–brain axis can modulate nutrient sensing and energy balance. As indirect evidence, bacterial SCFAs induce secretion of serotonin or PYY via GPCR activation on EECs. As such, microbial byproducts can affect functional expression of intestinal peptide hormones and potently modulate food intake leading to satiation. Moreover, dietary prebiotic components such as nondigestible carbohydrates change the ability of the gut microbiota to alter glucagon-like peptides, serotonin, or PYY via SCFAs and to promote satiety and weight loss.[38,49] Prebiotics increase the abundance of *Bifidobacterium* and *Lactobacillus*, or microbiota profiles with ability to modulate EEC differentiation and subsequently gut glucagon-like peptide production, ultimately leading to increased satiety, decreased energy intake, fat mass and body weight, suggesting the importance of "diet-microbiota–gut–brain" axis in the control of food intake.[50–52] Moreover, probiotic-producing SCFA and endocannabinoid stimulate production of glucagon-like peptides, which improves gut permeability and reduces low-grade inflammation and metabolic disorders in obesity and type 2 diabetes.[49]

2.5 IMPLICATIONS OF FOODBORNE PATHOGENS IN MUCOSAL EXPOSOME AND RELATED BIOMARKERS

Various types of effector molecules in the mucosa are altered by the invasiveness of the pathogens and their toxicity, linked to the gastrointestinal disorders including inflammatory diseases and epithelial cancers. Based on the mechanisms of actions of the foodborne pathogens, microbial invasion and toxin production can cause injuries in components of the mucosal exposome, including gastrointestinal barriers, composition, and diversity of gut microbiota, mucosal immunity, and mucosal neuroendocrine system.

2.5.1 IMPACTS ON THE GASTROINTESTINAL BARRIER AS THE TARGET OF GUT PATHOGENS

Many of gastroenteritis-causing pathogens may damage the epithelial integrity by altering the water balance of enterocytes, ultimately leading to severe diarrhea. Most well-known diarrhea-causing cholera enterotoxin and entero-toxigenic *Escherichia coli* (ETEC)-producing heat-labile toxin activate G-protein-coupled adenylate cyclase, resulting in increase of cyclic AMP. Moreover, production of cyclic GMP via activation of guanylate cyclase is increased by heat-stable toxin of ETEC. cAMP and cGMP then leads to activation of protein kinase G and A, respectively, which phosphorylate the cystic fibrosis transmembrane conductance regulator chloride channel proteins and lead to ATP-mediated efflux of chloride ions and secretion of H_2O, Na^+, K^+, and HCO_3^- into the intestinal lumen. Consequently, the entry of Na^+ and water into enterocytes is diminished. These combined effects result in rapid fluid loss from the intestine. Some foodborne bacteria and mold cause epithelial cytotoxicity through cytotoxins such as ribosome-inactivating toxins which specifically bind to the eukaryotic ribosomes and arrest the global protein translation in the enterocytes. For instance, shiga-toxin, shiga-like toxins, and trichothecene mycotoxins which contaminate diverse food stuffs can cause acute gastroenteritis with diarrhea in humans and animals via inactivation of epithelial ribosomal inactivation and subsequent apoptosis. In addition to actions of enterotoxins, bacterial colonization itself can cause injuries in the gut epithelia. Instead of active invasive infection, persistent colonization of enteropathogenic *E. coli* (EPEC) and diffusely adherent *E. coli* make histological changes in the epithelial surface structure via type III secretion system which delivers microbial metabolites, disrupting the cytoskeletal structure of the host enterocytes. Subsequently, loosening of the tight junction between enterocytes results in loss of epithelial polarity and cell death, which facilitates the influx of luminal microbes and severe inflammation. Moreover, mucosa-associated *E. coli* strains, including EPEC, are frequently observed on the intestinal mucosal surface in patients with chronic diseases such as IBD and colorectal cancer.[53,54] Enteropathogenic *E. coli* also suppresses the expression of DNA repair protein, and therefore, it has the potential to play a role in promoting intestinal cancer.[53,55]

2.5.2 IMPACTS ON THE MUCOSAL IMMUNITY AND EFFECTOR BIOMARKERS

Many gastroenteritis-causing pathogens can impair mucosal immunity. This disruption in the mucosal immune defense may allow more pathogens to spread systemically in host body. One prevalent mechanism is the dysfunction of mucosal innate immune defenses via host cell cytoxicity and immune evasion caused by foodborne pathogens including enteroinvasive

E. coli (EIEC), *Listeria monocytogenes*, and *Salmonella enterica*. Although they are engulfed by macrophages, they can survive during the macrophage migration from the mucosal area to the lymph nodes. They can escape or inhibit the intracellular lysophagosomal fusion and degradation process in the monocytes. Some foodborne pathogens secrete the cytotoxins including ribosome-inactivating toxins against the phagocytes and lymphocytes. In terms of clinical diagnosis, there are anemia, leukopenia, and thrombocytopenia in response to foodborne ribosome-inactivating foodborne toxins. Moreover, enterocyte cytokines can be considered as diagnostic biomarkers of mucosal immune toxicity of ribosome-inactivating microbial toxins.[56] As the hematological biomarker, heptoglobin is specifically increased by deoxynivalenol (DON), a trichothecene mycotoxin, and has been verified as an effector biomarker.[57] Heptoglobin is an important protein in acute hemolysis, playing important roles in adsorbing and recycling of the free hemoglobin.[56,57] Although heptoglobin and serum amyloid A are increased by DON for 1–8 days in mini-pig gavage model,[56] other biomarkers of extended-time exposure would be more useful predictors for the chronic impacts. For instance, immunoglobulin M, A, and G is decreased during the early time of exposure to trichothecenes,[57] IgA production is significantly increased in mucosa and spleen after dietary exposure for 8 weeks and excessive increase of IgA in blood forms immune complex, is deposited in glomerulus, and leads to IgA nephritis symptoms, which is similar to human Berger's disease as the autoimmune diseases.[58–60] In terms of regulation, only 1–10 ppm of dietary DON exposure in animal experiments can increase IgA as a subchronic effector biomarker of hypersensitivity. As these exposure levels are almost close to the regulation limit (1 ppm) in most countries,[61,62] it is necessary to reestablish safe levels of trichothecenes. Mucosal or serum IgA and IgA-immune complex thus are expected very sensitive effector biomarkers of chronic exposure in association with mucosal stress-related systemic diseases including IgA nephritis. In addition to the IgA-related hypersensitivity, hyper-responsiveness to self-antigens is associated with the autoimmune disorders, which can be observed in several gut infections by foodborne bacteria including *Campylobacter jejuni* and *Helicobacter pylori*. Guillain–Barre syndrome (GBS) is an acute neuromuscular paralysis as a serious sequelae of *Campylobacter* infection. An estimated 1 case of GBS occurs for every 1000 cases of Campylobacteriosis and up to 40% of patients with the autoimmune syndrome have evidence of recent *Campylobacter* infection, which is mechanistically due to the host antigens sequence homology with *C. jejuni* serogroup O19.[63] Another example of molecular mimicry is the LPS O antigen of 80% of *H. pylori* strains with homology with human Lewis serum antigen, contributing to the bacteria-induced gastritis and tumor formation.[64]

2.5.3 IMPACTS ON THE MUCOSAL NEUROENDOCRINE SYSTEM: BASED ON CASE EXAMPLES OF FOODBORNE TRICHOTHECENE MYCOTOXIN EXPOSURE

As speculated from the nomenclature of synonym vomitoxin for DON, this toxin was originally found to trigger vomiting and anorexia in swine, rapidly leading to weight loss. This toxicopathologic pattern can be commonly observed in animals exposed to 8-ketotrichothecene mycotoxins.[65,66] Mechanistically, mucosal or systemic inflammatory factors (IL-1β, IL-6, TNF-α, cyclooxygenase-2 (COX-2), microsomal prostaglandin synthase1 (mPGES-1)) triggering feed refusal are also increased by 8-ketotrichothecenes.[67] However, anorexia is not generally associated with systemic inflammatory factors. Instead, serotonin (monoamine 5-HT) is well-known neuro-endocrine biomarker of trichothecene exposure and related anorexia since antagonists of serotonin receptor suppresses feed refusal.[68–70] Although serotonin and other related metabolites can be effector biomarkers, they are temporarily elevated in acute exposure to DON and thus they are not enough to represent extended or chronic exposure to 8-ketotrichothecenes. Serotonin-independent serotonin receptor activation its upstream modulators including peptide $PYY_{3–36}$ also mediates DON-induced anorexia by regulating the food intake and energy metabolism.[71]

2.5.4 APPLICATIONS AND PERSPECTIVES OF THE EFFECTOR BIOMARKERS OF FOODBORNE MICROBIAL PATHOGENICITY

Foodborne-pathogen-mediated metabolites are crucial exposure biomarkers to indicate the luminal crosstalk between microbes and host. Indeed, the effector biomarkers of host molecules can be also applicable to develop quantitative bioassay such as the biomarker-linked reporter assay of the pathogenic agents, which would be supportive of the microbial and chemical-based analysis methods. Recently, a human-enterocyte-based biomarker reporter system has been developed to assess the early mucosal actions of mucosal xenobiotics as an alternative to animal experimentation,[72,73] which may address some ethical issues in the use of animal models. Moreover, development of biomimetic models close to physiological conditions would substantially contribute to reduction of animal use in the pathological assessment of the foodborne pathogens and toxins.

KEYWORDS

- mucosal exposome
- food toxicity biomarkers
- human microbiota
- mucosal immunity
- gastrointestinal neuroendocrine system

REFERENCES

1. Fiocchi, C. Towards a 'Cure' for IBD. *Dig. Dis.* **2012**, *30*, 428–433.
2. Putignani, L.; Dallapiccola, B. Foodomics as Part of the Host–Microbiota–Exposome Interplay. *J Proteomics* **2016**, *47*, 3–20.
3. Fiocchi, C. Integrating Omics: The Future of IBD? *Dig. Dis.* **2014**, *32* (Suppl. 1), 96–102.
4. De Rosa, V.; Galgani, M.; Santopaolo, M.; Colamatteo, A.; Laccetti, R.; Matarese, G. Nutritional Control of Immunity: Balancing the Metabolic Requirements with an Appropriate Immune Function. *Semin. Immunol.* **2015**, *27*, 300–309.
5. Donaldson, G. P.; Lee, S. M.; Mazmanian, S. K. Gut Biogeography of the Bacterial Microbiota. *Nat. Rev. Microbiol.* **2016**, *14*, 20–32.
6. Fallingborg, J. Intraluminal pH of the Human Gastrointestinal Tract. *Dan. Med. Bull.* **1999**, *46*, 183–196.
7. Press, A. G.; Hauptmann, I. A.; Hauptmann, L.; Fuchs, B.; Fuchs, M.; Ewe, K.; Ramadori, G. Gastrointestinal pH Profiles in Patients with Inflammatory Bowel Disease. *Aliment. Pharmacol. Ther.* **1998**, *12*, 673–678.
8. Hildebrandt, M. A.; Hoffmann, C.; Sherrill-Mix, S. A.; Keilbaugh, S. A.; Hamady, M.; Chen, Y. Y.; Knight, R.; Ahima, R. S.; Bushman, F.; Wu, G. D. High-Fat Diet Determines the Composition of the Murine Gut Microbiome Independently of Obesity. *Gastroenterology* **2009**, *137*, 1716–1724 e1711–1712.
9. Walker, A. W.; Ince, J.; Duncan, S. H.; Webster, L. M.; Holtrop, G.; Ze, X.; Brown, D.; Stares, M. D.; Scott, P.; Bergerat, A.; Louis, P.; McIntosh, F.; Johnstone, A. M.; Lobley, G. E.; Parkhill, J.; Flint, H. J. Dominant and Diet-Responsive Groups of Bacteria within the Human Colonic Microbiota. *ISME J.* **2011**, *5*, 220–230.
10. Tilg, H.; Moschen, A. R. Food, Immunity, and the Microbiome. *Gastroenterology* **2015**, *148*, 1107–1119.
11. Hartiala, J.; Bennett, B. J.; Tang, W. H.; Wang, Z.; Stewart, A. F.; Roberts, R.; McPherson, R.; Lusis, A. J.; Hazen, S. L.; Allayee, H.; Consortium, C. A. Comparative Genome-Wide Association Studies in Mice and Humans for Trimethylamine N-Oxide,

a Proatherogenic Metabolite of Choline and l-Carnitine. *Arterioscler. Thromb. Vasc. Biol.* **2014**, *34*, 1307–1313.

12. Wang, Z.; Klipfell, E.; Bennett, B. J.; Koeth, R.; Levison, B. S.; Dugar, B.; Feldstein, A. E.; Britt, E. B.; Fu, X.; Chung, Y. M.; Wu, Y.; Schauer, P.; Smith, J. D.; Allayee, H.; Tang, W. H.; DiDonato, J. A.; Lusis, A. J.; Hazen, S. L. Gut Flora Metabolism of Phosphatidylcholine Promotes Cardiovascular Disease. *Nature* **2011**, *472*, 57–63.

13. Koeth, R. A.; Wang, Z.; Levison, B. S.; Buffa, J. A.; Org, E.; Sheehy, B. T.; Britt, E. B.; Fu, X.; Wu, Y.; Li, L.; Smith, J. D.; DiDonato, J. A.; Chen, J.; Li, H.; Wu, G. D.; Lewis, J. D.; Warrier, M.; Brown, J. M.; Krauss, R. M.; Tang, W. H.; Bushman, F. D.; Lusis, A. J.; Hazen, S. L. Intestinal Microbiota Metabolism of l-Carnitine, a Nutrient in Red Meat, Promotes Atherosclerosis. *Nat. Med.* **2013**, *19*, 576–585.

14. Cook, S. I.; Sellin, J. H. Review Article: Short Chain Fatty Acids in Health and Disease. *Aliment. Pharmacol. Ther.* **1998**, *12*, 499–507.

15. Tan, J.; McKenzie, C.; Potamitis, M.; Thorburn, A. N.; Mackay, C. R.; Macia, L. The Role of Short-Chain Fatty Acids in Health and Disease. *Adv. Immunol.* **2014**, *121*, 91–119.

16. Hamer, H. M.; Jonkers, D.; Venema, K.; Vanhoutvin, S.; Troost, F. J.; Brummer, R. J. Review Article: The Role of Butyrate on Colonic Function. *Aliment. Pharmacol. Ther.* **2008**, *27*, 104–119.

17. den Besten, G.; van Eunen, K.; Groen, A. K.; Venema, K.; Reijngoud, D. J.; Bakker, B. M. The Role of Short-Chain Fatty Acids in the Interplay between Diet, Gut Microbiota, and Host Energy Metabolism. *J. Lipid Res.* **2013**, *54*, 2325–2340.

18. Macfarlane, G. T.; Macfarlane, S. Fermentation in the Human Large Intestine: Its Physiologic Consequences and the Potential Contribution of Prebiotics. *J. Clin. Gastroenterol.* **2011**, *45 Suppl.*, S120–127.

19. Bezirtzoglou, E. E. Intestinal Cytochromes P450 Regulating the Intestinal Microbiota and its Probiotic Profile. *Microb. Ecol. Health Dis.* **2012**, *23*.

20. Cheng, J.; Fang, Z. Z.; Kim, J. H.; Krausz, K. W.; Tanaka, N.; Chiang, J. Y.; Gonzalez, F. J. Intestinal CYP3A4 Protects against Lithocholic Acid-Induced Hepatotoxicity in Intestine-Specific VDR-Deficient Mice. *J. Lipid Res.* **2014**, *55*, 455–465.

21. Ahlawat, S.; Xie, F.; Zhu, Y.; D'Hondt, R.; Ding, X.; Zhang, Q. Y.; Mantis, N. J. Mice Deficient in Intestinal Epithelium Cytochrome P450 Reductase Are Prone to Acute Toxin-Induced Mucosal Damage. *Sci. Rep.* **2014**, *4*, 5551.

22. Zhu, Y.; Xie, F.; Ding, L.; Fan, X.; Ding, X.; Zhang, Q. Y. Intestinal Epithelium-Specific Knockout of the Cytochrome P450 Reductase Gene Exacerbates Dextran Sulfate Sodium-Induced Colitis. *J. Pharmacol. Exp. Ther.* **2015**, *354*, 10–17.

23. Maka, D. A.; Murphy, L. K. Drug–Nutrient Interactions: A Review. *AACN Clin. Issues* **2000**, *11*, 580–589.

24. Dolton, M. J.; Roufogalis, B. D.; McLachlan, A. J. Fruit Juices as Perpetrators of Drug Interactions: The Role of Organic Anion-Transporting Polypeptides. *Clin. Pharmacol. Ther.* **2012**, *92*, 622–630.

25. Kane, G. C.; Lipsky, J. J. Drug–Grapefruit Juice Interactions. *Mayo Clin. Proc.* **2000**, *75*, 933–942.

26. Koe, X. F.; Tengku Muhammad, T. S.; Chong, A. S.; Wahab, H. A.; Tan, M. L. Cytochrome P450 Induction Properties of Food and Herbal-Derived Compounds Using a Novel Multiplex RT-qPCR In Vitro Assay, a Drug–Food Interaction Prediction Tool. *Food Sci. Nutr.* **2014**, *2*, 500–520.

27. Hermann, R.; von Richter, O. Clinical Evidence of Herbal Drugs as Perpetrators of Pharmacokinetic Drug Interactions. *Planta Med.* **2012**, *78*, 1458–1477.

28. Lei, L.; Waterman, M. R.; Fulco, A. J.; Kelly, S. L.; Lamb, D. C. Availability of Specific Reductases Controls the Temporal Activity of the Cytochrome P450 Complement of *Streptomyces coelicolor* A3(2). *Proc. Natl. Acad. Sci. USA* **2004**, *101*, 494–499.

29. Sperry, J. F.; Wilkins, T. D. Presence of Cytochrome *c* in *Desulfomonas pigra*. *J. Bacteriol.* **1977**, *129*, 554–555.

30. Bernstein, C.; Holubec, H.; Bhattacharyya, A. K.; Nguyen, H.; Payne, C. M.; Zaitlin, B.; Bernstein, H. Carcinogenicity of Deoxycholate, a Secondary Bile Acid. *Arch. Toxicol.* **2011**, *85*, 863–871.

31. Saracut, C.; Molnar, C.; Russu, C.; Todoran, N.; Vlase, L.; Turdean, S.; Voidazan, S.; Copotoiu, C. Secondary Bile Acids Effects in Colon Pathology. Experimental Mice Study. *Acta Cir. Bras.* **2015**, *30*, 624–631.

32. Zelante, T.; Iannitti, R. G.; Cunha, C.; De Luca, A.; Giovannini, G.; Pieraccini, G.; Zecchi, R.; D'Angelo, C.; Massi-Benedetti, C.; Fallarino, F.; Carvalho, A.; Puccetti, P.; Romani, L. Tryptophan Catabolites from Microbiota Engage Aryl Hydrocarbon Receptor and Balance Mucosal Reactivity via Interleukin-22. *Immunity* **2013**, *39*, 372–385.

33. Ikuta, T.; Kurosumi, M.; Yatsuoka, T.; Nishimura, Y. Tissue Distribution of Aryl Hydrocarbon Receptor in the Intestine: Implication of Putative Roles in Tumor Suppression. *Exp. Cell. Res.* **2016**, *343* (2), 126–134.

34. Esser, C.; Rannug, A. The Aryl Hydrocarbon Receptor in Barrier Organ Physiology, Immunology, and Toxicology. *Pharmacol. Rev.* **2015**, *67*, 259–279.

35. Murray, I. A.; Patterson, A. D.; Perdew, G. H. Aryl Hydrocarbon Receptor Ligands in Cancer: Friend and Foe. *Nat. Rev. Cancer* **2014**, *14*, 801–814.

36. Opitz, C. A.; Litzenburger, U. M.; Sahm, F.; Ott, M.; Tritschler, I.; Trump, S.; Schumacher, T.; Jestaedt, L.; Schrenk, D.; Weller, M.; Jugold, M.; Guillemin, G. J.; Miller, C. L.; Lutz, C.; Radlwimmer, B.; Lehmann, I.; von Deimling, A.; Wick, W.; Platten, M. An Endogenous Tumour-Promoting Ligand of the Human Aryl Hydrocarbon Receptor. *Nature* **2011**, *478*, 197–203.

37. Thorburn, A. N.; Macia, L.; Mackay, C. R. Diet, Metabolites, and "western-lifestyle" Inflammatory Diseases. *Immunity* **2014**, *40*, 833–842.

38. Duca, F. A.; Lam, T. K. Gut Microbiota, Nutrient Sensing and Energy Balance. *Diabetes Obes. Metab.* **2014**, *16 Suppl 1*, 68–76.

39. Mace, O. J.; Tehan, B.; Marshall, F. Pharmacology and Physiology of Gastrointestinal Enteroendocrine Cells. *Pharmacol. Res. Perspect.* **2015**, *3*, e00155.

40. Bertrand, P. P. The Cornucopia of Intestinal Chemosensory Transduction. *Front. Neurosci.* **2009**, *3*, 48.

41. Steinert, R. E.; Beglinger, C. Nutrient Sensing in the Gut: Interactions between Chemosensory Cells, Visceral Afferents and the Secretion of Satiation Peptides. *Physiol. Behav.* **2011**, *105*, 62–70.

42. Worthington, J. J. The Intestinal Immunoendocrine Axis: Novel Cross-Talk between enteroendocrine cells and the Immune System during Infection and Inflammatory Disease. *Biochem. Soc. Trans.* **2015**, *43*, 727–733.

43. Cani, P. D.; Everard, A.; Duparc, T. Gut Microbiota, Enteroendocrine Functions and Metabolism. *Curr. Opin. Pharmacol.* **2013**, *13*, 935–940.

44. Carr, D. J.; Weigent, D. A.; Blalock, J. E. Hormones Common to the Neuroendocrine and Immune Systems. *Drug Des. Deliv.* **1989**, *4*, 187–195.

45. Luyer, M. D.; Greve, J. W.; Hadfoune, M.; Jacobs, J. A.; Dejong, C. H.; Buurman, W. A. Nutritional Stimulation of Cholecystokinin Receptors Inhibits Inflammation via the Vagus Nerve. *J. Exp. Med.* **2005,** *202*, 1023–1029.

46. Bauer, P. V.; Hamr, S. C.; Duca, F. A. Regulation of Energy Balance by a Gut–Brain Axis and Involvement of the Gut Microbiota. *Cell Mol. Life Sci.* **2016,** *73*, 737–755.

47. Emge, J. R.; Huynh, K.; Miller, E. N.; Kaur, M.; Reardon, C.; Barrett, K. E.; Gareau, M. G. Modulation of the Microbiota–Gut–Brain Axis by Probiotics in a Murine Model of Inflammatory Bowel Disease. *Am. J. Physiol. Gastrointest. Liver Physiol.* **2016,** *310*, G989–998.

48. Bercik, P.; Denou, E.; Collins, J.; Jackson, W.; Lu, J.; Jury, J.; Deng, Y.; Blennerhassett, P.; Macri, J.; McCoy, K. D.; Verdu, E. F.; Collins, S. M. The Intestinal Microbiota Affect Central Levels of Brain-Derived Neurotropic Factor and Behavior in Mice. *Gastroenterology* **2011,** *141*, 599–609, 609 e591–593.

49. Everard, A.; Cani, P. D. Diabetes, Obesity and Gut Microbiota. *Best Pract. Res. Clin. Gastroenterol.* **2013,** *27*, 73–83.

50. Paul, H. A.; Bomhof, M. R.; Vogel, H. J.; Reimer, R. A. Diet-Induced Changes in Maternal Gut Microbiota and Metabolomic Profiles Influence Programming of Offspring Obesity Risk in Rats. *Sci. Rep.* **2016,** *6*, 20683.

51. Reid, D. T.; Eller, L. K.; Nettleton, J. E.; Reimer, R. A. Postnatal Prebiotic Fibre Intake Mitigates Some Detrimental Metabolic Outcomes of Early Overnutrition in Rats. *Eur. J. Nutr.* **2015,** 55 (8), 2399–2409.

52. Yang, J.; Summanen, P. H.; Henning, S. M.; Hsu, M.; Lam, H.; Huang, J.; Tseng, C. H.; Dowd, S. E.; Finegold, S. M.; Heber, D.; Li, Z. Xylooligosaccharide Supplementation Alters Gut Bacteria in both Healthy and Prediabetic Adults: A Pilot Study. *Front. Physiol.* **2015,** *6*, 216.

53. Maddocks, O. D.; Short, A. J.; Donnenberg, M. S.; Bader, S.; Harrison, D. J. Attaching and Effacing *Escherichia coli* Downregulate DNA Mismatch Repair Protein In Vitro and Are Associated with Colorectal Adenocarcinomas in Humans. *PLoS One* **2009,** *4*, e5517.

54. Weber, P.; Koch, M.; Heizmann, W. R.; Scheurlen, M.; Jenss, H.; Hartmann, F. Microbic Superinfection in Relapse of Inflammatory Bowel Disease. *J. Clin. Gastroenterol.* **1992,** *14*, 302–308.

55. Maddocks, O. D.; Scanlon, K. M.; Donnenberg, M. S. An *Escherichia coli* Effector Protein Promotes Host Mutation via Depletion of DNA Mismatch Repair Proteins. *MBio* **2013,** *4*, e00152–00113.

56. Mikami, O.; Kubo, M.; Murata, H.; Muneta, Y.; Nakajima, Y.; Miyazaki, S.; Tanimura, N.; Katsuda, K. The Effects of Acute Exposure to Deoxynivalenol on Some Inflammatory Parameters in Miniature Pigs. *J. Vet. Med. Sci.* **2011,** *73*, 665–671.

57. Kim, E. J.; Jeong, S. H.; Cho, J. H.; Ku, H. O.; Pyo, H. M.; Kang, H. G.; Choi, K. H. Plasma Haptoglobin and Immunoglobulins as Diagnostic Indicators of Deoxynivalenol Intoxication. *J. Vet. Sci.* **2008,** *9*, 257–266.

58. Dewa, Y.; Kemmochi, S.; Kawai, M.; Saegusa, Y.; Harada, T.; Shimamoto, K.; Mitsumori, K.; Kumagai, S.; Sugita-Konishi, Y.; Shibutani, M. Rapid Deposition of Glomerular IgA in BALB/c Mice by Nivalenol and its Modifying Effect on High IgA Strain (HIGA) Mice. *Exp. Toxicol. Pathol.* **2011,** *63*, 17–24.

59. Goyarts, T.; Danicke, S.; Tiemann, U.; Rothkotter, H. J. Effect of the Fusarium Toxin Deoxynivalenol (DON) on IgA, IgM and IgG Concentrations and Proliferation of Porcine Blood Lymphocytes. *Toxicol. In Vitro* **2006,** *20*, 858–867.

60. Pestka, J. J.; Moorman, M. A.; Warner, R. L. Dysregulation of IgA Production and IgA Nephropathy Induced by the Trichothecene Vomitoxin. *Food Chem. Toxicol.* **1989,** *27,* 361–368.

61. Park, D. L.; Troxell, T. C. U.S. Perspective on Mycotoxin Regulatory Issues. *Adv. Exp. Med. Biol.* **2002,** *504,* 277–285.

62. Yazdanpanah, H.; Shafaati, A.; Foroutan, S. M.; Zarghi, A.; Aboul-Fathi, F.; Khoddam, A.; Shaki, F.; Nazari, F. Occurrence of Deoxynivalenol in Foods for Human Consumption from Tehran, Iran. *Iran. J. Pharm. Res.* **2014,** *13,* 87–92.

63. Nachamkin, I.; Allos, B. M.; Ho, T. *Campylobacter* species and Guillain-Barre Syndrome. *Clin. Microbiol. Rev.* **1998,** *11,* 555–567.

64. Wang, G.; Ge, Z.; Rasko, D. A.; Taylor, D. E. Lewis Antigens in *Helicobacter pylori:* Biosynthesis and Phase Variation. *Mol. Microbiol.* **2000,** *36,* 1187–1196.

65. Flannery, B. M.; Wu, W.; Pestka, J. J. Characterization of Deoxynivalenol-Induced Anorexia Using Mouse Bioassay. *Food Chem. Toxicol.* **2011,** *49,* 1863–1869.

66. Wu, W.; Flannery, B. M.; Sugita-Konishi, Y.; Watanabe, M.; Zhang, H.; Pestka, J. J. Comparison of Murine Anorectic Responses to the 8-Ketotrichothecenes 3-Acetyl-deoxynivalenol, 15-Acetyldeoxynivalenol, Fusarenon X and Nivalenol. *Food Chem. Toxicol.* **2012,** *50,* 2056–2061.

67. Girardet, C.; Bonnet, M. S.; Jdir, R.; Sadoud, M.; Thirion, S.; Tardivel, C.; Roux, J.; Lebrun, B.; Mounien, L.; Trouslard, J.; Jean, A.; Dallaporta, M.; Troadec, J. D. Central Inflammation and Sickness-Like Behavior Induced by the Food Contaminant Deoxynivalenol: A PGE2-Independent Mechanism. *Toxicol. Sci.* **2011,** *124,* 179–191.

68. Prelusky, D. B. The effect of Low-Level Deoxynivalenol on Neurotransmitter Levels Measured in Pig Cerebral Spinal Fluid. *J. Environ. Sci. Health B* **1993,** *28,* 731–761.

69. Prelusky, D. B. The Effect of Deoxynivalenol on Serotoninergic Neurotransmitter Levels in Pig Blood. *J. Environ. Sci. Health B* **1994,** *29,* 1203–1218.

70. Prelusky, D. B.; Rotter, B. A.; Thompson, B. K.; Trenholm, H. L. Effect of the Appetite Stimulant Cyproheptadine on Deoxynivalenol-Induced Reductions in Feed Consumption and Weight Gain in the Mouse. *J. Environ. Sci. Health B* **1997,** *32,* 429–448.

71. Wu, W.; Bates, M. A.; Bursian, S. J.; Flannery, B.; Zhou, H. R.; Link, J. E.; Zhang, H.; Pestka, J. J. Peptide YY3–36 and 5-Hydroxytryptamine Mediate Emesis Induction by Trichothecene Deoxynivalenol (Vomitoxin). *Toxicol. Sci.* **2013,** *133,* 186–195.

72. Yang, H.; Park, S. H.; Choi, H. J.; Do, K. H.; Kim, J.; An, T. J.; Lee, S. H.; Moon, Y. Mechanism-Based Alternative Monitoring of Endoplasmic Reticulum Stress by 8-Keto-Trichothecene Mycotoxins Using Human Intestinal Epithelial Cell Line. *Toxicol. Lett.* **2010,** *198,* 317–323.

73. Denizot, J.; Desrichard, A.; Agus, A.; Uhrhammer, N.; Dreux, N.; Vouret-Craviari, V.; Hofman, P.; Darfeuille-Michaud, A.; Barnich, N. Diet-Induced Hypoxia Responsive Element Demethylation Increases CEACAM6 Expression, Favouring Crohn's Disease-Associated *Escherichia coli* Colonisation. *Gut* **2015,** *64,* 428–437.

CHAPTER 3

TOXICITY OF INGESTED NANOMATERIALS

YI CAO[*]

Key Laboratory of Environment-Friendly Chemistry and Applications of Ministry of Education, Lab of Biochemistry, College of Chemistry, Xiangtan University, Hunan 411105, PR China

[*]*E-mail: caoyi39@xtu.edu.cn*

CONTENTS

ABSTRACT

With the rapid development of nanotechnologies, a number of engineered nanomaterials (NMs), particularly solid NMs like TiO_2, ZnO, Ag, and SiO_2 NMs, are increasingly used in food science as food additives or in food packaging. This raises the health concern about the adverse health effects of engineered NMs after oral exposure. The present book chapter discusses the potential toxicity of ingested NMs based on the most recent studies. It has been shown that following oral exposure, ZnO NMs, and Ag NMs to a lesser extent could be absorbed into gastrointestinal tract and accumulated into various organs such as liver, kidney, and spleen, whereas TiO_2 and SiO_2 NMs are hardly absorbed. A number of studies showed that ZnO NMs were the most toxic NMs, followed by Ag NMs to a lesser extent, whereas TiO_2 and SiO_2 NMs generally showed little to no toxic effects both in vitro and in vivo. This could be explained by the relatively high bioavailability and activity of ZnO and Ag NMs, whereas oral exposure to TiO_2 and SiO_2 NMs generally showed negligible bioavailability. In the future, it is necessary and urgent to further assess the toxicity of engineered NMs used in food, especially in the case of long-time exposure, to ensure the safe use of them. There is also a need to further study the mechanisms.

3.1 INTRODUCTION

According to Recommendation 2011/696/EU by the European Commission, a nanomaterial (NM), or nanoparticle (NP), was defined as "a natural, incidental or manufactured material containing particles, in an unbound state or as an aggregate and where, for 50% or more of the particles in the number size distribution, one or more external dimensions is in the size range 1–100 nm".[1] In daily life, human beings are exposed to naturally occurring ultrafine particles (particles with diameter ≤0.1 μm, the same name as NP) mainly from sources of combustion (gas stoves, diesel and gasoline engines, cooking, fires, and smoking) and heating elements (hair dryers, electric irons, space heaters).[2] It has long been shown that airway exposure to ultrafine particles could induce adverse health effects; however, recent studies by using laboratory animals (e.g., rats and mice) also showed that oral exposure to ultrafine particles as well as particulate matter with larger sizes could damage both the gastrointestinal tract (GIT) and the secondary organs, for example, bone marrow, liver, and lung.[3–6] These studies raised the health concern of NPs following oral exposure, which may be relevant for particle-contaminated food.[7]

With the rapid development of nanotechnology, engineered NMs are being produced and increasingly used for many commercial purposes. In food sciences, engineered NMs are used as food additives for color quality, nutritional improvement, and antibacterial effects.[8,9] Engineered NMs are also used in food packages, which may result in the contamination by NMs as they migrate from the packages into food.[10,11] By analog to naturally occurring ultrafine particles, this raises the health concern of ingested NMs. In this book chapter, the potential health effects of engineered NMs used as food additives or in packaging will be discussed.

3.2   NANOMATERIALS IN FOOD

There are many engineered NMs already present in foods that are sold in the market, and the use of NMs in food continues to increase especially in regards to solid NMs, namely titanium dioxide (TiO_2), zinc oxide (ZnO), silver (Ag), and silica dioxide (SiO_2) NPs. TiO_2 particles naturally occur as rutile, anatase, and/or brookite, and food-grade TiO_2 particles (referred as E171 in the food industry in EU) are added as food additives for color quality due to their natural white color. However, it should be noticed that food-grade TiO_2 particles contain only a portion of nanosized particles as TiO_2 NPs are colorless, whereas most of the TiO_2 particles used in food is of the pigment-grade form, because it is primary the particles with size range of 200–300 nm that show the desired light scattering.[12,13] Nevertheless, nanosized TiO_2 particles are some of the most commonly used NPs[14] and have been found in common food products, for example, candies, sweets, and chewing gums, which may count for <40% of the total food-grade TiO_2 particles.[15–18] Moreover, it has been shown that TiO_2 particles in chewing gums were released from the food readily. Over 90% of food-grade TiO_2 particles were released when a volunteer chewed a gum for 10 min.[15] Although it is difficult to estimate the total dietary intake of TiO_2 particles as the commercial products contain different amount of particles, this study indicated that consuming food-grade TiO_2 particle-containing products may result in the exposure of TiO_2 particles to human GIT easily.

ZnO NMs are added as food additives to supplement zinc. Zinc is one of the most important essential trace elements for human beings. More than 300 enzymes (e.g., zinc-figure proteins) contain zinc, and zinc deficiency has been implicated in the development of certain diseases, such as diabetes, infectious diseases (e.g., diarrhea, pneumonia, and malaria), and arthritis.[19] Compared with ZnO powder, oral administration of ZnO NMs could result in a higher accumulation of zinc ion at various organs (e.g., livers, pancreas, and kidneys), and some recent studies showed beneficial effects of orally administrated ZnO NMs to streptozotocin-induced Type 1 and 2 diabetic rats, although the beneficial effects are not yet proved in patients.[9,20] Another important application of ZnO NMs in food science is their strong antibacterial ability especially against Gram-positive bacteria; therefore, these NMs could be used as food additives or in packaging.[11,21] The antibacterial mechanisms of ZnO NMs are not fully known, but the formation of various reactive oxygen species (ROS) such as hydroxyl radicals ($\cdot$OH), singlet oxygen or superoxide anion ($O_2^{\cdot-}$), and hydrogen peroxide (H_2O_2), likely play a central role. The interaction between ZnO NMs and bacteria as well as the subsequent release of zinc ions may also be responsible for the antibacterial ability.[21–23]

A number of studies indicate antimicrobial activity of Ag NMs toward a broad range of microbes; therefore, Ag NMs (referred as E174 in the food industry in EU) are used in food science mostly for antimicrobial and sterilization purposes.[23–25] Most importantly, microorganisms are becoming increasingly resistant to antibiotics, whereas Ag NMs could effectively inhibit or kill microorganisms without the development of resistance; therefore, Ag NMs are increasingly used recently in food science as well as medicine.[25] Due to the health concern of Ag NMs, they are rarely added as food ingredients, but rather are used in food packaging or in the containers.[8,11,25] This may result in Ag NMs contamination when they migrate into food, but it is still a big challenge to measure the migration rate.[8,11] In addition, Ag NMs may also be added into animal feed to replace antibiotics, which may result in Ag residues in meat.[25]

SiO_2 NMs (referred as E551 in the food industry in EU) are used in clearing beers and wines, and as an anticaking and anticlumping agent (mostly amorphous form for this purpose).[8,9] The US Food and Drug Administration (FDA) and the Code of Federal Regulations Title 21 permit the use of SiO_2 in foods to not exceed 2% by weight. A recent survey found nano-sized SiO_2 particles (primary size range of 9–26 nm) from food products sold in USA, and the Si concentrations in 14 foods and consumer goods were found to be 0.13–1.63% by weight except for one toothpaste sample. Therefore, most food samples contained less than the permitted weight percentage set in the code of FDA.[26]

3.3   TOXICITY OF INGESTED NANOMATERIALS

The basic components of the GIT include the mouth, esophagus, stomach, small intestine, and large intestine. The small intestine, particularly the segments of the jejunum and ileum, is the site of most nutrient digestion as well as the absorption of particles/NPs.[13] Nanosized particles may cross the GIT and accumulate at the base of Peyer's patches, where microfold cells (M cells; cells lack microvilli and express strong cell junction to function as a barrier) transport microorganisms and particles from the gut lumen to immune cells, thus initiate the mucosal immune response.[27] As NPs are small, some of them may also be absorbed into the bloodstream to initiate a secondary effect. In the following section, the adverse health effects of the four solid NMs following oral exposure will be discussed, with the focus on GIT, but the secondary effects will also be mentioned. For the in vitro effects of the solid NMs, only studies using relevant cell lines derived from GIT will be discussed (the most popular cells used in vitro for this purpose are intestinal cells such as Caco-2 cells).

3.3.1   TiO$_2$ NANOMATERIALS

The absorption, distribution, and excretion of TiO$_2$ NMs following oral exposure in laboratory animals (rats or mice) have been determined in three independent studies, and results showed a very minor extent of absorption in GIT even after repeated exposure of very high concentrations of TiO$_2$ particles (up to 1041.5 mg/kg body weight in rats). Most of TiO$_2$ NMs were excreted by feces, and the accumulation of Ti in various organs was negligible.[3,28,29] Consistent with the animal data, oral exposure of human volunteers to different sizes of TiO$_2$ particles (i.e., 15, 100, and <5000 nm) showed no absorption regardless of particle sizes; the Ti content was not significantly elevated in urine or blood samples.[30] The very low absorption of TiO$_2$ NPs in vivo was also supported by in vitro studies, that no translocation was observed by using Caco-2 cells only, whereas there was some accumulation of NPs in M cells and mucus-secreting cells.[3,30,31] All of these studies indicated negligible bioavailability of TiO$_2$ NMs following oral exposure.

The toxicity of TiO$_2$ particles to animals after oral exposure has been conducted following Organization for Economic Cooperation and Development (OECD) test guideline, and results showed that single dose or repeated exposure (14, 28, and 90 days) to different forms and concentrations (up to 24,000 mg/kg body weight in rats) of TiO$_2$ particles showed little to no toxicity by the assessment of a variety of toxicological end points, which is in agreement with the low uptake of TiO_2 particles.[13,32] Oral exposure of different forms of TiO_2 particles (pigment-grade or nanosized) induced no genotoxicity to bone marrow cells either, which is likely to be associated with the negligible absorption of particles into blood or liver.[33] In addition, the exposure of fruitfly, *Drosophila melanogaster*, to food-grade TiO_2 particles (up to 2 mg/ml) showed little toxicity. These NPs did not affect the survival or fecundity, although the exposure was associated with signs of oxidative stress (downregulation of catalase and to a lesser extent superoxide dismutase 2).[34]

In contrast to the studies mentioned above, some studies showed systemic effects to orally exposed TiO_2 particles in laboratory animals. Five-day oral exposure to anatase TiO_2 NPs (0, 1, and 2 mg/kg body weight) in rats was associated with toxicity to endocrine and reproductive systems.[35] Gastrointestinal administration of TiO_2 NPs (0, 2, 10, and 50 mg/kg body weight) to rats for 30 and 90 days resulted in adverse cardiovascular effects.[36] Genotoxicity was observed in bone marrow cells and livers of orally exposed rats (0, 10, 50, and 200 mg/kg body weight for 30 days) or mice (0, 40, 200, and 1000 mg/kg body weight for 7 days).[37–41] Some studies also indicated adverse effects of ingested TiO_2 particles to central nervous systems.[42] With the low level of particle translocation and accumulation in secondary organs, these effects may be affected by mediators of oxidative stress and/or inflammation. The reason for inconsistent results obtained by different studies remains unclear, but it has been suggested that studies should rigorously follow standard protocols for toxicology studies for future risk assessments.[13,32]

The in vitro toxicity of TiO_2 particles has been extensively studied by using intestinal cells such as Caco-2 cells, and results generally indicated little to no effects of TiO_2 particles on cytotoxicity, oxidative stress (e.g., production of intracellular ROS), or inflammatory response (e.g., release of inflammatory cytokines) even after incubation with up to 200 µg/ml particles.[15,31,43–49] In addition, a recent study showed that 200 µg/ml TiO_2 particles were only minimally toxic to undifferentiated Caco-2 cells but not to differentiated ones, probably due to a much greater uptake of particles into the undifferentiated cells, which indicated a crucial role of differentiated status of Caco-2 cells on the safety assessment of particles. At concentrations lower than 200 µg/ml, no toxic effect was observed for neither type of cells.[31]

Overall, oral exposure to TiO_2 NMs is generally considered to be safe, and they have already been added as food additives in a number of commercial products. However, more work is still needed to assess the potential health effects of TiO_2 NMs to humans. For example, it was recently shown that exposure of food-grade TiO_2 as well as TiO_2 isolated from the candy coating of chewing gum (both of the samples contain a portion of nanosized TiO_2) elicited a bona fide biological response and resulted in disruption of the brush border to Caco-2$_{BBe1}$ human derived cell system, which was independent of particle sedimentation. At the lowest concentration (i.e., 350 ng/ml or 100 ng/cm²), approximately 42% of microvilli were lost after exposure.[16] A previous study showed that polystyrene NPs exposure affected iron uptake and transport due to the disruption of membrane of intestinal cells and remodeling of the intestinal microvilli.[27] Therefore, it may be necessary to further study if the loss of microvilli elicited by food-grade TiO_2 exposure will affect the absorption of nutrients.

3.3.2   ZnO NANOMATERIALS

Unlike TiO_2 particles, several in vivo studies showed that oral exposure of animals to ZnO NMs could lead to a high accumulation of zinc in various organs, such as GIT, liver, and kidney.[20,28,50,51] But the concentration of zinc may return to normal after a recovery time (e.g., 7-day[50] or 14-day[51] post treatment). Due to the relatively high bioavailability, oral exposure to high concentrations of ZnO NMs was shown to be toxic to animals. A recent study showed that exposure of Sprague–Dawley rats to positively charged ZnO NPs (500, 1000, and 2000 mg/kg/day for 14 days) by gavage showed adverse changes in clinical signs, mortality, body weight, food consumption, hematology, serum biochemistry, gross pathology, organ weight, and histopathology, with the spleen, stomach, and pancreas as the main target organs. At concentrations <500 mg/kg, no obvious effect was observed.[52] In another study, 2000 mg/kg ZnO NMs exposure in rats resulted in alteration of some hemato-biochemical parameters but with few histopathological lesions in livers and kidneys.[51] Moreover, repeated exposure of pregnant rats by gavage to positively charged ZnO NMs,[53] but not the negatively charged ones,[41] showed embryotoxicity at 400 mg/kg body weight. In contrast, at much lower concentrations (1, 3, and 10 mg/kg body weight) oral exposure to ZnO NMs in rats did not show any sign of toxicity and resulted in beneficial effects to streptozotocin-induced diabetes.[20]

In agreement with the in vivo data, extensive studies showed that ZnO NMs were toxic to intestinal cells, which can induce cytotoxicity, inflammatory response, oxidative stress, and genotoxicity.[43,44,48,49,55,56] The LD_{50} of ZnO NMs to Caco-2 cells was calculated as low as 0.431 µg/ml in one study,[56] whereas in another study, the LD_{50} was calculated as 28 µg/ml by using WST-1 assay (note that in that study, a high level of proteins was present during exposure, which may reduce the toxicity of ZnO NMs).[55] The sizes and shape are important in defining the toxic effects of ZnO particles, as nanosized particles were shown to be more toxic than larger particles.[57,58] The toxic effects were also confirmed by using 3D spheroid cell cultures, although 3D cultures were more resistant to ZnO NPs exposure compared with conventional cultures.[59] A number of studies showed that oxidative stress was important in mediating the toxic effects of ZnO NPs, as they can induce ROS (intracellular ROS and/or mitochondrial ROS) as well as changes in antioxidant systems (i.e., glutathione and antioxidant enzymes) in intestinal cells.[44,49,55,56,58,59] Another well-accepted mechanism is the internalization of ZnO particles and the subsequently release of Zn^{2+} inside cells, which can disrupt the cellular Zn^{2+} homeostasis and consequently lead to cell death.[21,60] However, it should be noticed that it requires the direct contact of ZnO NMs with the cells and the dissolution of particles inside the cells (e.g., in lysosomes), whereas the extracellular dissolution of ZnO particles is considered to play little role in determining cytotoxicity.[61,62]

To assess the toxicity of orally exposed ZnO NMs, it should be noticed that the interaction between nutrients and particles may affect the toxicological responses as revealed by recent studies. For example, it was shown that the presence of 300 μg/ml vitamin C enhanced the cytotoxicity of ZnO NMs to gastric epithelial cell line and neural stem cells, which is associated with the promoted dissolution of ZnO NMs to release zinc ions by vitamin C. Moreover, under in vivo conditions, repeated oral exposure to ZnO NMs (14 mg/kg body weight) plus vitamin C (50 mg/kg body weight) resulted in injury of the liver and kidneys in mice.[63] In another study, it was shown that the presence of saturated fatty acids (palmitic acid), but not the mixture of unsaturated fatty acids plus saturated fatty acids (palmitic acid/oleic acid 1:2), enhanced the cytotoxicity of ZnO NMs to Caco-2 cells, which is likely to be associated with the production of mitochondrial ROS but not lysosomal destabilization.[55] All of these studies highlighted the likelihood of synergistic toxicity in a complex system when considering the safety of ZnO NMs used in food.

Taken together, the uptake of ZnO NMs is high, which can lead to toxicological responses both in vivo and in vitro. This indicates that the use of ZnO NMs as food additives should follow the guideless to reduce the potential adverse health effects. In the United States, the Recommended Dietary Allowance of Zn was set at 8 mg for women and 11 mg for men. The tolerable upper intake level was set at 40 mg because higher levels can block the absorption of Cu.[64]

3.3.3 AG NANOMATERIALS

The biodistribution and toxicological response of Ag NMs following oral exposure to animals has been shown by recent studies. Oral gavage of different sizes (10, 75, and 110 nm) and concentrations (100, 200, and 400 mg/kg body weight) of Ag NMs to rats for 13 weeks showed size dependent (10 $\gg$ 75 > 110 nm) accumulation of Ag in various organs, such as liver, kidney, and spleen. In addition, there was gender difference that the accumulations Ag NMs were significantly higher in female rats, especially in the kidney, liver, jejunum, and colon. Despite the accumulation of Ag in organs, there was no significant changes in body weights, intakes of feed and water, alteration in blood and reproductive system or genotoxicity.[65] Similarly, intragastric administration of 12 nm Ag NMs to male rats after a single exposure (2000 mg/kg body weight) or multiple exposures over 30 days (250 mg/kg body weight) resulted in Ag accumulation in the liver, kidneys, spleen, stomach, and small intestine, with the liver and kidneys being the highest. However, >99% of Ag NMs were efficiently excreted from the organisms, and no sign of toxicity was observed after single or multiple dose exposure.[66] One study indicated that it was mainly Ag ions, and to a much lesser extent silver NPs, that passed the intestines and then formed Ag NPs in various organs of rats. Again, no hepatotoxicity or immunotoxicity was observed following repeated oral exposure to Ag NMs (<20 nm; 90 mg/kg body weight).[67]

In contrast to negative toxicological response discussed above, some studies also reported adverse health effects of Ag NMs after oral exposure. For example, 13-week repeated oral exposure of Ag NMs (11 nm; up to 1030.5 mg/kg body weight) to rats induced toxicity in liver and kidney, showed increases in serum alkaline phosphatase and calcium as well as lymphocyte infiltration, which could be explained by extensive systemic distribution of Ag NMs, whereas SiO_2 and Fe_2O_3 NPs showed no systemic distribution and did not induce toxicity.[68] In another study following OECD test guidelines, 90-day repeated oral exposure (56 nm; 90-day exposure to 30, 125, and 500 mg/kg) resulted in decreased body weight, slight liver damage as well as histopathologic changes in rats, along with a dose-dependent accumulation of silver in all tissues.[69] Loss of body weight and damage of microvilli as well as intestinal glands were also observed in orally exposed mice.[70] Genotoxicity was observed in bone marrow of rats receiving 5-day repeated oral exposure (10 nm; 5, 25, 50, 100 mg/kg body weight), which was associated with oxidative stress.[71] Inflammatory responses were observed both in intestinal cells as well as immune cells by some studies using orally exposed rats or mice.[25,72] The reason for the inconsistent data reported by different studies was unknown, but several factors, for example, size, dose, shape, and charge, may play an important role in defining the biological response to Ag NMs. Therefore, systematically assessing the full range of Ag NMs that are intended for the use in food according to size, shape, charge, and concentrations is required for future studies, and it should be conducted on case by case basis.[25]

A number of in vitro studies showed that Ag NM exposure was associated with toxic effects to intestinal cells, such as cytotoxicity, genotoxicity, and inflammatory response.[43,56,73-78] Some studies indicated that oxidative stress (i.e., increased production of intracellular ROS) was associated with the toxicity of Ag NMs in vitro,[73-75] whereas other studies did not find oxidative stress following Ag NM exposure.[43,56,78] It seems that the toxicity of Ag NMs in vitro could be mediated by both ROS-dependent and independent manner. The toxicity of Ag NMs in vitro was also assessed using advanced 3D coculture model in recent studies. In coculture consisting of Caco-2 cells, human macrophages and dendritic cells, Ag NMs (dose range 1.25–625 µg/cm^2) were shown to be more toxic compared with TiO_2 and Au NMs and that Ag NMs significantly induced cytotoxicity, disruption of barrier function, and release of inflammatory mediators. In addition, inflammation induced by cytokine exposure modulated the response of the model to Ag NMs.[79] In coculture model consisting of Caco-2 and HT29-MTX cells, Ag NMs (dose range 10–100 µg/ml) induced inflammatory response without ROS or cytotoxicity (assessed as metabolic activity), and the effect was reduced by the presence of mucus layer. Furthermore, compared with larger Ag particles (200 nm) or $AgNO_3$, Ag NPs (20 nm) resulted in higher release of IL-8 as well as different expression of proteins with a distinct pattern of cellular responses as revealed by proteomic analysis, which indicated a NP-specific role in determining the toxicological response.[80]

Overall, oral exposure to Ag NMs could be absorbed into GIT and accumulated in various secondary organs to a lower level. In vivo toxicity could be induced depending on the types and doses of Ag NMs, whereas in vitro toxicity could be induced by high concentrations of Ag NMs, both in ROS-dependent and independent manner. Besides the traditionally used toxicological end points, one possible toxicological effect of Ag NMs as well as other NMs with antimicrobial applications that may need extensive studies is the interaction of NMs with gut microbiota.[81] There are enormous amount of microorganisms living within the human gut (microbiota), and the alterations of gut microbiota have been implicated in the development of metabolic diseases such as obesity and diabetes.[82] For NMs with antimicrobial effects, it may be therefore necessary to examine their effects on gut microbiota.[81] For Ag NMs, it was only recently shown that oral exposure of Sprague–Dawley rats to different sizes of Ag NMs for 13 weeks resulted in changes of ileal–mucosal microbial populations, decreased expression of immuno-modulatory genes and an apparent shift in the gut microbiota toward greater proportions of Gram-negative bacteria, which in combination indicated that Ag NMs may modulate the gut-associated immune response and the overall homeostasis of the GIT.[83] It remains unclear if similar effects will be induced by Ag NMs in human gut.

3.3.4   SiO$_2$ NANOMATERIALS

The absorption and toxicity of SiO$_2$ NMs (average size 110 nm; 50 mg/kg body weight) via different routes was compared in mice, and results showed that most of the particles via oral exposure were excreted through feces, with only a small portion of ingested particles being absorbed into intestines as well as livers. Consistent with the low oral absorption, no adverse health effects was observed after oral exposure to particles, whereas intramuscular and hypodermic injection of particles caused inflammatory response around the injection sites.[84] Similarly, oral administration of different sizes of SiO$_2$ particles (70, 300, and 100 nm; 2.5 mg/mouse at the age of 6 weeks) to mice can lead to absorption of particles into intestines, which was enhanced by surface coating with carboxyl or amine groups. No significant changes in hematological, histopathological, and biochemical parameters were observed after exposure.[85] Repeated oral exposure (14 and 90 days; concentration up to 2000 mg/kg body weight) to SiO$_2$ particles (20 and 200 nm, negatively charged) did not induce toxicity to rats.[86] No obvious genotoxicity of SiO$_2$ particles was observed in various organs of orally exposed rats, although there was a weak increase in the percentage of micronucleated cells in the colon, indicating a minimally genetic damage.[87] In contrast, oral exposure of D. melanogaster to SiO$_2$ NMs (<30 nm; 1–100 µg/ml) resulted in oxidative stress, caspases activation, membrane destabilization, and mitochondrial membrane potential loss in the mid-gut cells, along with uptake of particles via endocytic vesicles and by direct membrane penetration.[88] This indicates that the responses to orally exposed SiO$_2$ NMs could be dependent on the species, but generally SiO$_2$ NMs showed good biocompatibility following oral exposure.

Consistent with the in vivo data, most of the in vitro studies showed little to no effects of different sizes of SiO$_2$ particles on oxidative stress, cytotoxicity, or genotoxicity to intestinal cells, even with very high concentrations (up to 150 µg/ml).[48,49,74,89–91] Significantly increased inflammatory response (i.e., expression of IL-8 mRNA and release of IL-8 proteins) was only observed in 20 µg/cm^2 SiO$_2$ NM exposed undifferentiated Caco-2 cells but not in the differentiated cells, whereas ZnO NMs at the same concentration induced the inflammatory response in both models.[92] Only very few studies showed the toxic effects of SiO$_2$ particles to intestinal cells. For example, a recent study showed that SiO$_2$—15 nm, but not that of 55 nm or the control quartz DQ12, induced cytotoxicity, genotoxicity, oxidative stress, and inflammatory response to Caco-2 cells, which indicated a role of particle sizes in defining the toxicological responses. However, most of the toxicological responses were observed at relatively high concentrations (≥32 µg/ml).[93]

Overall, most of the in vivo and in vitro studies showed little to no effects of SiO$_2$ NMs, and the use of them as food additives as well as in medicine is considered to be relatively safe. However, similar to what have been seen in food-grade TiO$_2$-exposed Caco-2$_{BBe1}$ human-derived cell system, food-grade SiO$_2$-disrupted microvilli even at very low concentrations (i.e., 1 µg/ml or 10 ng/cm^2 of epithelium), which may be associated with increased production of ROS.[26] This impact of food-grade SiO$_2$ on human microvilli and consequently the absorption of nutrition, especially during long-time exposure, may need to be further explored.

3.4 CONCLUSIONS

Engineered NMs are increasingly used in food science as food additives or in food packaging, and the use of them in food is still expected to be increasing in the future. Solid NMs, for example, TiO$_2$, ZnO, Ag, and SiO$_2$ NMs, are among the most popular engineered NMs used in food. Following oral exposure, ZnO NMs, and Ag NMs to a lesser extent, could be absorbed into GIT and accumulated into various organs such as liver, kidney, and spleen, whereas TiO$_2$ and SiO$_2$ NMs are hardly absorbed. A number of studies showed that ZnO NMs were the most toxic NMs, followed by Ag NMs to a lesser extent, whereas TiO$_2$ and SiO$_2$ NMs generally showed little to no toxic effects both in vitro and in vivo. This could be explained by the relatively high bioavailability and activity of ZnO and Ag NMs, whereas oral exposure to TiO$_2$ and SiO$_2$ NMs generally showed negligible bioavailability. In the future, it is necessary and urgent to assess the toxicity of engineered NMs used in food, especially in the case of long-time exposure, to ensure the safe use of them. There is also a need to further study the mechanisms. It is generally agreed that oxidative stress plays a crucial role in the toxic effects induced by the oral NP exposure, as NPs are biologically active which can result in oxidative damage. Secondary effects may be induced by the translocation and subsequently accumulation of NPs in indirectly exposed organs and/or the release of biological mediators from the contact sites to secondary organs.

KEYWORDS

- **nanomaterial**
- **nanoparticle**
- **trace element**
- **microorganisms**
- **exposure**

REFERENCES

1. European Commission. *Commission Recommendation of 18 October 2011 on the Definition of Nanomaterial Text with EEA Relevance*, 2011.
2. Wallace, L.; Ott, W. Personal Exposure to Ultrafine Particles. *J. Exposure Sci. Environ. Epidemiol.* **2011**, *21*, 20–30.
3. Brun, E.; Barreau, F.; Veronesi, G.; Fayard, B.; Sorieul, S.; Chaneac, C.; Carapito, C.; Rabilloud, T.; Mabondzo, A.; Herlin-Boime, N.; Carriere, M. Titanium Dioxide Nanoparticle Impact and Translocation through Ex Vivo, In Vivo and In Vitro Gut Epithelia. *Part Fibre Toxicol.* **2014**, *11*, 13.
4. Danielsen, P. H.; Loft, S.; Jacobsen, N. R.; Jensen, K. A.; Autrup, H.; Ravanat, J. L.; Wallin, H.; Moller, P. Oxidative Stress, Inflammation, and DNA Damage in Rats after Intratracheal Instillation or Oral Exposure to Ambient Air and Wood Smoke Particulate Matter. *Toxicol. Sci.* **2010**, *118*, 574–585.
5. Hadrup, N.; Sharma, A. K.; Poulsen, M.; Nielsen, E. Toxicological Risk Assessment of Elemental Gold Following Oral Exposure to Sheets and Nanoparticles—A Review. *Regul. Toxicol. Pharmacol.* **2015**, *72*, 216–221.
6. Moller, P.; Folkmann, J. K.; Danielsen, P. H.; Jantzen, K.; Loft, S. Oxidative Stress Generated Damage to DNA by Gastrointestinal Exposure to Insoluble Particles. *Curr. Mol. Med.* **2012**, *12*, 732–745.
7. Marano, F.; Guadagnini, R. Health Impacts of Nanomaterials. In: *Nanosciences and Nanotechnology: Evolution or Revolution?*; Lourtioz, J. M., Lahmani, M., Dupas-Haeberlin, C., Hesto, P., Eds.; Springer International Publishing: Cham, 2016; pp. 273–286.

8. Contado, C. Nanomaterials in Consumer Products: A Challenging Analytical Problem. *Front. Chem.* **2015**, *3*, 48.

9. Wang, H.; Du, L. J.; Song, Z. M.; Chen, X. X. Progress in the Characterization and Safety Evaluation of Engineered Inorganic Nanomaterials in Food. *Nanomedicine (Lond.)* **2013**, *8*, 2007–2025.

10. Arvanitoyannis, I. S.; Bosnea, L. Migration of Substances from Food Packaging Materials to Foods. *Crit. Rev. Food Sci. Nutr.* **2004**, *44*, 63–76.

11. Bumbudsanpharoke, N.; Ko, S. Nano-Food Packaging: An Overview of Market, Migration Research, and Safety Regulations. *J. Food Sci.* **2015**, *80*, R910–R923.

12. Braun, J. H. Titanium Dioxide: A Review. *J. Coat. Technol.* **1997**, *69*, 59–72.

13. Warheit, D. B.; Donner, E. M. Risk Assessment Strategies for Nanoscale and Fine-Sized Titanium Dioxide Particles: Recognizing Hazard and Exposure Issues. *Food Chem. Toxicol.* **2015**, *85*, 138–147.

14. Rollerova, E.; Tulinska, J.; Liskova, A.; Kuricova, M.; Kovriznych, J.; Mlynarcikova, A.; Kiss, A.; Scsukova, S. Titanium Dioxide Nanoparticles: Some Aspects of Toxicity/ Focus on the Development. *Endocr. Regul.* **2015**, *49*, 97–112.

15. Chen, X. X.; Cheng, B.; Yang, Y. X.; Cao, A.; Liu, J. H.; Du, L. J.; Liu, Y.; Zhao, Y.; Wang, H. Characterization and Preliminary Toxicity Assay of Nano-Titanium Dioxide Additive in Sugar-Coated Chewing Gum. *Small* **2013**, *9*, 1765–1774.

16. Faust, J. J.; Doudrick, K.; Yang, Y.; Westerhoff, P.; Capco, D. G. Food Grade Titanium Dioxide Disrupts Intestinal Brush Border Microvilli In Vitro Independent of Sedimentation. *Cell Biol. Toxicol.* **2014**, *30*, 169–188.

17. Weir, A.; Westerhoff, P.; Fabricius, L.; Hristovski, K.; von Goetz, N. Titanium Dioxide Nanoparticles in Food and Personal Care Products. *Environ. Sci. Technol.* **2012**, *46*, 2242–2250.

18. Yang, Y.; Doudrick, K.; Bi, X.; Hristovski, K.; Herckes, P.; Westerhoff, P.; Kaegi, R. Characterization of Food-Grade Titanium Dioxide: The Presence of Nanosized Particles. *Environ. Sci. Technol.* **2014**, *48*, 6391–6400.

19. Haase, H.; Overbeck, S.; Rink, L. Zinc Supplementation for the Treatment or Prevention of Disease: Current Status and Future Perspectives. *Exp. Gerontol.* **2008**, *43*, 394–408.

20. Umrani, R. D.; Paknikar, K. M. Zinc Oxide Nanoparticles Show Antidiabetic Activity in Streptozotocin-Induced Type 1 and 2 Diabetic Rats. *Nanomedicine (Lond.)* **2014**, *9*, 89–104.

21. Shi, L. E.; Li, Z. H.; Zheng, W.; Zhao, Y. F.; Jin, Y. F.; Tang, Z. X. Synthesis, Antibacterial Activity, Antibacterial Mechanism and Food Applications of ZnO Nanoparticles: A Review. *Food Addit. Contam Part A Chem. Anal. Control Exposure Risk Assess.* **2014**, *31*, 173–186.

22. Amna, S.; Shahrom, M.; Azman, S.; Kaus, N. H. M.; Ling, C. A.; Siti Khadijah, M. B.; Habsah, H.; Dasmawati, M. Review on Zinc Oxide Nanoparticles: Antibacterial Activity and Toxicity Mechanism. *Nano-Micro Lett.* **2015**, *7*, 219–242.

23. Dizaj, S. M.; Lotfipour, F.; Barzegar-Jalali, M.; Zarrintan, M. H.; Adibkia, K. Antimicrobial Activity of the Metals and Metal Oxide Nanoparticles. *Mater. Sci. Eng., C: Mater. Biol. Appl.* **2014**, *44*, 278–284.

24. Antony, J. J.; Sivalingam, P.; Chen, B.; Toxicological Effects of Silver Nanoparticles. *Environ. Toxicol. Pharmacol.* **2015**, *40*, 729–732.

25. Gaillet, S.; Rouanet, J. M. Silver Nanoparticles: Their Potential Toxic Effects after Oral Exposure and Underlying Mechanisms—A Review. *Food Chem. Toxicol.* **2015**, *77*, 58–63.

26. Yang, Y.; Faust, J. J.; Schoepf, J.; Hristovski, K.; Capco, D. G.; Herckes, P.; Wester-hoff, P. Survey of Food-Grade Silica Dioxide Nanomaterial Occurrence, Characterization, Human Gut Impacts and Fate Across its Lifecycle, *Sci. Total Environ.* **2016**, *565*, 902–912.

27. Mahler, G. J.; Esch, M. B.; Tako, E.; Southard, T. L.; Archer, S. D.; Glahn, R. P.; Shuler, M. L. Oral Exposure to Polystyrene Nanoparticles Affects Iron Absorption. *Nat. Nanotechnol.* **2012**, *7*, 264–271.

28. Cho, W. S.; Kang, B. C.; Lee, J. K.; Jeong, J.; Che, J. H.; Seok, S. H. Comparative Absorption, Distribution, and Excretion of Titanium Dioxide and Zinc Oxide Nanoparticles after Repeated Oral Administration. *Part Fibre Toxicol.* **2013**, *10*, 9.

29. Geraets, L.; Oomen, A. G.; Krystek, P.; Jacobsen, N. R.; Wallin, H.; Laurentie, M.; Verharen, H. W.; Brandon, E. F.; de Jong, W. H. Tissue Distribution and Elimination after Oral and Intravenous Administration of Different Titanium Dioxide Nanoparticles in Rats. *Part Fibre Toxicol.* **2014**, *11*, 30.

30. Jones, K.; Morton, J.; Smith, I.; Jurkschat, K.; Harding, A. H.; Evans, G. Human In Vivo and In Vitro Studies on Gastrointestinal Absorption of Titanium Dioxide Nanoparticles. *Toxicol. Lett.* **2015**, *233*, 95–101.

31. Song, Z. M.; Chen, N.; Liu, J. H.; Tang, H.; Deng, X.; Xi, W. S.; Han, K.; Cao, A.; Liu, Y.; Wang, H. Biological Effect of Food Additive Titanium Dioxide Nanoparticles on Intestine: An In Vitro Study. *J. Appl. Toxicol.* **2015**, *35*, 1169–1178.

32. Warheit, D. B.; Brown, S. C.; Donner, E. M. Acute and Subchronic Oral Toxicity Studies in Rats with Nanoscale and Pigment Grade Titanium Dioxide Particles. *Food Chem. Toxicol.* **2015**, *84*, 208–224.

33. Donner, E. M.; Myhre, A.; Brown, S. C.; Boatman, R.; Warheit, D. B. In Vivo Micronucleus Studies with 6 Titanium Dioxide Materials (3 Pigment-Grade & 3 Nanoscale) in Orally-Exposed Rats. *Regul. Toxicol. Pharmacol.* **2016**, *74*, 64–74.

34. Jovanovic, B.; Cvetkovic, V. J.; Mitrovic, T. L. Effects of Human Food Grade Titanium Dioxide Nanoparticle Dietary Exposure on *Drosophila melanogaster* Survival, Fecundity, Pupation and Expression of Antioxidant Genes. *Chemosphere* **2016**, *144*, 43–49.

35. Tassinari, R.; Cubadda, F.; Moracci, G.; Aureli, F.; D'Amato, M.; Valeri, M.; De, B. B.; Raggi, A.; Mantovani, A.; Passeri, D.; Rossi, M.; Maranghi, F. Oral, Short-Term Exposure to Titanium Dioxide Nanoparticles in Sprague–Dawley Rat: Focus on Reproductive and Endocrine Systems and Spleen. *Nanotoxicology* **2014**, *8*, 654–662.

36. Chen, Z.; Wang, Y.; Zhuo, L.; Chen, S.; Zhao, L.; Luan, X.; Wang, H.; Jia, G. Effect of Titanium Dioxide Nanoparticles on the Cardiovascular System after Oral Administration. *Toxicol. Lett.* **2015**, *239*, 123–130.

37. Chen, T.; Yan, J.; Li, Y. Genotoxicity of Titanium Dioxide Nanoparticles. *J. Food Drug Anal.* **2014**, *22*, 95–104.

38. Chen, Z.; Wang, Y.; Ba, T.; Li, Y.; Pu, J.; Chen, T.; Song, Y.; Gu, Y.; Qian, Q.; Yang, J.; Jia, G. Genotoxic Evaluation of Titanium Dioxide Nanoparticles In Vivo and In Vitro. *Toxicol. Lett.* **2014**, *226*, 314–319.

39. Grissa, I.; Elghoul, J.; Ezzi, L.; Chakroun, S.; Kerkeni, E.; Hassine, M.; El, M. L.; Mehdi, M.; Ben, C. H.; Haouas, Z. Anemia and Genotoxicity Induced by Sub-chronic Intragastric Treatment of Rats with Titanium Dioxide Nanoparticles. *Mutat. Res. Genet. Toxicol. Environ. Mutagen.* **2015**, *794*, 25–31.

40. Mohamed, H. R. Estimation of TiO(2) Nanoparticle-Induced Genotoxicity Persistence and Possible Chronic Gastritis-Induction in Mice. *Food Chem. Toxicol.* **2015**, *83*, 76–83.

41. Sycheva, L. P.; Zhurkov, V. S.; Iurchenko, V. V.; Daugel-Dauge, N. O, Kovalenko, M. A.; Krivtsova, E. K.; Durnev, A. D. Investigation of Genotoxic and Cytotoxic Effects of Micro- and Nanosized Titanium Dioxide in Six Organs of Mice In Vivo. *Mutat. Res.* **2011,** *726*, 8–14.

42. Czajka, M.; Sawicki, K.; Sikorska, K.; Popek, S.; Kruszewski, M.; Kapka-Skrzypczak, L. Toxicity of Titanium Dioxide Nanoparticles in Central Nervous System. *Toxicol. In Vitro* **2015,** *29*, 1042–1052.

43. Abbott Chalew, T. E.; Schwab, K. J. Toxicity of Commercially Available Engineered Nanoparticles to Caco-2 and SW480 Human Intestinal Epithelial Cells. *Cell Biol. Toxicol.* **2013,** *29*, 101–116.

44. De, A.; Barone, I. F.; Zijno, A.; Bizzarri, L.; Russo, M. T.; Pozzi, R.; Franchini, F.; Giudetti, G.; Uboldi, C.; Ponti, J.; Rossi, F.; De, B. B. Comparative Study of ZnO and TiO(2) Nanoparticles: Physicochemical Characterisation and Toxicological Effects on Human Colon Carcinoma Cells. *Nanotoxicology* **2013,** *7*, 1361–1372.

45. Fisichella, M.; Berenguer, F.; Steinmetz, G.; Auffan, M.; Rose, J.; Prat, O. Intestinal Toxicity Evaluation of TiO$_2$ Degraded Surface-Treated Nanoparticles: A Combined Physico-chemical and Toxicogenomics Approach in Caco-2 Cells. *Part Fibre Toxicol.* **2012,** *9*, 18.

46. Gerloff, K.; Fenoglio, I.; Carella, E.; Kolling, J.; Albrecht, C.; Boots, A. W.; Forster, I.; Schins, R. P. Distinctive Toxicity of TiO$_2$ Rutile/Anatase Mixed Phase Nanoparticles on Caco-2 Cells, Chem. Res. Toxicol. **2012,** *25*, 646–655.

47. Koeneman, B. A.; Zhang, Y.; Westerhoff, P.; Chen, Y.; Crittenden, J. C.; Capco, D. G. Toxicity and Cellular Responses of Intestinal Cells Exposed to Titanium Dioxide. *Cell Biol. Toxicol.* **2010,** *26*, 225–238.

48. McCracken, C.; Zane, A.; Knight, D. A.; Dutta, P. K.; Waldman, W. J. Minimal Intestinal Epithelial Cell Toxicity in Response to Short- and Long-Term Food-Relevant Inorganic Nanoparticle Exposure. *Chem. Res. Toxicol.* **2013,** *26*, 1514–1525.

49. Setyawati, M. I.; Tay, C. Y.; Leong, D. T. Mechanistic Investigation of the Biological Effects of SiO(2), TiO(2), and ZnO Nanoparticles on Intestinal Cells. *Small* **2015,** *11*, 3458–3468.

50. Konduru, N. V.; Murdaugh, K. M.; Sotiriou, G. A.; Donaghey, T. C.; Demokritou, P.; Brain, J. D.; Molina, R. M. Bioavailability, Distribution and Clearance of Tracheally-Instilled and Gavaged Uncoated or Silica-Coated Zinc Oxide Nanoparticles. *Part Fibre Toxicol.* **2014,** *11*, 44.

51. Srivastav, A. K.; Kumar, M.; Ansari, N. G.; Jain, A. K.; Shankar, J.; Arjaria, N.; Jagdale, P.; Singh, D. A Comprehensive Toxicity Study of Zinc Oxide Nanoparticles versus their Bulk in Wistar Rats: Toxicity Study of Zinc Oxide Nanoparticles. *Hum. Exp. Toxicol.* **2016,** *35*, 1286–1304.

52. Ko, J. W.; Hong, E. T.; Lee, I. C.; Park, S. H.; Park, J. I.; Seong, N. W.; Hong, J. S.; Yun, H. I.; Kim, J. C. Evaluation of 2-Week Repeated Oral Dose Toxicity of 100 nm Zinc Oxide Nanoparticles in Rats. *Lab Anim. Res.* **2015,** *31*, 139–147.

53. Hong, J. S.; Park, M. K.; Kim, M. S.; Lim, J. H.; Park, G. J.; Maeng, E. H.; Shin, J. H.; Kim, M. K.; Jeong, J.; Park, J. A.; Kim, J. C.; Shin, H. C. Prenatal Development Toxicity Study of Zinc Oxide Nanoparticles in Rats. *Int. J. Nanomed.* **2014,** *9* (Suppl 2.), 159–171.

54. Hong, J. S.; Park, M. K.; Kim, M. S.; Lim, J. H.; Park, G. J.; Maeng, E. H.; Shin, J. H.; Kim, Y. R.; Kim, M. K.; Lee, J. K.; Park, J. A.; Kim, J. C.; Shin, H. C. Effect of Zinc Oxide Nanoparticles on Dams and Embryo-Fetal Development in Rats. *Int. J. Nanomed.* **2014,** *9* (Suppl. 2), 145–157.

55. Cao, Y.; Roursgaard, M.; Kermanizadeh, A.; Loft, S.; Moller, P. Synergistic Effects of Zinc Oxide Nanoparticles and Fatty Acids on Toxicity to Caco-2 cells. *Int. J. Toxicol.* **2015**, *34*, 67–76.

56. Song, Y.; Guan, R.; Lyu, F.; Kang, T.; Wu, Y.; Chen, X. In Vitro Cytotoxicity of Silver Nanoparticles and Zinc Oxide Nanoparticles to Human Epithelial Colorectal Adenocarcinoma (Caco-2) Cells. *Mutat. Res.* **2014**, *769*, 113–118.

57. Cepin, M.; Hribar, G.; Caserman, S.; Orel, Z. C. Morphological Impact of Zinc Oxide Particles on the Antibacterial Activity and Human Epithelia Toxicity. *Mater. Sci. Eng., C: Mater. Biol. Appl.* **2015**, *52*, 204–211.

58. Kang, T.; Guan, R.; Chen, X.; Song, Y.; Jiang, H.; Zhao, J. In vitro Toxicity of Different-Sized ZnO Nanoparticles in Caco-2 Cells. *Nanoscale Res. Lett.* **2013**, *8*, 496–498.

59. Chia, S. L.; Tay, C. Y.; Setyawati, M. I.; Leong, D. T. Biomimicry 3D Gastrointestinal Spheroid Platform for the Assessment of Toxicity and Inflammatory Effects of Zinc Oxide Nanoparticles. *Small* **2015**, *11*, 702–712.

60. Saptarshi, S. R.; Duschl, A.; Lopata, A. L. Biological Reactivity of Zinc Oxide Nanoparticles with Mammalian Test Systems: An Overview. *Nanomedicine (Lond.)* **2015**, *10*, 2075–2092.

61. Gilbert, B.; Fakra, S. C.; Xia, T.; Pokhrel, S.; Madler, L.; Nel, A. E. The Fate of ZnO Nanoparticles Administered to Human Bronchial Epithelial Cells. *ACS Nano* **2012**, *6*, 4921–4930.

62. Moos, P. J.; Chung, K.; Woessner, D.; Honeggar, M.; Cutler, N. S.; Veranth, J. M. ZnO Particulate Matter Requires Cell Contact for Toxicity in Human Colon Cancer Cells. *Chem. Res. Toxicol.* **2010**, *23*, 733–739.

63. Wang, Y.; Yuan, L.; Yao, C.; Ding, L.; Li, C.; Fang, J.; Sui, K.; Liu, Y.; Wu, M. A Combined Toxicity Study of Zinc Oxide Nanoparticles and Vitamin C in Food Additives. *Nanoscale* **2014**, *6*, 15333–15342.

64. Institute of Medicine, Dietary Reference Intakes for Vitamin A, Vitamin K, *Arsenic, Boron, Chromium, Copper, Iodine, Iron, Manganese, Molybdenum, Nickel, Silicon, Vanadium, and Zinc.* The National Academies Press: Washington, DC, 2001.

65. Boudreau, M. D.; Imam, M. S.; Paredes, A. M.; Bryant, M. S.; Cunningham, C. K.; Felton, R. P.; Jones, M. Y.; Davis, K. J.; Olson, G. R. Differential Effects of Silver Nanoparticles and Silver Ions on Tissue Accumulation, Distribution, and Toxicity in the Sprague Dawley Rat Following Daily Oral Gavage Administration for 13 Weeks. *Toxicol. Sci.* **2016**, *150*, 131–160.

66. Hendrickson, O. D.; Klochkov, S. G.; Novikova, O. V.; Bravova, I. M.; Shevtsova, E. F.; Safenkova, I. V.; Zherdev, A. V.; Bachurin, S. O.; Dzantiev, B. B. Toxicity of Nanosilver in Intragastric Studies: Biodistribution and Metabolic Effects. *Toxicol. Lett.* **2016**, *241*, 184–192.

67. van der Zande, M.; Vandebriel, R. J.; Van Doren, E.; Kramer, E.; Herrera, R. Z.; Serrano-Rojero, C. S.; Gremmer, E. R.; Mast, J.; Peters, R. J.; Hollman, P. C.; Hendriksen, P. J.; Marvin, H. J.; Peijnenburg, A. A.; Bouwmeester, H. Distribution, Elimination, and Toxicity of Silver Nanoparticles and Silver Ions in Rats after 28-Day Oral Exposure. *ACS Nano* **2012**, *6*, 7427–7442.

68. Yun, J. W.; Kim, S. H.; You, J. R.; Kim, W. H.; Jang, J. J.; Min, S. K.; Kim, H. C.; Chung, J.; Jeong, B. C.; Kang, D. H.; Che, J. H. Comparative Toxicity of Silicon Dioxide, Silver and Iron Oxide Nanoparticles after Repeated Oral Administration to Rats. *J. Appl. Toxicol.* **2015**, *35*, 681–693.

69. Kim, Y. S.; Song, M. Y.; Park, J. D.; Song, K. S.; Ryu, H. R.; Chung, Y. H.; Chang, H. K.; Lee, J. H.; Oh, K. H.; Kelman, B. J.; Hwang, I. K.; Yu, I. J. Subchronic Oral Toxicity of Silver Nanoparticles. *Part Fibre Toxicol.* **2010**, *7*, 20.

70. Shahare, B.; Yashpal, M. Toxic Effects of Repeated Oral Exposure of Silver Nanoparticles on Small Intestine Mucosa of Mice. *Toxicol. Mech. Methods* **2013**, *23*, 161–167.

71. Patlolla, A. K.; Hackett, D.; Tchounwou, P. B. Genotoxicity Study of Silver Nanoparticles in Bone Marrow Cells of Sprague–Dawley Rats. *Food Chem. Toxicol.* **2015**, *85*, 52–60.

72. Lappas, C. M. The Immunomodulatory Effects of Titanium Dioxide and Silver Nanoparticles. *Food Chem. Toxicol.* **2015**, *85*, 78–83.

73. Bohmert, L.; Niemann, B.; Lichtenstein, D.; Juling, S.; Lampen, A. Molecular Mechanism of Silver Nanoparticles in Human Intestinal Cells. *Nanotoxicology* **2015**, *9*, 852–860.

74. McCracken, C.; Zane, A.; Knight, D. A.; Hommel, E.; Dutta, P. K.; Waldman, W. J. Oxidative Stress-Mediated Inhibition of Intestinal Epithelial Cell Proliferation by Silver Nanoparticles. *Toxicol. In Vitro* **2015**, *29*, 1793–1808.

75. Miethling-Graff, R.; Rumpker, R.; Richter, M.; Verano-Braga, T.; Kjeldsen, F.; Brewer, J.; Hoyland, J.; Rubahn, H. G.; Erdmann, H. Exposure to Silver Nanoparticles Induces Size- and Dose-Dependent Oxidative Stress and Cytotoxicity in Human Colon Carcinoma Cells. *Toxicol. In Vitro* **2014**, *28*, 1280–1289.

76. Sahu, S. C.; Roy, S.; Zheng, J.; Ihrie, J. Contribution of Ionic Silver to Genotoxic Potential of Nanosilver in Human Liver HepG2 and Colon Caco2 Cells Evaluated by the Cytokinesis-Block Micronucleus Assay. *J. Appl. Toxicol.* **2016**, *36*, 532–542.

77. Sahu, S. C.; Roy, S.; Zheng, J.; Yourick, J. J.; Sprando, R. L. Comparative Genotoxicity of Nanosilver in Human Liver HepG2 and Colon Caco2 Cells Evaluated by Fluorescent Microscopy of Cytochalasin B-Blocked Micronucleus Formation. *J. Appl. Toxicol.* **2014**, *34*, 1200–1208.

78. Sahu, S. C.; Zheng, J.; Graham, L.; Chen, L.; Ihrie, J.; Yourick, J. J.; Sprando, R. L. Comparative Cytotoxicity of Nanosilver in Human Liver HepG2 and Colon Caco2 Cells in Culture. *J. Appl. Toxicol.* **2014**, *34*, 1155–1166.

79. Susewind, J.; de Souza Carvalho-Wodarz, C.; Repnik, U.; Collnot, E. M.; Schneider-Daum, N.; Griffiths, G. W.; Lehr, C. M. A 3D Co-culture of Three Human Cell Lines to Model the Inflamed Intestinal Mucosa for Safety Testing of Nanomaterials. *Nanotoxicology* **2016**, *10*, 53–62.

80. Georgantzopoulou, A.; Serchi, T.; Cambier, S.; Leclercq, C. C.; Renaut, J.; Shao, J.; Kruszewski, M.; Lentzen, E.; Grysan, P.; Eswara, S.; Audinot, J. N.; Contal, S.; Ziebel, J.; Guignard, C.; Hoffmann, L.; Murk, A. J.; Gutleb, A. C. Effects of silver Nanoparticles and Ions on a Co-culture Model for the Gastrointestinal Epithelium, *Part Fibre Toxicol.* **2016**, *13*, 9.

81. Pietroiusti, A.; Magrini, A.; Campagnolo, L. New Frontiers in Nanotoxicology: Gut Microbiota/Microbiome-Mediated Effects of Engineered Nanomaterials. *Toxicol. Appl. Pharmacol.* **2016**, *299*, 90–95.

82. Gregor, M. F.; Hotamisligil, G. S. Inflammatory Mechanisms in Obesity. *Annu. Rev. Immunol.* **2011**, *29*, 415–445.

83. Williams, K.; Milner, J.; Boudreau, M. D.; Gokulan, K.; Cerniglia, C. E.; Khare, S. Effects of Subchronic Exposure of Silver Nanoparticles on Intestinal Microbiota and Gut-Associated Immune Responses in the Ileum of Sprague–Dawley Rats. *Nanotoxicology* **2015**, *9*, 279–289.

84. Fu, C.; Liu, T.; Li, L.; Liu, H.; Chen, D.; Tang, F. The Absorption, Distribution, Excretion and Toxicity of Mesoporous Silica Nanoparticles in Mice Following Different Exposure Routes. *Biomaterials* **2013**, *34*, 2565–2575.

85. Yoshida, T.; Yoshioka, Y.; Takahashi, H.; Misato, K.; Mori, T.; Hirai, T.; Nagano, K.; Abe, Y.; Mukai, Y.; Kamada, H.; Tsunoda, S.; Nabeshi, H.; Yoshikawa, T.; Higashisaka, K.; Tsutsumi, Y. Intestinal Absorption and Biological Effects of Orally Administered Amorphous Silica Particles. *Nanoscale Res. Lett.* **2014**, *9*, 532–539.

86. Kim, Y. R.; Lee, S. Y.; Lee, E. J.; Park, S. H.; Seong, N. W.; Seo, H. S.; Shin, S. S.; Kim, S. J.; Meang, E. H.; Park, M. K.; Kim, M. S.; Kim, C. S.; Kim, S. K.; Son, S. W.; Seo, Y. R.; Kang, B. H.; Han, B. S.; An, S. S.; Lee, B. J.; Kim, M. K. Toxicity of Colloidal Silica Nanoparticles Administered Orally for 90 Days in Rats. *Int. J. Nanomed.* **2014,** *9* (Suppl. 2), 67–78.

87. Tarantini, A.; Huet, S.; Jarry, G.; Lanceleur, R.; Poul, M.; Tavares, A.; Vital, N.; Louro, H.; Joao, S. M.; Fessard, V. Genotoxicity of Synthetic Amorphous Silica Nanoparticles in Rats Following Short-Term Exposure. Part 1: Oral Route. *Environ. Mol. Mutagen* **2015,** *56*, 218–227.

88. Pandey, A.; Chandra, S.; Chauhan, L. K.; Narayan, G.; Chowdhuri, D. K. Cellular Internalization and Stress Response of Ingested Amorphous Silica Nanoparticles in the Midgut of *Drosophila melanogaster. Biochim. Biophys. Acta* **2013**, *1830*, 2256–2266.

89. Contado, C.; Mejia, J.; Lozano, G. O.; Piret, J. P.; Dumortier, E.; Toussaint, O.; Lucas, S. Physicochemical and Toxicological Evaluation of Silica Nanoparticles Suitable for Food and Consumer Products Collected by Following the EC Recommendation. *Anal. Bioanal. Chem.* **2016**, *408*, 271–286.

90. Sergent, J. A.; Paget, V.; Chevillard, S. Toxicity and Genotoxicity of Nano-SiO$_2$ on Human Epithelial Intestinal HT-29 Cell Line. *Ann. Occup. Hyg.* **2012**, *56*, 622–630.

91. Yang, Y. X.; Song, Z. M.; Cheng, B.; Xiang, K.; Chen, X. X.; Liu, J. H.; Cao, A.; Wang, Y.; Liu, Y.; Wang, H. Evaluation of the Toxicity of Food Additive Silica Nanoparticles on Gastrointestinal Cells. *J. Appl. Toxicol.* **2014**, *34*, 424–435.

92. Gerloff, K.; Pereira, D. I.; Faria, N.; Boots, A. W.; Kolling, J.; Forster, I.; Albrecht, C.; Powell, J. J.; Schins, R. P. Influence of Simulated Gastrointestinal Conditions on Particle-Induced Cytotoxicity and Interleukin-8 Regulation in Differentiated and Undifferentiated Caco-2 Cells. *Nanotoxicology* **2013**, *7*, 353–366.

93. Tarantini, A.; Lanceleur, R.; Mourot, A.; Lavault, M. T.; Casterou, G.; Jarry, G.; Hogeveen, K.; Fessard, V. Toxicity, Genotoxicity and Proinflammatory Effects of Amorphous Nanosilica in the Human Intestinal Caco-2 Cell Line. *Toxicol. In Vitro* **2015**, *29*, 398–407.

CHAPTER 4

SAFETY EVALUATION OF CHEMISTRIES USED IN THE FOOD AND BEVERAGE PROCESSING AND PACKAGING INDUSTRIES

MARTIN HOAGLAND[1], KATHRYN SANDE[2], ALISON BEHLING[3], and NATHAN PECHACEK[2*]

[1]Nalco Water, An Ecolab Company, Naperville, IL, United States

[2]Ecolab, Eagan, MN, United States

[3]Ecolab, St. Paul, MN, United States

[*]Corresponding author. E-mail: Nathan.Pechacek@ecolab.com

CONTENTS

ABSTRACT

This chapter highlights some of the methods used to assess the risk of dietary exposure to various chemicals used in food and beverage processing, public eating facilities, and packaging industries using the perspective of US federal agencies regarding human dietary exposure and safety. These methods can be used to evaluate the safety of these chemicals prior to registration with regulatory agencies or as part of a product registration. In addition, these methods can also be used to determine the potential health hazards of foods which have been accidentally contaminated by chemical products which have been misused during processing, packaging, cleaning, or sanitizing.

4.1   INTRODUCTION

Many chemicals may come into contact with food during its production, processing, and packaging prior to consumption. Some of this contact is purposeful, whereas others are inadvertent. Some of these chemistries may become incorporated into the food, whereas others dissipate prior to consumption. Regardless of the intent or ultimate fate of these chemistries in food, they fall under regulatory scrutiny and require a thorough safety evaluation prior to their use on or around food.

This chapter focuses on how safety is evaluated for chemicals used for cleaning, sanitizing, and during the processing and packaging of food, as well as in the preparation of food in public eating facilities such as cafeterias and restaurants. Examples of such chemicals include products that are intended for direct food contact such as animal carcass antimicrobial washes and those used to clean and sanitize the surfaces of freshly harvested produce. Another example includes chemicals used to clean and sanitize the interior of beverage containers such as bottles and cans prior to filling. Products used for cleaning and sanitizing in public eating facilities also come in contact with food through use on utensils, cutting boards, and other food-preparation surfaces. Equipment maintenance products used in the food-processing industry such as conveyor belt lubricants may have indirect or inadvertent food contact. Lastly, some chemicals are used to clean and sanitize food-processing equipment following use and prior to further use. Examples are chemicals used to clean and sanitize holding vessels, mixing containers, transfer lines, and other types of equipment and machinery used in the food and beverage-processing facilities. As noted, this chapter will focus on the processing and packaging aspect of food using a US regulatory lens and will not address chemicals that come into contact with food during its production. This latter category include agricultural fertilizers and pesticides used to treat food crops and veterinary medicines used to treat food production animals, to name just a few examples.

Substances that can become a part of food or come in contact with food are regulated in the United States primarily by the US Food and Drug Administration (FDA) and US Environmental Protection Agency (EPA). The FDA regulates food additives whether such comes into direct contact with food as per the previously mentioned example of carcass wash chemicals or through indirect exposure addition to food, as per food-packaging materials or food-equipment maintenance materials in which trace levels of material may migrate to the food through physical contact and diffusion. Food contact antimicrobials are regulated by both the FDA and the EPA; however, the applications are divided as indicated in Table 4.1. The FDA regulates antimicrobials used in meat and poultry washes, as well as those used on processed foods or in food-packaging material. The EPA regulates antimicrobials as pesticides for applications on raw agricultural commodities and food contact hard surface sanitization. The US Department of Agriculture (USDA) can also have a role in the regulation of antimicrobials, in particular those used directly on meat and poultry; however, while registration of these materials is required through the USDA Food Safety and Inspection Service, notification to the FDA is also required through the submission of a Food Contact Notification (FCN) or Food Additive Petition (FAP).

The FDA and EPA use dietary risk assessments to evaluate the hazards posed by direct and indirect food contact substances (FCSs) and follow the principles of risk assessment including hazard identification, dose–response assessment, exposure assessment, and risk characterization.[11,14] The purpose of these premarket assessments is to determine whether the regulated use of the FCS is safe for humans, companion animals, and food-producing animals as well as the environment. The scope of this chapter is to describe the dietary-risk assessment methods used by the FDA and EPA to evaluate the risk posed to humans from food-contact substances.

4.2   FOOD ADDITIVES

4.2.1   BACKGROUND

The FDA is tasked with ensuring the foods we eat are safe. With this mission, the FDA conducts premarket reviews of food additives to prevent unsafe

TABLE 4.1 US Food and Drug Administration (FDA) and Environmental Protection Agency (EPA) Jurisdiction of Antimicrobials that Come in Contact with Food.[5]

Food contact use	FDA	EPA
In/on processed food or beverages[a]	X	
Application to process water in a food-processing facility[a]	X	
- When water is vehicle for transporting the antimicrobial to food	X	
In/on animal feed	X	
Application to animal drinking water		X
Pre and/or postharvest field use on crops		X
By consumers on raw agricultural commodities:		X
- Home produce washes, home gardens		
Application to process water for postharvest use (field washing) of raw agricultural commodities		X
Treatment of permanent or semipermanent food contact surfaces (sanitizers)		X
In the production of food contact articles, other than food packaging, where the antimicrobial is intended to have an ongoing effect on the article's food contact surface or in food that may contact the article		X
Application to process water in a food-processing facility to control a pest in the water	X	X
-e.g. pulp and paperboard use, use in cane-sugar and beet-sugar mills		
Treatment of raw agricultural commodities in a food-processing facility	X	X
Production of food packaging	X	X
Production of food-contact articles other than food packaging; no intended effect on the surface of the article	X	X

[a]Antimicrobials applied to raw meat tissue such as poultry, beef, or pork are reviewed for safety by the US FDA but must also meet regulatory requirements of USDA Food Safety Inspection Service (FSIS).

products from reaching the food supply. A food additive is legally defined as any substance whose intended use results or may reasonably be expected to result, directly or indirectly, in its becoming a component or otherwise affecting the characteristics of any food if such substance is not generally recognized as safe (GRAS).[10]

Traditionally, substances added to foods were distinguished based on direct or indirect addition to food. Direct food additives include those substances intentionally added to food that impart an ongoing technical effect on food. For example, xanthan gum is used to enhance texture. Other examples of direct additives include preservatives and artificial sweeteners. As previously described, indirect additives include food-packaging materials that are not intentionally added to food meet the legal definition of food in the United States and are considered food additives based on the likely migration to food through direct contact. The components of paper and plastic food packaging that contact food are examples of indirect food additives. Other examples include processing aids including flocculants and clarifying agents which are used in water and food-treatment processes to concentrate small particles for aggregation or removal, boiler additives which include anticorrosive compounds used to prevent mechanical failure of the steam-containing boilers and lines, rinse aids used in dish machines to help the sheeting of water off dishes, and manufacturing materials with incidental food contact such as solvents, lubricants, and release agents commonly applied to manufacturing equipment (e.g., extruders, gaskets). Since the advent of the FDA FCN program in 2000, use of the term "FCS" is preferred and replacing the term "indirect food additive" in reference to incidental food additives.

4.2.2 PATHWAYS TO REGULATORY COMPLIANCE

Foods containing unauthorized substances may be considered adulterated by the FDA and present potential health risks to the public. Given the regulatory authority of the FDA, compliance is essential to introduce safe food products including FCSs to commerce, ensure continued business operations, and promote consumer confidence. Various pathways for regulatory compliance exist to introduce food products and FCSs including:

- *FAP*: Commonly used for direct food additives, the sponsor may submit an FAP to the FDA for premarket review. This petition should contain the necessary information to support the safety of the proposed food additive and use conditions. The cumulative human dietary concentration (DC) exposure to the intended food additive are typically greater than 1 part per million (ppm) or 1 mg/kg food. That is, if the daily intake of the proposed food additive for all uses exceed 1 mg/kg food, the FDA recommends submission of an FAP. However, antimicrobial agents are reviewed at one-fifth of the level of standard food additives. As such, submission of an FAP is recommended when cumulative dietary exposures to an antimicrobial exceed 200 μg/kg food. Information to support the safety of the petitioned use at the dietary exposure levels may include human clinical trial testing and epidemiology studies, in addition to meeting the recommended testing guidelines for the lower tiers of exposure (Table 4.2). A petitioned food additive that is approved by the FDA results in listing within Section 21 of the Code of Federal Regulations (CFR) and the approved food additive then must be used within the limitations set by its established regulation. As food additives listed within 21 CFR may be used without notification to the FDA, successfully petitioned food additives do not result in exclusivity for the petitioner. Examples of direct food additives include the vitamin folic acid and tertiary butylhydroquinone, an antioxidant and preservative.

- *Threshold of regulation (TOR) exemption*: A rare option, the manufacturer or supplier may request exemption of regulation for food-contact articles based on an incremental DC of less than 0.5 parts per billion (ppb) or 0.5 μg/kg food with the intended use. Alternatively, the FDA may also permit a TOR exemption in cases in which the use of a regulated direct food additive is ≤1% of the acceptable dietary intake (ADI). In the former case for FCS, such exposure levels are considered de minimis in nature and no toxicological testing is recommended at such levels. That is, exposure levels are low such that should the FCS be potentially hazardous, the risk is mitigated (i.e., below the Toxicological Threshold of Concern). However, information should be submitted that allows for evaluation of all potential materials that may migrate to food from the FCS including impurities and decomposition chemistries. This information may include manufacturing methods or material composition (e.g., levels of heavy metals within elastomers or plastics). Similar to FAPs, TORs are not exclusive to the supplier or manufacturer and anyone may use the listed material under the specified use conditions without notification to the FDA. Examples of materials authorized for use through TOR exemptions include antioxidants used at low levels in polymeric plastics to prevent discoloration and lubricants used for conveyance of beverage containers. Interestingly, houseware articles including those food-contact articles used in the home or restaurant (e.g., alloys and ceramics) may be the subject of TOR exemptions. For example, an antimicrobial formulation of silver and cetyltrimethylammonium bromide is permitted for use in composite countertops at up to 1% by weight. The FDA maintains the database for such exemptions within 21 CFR 170.39 (threshold of regulation for substances used in food-contact articles).

- *GRAS affirmation:* Within the definition of food additives lies the "GRAS" exemption: a food substance that is reviewed by experts to be GRAS for a particular use in food is determined to not be a food additive and thus not subject to regulatory authority of the FDA. This process may be completed outside notification to the FDA, referred to as a self-GRAS determination. However, the FDA does support a voluntary GRAS Notification program that allows for agency review of substances for the intended use conditions. Examples of substances effectively notified to the FDA GRAS Notification program include steviol glycosides as a sugar replacement based on historical use, citrus fiber for use as a texturizer in yogurt, and breadfruit flour for use as a gluten-free substitute in baked goods. Companies may choose the self-GRAS determination route to quickly meet business needs and to protect confidential business information, whereas avoiding public notification (e.g., publication in the CFR and internet). This route may include third-party legal partner opinion letters of GRAS determination, which provide legal protection as well as an outside opinion regarding the determination of safety. Affirmations of GRAS status can be evaluated based on historical use or safety information. Food substances affirmed as GRAS by the FDA are listed within 21 CFR and may be readily used under the stated conditions without notification.

- *FCN:* The sponsor may notify the FDA for intent to use a FCS through submission of an FCN for review. Upon submission of a complete package, the FDA is mandated by law to complete its review within 120 days. Deadlines for other regulatory pathways of food additives may be extended. The FCN program provides exclusivity to the notifier though effective notifications are posted online and competitors commonly are quick to submit "me too" notifications. Such latter notifications are still required to address potential concerns including differences in purity profile. The review of these notifications by the FDA typically extend the full review period of 120 days. The safety information necessary to support the notified use is based on the cumulative exposures of the FCS and constituent materials that may migrate to food. As part of the review, the FDA determines the cumulative estimated daily intake (cEDI) exposure levels for notified and regulated uses, and whether the notification will be substitutive in nature or subsumed by the previously notified and regulated use levels. Should such increases in exposure levels surpass the tiered safety exposure levels (Table 4.2), FDA may require additional toxicology testing to evaluate safety. Absence of such information may result in safety concerns on behalf of the FDA and a recommendation to withdraw the petition. Failure to withdraw may then lead to objection by the agency on safety concerns. In these cases, submitted materials are no longer confidential and are subject to public requests for information via the Freedom of Information Act.

TABLE 4.2 Tiered Safety Guidance.[9]

Exposure level		Minimum testing recommendation
DCa (ppb)	EDIb (μg/person /d)	
≤ 0.5	≤ 1.5	For a single use or incremental exposure
		No safety studies recommended
		Review of the potential carcinogenicity
$0.5 < X \leq 50$	$1.5 < X \leq 150$	For a cumulative exposure
		Review of the potential carcinogenicity
		Test for mutagenicity (i.e., bacterial reversion "Ames" assay)
		Test for clastogenicity
		In vitro chromosome aberration
		Mouse lymphoma $tk^{\pm}$ assay
$50 < X \leq 1000$	$150 < X \leq 3000$	For a cumulative exposure, the above recommendations for $0.5 < X \leq 50$ ppb
		In addition
		Two subchronic oral toxicity tests (one rodent, one nonrodent)
		Specialized testing recommendations based on structure similarities to known toxicants
>1 ppm	>3 mg/person/day	For a cumulative exposure
		Submission of a food additive petition

aDietary concentration, ppb = μg substance/kg food.
bEstimated daily intake: μg substance/person/day.

4.2.3   EXPOSURE CALCULATIONS

DC exposure levels indicate the intended use level within food and may be expressed as in various units of measure (e.g., mg/kg, µg/g, or ppm). As food additives may only be used in certain foods, this information may be used to revise dietary exposures. For example, a food additive may be used only with potatoes. If the average and high rates of intake of potatoes is 75 and 300 g/day, respectively, the dietary exposure level may be adjusted accordingly in place of applying the general use level to all foods (i.e., 3000 g food/liquid per day). Similar adjustments may be appropriate based on intended use conditions and consumption data (e.g., only in soups, only in frozen foods). The FDA provides consumption factors (CFs) for common FCSs.[9] For example, the agency's glass CF of 0.1 estimates the highest level of migration into food and/or liquid consumed daily that contacts glass is 10% of the 3000 g food/liquid daily diet.

With preliminary exposure assessments of FCS, one may determine the worst-case exposure level by assuming 100% migration ($\langle M \rangle$) of FCSs into food. Peroxide-based chemistries are commonly employed in meat and poultry-processing plants. With poultry washes, products containing 50% hydrogen peroxide with the end-use level of 1 ppm may be calculated in the following hypothetical example:

$$\text{Exposure } (E) = \text{Product use level (mg / kg, µg / kg or ppm)}$$
$$\times \text{ Level of leachable chemistry in product}$$
$$E = 1 \text{ ppm} \times 0.5$$
$$E = 0.5 \text{ ppm } (0.5 \text{ mg/kg})$$

Based on the publically available literature, one can reasonably assume that the average per capita consumption of poultry is 0.1 kg/day. Should the notified use capture 100% of the chicken market and only be used with chicken, the worst-case exposure level would be 0.05 ppm (i.e., downwardly adjust the 0.5 ppm exposure estimate by an order of magnitude given the 0.1 kg/day average consumption estimate). The FDA assumes that the average adult weighs 60 kg and consumes 3 kg food/liquid per day. Using an additional assumption that dietary exposure to the hydrogen peroxide chemistry only occurs through consumption of chicken, the DC and estimated daily intake (EDI) can then be determined:

$$DC = E/3 \text{ kg food/liquid per day} = 0.05 \text{ mg/3 kg} = 0.0167$$
$$ppm = 16.7 \text{ ppb (or } \mu g/kg)$$

$$EDI = 16.7 \ \mu g/kg \times 3 \text{ kg food/liquid per day}$$
$$= 50 \ \mu g/person/day \text{ or } 0.83 \ \mu g/kg \text{ bw/day.}$$

With rapidly decaying chemistries (e.g., halogenated dimethylhydan-toins), safety evaluations focus not on the starting materials but on the resulting reaction products and potential chemistries humans are exposed to through the notified use conditions (e.g., dimethylhydantoin, bromide, bromoform). In the case of chemistries that break down to inert compounds (e.g., hydrogen peroxide to oxygen and water, or peroxyacids to acetic acid and water), safety reviews focus on the other components of the products (e.g., stabilizers including 1-hydroxyethane 1,1-diphos-phonic acid (HEDP)).

The exposure level of chemistries used in the manufacture of paper for food packaging, including antimicrobials used to control microorganism growth (i.e., slimicides), may be calculated using the conservative retention level of 2% as prescribed by the FDA for wet-end-use applications with the default paper weight of 50 mg/in² and the standard assumption of 10 g food/in². For example,

$$DC = 50 \text{ mg}/\text{in}^2 \times (1/10 \text{ g food}/\text{in}^2) \times \text{Product level}$$
$$\times \text{Level of Leachable Chemistry} \times 0.02 \times CF \times \langle M \rangle$$

$$DC = 50 \text{ mg}/\text{in}^2 \times (1/10 \text{ g food}/\text{in}^2) \times 0.125\% \text{ dry weight fiber}$$
$$\times 100\% \times 0.02 \times 0.1 \times 100\%$$

$$DC = 1.25 \times 10^{-5} \text{ mg}/\text{g food} = 1.25 \times 10^{-8} \text{ g}/\text{g food} = 12.5 \text{ ppb}$$

$$EDI = 12.5 \text{ ppb} \times 3 \text{ kg/day} = 37.5 \ \mu g/person/day \text{ or } 0.63 \ \mu g/kg \text{ bw/day}$$

As coatings are not diluted as materials applied on the wet-end, the 2% retention level factor is not applied. However, should materials be applied that are volatile, off-gassing may substantially minimize the potential expo-sure level.

Should worst-case exposure levels present concern or require the support of toxicological testing, migration studies may be conducted using food simulants to present more definitive dietary exposure levels of FCS mate-rials. As the types of foods that contact the different types of packaging varies, food-type distribution factors (f_T) are used by the FDA to account for the variable nature of food-types contacting each food-contact article. Such distribution factors were determined by the FDA based on the relative level of food-type (i.e., aqueous, acidic, alcoholic, and fatty) that contacts each packaging material. Diffusion of FCS materials to food is dependent on solubility. Migration studies that determine the level to which a material migrates will rely on the use of different solvents to mimic various food stuffs. The currently recommended solvents are 10% aqueous ethanol for aqueous and acidic foods, 50% aqueous ethanol for alcohol, and Miglyol 812 for fatty foods. The recommended volume of solvent to be used in migration studies is 10 ml in^2. The analytical detection method for the migrants is not stringent (e.g., high-pressure liquid chromatography gas chromatography-mass sprectrophotometry); however, the method should be sufficiently sensitive. In addition, standard curves for the FCS should be generated for quantification purposes of the samples.

$$\langle M \rangle = f_{\text{aqueous and acidic}}(M_{\text{10\% ethanol}}) + f_{\text{alcohol}}(M_{\text{50\% ethanol}}) + f_{\text{fatty}}(M_{\text{fatty}})$$

$$\text{EDI} = 3 \text{ kg food/person/day} \times \langle M \rangle \times \text{CF}$$

where CF is the consumption factor (unitless) and $\langle M \rangle$ is the migration (unitless).

With the hypothetical example of a hydrophilic bleaching agent for use in uncoated paper, the migration $\langle M \rangle$ of the bleaching agent from sample paper was 1.0, 0.5, and 0.05 mg/kg for 10% aqueous ethanol, 50% aqueous ethanol, and Miglyol 812, respectively. Using the FDA's food-type distribution factors and CFs for uncoated and clay-coated paper (f_{T} of 0.58, 0.01, 0.01, and 0.41 for aqueous, acidic, alcoholic, and fatty foods; and CF of 0.1), the EDI may be calculated:

$$\langle M \rangle = 0.58 \, (1.0 \text{ mg/kg}) + 0.01(0.5 \text{ mg/kg}) + 0.41(0.05 \text{ mg/kg})$$

$$\langle M \rangle = 0.61 \text{ mg/kg}$$

$$\text{EDI} = 3 \text{ kg food/person/day} \times 0.61 \text{ mg/kg food} \times 0.1$$

$$\text{EDI} = 0.18 \text{ mg/person/day}$$

$$= 0.18 \text{ mg/person/day} \times 1 \text{ person/60 kg}$$

$$= 0.0031 \text{ mg/kg/day} \, (3.1 \text{ µg/kg bw/day})$$

As polymeric materials migrate poorly and have low bioavailability, migration samples should be evaluated for the presence of low molecular oligomers (i.e., molecular weight of less than 1000 Da) and unreacted monomers.

Whereas the above calculations assume 100% market capture, should exposure levels present concern with migration testing, one may propose a market volume limitation (MVL) as an alternative CF whereby a manufacturer limits production levels to capture a fraction of the market. The amount of paper consumed within the United States is 80 million tons per year, 54% of which is used for packaging. Conservatively assuming that 100% of the manufactured paper is bleached and 50% of the paper is used in food-contact applications, the US market of food contact bleached paper is 22 million tons. If the previous example of the paper bleaching agent is to be used at 500 mg/kg dry weight paper, 100% market capture would require manufacture of 11,000 t of the bleaching agent. Through the MVL, the manufacturer reports maximum production levels of 11 t/year. The MVL is determined as follows:

$$MVL = 11 \text{ tons maximally produced} /11,000 \text{ t}$$

$$=0.001$$

This MVL may then be applied to the previously calculated EDI:

$$EDI = 0.18 \text{ mg/person/day} \times 0.001$$

$$=0.00018 \text{ mg/person/day } (0.18 \text{ µg/person/day})$$

4.3 TOXICOLOGY

4.3.1 FDA-TIERED SAFETY GUIDANCE

While food additives may be authorized by the FDA through several ways, it is important to note the safety paradigm used to evaluate safety for each is the same (Table 4.2).[9] However, based on the intended biological activity of antimicrobials, the testing guidance tiers set by the FDA are one-fifth that of the standard food additive. For threshold toxicants, the agency reviews food additives based on the ADI and the cEDI. Traditionally, safety concerns of food additives are negligible when the cEDI < ADI; however, the FDA recommends consultation should the estimated exposure approach the identified point of departure (POD) (e.g., no observed adverse effect level—NOAEL, lowest observed adverse effect level—LOAEL, benchmark dose level—BMDL) for the chemical of interest.

Regarding polymeric materials, the safety of exposures to such FCS at DC exposure levels <50 ppb is based on toxicology information of the constituent monomers, a conservative worst-case approach. Reactive double bonds present within monomers are well established to be consumed within the polymerization process. As such, these structural alerts do not present a safety concern for the polymeric materials. Recent focus on the low bioavailability assumption of polymeric materials is currently under review by the FDA. Notifiers should address the safety of exposures of oligomeric materials whose molecular weight is <1000 Da.

Unavoidable carcinogenic impurities (e.g., acrylamide, styrene, oligomeric siloxanes) may be present within FCS. Risk assessments of these exposures are evaluated by the FDA using worst-case lifetime cancer risk (LCR) analysis, wherein the traditional threshold of concern for excess cancer cases in the population is 10^{-6} for cumulative exposure levels and 10^{-8} for incidental exposures. For example, the unit risk factor (URF) used by the FDA for acrylamide is 0.72 (mg/kg bw/day)$^{-1}$. In the following example, the dietary exposure level associated with a novel polyacrylamide FCS is 1 ppb (e.g., 1 µg/kg food). With the residual acrylamide level of 250 ppm within the FCS and assuming 100% migration of the acrylamide monomer, the worst-case DC level is 0.25×10^{-12} (e.g., 0.25 ng/kg food). Converting this value for the LCR analysis,

$$0.25 \text{ ng/kg food} \times 3 \text{ kg food/day} = 0.75 \text{ ng/person/day}$$

$$0.75 \text{ ng/person/day} \times 1 \text{ person/60 kg bw} = 0.0125 \text{ ng/kg bw/day}$$

$$0.0125 \text{ ng/kg bw/day} = 1.25 \times 10^{-8} \text{ mg/kg bw/day}$$

$$LCR = URF \times E$$

$$LCR = 0.72 \text{ (mg/kg bw/day)}^{-1} \times 1.25 \times 10^{-8} \text{ mg/kg bw/day}$$

$$LCR = 9.0 \times 10^{-9} \text{ (which is } <1 \times 10^{-8})$$

Traditionally, the risk associated with acrylamide exposure from this hypothetical FCS would be considered tolerable by the FDA. Carcinogen URF values are available through the FDA.

The FDA recently updated the tiered safety guidance to address specific susceptible subpopulations within its infant safety guidance. EDI thresholds were previously reported using the traditional assumptions of 60 kg as the weight of an adult and 3 kg of food/liquid diet per day. However, these assumptions are not representative of infant body weight and food consumption patterns. The updated tiers include thresholds based on infant safety guidance and the average infant intake of 0.9 kg food/liquid per day and body weight of 6.3 kg.[12]

4.4   FOOD CONTACT SURFACE ANTIMICROBIALS

Antimicrobials are regulated in the United States by the FDA, EPA, and in some instances, the USDA. Table 4.1 summarizes the various antimicrobial use patterns and the respective regulatory jurisdictions of the federal agencies. Some antimicrobials fall under dual jurisdiction between both the EPA and FDA, which is also highlighted in the table. FDA regulates and evaluates the safety of antimicrobials as they would other FCSs. As FDA regulation of FCSs has been addressed above, the focus of this section will be on EPA's approach to regulating food contact antimicrobials.

4.4.1   EPA REGULATION OF ANTIMICROBIALS

The EPA regulates antimicrobials as pesticides under the Federal Insecticide Fungicide and Rodenticide Act (FIFRA). Antimicrobial pesticides are a diverse class of substances that are quite different from conventional pesticides.[2] Conventional pesticides are often added directly to food crops to treat pests, whereas antimicrobial pesticides are typically indirect FCSs that come in contact with surfaces that then contact food, such as countertops or food-processing equipment. Antimicrobials may also come in direct contact with food through the treatment of raw or processed agricultural commodities in postharvest applications or in food-processing facilities and from consumer produce washes.

- Antimicrobials are defined by the EPA as pesticides that:
- disinfect, sanitize, reduce, or mitigate growth or development of microbiological organisms; or
- protect inanimate objects, industrial processes or systems, surfaces, water, or other chemical substances from contamination, fouling, or deterioration caused by bacteria, viruses, fungi, protozoa, algae, or slime;
- and are exempt from or not subject to a tolerance or a food additive regulation and are registered through the EPA Antimicrobial Division (AD). They are subject to the registration requirements described in FIFRA section 3(h).[2]

FDA regulates antimicrobial residues that may occur on processed food through the indirect food additives process. These uses are also considered pesticide uses subject to registration under FIFRA. However, a food additive clearance must be obtained from FDA before EPA will approve the application.

Every ingredient in EPA-registered antimicrobial products must have the proper registration. The AD within the EPA Office of Pesticide Programs (OPP) oversees the registration of active ingredients, other ingredients, and end-use products that fall within antimicrobial use patterns. An active ingredient is the substance in a product designed to control pests. All nonactive ingredients (i.e., not designed to control pests) are considered "other" ingredients. Note that prior to 1997, "inert" was the preferred terminology for nonactive ingredients but that term is phased out in preference to the more neutral term "other".[3] Other ingredients are added to antimicrobial products to aid in product usability and performance. Active ingredients may also be registered by a different division within OPP but still be used in an antimicrobial end-use product. An antimicrobial end-use product containing an active ingredient registered by a different division would still be registered by AD based on the use pattern.

4.4.2 EPA FOOD CONTACT ANTIMICROBIAL DIETARY RISK ASSESSMENT

Active ingredients and end-use products registered through AD must meet data requirements outlined under 40 CFR 158W.[4] They must also be evaluated for safety. All nonactive ingredients (other ingredients) used in food contact hard surface sanitizers used in public eating facilities and food-processing plants must also be assessed for safety and approved under 40 CFR 180.940 (previously 21 CFR 178.1010).

Antimicrobials used on food contact hard surfaces are referred to as food contact surface sanitizers (FCSS). The FCSS are used on surfaces that touch food such as food-processing equipment, utensils, and countertops. These FCSS are considered indirect FCSs. The EPA evaluates FCSS using one of three scenarios depending on the use pattern of the ingredient or end-use product: (1) public eating facilities, (2) food-processing facilities, including dairy, and (3) food-processing facilities, excluding dairy. The inert ingredient clearances for FCSS outlined in 40 CFR 180.940 correspond to the (1), (2), and (3) uses described above.

The FDA originally developed the dietary risk assessment guidelines for FCSS[8] as prior to 1996, these materials were regulated by FDA under 21 CFR 180.1010. The FDA model was adopted by EPA when the jurisdiction to regulate FCSS transferred from FDA to EPA in 1996 through the Food

Quality Protection Act (FQPA).[13] It is worth noting that the FCSS exposure model for FCSS is the same model still used by the FDA to evaluate exposure to substances used in rinse additives (i.e., substance that enhances water removal from utensils, dishes, bowls, and glassware).

The purpose of an FCSS dietary risk assessment is to estimate human exposure to residues that could transfer from a treated surface into food. Those residues in food are then evaluated to determine whether the exposure poses a health risk. An EDI is the concentration of an FCSS residue that has transferred from a surface into food. EDIs are often conservative estimates based on worst case theoretical exposure assumptions and therefore result in much higher estimates of exposure than would be expected if actual residues in food were measured. For example, unless chemical specific residue data and transfer data are available, it is assumed that 100% of an antimicrobial residue from a food contact surface will transfer into food. An EDI is estimated using three types of information:

1. The residual solution or residue concentration left on the surface;
2. the mass or volume of food contacting a treated surface; and
3. estimate of an individual's daily intake of foods assumed to have come into contact with a treated surface.

The exposure scenario outlined below is considered the broadest use scenario used by EPA for FCSS that are used in institutional settings (i.e., public eating facilities). The broadest use scenario is applicable for products that will be used in such facilities and that will not be rinsed further. This scenario is used to establish 40 CFR 180.940(a) clearances. An approval based on this approach also applies to approvals for uses of an FCSS in food-processing environments. However, the reverse is not true. Approvals based on the exposure scenario used for food-processing facilities cannot be used for an FCSS in a public eating establishment.

An EDI is used to calculate a human dietary dose, which is derived by dividing the residue of a substance in food by human body weight. The calculations used to derive the EDI and daily dietary dose (DDD) are shown below.

$$\text{EDI (mg/person/day)} = AR \times RS \times SA \times F \times 1 \times 10^{-6}$$

$$\text{DDD (mg/kg/day)} = (AR \times RS \times SA \times F \times 1 \times 10^{-6})/BW$$

where AR is the application rate (ppm or mg/kg); RS is the residual solution (mg/cm^2); SA is the surface area (SA) of the treated surface which comes into contact with food (cm^2); F is the fraction of the pesticide transferred or migrated to food (unitless); BW is the body weight (kg); $1 \times 10^{-6} = 1$ kg/1000,000 mg.

It is common to use the following default assumptions when calculating an EDI for an FCSS:

- A residual solution of 1 mg/cm^2 when substance-specific residue data are not available,
- 4000 cm^2 of SA of the treated surface that comes in contact with food,
- 100% transfer of the FCSS to food, although a value less than 100% may be used when there is a reasonable justification to use a lesser value, and
- body weights of 80 kg for adults, 15 kg for children, and 10 kg for infants.

Hazard identification is then conducted on all FCSS of interest to identify a POD, which represents the dose descriptor (usually NOAEL, LOAEL, or BMDL) indicative of the potency of the hazard and which serves as the comparison point for estimated exposures. The POD for each substance is then compared to the DDD derived for each substance. The DDD is divided by the POD to determine the margin of exposure (MOE). An acceptable MOE is dependent on several factors. A value of 100 is often considered an acceptable MOE, but it could be higher or lower depending on factors, such as (1) the severity of the adverse effect that serves as the basis for the POD, (2) the quality of the toxicity database, (3) whether there is evidence that animals and humans metabolize a substance in a similar manner, or (4) whether there is a need for a larger margin of safety due to concerns about sensitive populations. Potential sensitivity of infants and children may be addressed through an additional tenfold (10×) safety factor referred to as the FQPA safety factor.[13] The FQPA safety factor is usually added when an adverse effect(s) observed in a toxicology study is one that infants and children are known to be particularly vulnerable or sensitive (e.g., neurological effects). In addition, it may be applied when there is concern about the sensitivity of infants and children from exposure to a pesticide, but there is an incomplete database available to fully evaluate the risk posed by the exposure.

Another way the EPA evaluates whether an estimated exposure to an FCSS is acceptable is by comparing the DDD to a toxicological benchmark referred to as the population-adjusted dose (PAD). Comparing an estimated exposure to the PAD is similar to using an MOE approach to determine whether a potential exposure poses an increased health risk. The EPA considers a DDD to be acceptable when it is less than the PAD and where the DDD is expressed as a percent of the PAD (%PAD). A value less than 100% of the PAD is considered acceptable.

The EPA has used the FCSS exposure scenario described in this chapter in dietary risk assessments for various antimicrobials including quaternary ammonium compounds (alkyl dimethyl benzyl ammonium chloride known and didecyl dimethyl ammonium chloride), chlorine dioxide, and pine oil. Table 4.3 is an example of an EPA dietary risk assessment to evaluate risk from indirect contact to ADBAC residues when used as a sanitizer or disinfectant on hard nonporous surfaces in food-handling establishments.

TABLE 4.3 Alkyl Dimethyl Benzyl Ammonium Chloride (ADBAC) Dietary Risk Assessment for Food Contact Surface Sanitizers (FCSS) Used in Public Food Handling Establishments.[6]

Exposure and hazard parameters	Value
Maximum application rate of ADBAC (ppm)	200
Residue solution concentration (mg/cm^2)	1
Surface area of the treated surface which comes into contact with food (cm^2)	4000 (food utensils only)
Transfer rate	10% (EPA assumption)
Body weight	
Adult (kg)	70 kg males/60 kg females
Child (kg)	15 kg
Note: The default body weight values may have changed since the assessment was conducted in 2006	
Point of departure (NOAEL)	44 mg/kg/day
Total UF	Total UF = 100
	10× for interspecies extrapolation
	10× for intraspecies variation
FQPA safety factor	1×
PAD	0.44 mg/kg/day
PAD = NOAEL/UF	
Exposure estimates and %PAD for FCSS uses	
Total EDI for all FCSS uses combined (see equation above)	
Adult males	0.0815 mg/day
Adult females	0.0815 mg/day
Child	0.0815 mg/day

TABLE 4.3 *(Continued)*

Exposure and hazard parameters	Value
DDD = EDI/body weight (see equation above)	
Adult males	
Adult females	0.00116 mg/kg/day
Child	0.00136 mg/kg/day
	0.00543 mg/kg/day
% PAD	
%PAD = (DDD/PAD) × 100%	
Adult males	0.26
Adult females	0.31
Child	1.23

UF, uncertainty factors; *PAD*, population adjusted dose; *EDI*, estimated daily intake; *DDD*, daily dietary intake; *FCSS*, food contact surface sanitizers; *NOAEL*, no-observed adverse-effect level.

The model for evaluating dietary exposure for FCSS used in food-processing facilities, but not public eating facilities, is currently being revised. The new model will evaluate the transfer of FCSS from food-processing equipment to liquid-like foods and it will incorporate updated food consumption rates based on the Center for Disease Control (CDC) National Human and Nutrition Examination Survey data.[1]

4.5   EVALUATING RISKS POSED BY CHEMICAL PRODUCT MISUSE

4.5.1   CASE STUDIES WITH FOOD ADDITIVES/PROCESSING AIDS

4.5.1.1   USE OF A FOOD ADDITIVE ABOVE THE REGULATED MAXIMUM LIMIT

If a food additive is used over the regulated maximum limit, there are few options available to the food processor. Although a dietary exposure assessment could be done in this situation, it may not be used to permit release of the food product for sale. The product would be considered adulterated and/or misbranded regardless of the outcome of the risk assessment.

4.5.1.2   PROCESSING AID USED ABOVE THE REGULATED MAXIMUM LIMIT

If a processing aid is used above a regulated maximum limit, there are options for a risk assessment to determine whether it is appropriate to release the food product for consumption. This is predicated on the processing aid not having an ongoing technical effect in the food and the components of the processing aids are present only at negligible levels.

As an example, a peroxyacid-based antimicrobial food treatment cleared under 21 CFR 173.370(b)(2) is limited to use as an antimicrobial agent on poultry carcasses, parts and organs at a maximum concentration of 220 ppm (mg/l) peroxyacetic acid (POAA, CASRN 79-21-0), 110 ppm hydrogen peroxide (H_2O_2, CASRN 7722-84-1), and 13 ppm 1-hydroxyethylidene-1,1-diphosphonic acid (CASRN 2809-21-4). For a theoretical situation of product overapplication, it is assumed that the maximum concentrations of POAA, H_2O_2, and HEDP were all used at twice the permitted maximum (i.e., 440 ppm POAA, 220 ppm H_2O_2, 26 ppm HEDP).

In the theoretical situation noted above, it is reasonable to conclude that POAA will quickly degrade to acetic acid, water, and oxygen (i.e., the components of its equilibrium chemistry) and residue analysis has been conducted that demonstrates this is the case. Hydrogen peroxide is also known to rapidly degrade to water and oxygen. The HEDP will likely remain on the poultry unless it is washed off, and while this is the best corrective action, the overapplication may not be recognized before the poultry has been processed further (e.g., ground poultry), in which case rinsing is impractical.

The question now before the food-processing facility is whether the level of chemical overapplication necessitates disposal of the poultry product. This is where a dietary exposure assessment can be used to determine whether exposure to HEDP, the only remaining substance from the anti-microbial product that is of a potential health consequence, would pose an unacceptable human health risk if the poultry product is consumed by the public. Given the rapid dissipation of POAA and H_2O_2, and the relatively innocuous breakdown products of these substances, further evaluation is not warranted for POAA and H_2O_2. The estimated dietary exposure of HEDP is then compared with an ADI, oral reference dose (RfD), or other health-based guidance value to determine if the dietary exposure is below the already established safe consumption level.

Assumptions used for a theoretical product overapplication:

- Application rate of 1 fluid ounce/pound food
- 100% uptake HEDP
- Four-ounce (4 oz. = 113 g) serving size
- 1 Fluid ounce = 0.0296 liters (l)
- 1 pound = 454 g
- ADI/RfD of HEDP = 0.05 mg HEDP/kg bw/day[7]

$$\frac{26 \text{ mg HEDP}}{\text{liter solution}} \times \frac{1 \text{ fl oz solution}}{\text{pound turkey}} \times \frac{0.0296 \text{ liter}}{1 \text{ fl. oz.}} \times \frac{1 \text{ pound}}{454 \text{ grams}}$$

$$\times \frac{113 \text{ gram turkey}}{\text{person} / \text{day}} = \frac{0.19 \text{ mg HEDP}}{\text{person} / \text{day}}$$

For a 15-kg child, the dietary exposure would be

$$(0.19 \text{ mg HEDP} / \text{day}) / (15 \text{ kg}) = 0.01 \text{ mg HEDP} / \text{kg bw} / \text{day}$$

As demonstrated for this theoretical example of over application of a peroxy-acid-based antimicrobial food treatment, the conservative, yet reasonable, dietary exposure scenario for a sensitive population (e.g., young child) determined that the acute (i.e., short-term) dietary exposure to HEDP is approximately 20% of the chronic RfD (i.e., acceptable daily intake) established by the US EPA.

4.5.2 CONTAMINATION ISSUES

4.5.2.1 USE OF FOOD CONTACT SURFACE SANITIZER DIRECTLY ON FOOD

Although such a scenario is undesirable, the components of food contact surface sanitizers have been evaluated by the EPA or other regulatory authorities and a suitable dietary exposure assessment could be calculated based on the concentration of active and other ingredients applied to the food. The exposure assessment of the individual components of the sanitizer would be similar to the example shown above.

4.5.2.2 USE OF FOOD CONTACT SURFACE SANITIZER (FCSS) ABOVE LABEL MAXIMUM LIMIT OR WITHOUT DRAINING AND/ OR RINSING

The methods to determine EDI of an FCSS residue that has transferred from a surface into food were described previously using three types of information:

1. The residual solution or residue concentration left on the surface, typically assumed to be 1 mg solution/cm²;
2. the mass or volume of food contacting a treated surface;
3. estimate of an individual's daily intake of foods assumed to have come into contact with a treated surface; and
4. concentration of FCSS component in the solution (mg substance/kg solution).

In this situation, the critical components of an evaluation include the amount of food product affected (mass or volume), the SA of the actual food-processing equipment treated, and the serving size of the food product when consumed. The dietary exposure would then be determined assuming 100% transfer of the residue to the food product using the equation below.

$$\frac{mg\ substance}{person\ /\ day} = \frac{mg\ substance}{kg\ solution} \times \frac{mg\ solution}{cm^2} \times SA(cm^2)$$

$$\times \frac{kg\ food}{person\ /\ day} \times \frac{kg\ solution}{10^6\ mg\ solution}$$

4.5.2.3 USE OF EQUIPMENT CLEANERS WITHOUT RINSING

The components of most industrial cleaners are well known toxicologically, and at the dilutions commonly used, are of negligible toxicological concern. In addition, many ingredients such as sodium hydroxide, sodium carbonate, phosphoric acid, and sodium hypochlorite have clearance as food ingredients or processing aids. Despite the food regulatory clearances of the major components of equipment cleaners, there are likely other components in the cleaner which may not have such clearances and this intended use (i.e., cleaning) presupposes that the chemical ingredients of cleaners will not become a component of the food product; hence, in this scenario, the components of the cleaning solution would be considered a contaminant of the food. The question then becomes: Does the level of contamination cause the food product to be adulterated?

A conservative dietary exposure can be calculated provided the concentration of the cleaning product components are known, the volume of cleaning solution used, total quantity (e.g., batch size) of food product involved in the incident, and the serving size of the food product using models and methods as described above. The concentration of hazardous components of cleaning products can be found on the Safety Data Sheet for the cleaning product; the complete cleaning product formula information is also known by the manufacturer of the cleaning product and can be used to determine the concentration of components. The food manufacturer would have information on the amount of food affected and the serving size. This information would need to be available to the assessor regardless of whether the food manufacturer, the cleaning product manufacturer, or a third party completes the dietary exposure assessment.

Typically the biggest issue in the scenario of misuse of a cleaning product is whether the potential contaminants will change the flavor, aroma, or appearance of the food. The FDA and USDA consider foods with altered organoleptic properties as adulterated and would cause the food product to be recalled (i.e., remove a product from the market) and/or condemned (i.e., determine that a product is unfit to release to the market).

4.6   CORRECTIONS AND CORRECTIVE ACTIONS

4.6.1   RINSE THE CONTAMINATED FOOD WITH POTABLE WATER

This is most helpful in meat-processing facilities when antimicrobials are misused. The best correction is to rinse the antimicrobial off the carcass or primal cuts and retreat with the antimicrobial. If the discovery is not made until after the meat is further processed, then a risk assessment or dietary exposure assessment should be conducted to determine if the product must be discarded, may be reconditioned, or is still acceptable for human consumption.

If a food contact surface sanitizer or disinfectant is used above its maximum limit for a no-rinse application, the equipment, utensils, and other food contact surfaces can be rinsed with potable water to remove excess sanitizer prior to processing food. It is also useful to use a portion of a food to "clean out" filling equipment, which is then discarded, prior to packaging the remaining food for retail sale.

4.6.2 USE OF VALIDATED CLEANING AND SANITIZING METHODS

After an incident occurs, it is critical to determine the root cause and make corrections to the facility Sanitation Standard Operating Procedures, the Food Safety Plan, or both to ensure that the correct process is accurately documented and that the food manufacturing personnel are properly trained on the procedures to prevent chemical contamination of the food product.

KEYWORDS

- **dietary exposure**
- **chemicals**
- **packaging industries**
- **potential health hazards**
- **chemical products**

REFERENCES

1. Centers for Disease Control (CDC). *National Health and Nutrition Examination Survey*, 2016. http://www.cdc.gov/nchs/nhanes/ (accessed May 17, 2016).
2. Environmental Protection Agency (EPA). Chapter 4—Additional Considerations for Antimicrobial Products. *Pesticide Registration Manual*, 2016. https://www.epa.gov/pesticide-registration/pesticide-registration-manual-chapter-4-additional-considerations#whatis (accessed May 17, 2016).
3. EPA. *PRN 97-6: Use of Term "Inert" in the Label Ingredients Statement*, 2016. https://www.epa.gov/pesticide-registration/prn-97-6-use-term-inert-label-ingredients-statement#policy (accessed May 17, 2016).
4. EPA. *EPA Data Requirements for Registration of Antimicrobial Pesticides: Part 158W*, 2016. https://www.epa.gov/pesticide-registration/epa-data-requirements-registration-antimicrobial-pesticides-part-158w (accessed May 17, 2016).
5. EPA. Chapter 18—Other Federal or State Agency Requirements, *Pesticide Registration Manual*, 2016. http://www.epa.gov/pesticide-registration/pesticide-registration-manual-chapter-18-other-federal-or-state-agency#fda (accessed May 20, 2016).

6. EPA. *Memorandum: Dietary Risk Assessment for Alkyl Dimethyl Benzyl Ammonium Chloride (ADBAC) for Reregistration Eligibility Decision (RED) Process*, 2006.

7. EPA. Hydroxyethylidine Diphosphonic Acid—Exemption from the Requirement of a Tolerance. *Fed. Regist.* May **1998,** *63* (99), 28253–28258. FR Doc. 98-13603.

8. Food and Drug Administration (FDA). *Sanitizing Solutions: Chemistry Guidelines for Food Additive Petitions.* Chemistry Review Branch, Office of Pre-Market Approval, Center for Food Safety and Applied Nutrition: Washington DC, 1993.

9. FDA. *Guidance for Industry and Other Stakeholders. Toxicological Principles for the Safety Assessment of Food Ingredients (Redbook 2000)*, 2000. http://www.fda.gov/downloads/Food/GuidanceRegulation/UCM222779.pdf (accessed May 27, 2016).

10. FDA. *Determining the Regulatory Status of a Food Ingredient*, 2016. http://www.fda.gov/Food/IngredientsPackagingLabeling/FoodAdditivesIngredients/ucm228269.htm (accessed May 17, 2016).

11. National Academy of Science (NAS). *Science and Decisions: Advancing Risk Assessment.* National Academy Press: Washington, DC, 2009.

12. Neal-Kluever, A.; Aungst, J.; Gu, Y.; Hatwell, K.; Muldoon-Jacobs, K.; Liem, A.; Ogungbesan, A.; Shackelford, M. Infant Toxicology: State of the Science and Considerations in Evaluation of Safety. *Food Chem. Toxicol.* **2014,** *70,* 68–83.

13. US Government Printing Office (GPO). *Food Quality Protection Act of 1996*, 1996. https://www.gpo.gov/fdsys/pkg/PLAW-104publ170/pdf/PLAW-104publ170.pdf (accessed May 17, 2016).

14. World Health Organization (WHO). *Principles and Methods for the Risk Assessment of Chemicals in Food.* A Joint Publication of the Food and Agriculture Organization of the United Nations and the World Health Organization, Environmental Health Criteria 240: Geneva, Switzerland, 2009.

CHAPTER 5

DEVELOPMENTAL NEUROTOXICITY CONSIDERATIONS FOR FOOD ADDITIVE SAFETY

YEN-CHING WU, SHRUTI V. KABADI, and APRIL NEAL-KLUEVER*

FDA/CFSAN/OFAS/DFCN, 5100 Paint Branch Pkwy, HFS 275, College Park, MD 20740, United States

*Corresponding author. E-mail: April.kluever@fda.hhs.gov

CONTENTS

ABSTRACT

Developmental exposures to chemicals have the potential to induce adverse neurobehavioral effects. Developmental neurotoxicity testing (DNT) has added value to a risk or safety assessment by assessing the potential risk of chemical exposure to induce such effects in humans. In particular, DNT testing can provide relevant safety information for food additives and other chemicals that are intended for use by individuals during early life stages. In this chapter, we summarized the events that led to the establishment of the various current regulatory DNT guidelines, including international collaborative efforts that validated the reliability and sensitivity of measurements used to test for DNT potential. We also reviewed the current approach to DNT testing used by the US EPA and the US FDA CFSAN. We highlighted the underlying reasons why the nervous system is especially sensitive to chemical toxicity during early life stages. We then followed this discussion with an overview of the behavioral assays and histological assessments commonly used in DNT testing, including some of the current challenges and future directions.

5.1   INTRODUCTION

Developmental exposure to various types of chemicals, including industrial and environmental chemicals, pesticides, or drugs has the potential to induce adverse neurological effects in humans or animals.[97,143] However, there are significant challenges to the study of developmental neurotoxicity testing (DNT). For example, some adverse neurological outcomes, such as an increased susceptibility to neurodegeneration, may be latent, and the clinical symptoms may take years to decades to manifest after initial chemical exposure.[58,144] In addition, subtle effects, such as a minor deficit in intelligence quotient (IQ), may be undetectable on an individual basis and can only be appreciated at the population level.[16] To illustrate this concept, relatively low exposures to developmental neurotoxicants such as lead, organophosphates, or methyl mercury may shift the population IQ distribution toward lower IQ scores while only impacting individual IQ scores by less than ten points.[16,88] Although the individual burden from such a shift may be considered clinically negligible, the population burden of a shift toward lower IQ scores can have profound societal costs.[88,142] Due to the difficulty in detecting adverse neurological events, a limited number of chemicals have been universally accepted to present DNT hazards to both laboratory animals and humans.[97,98]

These confirmed DNT toxicants include lead, organic mercury, polychlori-nated biphenyls, toluene, valproic acid and other anticonvulsants, ethanol, and retinoic acid.[97,106] In contrast, many more chemicals are known to elicit DNT effects in laboratory animals but are only suspected (but not proven) human neurotoxicants,[97] such as the herbicide paraquat,[90] nicotine,[205] and aspirin.[106]

Regulatory agencies face the challenge of determining whether a chem-ical is safe for a particular use. The data required to make such a safety deter-mination includes exposure route, amount of exposure, duration of expo-sure, and end points of toxicological concern. This chapter focuses on the end point of DNT. We discuss the current state of DNT by first describing the evolution of regulatory DNT test guidelines and the currently available test methods regarding their utility and limits to analyze neurotoxic potential. We provide the context in which regulatory agencies must make a determi-nation regarding the safe use of a chemical and provide a basic overview of neural development from both anatomical and behavioral perspectives. This overview provides the background for our discussion of the regulatory toxicology assays that are commonly used to assess DNT effects in animals. Finally, we conclude by briefly highlighting some of the challenges of DNT testing and describe emerging approaches that are aimed to resolve some of these problems.

5.2 REGULATORY DNT GUIDELINES

The first DNT guideline was published by the US EPA in 1991 as a direct outcome of the landmark 1983 National Research Council report, "Risk Assessment in the Federal Government."[129,170] The US EPA's Office of Prevention, Pesticides, and Toxic Substances (OPPTS) reissued the 1991 DNT guideline in 1998 and designated it as OPPTS 870.6300[224] with minor revisions, as part of an effort by the United States to harmonize with its own program offices as well as internationally with the Organization for Economic Cooperation and Development (OECD).[148]

The measures used to test for DNT potential were established by several large-scale interlaboratory collaborative studies that validated and tested their reliability and sensitivity. Those collaborations include the US Collabora-tive Behavioral Teratology Study (1978–1984), the Cincinnati Test Protocol (1984), the Collaborative Studies of the Japanese Teratology Society (1982–1985), the European Interlaboratory Collaborative Study (1985–1988), the Williamsburg Workshop (1989), and several others.[7,148]

The US EPA's DNT guideline served as the foundation for the OECD Test Guideline (TG) 426, a DNT guidance document initiated by the OECD Working Group on Reproductive and Developmental Toxicity to accommodate the regulatory needs of the OECD countries.[172] Existing differences between the EPA and OECD guidelines are reviewed by Aoyama and colleagues.[7] The OECD TG 426 is recognized by the scientific community as a well-designed test to characterize potential DNT hazards of particular chemicals to human health.[148,230] Since the OECD TG 426 protocol was published 10 years after the latest revision of the EPA OPPTS 870.6300 protocol, its content reflects more current recommendations based on more recent scientific discourse (e.g., longer dosing period, consideration of directing dosing to pups).[148,230]

The US EPA OPPTS 870.6300 and OECD TG 426 protocols are considered "stand-alone" DNT protocols because their primary objective is to detect potential DNT. This is in contrast to another existing option where assessments of DNT are incorporated into other study types (e.g., reproductive assays) by the addition of a designated DNT cohort as a flexible study cassette. For example, in the Extended One-generation Reproductive Toxicity Study (EOGRTS) OECD TG 443, the first filial generation (F_1) pups are assigned to three different cohorts that evaluate distinct sets of parameters.[173] Cohort 1 animals evaluate reproductive/developmental end points, while Cohort 2 and 3 animals are dedicated to developmental neurotoxicity and immunotoxicity assessments, respectively. The National Toxicology Program (NTP) designed a Modified One Generation (MOG) reproductive study that also uses a dedicated DNT testing cohort.[171,*] In the MOG study, weaned offspring are assigned to up to four cohorts (i.e., subchronic toxicity, teratology, breeding and littering, and optional DNT/developmental immunotoxicity cohorts) that each evaluate distinct parameters, which are considered interchangeable cassettes. The DNT cohort assesses both physical and developmental landmarks and captures neurobehavioral end points adapted from the OECD TG 426. A subset of animals from the subchronic cohort is also set aside for neurohistopathology assessment. The MOG also examines target organ toxicity and is used establish the appropriate dose levels for a subsequent cancer bioassay.[80]

The various existing regulatory guidelines that assess DNT differ in their testing approaches (e.g., treatment duration, testing age, type of end points measured). A summary table (Table 5.1) is provided to highlight some of

---

*The MOG guidance document can be accessed at https://ntp.niehs.nih.gov/ntp/test_info/mog_guidance_508.pdf.

the recommendations made by these various DNT testing guidelines. For example, the EOGRTS captures both neurospecific (e.g., motor activity, acoustic startle response) and general developmental landmark (e.g., vaginal patency, pup weight) end points in the offspring but assesses each using two separate animal cohorts. In contrast, other protocols use one animal cohort to capture both sets of end points. Developmental landmarks are collected in conjunction with neurospecific end points to account for potential chemical effects on the neuroendocrine axis (i.e., hypothalamus–pituitary axis and hypothalamus–thyroid axes, see Section 4).

DNT guidelines continue to evolve as analyses on the utility of current DNT protocols signal a need for improvements. Such recommendations for a change in the current DNT guidelines will be generally discussed in Section 5.5.4 of the book chapter. For additional details on current specific DNT guidelines, please consult the individual referenced guidelines as needed.

5.3 REVIEW OF REGULATORY APPROACHES TO DNT TESTING

5.3.1 THE US EPA

The US EPA recommends specific methods, or test guidelines, to generate data in support of the registration of a pesticide under the Federal Insecticide Fungicide Rodenticide Act (FIFRA; 7 United States Code Section 136, hereafter designated as USC), the setting of a tolerance or tolerance exemption for pesticide residues under section 408 of the Federal Food, Drug, and Cosmetic Act (FFDCA), or the decision-making process supporting potential regulation of an industrial chemical under the Toxic Substances Control Act (TSCA; 15 USC 2601).[†] In particular, DNT testing is a required component of pesticide registration under FIFRA (40 CFR 158.500). The FIFRA requires that the DNT study use a weight of evidence approach considering mechanistic information regarding potential neurotoxicological effects in humans or animals.[‡] The US EPA accepts either the submission of an OPTTS 870.6300 or OECD 426 guideline study to satisfactorily assess the potential DNT hazard for pesticides.[149] Consequently, the Office of Pesticide Programs (OPP) at the US EPA currently houses the largest collection of completed DNT studies, with over 100 studies as of early 2006.[148]

[†]https://www.epa.gov/test-guidelines-pesticides-and-toxic-substances/about-test-guidelines-pesticides-and-toxic.

[‡]http://www.ecfr.gov/cgi-bin/text-idx?SID=8fd5cf01b197f999dd333ea416dea9b5&mc=true&node=se4 0.24.158_1500&rgn=div8.

5.3.2 THE US FDA CENTER FOR FOOD SAFETY AND APPLIED NUTRITION

A key mission of the US Food and Drug Administration (US FDA) Center for Food Safety and Applied Nutrition (CFSAN) is to protect the public health by assuring the safety of the food supply. It is the primary responsibility of the Office of Food Additive Safety (our office is commonly referred to as OFAS) in the FDA CFSAN to evaluate the safety of substances added to food (i.e., food additives). The FDA's authority to regulate substances added to food derives from the 1958 Food Additives Amendment to the FFDCA. Food additives are broadly defined as "any substance the intended use of which results or may reasonably be expected to result, directly or indirectly, in its becoming a component or otherwise affecting the characteristics of any food." (FFDCA, Section 201).The standard of safety for all added substances to food is that there is a reasonable certainty in the minds of competent scientists that the substance is not harmful under the intended conditions of use (codified in 21 Code of Federal Regulations 170.3(i), hereafter, designated as the 21 CFR). A determination that a reasonable certainty of no harm will result from the proposed use(s) of food ingredients is required before manufacturers and distributors can market them in the United States (see sections 70.3 and 170.3 of the 21 CFR). The determination of reasonable certainty of no harm is contingent upon the results of toxicological studies, exposure information, and other types of information. Safety factors are used in extrapolating from animal models to humans. FDA provides recommendations for the conduct of specific toxicological tests that may be used to demonstrate the safety of the substance added to food under the intended conditions of use. Those tests are recommended based on the estimated cumulative exposure to the food additive as well as its known potential human health hazards or alerts based on chemical structure. FDA's guidance on the general approach to the safety assessment of food additives can be accessed through the online publication, "Toxicological Principles for the Safety Assessment of Direct Food Additives and Color Additives Used in Food," which is referred to as "Redbook."[§] Due to the significant challenges in identifying human neurotoxicants, regulatory agencies such as the US FDA CFSAN adopt an integrative approach to screen for potential human DNT hazards, using studies from laboratory animals, structural and physicochemical information of substances, and mechanistic information obtained from in vitro studies.

[§]http://www.fda.gov/Food/GuidanceRegulation/GuidanceDocumentsRegulatoryInformation/IngredientsAdditivesGRASPackaging/ucm2006826.htm.

In contrast to the US EPA, the US FDA CFSAN adopts a tiered-testing approach to assess the neurotoxic potential of chemicals, such that stand-alone DNT studies are not conducted unless warranted as described in Redbook 2000 (Chapter IV.C.10: Neurotoxicity Studies). The DNT testing may be recommended if relevant information (e.g., systemic toxicity tests, the public literature, and/or in silico methods such as structure–activity relationship models (SAR)) indicate a concern for DNT.[207] Although the US FDA CFSAN does not have its own formalized DNT study protocol, it may recommend that sponsors follow one of the guidelines or protocols from reputable sources, such as those identified in Table 5.1, stipulating that the approach used must satisfactorily test for the DNT hazard. DNT testing may be conducted as a stand-alone study or as an added specialized cohort of study animals in another developmental toxicity study (e.g., a two-generational developmental and reproductive study).

5.3.3 SUMMARY

Two approaches to regulatory DNT testing have been described, with the US EPA employing an approach that requires the submission of DNT test data for pesticide registration, and the US FDA CFSAN not requiring the submission of DNT test data by default but recommending such testing on a case-by-case basis. There are specific scientific differences that drive these two approaches. For example, pesticides contain active ingredients that prevent, destroy, repel, or mitigate a pest. Therefore, these active ingredients in pesticides are designed to exert adverse biological activity and some are specifically known to exert neurotoxic activity in mammalian species. The organophosphates and pyrethroids insecticides are two large classes of pesticides that demonstrate DNT potential.[198,225] Therefore, these chemicals present alerting information suggestive of a potential DNT hazard. In contrast, food additives regulated by the US FDA CFSAN are generally intended to have a beneficial, neutral, or lack of effect on the food and consumers. The US FDA CFSAN provides additional guidance and considerations for addressing the relatively small group of food additives (e.g., antimicrobials) designed to exert adverse biological activity. Food additives generally exhibit substantially different physicochemical properties from pesticides. Therefore, in evaluating food additives, a weight-of-evidence approach is warranted for recommending specialized toxicity testing such as the DNT assay.

TABLE 5.1 Comparison of Recommendations by Various Regulatory DNT Guidelines.

Protocols	EPA OPPTS 870.8600 DNT screen	EPA OPPTS 870.6300 DNT study	OECD TG 426 DNT study	NTP: DNT arm of MOG	OECD TG 443 EO-GRTS (Cohort 2)
Date adopted	June 1996	June 1996	October 16, 2007	September 24, 2015 (draft)	July 28, 2011
Animal subject	Sprague–Dawley rats	Rats but not the Fischer 344 strain. Justification needed for use of other mammalian species	Rats. Justification needed for use of other mammalian species	Sprague–Dawley rats	Rat is preferred. Justification needed for use of other mammalian species
Doses	3 plus a control	3 plus a control	At least 3 and a control	3 plus a control	3 plus a control
Treatment period	GD 6—weaning (i.e., PND 21)	GD 6—PND 10	At a minimum, GD 6—PND 21	GD 6—weaning (i.e., PND 28) or study termination	2 weeks premating through weaning (i.e., PND 28)
Administration route	Oral intubation	Oral administration or case-by-case	Oral but also considers other route most relevant to human exposure and available metabolism and distribution information test animals	Oral administration	Route most relevant to human exposure although protocol is desig ned for test substance through the diet
Direct dosing to weanlings	No	No	No, but direct dosing to pups should be considered based on exposure and pharmacokinetic information	Dietary sources from PND 28 to PND 72	Dietary sources from PND 28 to study termination
No. of animals for behavioral tests	Not specified	20/Sex/Dose	20/Sex/Dose (except cognitive function tests, which can range from 10/sex/dose or higher, depending on the test's sensitivity)	20/Sex/Dose	10/Sex/Dose

TABLE 5.1 *(Continued)*

Protocols	EPA OPPTS 870.8600 DNT screen	EPA OPPTS 870.6300 DNT study	OECD TG 426 DNT study	NTP: DNT arm of MOG	OECD TG 443 EO-GRTS (Cohort 2)
Detailed clinical observations	Maternal: daily prior to treatment, then daily GD 6—PND 21. Offspring: during weighing/behavioral testing	Maternal: daily prior to treatment, then daily GD 6—PND 10. Offspring: during weighing/behavioral testing	Maternal: at least twice during gestation and lactation (at least 10 dams/dose). Offspring: weekly and then every 2 weeks postweaning (at least 1 pup/sex/litter	Maternal: at least twice during gestation and lactation (at least 20 dams/dose). Offspring: daily (at least 1 pup/sex/litter)	Maternal and offspring: weekly (postweaning)
Motor activity	PNDs 13, 17, 21, 45(±2 days), 60(±2 days)	PNDs 13, 17, 21, and 60(±2 days)	1–3 times prior to weaning (e.g., PND 13, 17), day of weaning (PND 21), and at adult age (e.g., PND 60–70)	~PNDs 31–33 and ~PNDs 60–70	PNDs 63–75
Motor function test (e.g., strength, coordination)	Not discussed	Not discussed	Once in adolescence (e.g., PND 25 ± 2) and once in adulthood (e.g., PND 60–70)	~PNDs 31–33 and ~PNDs 60–70	PNDs 63–75 and also includes other FOBs[a]
Sensory function test	Acoustic startle test on PNDs 22 and 60	Acoustic startle habituation test on PNDs 22 and 60 ± 2 days	Open choice for specific test on PND 25 ± 2 and PNDs 60–70	Acoustic startle test ~PNDs 31–33 and ~PNDs 60–70	Acoustic startle on PND 24 ± 1, additional tests as part of FOBs[a] or as needed
Cognitive function test	Active avoidance test (an associative learning test) beginning on PNDs 60–61	Associative learning and memory task (open choice for specific test) on PNDs 21–24 and at adulthood (PND 60 ± 2 days)	Associative learning and memory (open choice for specific test) during adolescence (e.g., PND 25 ± 2) and during adulthood (PND 60 and older)	Associated learning and memory: Morris Water Maze preferred, otherwise passive or active/conditioned avoidance test, at ~PNDs 60–70	As needed

TABLE 5.1 *(Continued)*

Protocols	EPA OPPTS 870.8600 DNT screen	EPA OPPTS 870.6300 DNT study	OECD TG 426 DNT study	NTP: DNT arm of MOG	OECD TG 443 EO-GRTS (Cohort 2)
No. of animals for pathology	At least 6/dose	6/Sex/dose	10/Sex/Dose	10/Sex/Dose	10/Sex/Dose
Developmental landmarks	Vaginal opening, testes descent	Vaginal opening, preputial separation	Vaginal opening, preputial separation, eye opening, pinna unfolding, incisor eruption	Vaginal opening, preputial separation, testis descent, male areola nipple retention, anogenital distance	End points only collected in a separate cohort (i.e., Cohort 1[b])
Neuropathology	At weaning and following last behavioral measure in adulthood (perfusion fixation) including brain and spinal cord	PND 11 (immersion fixation) brain tissue evaluated and at study termination (perfusion fixation) and tissue and peripheral nervous tissue evaluated	PNDs 11–22 (immersion or perfusion fixation) brain tissue evaluated, PND ~70 (perfusion fixation) brain tissue and peripheral nervous tissue evaluated	PNDs 31–33 (immersion fixed) brain tissue evaluated, and study termination (~PNDs 90–100, perfusion fixed) both central and peripheral nervous system tissues will be evaluated	Neurohistopathology at PNDs 21–22 (optional perfusion fixation), PNDs 76–90 (uses high dose and controls only, perfusion fixed), brain histopathology, including spinal cord and optic nerve
Brain weight measure	At weaning and at study termination	Prefixation on PND 11 and at study termination	Postfixation PNDs 11–22 and prefixation ~PND 70, optional: prefixation PNDs 11–22	PNDs 31–33 and during study termination (~PNDs 90–100)	PNDs 21 or 22 and 76–90
Brain morphometric analysis	No	Yes	Yes	Not discussed	Yes

[a]FOBs: functional observational batteries (Appendix A, OECD TG 443): Home cage and open field (e.g., bizarre behavior, piloerection), manipulative (e.g., ease of removal, muscle tone, touch response, tail pinch response), and physiologic (e.g., temperature, body weight, pupil size).

[b]Cohort 1 of the OECD TG 443—EOGRTS captures the following developmental landmarks: anogenital distance, presence of male areola nipple retention, vaginal opening, preputial separation, and the presence of genital organ abnormalities.

The tiered-testing approach to DNT testing is not unique to the US FDA CFSAN and has also been adopted by the European Food Safety Authority (EFSA) for the toxicological testing of food additives for market in Europe. The EFSA's tiered approach to toxicity testing includes three tier levels. Tier 1 testing includes a modified 90-day toxicity test that allows for identification of chemicals with the potential to cause neurotoxic, immunologic, or reproductive organ effects. Those effects, if observed, warrant additional evaluation at higher tiers under EFSA's approach. In that paradigm, a DNT cohort to an EOGRTS would be conducted as a Tier 2 test if results from a Tier 1 study indicated a DNT concern. Further, a stand-alone OECD TG 426 assay may be conducted as a Tier 3 study if data from the lower tiers are suggestive of a DNT concern.[69]

5.4 LITERATURE REVIEW OF NEUROLOGICAL AND BEHAVIORAL DEVELOPMENT

The assays contained in the various DNT test guidelines typically range from behavioral assessments to histological neuropathology. Modification of animal behavior or the presence of neurological lesions after chemical exposure may indicate a disruption to a neural developmental event or pathway. This disruption in animals may further suggest that an analogous or parallel event could also be adversely affected in exposed humans.[213] Therefore, results from laboratory animal testing help risk or safety assessors determine whether a chemical might produce adverse effects in humans and the dose at which effects might occur.[81,148] To give our readers a better understanding of why the developing human brain may be particularly susceptible to neurotoxicity, we next provide a brief overview of the major human neural and behavioral developmental milestones that span from birth through adolescence.

5.4.1 OVERVIEW OF STRUCTURAL NEURODEVELOPMENT

Human neural development is a complex process that initiates from a single layer of embryonic cells a few weeks after fertilization and then takes decades to complete. During gestation, the nervous system transitions from a smooth, folded tube into a complex structure with characteristic sulci and gyri. This transition progresses through a dramatic sequence of important developmental events. The developing brain may be particularly susceptible to neurotoxicity because disruptions to critical events may produce adverse neurological effects that have little potential for remediation. Those critical events have been well defined by others, and include neurulation, neurogenesis, and gliogenesis, as described below.[4,26,146] Since each of these events represents a substantial body of literature, we provide key references for additional reading. Where appropriate, examples of adverse effects associated with defects in these processes are included. Adverse neurodevelopmental conditions that are associated with chemical exposure include mental retardation, autism spectrum disorder (ASD), and neural tube defects.[245]

5.4.1.1 NEURULATION

Neurulation encompasses the initial formation of neural tissue, which in humans begins around 17–18 days after fertilization and completes about 8–10 days later.[101] During neurulation, epithelial cells in the embryonic ectoderm undergo extensive, coordinated morphometric changes involving elevation and bending, apposition and fusion, and remodeling. Those steps ultimately result in the transition of the neuroepithelium from a flat neural plate to a tube with two continuous epithelial layers.[101,240] This complex process is accomplished through an intricate system of cues provided by cell-extrinsic factors such as morphogens (e.g., bone morphogenetic protein) and growth factors (e.g., sonic hedgehog), as well as cell-intrinsic factors such as genetic programing.[240] Disruption of neurulation can give rise to neural tube defects, such as spina bifida or anencephaly.[102,235]

5.4.1.1.1 Neurogenesis and Gliogenesis

After neurulation, extrinsic signaling molecules and cell-intrinsic factors signal the first waves of neurogenesis, when new neurons or neuronal precursors from progenitor cells are created. This event initiates around embryonic day 25 in humans and neurogenesis begins as a symmetrical division of neural epithelial cells in the ventricular zone.[135] These symmetrical divisions increase the cellular number of neural epithelial cells while simultaneously increasing brain volume. The neural epithelial cells exhibit a bipolar orientation with one process in contact with the pial surface of the brain and one process in contact with the ventricular region of the brain. As brain volume expands, the apical process of the neural epithelial cell extends to maintain contact with the pial membrane. That long, apical process subsequently serves as the guidance structure for migrating daughter cells formed from neural epithelial cell division.[136]

The brain is composed of neurons as well as other cell types, collectively described as glia. Glial cells, such as astrocytes, oligodendrocytes, and microglia, are responsible for diverse functions in the human nervous system, such as providing trophic support for neurons, specialized facilitative roles in neurotransmission, and providing immune functions for the brain. Macroglia (i.e., astrocytes and oligodendrocytes) derive from the same pool of progenitor neural epithelial cells; however, different extrinsic and cell-intrinsic factors are involved to stimulate genesis of discrete pools of those glial cells.[73,186] In contrast, microglia derive from other pools of progenitor cells, such as the yolk sac and fetal liver, during early embryonic development. Later in development a second wave of microglia progenitors originates from the bone marrow.[36,50,89] The invasion of microglia progenitors is complete prior to birth in humans and in the early postnatal period (postnatal days (PNDs) 0–15) in rodents. Upon colonization of neural tissue, the microglia progenitors differentiate, possibly through cues given by astrocytes, into the various populations of mature microglia.[50]

Toxicants that suppress neurogenesis include sedative and antiepileptic drugs,[214] ethanol,[155] and methyl mercury.[76] Although effects on gliogenesis have been less well studied in DNT studies, there is some evidence that exposure to the organophosphate chlorpyrifos may impair gliogenesis.[78,87]

5.4.1.1.2 *Migration*

Following neurogenesis, cells migrate to carefully selected positions to facilitate circuit formation. Neuronal migration follows two main migration models, radial and tangential migration.[136] In both humans and rodents, neuronal migration is generally completed prior to birth, with the exception of a few regions of the brain (e.g., the subventricular zone, olfactory bulb, cerebellum, etc.).[35,135] Glial cell migration continues well into the postnatal period. In contrast to neurons, glial cell progenitors continue to proliferate during migration.[35] Defective migration has been proposed as a key mechanism underlying adverse events due to developmental exposure to methyl mercury.[10,38]

5.4.1.1.3 Differentiation and Synaptogenesis

Neural cells undergo the process of differentiation, becoming specialized in performing a specific function, as they reach their varying final destinations in the brain. Local cues in the extracellular environment drive neuronal and glial differentiation and also direct the branching and outgrowth of neuronal axon and dendrites.[37] During synaptogenesis, the axons and dendrites mature and develop the capacity to transmit and receive neurotransmitter signal, respectively, once the axon of one neural cell contacts the dendrite(s) of another cell.[193] Synaptogenesis begins before gestation week 27 in humans and peaks during the postnatal time period.[26,114] Impairment of synaptogenesis by exposures to chemicals, such as lead, during development has been implicated in adverse effects on learning and memory.[10,167,229]

5.4.1.1.5 Neuroapoptosis

Apoptosis plays several essential roles that ensure proper functioning of the brain, including refinement of progenitor pools, error correction, and optimization of connectivity.[29] Developmental apoptosis in the nervous system is very high and is estimated to eliminate as many as 50% of neurons after synaptogenesis.[56] Deficient apoptosis during development has been implicated in several adverse outcomes, including neural tube closure defects[241] and outgrowth malformations.[137] In contrast, excessive apoptosis during brain development may adversely impact learning and memory later in life.[48] Anesthetics, such as isoflurane, that interact with the N-methyl-D-aspartate receptor or γ-amino butyric acid receptor appear to increase the rate of neuroapoptosis, particularly during developmental exposures,[117,119,190] which can lead to short-term behavioral changes in motor coordination related to cerebellar function.[21]

5.4.1.1.6 Myelination

Neurons have the capability to transmit or conduct electrical signals (i.e., action potentials) along axons across great distances. Neurons that require fast conduction of action potentials require myelination of their axons by oligodendrocytes. In humans, myelination begins around mid-gestation but continues for several years after birth.[118] Macrostructually, myelination can be visualized by the formation of white matter tracts, which are established by mid-gestation in the human fetal brain.[113] However, proper formation of the white matter tracts requires additional optimization that does not complete until several years after birth.[156] Brain myelination progresses rapidly after birth[63] and approaches adult levels at different ages, with some major white matter tracts (e.g., the corpus callosum) reaching adult appearance much earlier than others (e.g., the internal capsule).[63,110] Exposure to several chemicals including organotin[104,212] and ethanol[221] impairs myelination and are associated with the onset of long-term neurologic dysfunction.

5.4.1.1.7 Competitive Elimination

During the process of competitive elimination, also known as synaptic pruning, the number of synapses is reduced to refine neuronal networks. This essential event is required to optimize connectivity and ensure efficient neurotransmission. This process initiates in humans around the time of birth and completes by sexual maturation.[26] There is some evidence suggesting that impaired competitive elimination may play a role in the etiology of ASD.[65]

5.4.1.1.8 Development of Barrier Systems in the Brain

The brain and spinal cord are considered privileged regions due to the presence of several barrier mechanisms that separate systemic serum and blood from cerebral spinal fluid and the tissues of the nervous system. Together, these barrier mechanisms regulate the internal environment of the central nervous system (CNS) and have been termed the blood–brain barrier (BBB), blood–cerebral spinal fluid barrier (BCB), and blood–arachnoid barrier. These barriers initiate formation when the first blood vessels invade the neural tube during the early stages of embryonic neurodevelopment.[74] Effective barrier function between systemic circulation and the nervous system includes (1) the presence of tight junctions between endothelial (BBB) or epithelial (BCB) cells, and (2) expression of transporters that limit influx or efflux of large and small molecules.[189]

Tight junctions in brain barriers form early in embryonic development and are functional prior to birth in humans and laboratory animals.[12,158,159] The influx of small and/or hydrophilic molecules appears to be largely mediated by solute-linked carrier transporters while efflux is largely mediated by adenosine triphosphate-binding cassette transporters in the BBB and BCB, which together provide essential nutrients, ions, and other molecules required for brain function.[189] Differences in transport properties in the developing brain from the adult may result in a different toxicokinetic (TK) and/or toxicodynamic profile for a neurotoxicant. However, it is important to emphasize that the developing BBB is fully functional and is considered as a "dynamic barrier, different from the adult, and adapted to the specific requirements and environment of the early developing nervous system."[72,120,121] Neurotoxicants can promote neurotoxicity by disrupting the BBB (e.g., nitrobenzene, Cyclosporin A) or by altering its regulatory function (e.g., aluminum, lead).[247]

5.4.2   OVERVIEW OF BEHAVIORAL DEVELOPMENT

In the same way that structural development of the nervous system can take several years to complete, behavioral development is also a protracted process. Behavior is an integrated readout of the concerted activity of various brain circuits that is mediated by the function and coordinated efforts of cells located within different brain regions that respond to cues in the external and internal environments. For that reason, the neurodevelopment of individuals can be monitored by their external behavior. In this way, any potential disruptive neurotoxic effect that occurs during neurodevelopment may be detected by any behavioral changes that deviate from the norm. For example, hippocampal lesions that are present during postnatal week 2 in monkeys can result in severe memory deficits when assessed later in adulthood.[8] Alternatively, behavioral deficits can also result from nonphysical insults caused by exposure to toxic chemicals. For example, a single oral exposure to the organophosphate insecticide chlorpyrifos in neonate mice on PND 10 led to transient decreases in neural proteins within 24 h of treatment, an effect which did not extend into adulthood, yet still led to altered spontaneous behavior in treated animals as adults.[145]

The brain develops in a sequential fashion and is organized hierarchically from the least (e.g., brain stem) to most complex (e.g., limbic and cortical areas).[177] Cellular events that underlie brain maturation are previously described and include synaptogenesis, myelination, neuronal differentiation, as well as neurotransmitter modulation and contributions made by glia.[55] Because different brain regions and neural circuits develop and mature at different rates, distinctive neural systems have different periods of vulnerability, which span from the gestational period through adolescence.[2]

Overall, parts of the brain that are associated with more basic behaviors (e.g., motor and sensory systems) mature earlier, followed by areas that process spatial orientation, speech, and language. Regions that regulate the most complex executive functions, such as attention and motor coordination, mature last.[92] For example, synaptogenesis occurs at a maximal rate in brain areas that process visual and auditory information a few months following birth, reaching maximal density at 3–4 months of age[114,134] but proceeds more slowly in other areas such as the prefrontal cortex (PFC) that controls memory and decision-making;[134] the functional development of the PFC typically spans from 6 months of gestation to at least 15 months after birth.[114] Similarly, the order that the brain becomes myelinated follows the same regional pattern as synaptogenesis. This section describes some of the major behavioral milestones that loosely correspond with these asynchronous neural developmental processes.

5.4.2.1 PRIMITIVE REFLEXES

The brainstem is one of the first areas of the CNS to become functionally mature and facilitates an infant's breathing following birth. It also mediates the involuntary stereotypic movements of full-term newborn babies that are collectively termed "primitive reflexes," as described below.

5.4.2.1.1 Neonatal Grasp Reflex

This includes the Palmar and plantar grasp reflex, which both appear at 25 weeks in utero and disappear by 6–12 months.[84,244] The Palmar grasp reflex occurs when an object placed in a baby's causes the baby to grasp the object. The plantar grasp reflex is elicited by the stroking of the sole of the foot, which triggers the infant to extend their big toe. This response is distinct from adults, who instead draw the toe downward.[83,85]

5.4.2.1.2 Moro Reflex

This reflex is characterized by the neonate throwing out his/her arms and legs and then pulling them toward the body, as a startle response to sudden movement or loud noise.[157] This reflex emerges at 8–9 weeks in utero and disappears by 6 months after birth.[85]

5.4.2.1.3   Tonic Reflex

The asymmetric tonic neck reflex or "fencing posture" is observed by the extension of an infant's arm and leg toward the side that his/her head turns, accompanied by the bending of limbs on the opposing side. This reflex develops 18 weeks in utero and disappears when the infant is roughly 4 months of age.[17]

5.4.2.1.4   Babkin Reflex

Otherwise known as the hand–mouth reflex, this reflex is characterized by the flexion and rotation of the head, opening of the mouth, or a combination of these responses when pressure is applied to both palms of the infant. It can develop as early as 26 weeks in utero and disappears 5 months after.[86,174]

5.4.2.1.5   Galant Reflex

Also known as the truncal incurvation reflex, this reflex is characterized by the lateral flex of an infant in a prone position toward the side where his/her back is stroked.[243] This reflex is present at birth and disappears after 4 months.

The presence of primitive reflexes early in life is indicative of proper neural development.[133] As infancy progresses, primitive reflexes are replaced by voluntary motor movements (e.g., rolling over, sitting, standing, and walking) due to ongoing maturation of the cerebral cortex and the emergence of cortical inhibition.[108] Therefore, the dysfunctional or delayed onset of primitive reflexes or the persistence of these reflexes past the first year after birth is indicative of abnormal brain development. Dysfunction in the onset of primitive reflexes occurred in infants that experienced congenital poisoning from long-term exposure to methylmercury (MeHg) in Minamata City, Japan, in 1956, which provided the first evidence that MeHg is a developmental neurotoxicant.[38,105]

5.4.2.2   SENSORY SYSTEM DEVELOPMENT

Infancy is also characterized by the functional onset of various sensory systems as the primary cortical areas of the brain mature. Proper development of those systems are both time- and activity-dependent and rely on external cues, such that the functional capability of a sensory system can be lost if that sensory experience is disrupted during development.[177] For example, prenatal exposure to nicotine has adverse effects on the developing sensory cortex; nicotine impairs visuospatial function and auditory processing.[20,107]

1. **Somatosensory (touch)**: The infant somatosensory system is fully developed during gestation, which is demonstrated by the ability of neonates to direct their attention to the body part being touched.[236]

2. **Gustatory**: The gustatory system is functionally intact at birth such that an infant can discriminate the varying degrees of sweetness but does not have the aversion to bitter stimuli until a few days after.[236] The gustatory system continually changes over the course of fetal development through adulthood, such that the development of salt discrimination does not occur until four years of age.

3. **Olfactory**: The infant olfactory system is fully functional at birth and permits the infant to respond to both maternal odors as well as airborne stimulants. For example, 2-week-old breast-fed infants preferentially orient toward the axillary odor of their mother over that of another mother,[66] which implies that olfactory discrimination is already developed at this age.

4. **Visual**: The visual system of a full-term newborn is immature at birth and cannot detect color, although it is capable of sensing patterns, movement, and different light intensities.[100] An infant's initial visual acuity is approximately 20/600 but reaches 20/20 by the time he/she reaches 4–6 months of age.[23] In addition, infants begin to discriminate the color red at 2–3 months of age,[100] develop flicker sensitivity at 3 months, and establish binocular function at 3–7 months, whereas contrast sensitivity matures several years after birth.[23] The critical period of visual system development is between 20 weeks' gestational age and 2–3 years postnatally, with critical components of the visual system completing during the first year.[100] Importantly, routine exposure to external stimulation such as indirect light on objects, novelty or change, and movement is needed for continued postnatal visual development.

5. **Auditory**: The development of the auditory system begins with the formation of the structural parts of the ear (e.g., cochlea of the middle ear) during the first 20 weeks of gestation and the neurosensory part of the auditory system from 25 weeks' gestation to 5–6 months of

age.[99] The auditory system becomes functional at around 25–29 weeks of gestation when the cochlea and the auditory cortex in the temporal lobe of the brain connect. This connection requires external stimulation (e.g., spoken speech, music, and meaningful sounds from the environment) during 28–40 weeks of gestation in utero and also several years after birth to properly fine tune. However, if background noises in the environment are greater than 80 decibels (dB) in utero or 60 dB following birth, language development may be delayed, because the infant may not develop necessary circuits for recognition of pitch, speech, phonemes, and other characteristics of the mother's voice. Human hearing is not fully developed until the first 5–10 years of age.[187]

5.4.2.3 LANGUAGE ACQUISITION

Having a functional auditory system is essential to language development, since children who experience deafness prior to the age of three demonstrate a significant delay in the mastery of language in all its forms (i.e., written, read, spoken, and signed).[217] Shortly after birth, infants are able to discern different sounds that are present in languages of the world.[39] The first year after birth marks a prelinguistic period where infants develop the use of communicative gestures (e.g., pointing) and babble. Language comprehension also occurs at this time as infants begin to recognize their own name and associate objects with their names. The prelinguistic period ends when the child utters their first word, which typically occurs at the first year after birth. Infants in their first year become progressively more sensitive to speech sound contrasts related to their native language and insensitive to unimportant phonetic contrasts.[138] Language development continues throughout childhood and is greatly influenced by cognition and attention.[41,216] For example, better visual recognition memory is associated with better language skills (e.g., comprehension, language expression) in individuals ranging from the preschool years through adulthood.[185] The inability or delay in developing language apprehension is observed in certain disorders such as autism, attention deficit hyperactivity disorder (ADHD), auditory processing disorder, and developmental verbal dyspraxia. Prenatal exposure to neurotoxicants, such as ethanol[1] and cocaine,[162] are associated with developmental delays in language acquisition and comprehension.

5.4.2.4 DEVELOPMENT OF MEMORY AND COGNITIVE CONTROL

The frontal brain regions that support higher order executive functions such as cognition, attention, cognitive control, and memory are the last to reach maximal synaptic density and maturity. Unlike the visual and auditory cortex, which mature during the first three months and the first 3 years after birth, respectively, the frontal areas and cortico-cortical connections that encompass cognitive systems are only partially present in infants[11] and continue to mature through puberty.[55] Due to the protracted maturation of this region of the brain, it may exhibit increased vulnerability to neurotoxicant exposure.

Working memory is one of several memory systems that underlie memory that refers to the temporary storage and manipulation of information, such as visual images and verbal cues. Memory is the process of encoding, storing, and retrieving information, consists of three stages (i.e., sensory, short term, and long term) depending on how long the information is stored and comprises multiple memory systems that develop over different time courses and require different neurological substrates.[168] Working memory allows for the recognition of change in the environment, in terms of new location or person, and initially develops in infants around the age of 7–10 months.[61] As a result, separation anxiety sets in around that time, as working memory consolidation plays an important role in the formation of separation anxiety function.[108] Impairments in working memory function are observed in animal models of ADHD.[59] Furthermore, early exposure to neurotoxicants, such as lead[15,31] and cannabinoid compounds,[238] cause long-term impairments in learning and working memory function.

Compared with adults, children show decreased engagement of brain areas that are needed for manipulating items in working memory (i.e., dorsolateral prefrontal cortex (DLPFC) and the parietal cortex) and fail to activate other areas (i.e., ventrolateral PFC) that are needed to control one's thought and behavior, such as inhibiting the tendency to respond inappropriately.[27] Progress in both working memory and self-control occur as substantial experience-dependent synaptic pruning takes place from late childhood into adolescence,[114] in conjunction with increased myelination and increased neuronal connectivity within a brain region and across networks.[27] As a result, adults more effectively recruit neural resources and only activate brain areas that are needed to perform a task, whereas children and adolescents activate a broader region of the brain when performing the same task.[134] Abnormal maturation of connections within the DLPFC in late childhood and adolescence may underlie pathophysiology of neuropsychiatric disorders and impaired development of control processes.[11]

5.4.2.5 ONSET OF COGNITIVE AND SEXUAL MATURATION

The transition from a child into a sexually mature adult occurs during a specific time and requires the maturation of cognitive behavior (adolescence), the activation of hypothalamic–pituitary–gonadal (HPG) axis, and the hypothalamic–pituitary–adrenal (HPA) axis that culminates in puberty.[103,202,203] All of those events occur by distinct neural processes that require different neurobiological systems and distinct timing but are linked and contribute to behavior that typifies adolescence, which are described below.

The processes that mediate maturation of cognitive behavior include structural changes, such as reorganization of the neural circuitry that exists between different brain areas, as well as changes in levels of neurochemicals that support neural communication.[134] In addition, ongoing pruning of synaptic connections and increased myelination of neurons in certain brain regions raises the efficiency and speed of flow of information between neurons that may be located far apart from one another.[153] Those structural and developmental changes occur during adolescence in subcortical regions involved with emotion (i.e., amygdala) and reward (i.e., striatum), such that adolescents exhibit heightened activation of these brain areas when compared with adults.[134] In adolescents the brain area responsible for regulating cognitive control and goal-oriented behavior (i.e., PFC) is immature relative to the regions that mediate the reward system (i.e., striatum), and this is postulated to account for their increased susceptibility to risky behavior.[11,33] Overall, changes in activation of these subcortical regions result in adolescents who exhibit enhanced risk-taking behavior, heightened emotions, and impulsivity.[14,194]

Gonadal maturation initiates when a cue in the environment signals to the brain to secrete gonadotropin-releasing hormone (GnRH) from the hypothalamus that directs the synthesis and secretion of the pituitary gonadotropins, luteinizing hormone, and follicle stimulating hormone, which together promote the production and release of steroid hormones from the gonads of both sexes.[203] This is otherwise known as the HPG axis. Steroid hormones (e.g., testosterone and estradiol) then bind existing steroid receptors in the brain and restructure the brain in a manner as described above.[134,222] For example, sex hormones are involved in myelination and can organize structural connections between different brain regions.[176] In addition, decreased cortical gray matter is associated with increased levels of estradiol and testosterone in girls and boys, respectively.[175] In binding to the brain, steroid hormones can also increase the sexual activity and interest of individuals[203] and affect their processing of affective and social stimuli,[33] as well as regulate the release of GnRH and the overall activity of the HPG axis.[222]

The HPG axis is also linked to the HPA axis, which controls stress responses and sleep and regulates mood and energy use through the actions of cortisol. Adolescents exhibit higher baseline and stronger cortisol response to performance and social stress than younger children, which can inhibit the HPG axis at all the distinct stages of hormone production.[222] Conversely, the HPG axis can also alter HPA activity since estrogen can regulate glucocorticoid and mineralocorticoid receptors, which bind cortisol and inhibit the HPA axis.

A child's transition into a sexually mature adult is regulated by the neuroendocrine system, which can be affected by environmental factors such as the nutritional and health status of the individuals, as well as exposure to environmental chemicals. Several epidemiological studies identify a relationship between perinatal or pubertal exposure to environmental pollutants and pubertal development.[195] For example, two studies[196,239] observed that high blood lead levels in females were associated with a delay in menarche, although only one of these studies[196] also noted a significant delay in their breast development. Conversely, high level exposure to polybrominated biphenyls (PBBs) in utero and through lactation was linked to females exhibiting an earlier onset of menarche compared with females that received less PBB exposure or who were not breastfed.[19]

5.4.3 LITERATURE SUMMARY

At birth, the newborn human brain has the overall architecture of an adult brain but continues to mature throughout life with different areas developing at different rates. Consequently, there are a series of distinct critical windows of development for different brain regions, which if disrupted during development, can result in loss or altered brain function that cannot be remedied. Moreover, areas of the brain such as the PFC have a protracted developmental time scale that makes the human brain from infancy through adolescence especially vulnerable to the adverse effects of neurotoxicants compared to adults. The behavioral milestones from infancy through adolescence loosely mirror the ontogeny of neural development, with maturation of the primary cortices first, followed by development of brain areas that serve more complex functions, and the strengthening of connections within and between neural circuits. Consequently, changes in neural development may be tracked to some extent by changes in behavioral function.

5.5 EXPERIMENTAL METHODOLOGY

Having provided overviews for both neural and behavioral development in humans, we will now discuss common strategies employed by laboratory researchers to measure chemical effects on the nervous system of experimental animals, which include behavioral assessments and neuroanatomical evaluations. These methods are selected for a study on a case-by-case basis depending on the chemical and within the context of DNT regulatory guidelines at different regulatory organizations.

5.5.1 ANIMAL MODELS

While, conceptually, the optimum means of identifying chemicals that are potential human developmental neurotoxicants would entail assessing the effects directly in humans, in practicality the obvious ethical concerns preclude the conduct of human studies for such DNT evaluation. As a feasible alternative, laboratory animal models are used to effectively and reliably identify chemical substances with a potential to elicit developmental neurotoxic effects in humans. In selecting an appropriate animal model, several factors are considered: (1) the degree of similarity in the TK and developmental profile of the animal to humans, (2) the ease of handling and working with the animal, (3) the financial cost and time required to breed and maintain the animals, and (4) the existence of established reliable tests that can assess the integrity of the CNS of the test animal. Numerous articles have been published comparing the neurodevelopment and behavioral ontogeny of humans to various laboratory animal species.[8,106,184,211,236,237]

Rodents are considered the model of choice for DNT evaluation for several reasons. For example, like humans, the brain maturational sequence in the rodent initiates earlier in the hindbrain area before the forebrain region.[184] Moreover, the order in which the sensory reflexes are developed in rodents also parallels those of humans.[236] Importantly, rodents also exhibit clear effects in response to established developmental human neurotoxicants; both humans and rodents exhibited alterations in motor development, motor control, and cognitive function following exposures during development to methylmercury, lead, ethanol, and polychlorinated biphenyls.[81] From a logistical stand point, rodents are ideal for animal toxicity testing because they (1) are small, easily housed, and maintained, (2) adapt well to new environments, (3) are relatively inexpensive, (4) have a life span of 2–3 years, which enables researchers to perform lifetime toxicity studies,

(5) have genetic, biological, and behavioral characteristics that are well-understood by researchers, and (6) can model many symptoms of human conditions, particularly behavioral abnormalities.[25,227] In contrast, nonhuman primates are generally not used for DNT testing for many reasons, including the inconvenience and high cost associated with their housing and maintenance and the more stringent ethical concerns surrounding working with primates. However, nonhuman primates may sometimes be used for certain special behavioral tests because of their similarity to humans.

The validity of an animal model for behavioral assessments depends on the following factors: (1) the ability to demonstrate intensity- or severity-dependent behavioral changes, (2) the sensitivity to distinguish between behavioral outcomes observed in control versus treated groups, and (3) the efficiency to determine pharmacological, toxicological, and behavioral consequences of test compounds.

5.5.2   BEHAVIORAL   TESTS

A variety of behavioral tests has been developed to evaluate the effects of administered test substances on nervous system function.[67,125,163,219] In offspring, the neurobehavioral developmental effects of a neurotoxicant are assessed by evaluating the following parameters: (1) somatic growth and maturation (e.g., developmental landmarks), (2) reflex ontogeny, (3) sensory function, (4) motor function, (5) general motor activity and emotionality levels, and (6) memory and learning.[67] Evaluation is primarily based on quantitative measurements (e.g., velocity, distance, and time) as well as categorical parameters (e.g., strategies used and preference for a particular object, location, or environment). Here, we describe some functional assays that range from routinely conducted screening batteries to more refined procedures that further characterize a hazardous effect. These assays are not mandatory, which gives investigators the flexibility to optimally choose assays that are better suited for studying a particular chemical based on prior knowledge of its toxicological properties.

5.5.2.1   SCREENING TESTS FOR HAZARD IDENTIFICATION

The functional observation battery (FOB) represents a series of short tests that are routinely conducted in the home cage or open field. These tests cover a broad range of end points aimed at hazard identification and may also

identify dose levels at which adverse effects occur. Some examples of functional domains that are covered by the FOB and the underlying end points
that evaluate them, as adapted from Tilson[220] and Moser,[163] include: (1) autonomic function (e.g., lacrimation, salivation, defecation, urination), (2) CNS
activity (e.g., rearing, home cage posture), (3) CNS excitability (e.g., ease
of removal, handling reactivity, arousal, vocalizations), (4) neuromuscular
function (e.g., gait score, landing foot splay, righting reflex, grip strength),
(5) sensorimotor function (e.g., tail pinch response, approach response), and
(6) physiologic function (e.g., body weight and temperature). The FOBs can
be tailored to the needs of the investigator, who can select assays that they
identify as most relevant.

Three common reflexes that are routinely tested as part of the FOB
include the surface righting reflex, the negative geotaxis, and the acoustic
startle reflex. The surface righting reflex is demonstrated when an animal
that is held with its back on a surface or 12 in above surface is released, and
then is able to right itself, such that all four of its feet are in contact with the
surface. It is graded as follows: 1 = normal, immediate righting, 2 = slow or
poorly coordinated righting, and 3 = very slow or fails to achieve righting
within approximately 5 s.[151] The surface-righting reflex emerges in the rat
ranging from PNDs 1–3 and matures during its first week of development.[236]
The negative geotaxis (inclined plane) test examines the ability of rodents to
maintain their position on a stationary inclined plane or angle board.[64,164,242]
Sensorimotor function is measured by the angle at which the animal first
falls (threshold angle), the total number of falls on the inclined plane after
repeated trials, and the mean latencies to fall. Rats achieve this ability around
PND 9–11.[236] The acoustic or auditory startle reflex (ASR) is a stereotyped
motor response to a sudden intense auditory stimulus that is characterized by
contractions of the major muscles of the body, whereby both the forepaws
and hind paws of the rodent are initially extended and then flexed and result
in the animal in a hunched position. Because rats are born nonresponsive to
sound, the ASR is not functional until PND 12.[236] The ASR test evaluates
the rodent's sensorimotor process and can be further modified to address the
engagement of higher brain centers.[47] For example, prepulse inhibition (PPI)
of the startle reflex is a phenomenon whereby presentation of lower intensity
acoustic stimuli preceding the acoustic startle stimulus attenuates the ASR.
PPI is considered an operational measure of sensorimotor gating and is regulated by the forebrain neural circuitry.

Spontaneous motor activity is another screening test that is often evaluated in an open field and serves as an early indicator of a direct effect of a
chemical to the nervous system. The detection of neurotoxicity by this test

is not limited to the motor system but can also reflect potential effects on the animal's arousal, motivational state, diurnal activity, and sensorimotor function.[219] As a screening measure, this test is conducted in animals that undergo other further observational and functional procedures. They are assessed in terms of their distance traveled, the velocity of their movements, and their time of immobility versus activity as a means to evaluate motor function.[115,116] However, it can also serve as a standalone study and be modified for further characterization (e.g., evaluation of habituation following repeated testing). For example, in a modified version of this test, the apparatus is divided into different zones, such as central and peripheral compartments[139] or light and dark chambers,[131] and the behavior in each zone is captured separately (i.e., number of line crossings) to further determine place preference. The time spent in certain areas of the open field and defecation rate are both measures of emotional reactivity in the animal to the new environment.[67]

The ability to detect toxicity in screening studies heavily depends on the training of the investigator (e.g., ease of handling laboratory animals, knowledgeable of the basics of experimental design and good laboratory techniques), so that they can identify and interpret behavioral changes.[40] Because screening tests broadly assess neurological function and animal behavior are not under procedural control and widely vary, data from these studies are more variable and more difficult to interpret. On the other hand, the benefits of screening tests include the ability to retest an animal to determine an effect's onset, progression, duration, and potential reversibility.[163]

5.5.2.2   TESTS FOR HAZARD CHARACTERIZATION

Tests dedicated to hazard characterization are more costly to conduct than screening assays (e.g., time and cost to execute) but can also provide more information in regards to dose–response relationships. Furthermore, since these tests are designed to address specific functional effects, and animal subjects are trained and are under procedural control, the variability in data is reduced, which makes results easier to interpret.[163] While these studies are dedicated to evaluating distinct functional systems and appear straightforward, interpreting the data can be difficult. For example, sensory processing is a confounder of cognitive tests; results from sensory testing can influence the interpretation of cognitive outcomes.[40] Therefore, the conduct of multiple behavioral tests or follow-up specialized testing is often needed to identify the mechanism of action of an adverse effect.

In this section, we highlight some well-validated DNT tests for hazard characterization, which also have broad applicability beyond evaluating DNT. Many of the tests are also used to reveal functional deficits in animal models of, for example, CNS injury, neurologic disease, neurodevelopmental disorders, and evaluate the efficacy of neuroprotective effects of drugs.

5.5.2.2.1 Sensorimotor Tests

5.5.2.2.1.1 Beam Walk Test

This assay is used to detect sensorimotor and locomotor dysfunction.[79,93–95,122] The beam walk test commonly evaluates rodents for their ability to stay upright and walk across an elevated narrow beam to a safe platform. Fine motor function and coordination are assessed using parameters such as the time required to walk across the beam and the number of foot faults or slips. More recently, modified forms of the beam walk task have been developed to enable detection of more subtle impairments in fine motor coordination. Those include the introduction of hurdles (e.g., pegs) in the path of beam traversal[201] or a reduction in beam thickness.[197]

5.5.2.2.1.2 Rotarod Test

This is a routinely recommended DNT study that evaluates motor coordination as an indicator of midbrain and cerebellar dysfunction in rodents. Specifically, it examines the ability of a rat or mouse to balance itself on a rotating rod, beam, or pole that can be held at a constant or variable speed over time.[32,34,141,192,199] Fine motor coordination is measured by the amount of time the animal takes to balance itself or until it falls off the rotating beam. Moreover, the rotarod test is also designed to determine the degree of motor skill learning, which improves over repeated trials in healthy animals.

5.5.2.2.2 Cognitive Tests

5.5.2.2.2.1 Morris Water Maze Task

This test of spatial learning is a highly reliable and reproducible assessment that serves as the "gold standard" for assessing learning and memory function in rodents, and primarily reveals impairments in spatial and reference memory.[161,180,192,231,232] Consequently, this test is routinely conducted by investigators when evaluating short-term memory of test subjects. The test uses a circular pool that is divided into four imaginary quadrants, where one quadrant contains a hidden platform. Surrounding the pool are distinct visual cues displayed on the walls. Animals are trained daily over several trials (e.g., four trials per day for 4–5 days with trial durations set at 60 or 90 s) to find the hidden platform. For each trial, the subject is introduced from different starting locations. The degree that an animal learns the location of the platform is measured by how quickly it is able to locate the platform as it progresses through the trials, and the amount of time it spends in the target quadrant, after the platform is removed. All trials in the Morris Water Maze (MWM) end points are typically captured using a video tracking software. Performance is assessed in terms of navigational parameters, such as latency to find the platform, total distance traveled, velocity, time spent in different quadrants, etc.[24,122]

Since the MWM test design is highly dependent on visual acuity of the animal, a "visual probe" or "cue" test is conducted following the probe trial, to evaluate the test subject's vision. This involves placing a visible flag on the platform that is then placed back in the pool and measuring the latency for the test subject to locate the flagged platform. Aside from vision as a confounding variable, other confounders can include deficits in motor behavior or coordination, which can increase the animal's latency to find the platform. These can be further addressed by conducting additional testing to verify whether deficits exist in those other functional domains.

5.5.2.2.2.2 Appetitive Position Discrimination and Delayed Alternation T-Maze Tests

These tests use appetitive motivation as a means to evaluate cognition. In the "position discrimination" task, rodents are given a reward (e.g., food, sucrose water) consistently across trials in one arm and learn to choose the baited arm.[52] The latency to find the reward over a certain number of trials on a given day or over several days provides a measure of learning and memory. In the "delayed alternation" task, one trial contains two runs. In the first run, one arm is initially blocked off to force an animal to enter an alternate baited arm (forced run). The animal is then removed and subsequently returned, but then has the option of choosing between both arms (free-choice run). The animal is only rewarded if it chooses the previously unbaited arm.[45,54] The test apparatus for these tests can vary in shape and size. For example, it can be a simple 3-arm maze that is in a "T" or "Y" shape or a circular shaped 8-arm radial maze that has a central platform with 8 arms radiating from the center.[68,192,200] Similar to the MWM task, the location of the reward can be switched between trials to determine arm preferences and to evaluate "reference" and "working" memory functions.[68,200] This test is routinely recommended by standardized DNT guidelines as an option for evaluating cognition.

5.5.2.2.2.3   Novel Object Recognition Test

Unlike the MWM and the appetitive position discrimination and delayed alternation T-maze tests, the novel object recognition (NOR) does not rely on negative or positive reinforcement to evaluate recognition memory.[6] For the NOR test, an animal is familiarized to two identical objects and after a well-defined trial interval, tasked to notice a novel object which has replaced one of the identical objects. This test assesses "retention" memory based on the duration the test subject spends exploring the novel versus old/familiar object. Furthermore, it evaluates the ability of the animal to distinguish between the two objects, which is calculated as the "discrimination index".[18,122,123] Modifications to this protocol include changing the location or position of the two identical objects to enable assessment of "spatial" memory,[62] raising the level of difficulty by performing more trials, or introducing more novel objects.[130]

5.5.2.2.2.4   Passive/Active Avoidance Tests

Based on the principle of fear conditioning,[152,183] these tests follow a behavioral paradigm where rodents learn responses that will enable them to avoid an aversive stimulus.[165,166] Two forms of fear conditioning exist: passive avoidance and active avoidance. In the former, an animal is trained to avoid an aversive stimulus by not performing a certain behavior. For example, a rodent that is placed in a two-compartment chamber that has both well-lit and dark areas naturally prefers the dark side. However, when a foot shock is introduced every time the rodent enters the dark side, it learns to avoid entry. The latency to enter the dark side is a measure of memory of learned avoidance; the longer it takes the animal to re-enter the dark side, the greater the memory is attributed to that animal. However, this interpretation can be confounded by potential sensory processing deficits, which may make it difficult for the rodent to distinguish between shocked and nonshocked compartments. In active avoidance, an animal is trained to move over a barrier, when a signal that typically precedes an aversive stimulus is presented. For example, a rodent that is placed into a shuttle box with two well-lit chambers is trained to locate to the opposite side of the chamber where a cue is presented, which the rodent has learned, signals an oncoming shock. Its avoidance based on the cue alone is considered active avoidance.[248] The conduct of both the passive and active avoidance tests can be automated and computer controlled to avoid experimenter error. The passive avoidance test is more routinely conducted to evaluate DNT since it only requires one trial to learn, whereas active avoidance tests require multiple learning trials.

5.5.2.2.3 Emotional and Social Behavioral Tests

Tests that evaluate the emotional and social behavior of rodents are not currently included in DNT guidelines. However, the information they provide can further characterize a hazard and therefore is considered on a case-by-case basis as optional specialty testing.

5.5.2.2.3.1 Elevated Plus Maze

This test is used to screen the antianxiety effects of pharmacological agents and steroid hormones and to better understand the mechanisms and brain regions involved in anxiety-related behavior. The test uses an elevated plus maze, where rodents are placed on an apparatus with 4-arms (two open and two enclosed).[139] Evaluation is based on the test subject's approach-avoidance mechanism: the animal's preference for dark and enclosed spaces and fear of heights and open space relative to its motivation to explore the environment.[233]

5.5.2.2.3.2 Stress-Based Tests

These tests commonly subject rodents to short-term inescapable stress, who are then evaluated for their extent of immobility versus active movement. Extended periods of immobility with a few attempts to struggle are interpreted as depressive-like behavior. One classic example is the tail suspension test,[215] which involves suspending a mouse by the tail from a lever and recording its struggling movements. Another test is the Porsault's swim or forced swim test,[204] in which rodents swim in a water-filled cylinder without an escape platform for an extended period of time. Both activity (i.e., swimming, climbing, or struggle) and passivity (i.e., immobility or floating) data are collected. A substance is considered an antidepressant if it reduces an animal's period of immobility in the tail suspension and forced swim tests.

5.5.2.2.3.3   Social Interaction-Based Tests

Rodents socialize by olfactory interaction and vocal communication. Olfactory interaction is measured based on the amount of time a rodent spends sniffing olfactory stimuli from novel animals or the amount of time it takes to distinguish between novel versus familiar animals, through olfactory cues in an open field setting.[9,154,160] Vocal communication is assessed via ultrasonic vocalization recordings emitted during social interactions[150,191] and is particularly observed as an early communicative behavior of the pup–mother dyad. Despite the existence of tools to evaluate social interactions in rodents, social behavior is typically not recommended by the standardized DNT guidelines. Reasons for its exclusion include the following: (1) rodents are typically not given the opportunity to socialize in a laboratory setting, (2) existing tests are not as well standardized as others (e.g., motor activity assays), and (3) the evaluation of social interaction-based tests cannot be automated but require the observer to be trained and able to record the behavior of two or more animals concurrently.[40]

5.5.3   HISTOPATHOLOGICAL ASSESSMENT

The evaluation of histopathological changes is an important complement to behavioral observation as it may help identify the underlying mechanism(s) for a particular neural dysfunction. Histopathological assessments require the expertise of a pathologist to plan and execute the histopathological studies, as well as evaluate and interpret the resultant data. The selection of histological measures for a study depends on the study's objective and the potential neural defect under question. Generally, at least seven representative histological sections for each dose group should be examined microscopically by a trained pathologist for evidence of neuropathological damage.[181,182] Accuracy in histological assessments largely reflects the quality of the tissue samples, which depend on proper storage, handling, and preparation steps (e.g., fixation, sectioning, and staining). Histopathological exams can be performed using qualitative or quantitative (morphometric) measures via microscopy.[22] Microscopic examinations are performed to identify regions within the CNS and peripheral nervous system (PNS) that exhibit neuropathological alterations, especially indications of developmental damage to the nervous system, the types of neuropathological alterations (e.g., neuronal vacuolation, apoptosis) resulting from exposure to the test substance and to determine the degree of severity of the abnormalities.[127,188] Evaluation of the PNS routinely include sampling of nerves of the hind limb (e.g., sciatic and tibial), the spinal cord (e.g., cervical, thoracic, and lumbar) and the eyes (e.g., optic nerve), and the dorsal root ganglia.[22,172] For qualitative analysis, sections from the high-dose group are typically first compared with those of the control. Only when neuropathological alterations are observed from this initial comparison will evaluations of other dose groups take place.

As stated in OECD TG 426, significant changes that are indicative of developmental insult include but are not restricted to

- "alterations in the gross size or shape of the olfactory bulbs, cerebrum, or cerebellum;
- alterations in the relative size of various brain regions, including decreases or increases in the size of regions resulting from the loss or persistence of normally transient populations of cells or axonal projections (e.g., external germinal layer of cerebellum, corpus callosum);
- alterations in proliferation, migration, and differentiation, as indicated by areas of excessive apoptosis or necrosis, clusters, or dispersed populations of ectopic, disoriented, or malformed neurons or alterations in the relative size of various layers of cortical structures;
- alterations in patterns of myelination, including an overall size reduction or altered staining of myelinated structures;
- evidence of hydrocephalus, in particular, enlargement of ventricles, stenosis of the cerebral aqueduct, and thinning of the cerebral hemispheres."[172]

The observation of neuropathological alterations is followed by a subjective diagnosis to examine dose–response relationships and considers all regions of the CNS where histopathological abnormalities are reported. Accurate and reliable statistical analysis must be performed to determine the statistically significant effects between treatment groups. In such cases, morphometric or stereological evaluation is preferred over a qualitative or semiquantitative assessment to confirm the degree of neuropathological damage. Here, we describe some of the different histopathological evaluations that histologists conduct to determine the extent of neuropathological damage.

5.5.3.1   CELLULAR CHANGES—NEURONAL

Cellular alterations such as cell degeneration, death, and neuronal vacuolation are commonly assessed to determine the severity of cell injury or damage. Cell death can be identified as apoptosis or necrosis, depending on the progression of the cell death. Appropriate staining procedures are selected depending on the type of cell death to be examined. In addition, any kind of hypoplasia (i.e., decrease in neuronal cell number due to an impact on development such as death of neuronal precursors), hyperplasia (i.e., abnormal neuronal proliferation), heterotopiae (i.e., misplaced neuronal cells due to disturbances in early neuronal cell migration and terminal differentiation), and hypomyelination are also noted.[126]

Commonly used staining procedures for assessing cellular changes include hematoxylin and eosin (H & E) and cresyl violet/Nissl staining. Depending on the objective of the study, specialized staining procedures can be used for more in-depth investigation of the cellular alterations. For example, Fluorojade staining is used for estimating the extent of neurodegeneration by counting the number of neurodegenerating cells,[70] whereas the terminal deoxynucleotidyl transferase-mediated dUTP (deoxyuridine triphosphate) nick end-labeling (TUNEL) assay is used to specifically detect apoptotic cells in tissue sections.[140] To evaluate signs of hypomyelination, the classical cresyl violet stain, the Klüver Barrera technique which stains complex lipids, and immunohistochemistry against the myelin basic protein can all be used.[127]

5.5.3.2   CELLULAR CHANGES—NONNEURONAL

Tissue-related alterations may not be accompanied by cell death or degeneration. Tissue alterations are often characterized based on the type of tissues or cells that undergo transformation, due to an insult or other sources. Inflammation is often investigated in neurotoxicity studies, since chronic inflammation often provides a mechanistic link between neurodegeneration or cell death and behavioral dysfunction. Therefore, most of the immunologic-related changes that are explored in such studies involve inflammatory cells such as astrocytes and microglia or pro-inflammatory markers such as interleukins, tumor necrosis factor- alpha, etc. Tissue-related changes that are commonly examined in neurotoxicity studies to estimate the extent of neuropathological damage are summarized below.

5.5.3.2.1   Microglial Activation

Microglia are one class of glial cells that act as the first and main line of active immune defense in the CNS. Inflammation via activation of microglia is a hallmark of CNS pathology.[60] Previous studies suggest that sustained microglial activation after a neuropathological insult may contribute to chronic neuronal cell loss and associated behavioral dysfunction.[30,147] Microglia are classified based on their morphological features (ramified, hypertrophic, and bushy) and activation status (resting or active).[49,124,210,246] Commonly used staining procedures for the characterization of microglia include markers such as ionized calcium-binding adaptor molecule 1 (Iba 1).[209]

5.5.3.2.2 Astrocytosis or Astrogliosis

Astrocytes or astroglia are star-shaped glial cells in the CNS that provide energy metabolites to neurons, maintain the homeostasis of extracellular fluid, ions, and transmitters, regulate blood flow, and regulate synaptic function and remodeling.[96] Similar to microglia, astrocytes are classified based on their morphological features and activation status. Astrocytes can undergo alterations in expression (e.g., proliferation) and morphology, otherwise known as astrocytosis or astrogliosis, as a response to neuropathological insult.[208] Commonly used staining procedures for the characterization of astrocytes include markers such as Glial Fibrillary Acidic Protein and glutamine synthetase and S100β,[209] which are identified using immunofluorescence and quantified using tools such as unbiased stereology.

5.5.3.2.3 Cystic Formation

Cerebral cystic lesions are abnormal cavities that disrupt the continuity of the brain parenchyma. These lesions can vary depending on their content, their location, and their origin.[91] Those that arise from within the CNS can be static and derive from infarcts or other destructive processes, such as BBB breakdown, extravasation of plasma constituents, and edema formation.[82] Other lesions can be progressive (e.g., arachnoid cysts, ependymal cysts) or form from infectious processes.[109] Because the CNS has poor regenerative capacity, large areas of necrotic brain tissue are not replaced but filled with fluid or other material. Recently, the societies of toxicologic pathology from Europe, Great Britain, Japan, and North America established a joint initiative called the INHAND PROJECT (International Harmonization of Nomenclature and Diagnostic Criteria for Lesions in Rats and Mice) to harmonize the diagnostic nomenclature for proliferative and nonproliferative lesions in rodents. As a result, a set of standardized terms for classifying lesions that originate from toxicant induced, spontaneous, frequent, and age-related changes in the CNS and PNS, as well as descriptions of pertinent special diagnostic techniques, is available.[126]

5.5.4 METHODOLOGICAL CHALLENGES

Despite the existence of multiple DNT guidelines that vary in their testing design, all current DNT protocols have some shortcomings. Many reviews have been dedicated to evaluating the utility of current DNT guidelines, their limitations, and emerging technologies that can improve DNT study design.[40,148,206] In this section, we will briefly highlight some of the major challenges in DNT testing and cite pertinent resources that provide additional information.

5.5.4.1 PROTOCOL DESIGN

5.5.4.1.1 Guideline Selection

There may be limited prior knowledge of toxicological properties of chemical to be tested for DNT potential. Therefore, careful consideration is warranted in selecting which test guideline would best characterize the hazard of concern. In some cases, a standalone study (e.g., the OECD TG 426) may be ideal, but in other cases, it may be sufficient to characterize the DNT hazard as part of a larger study with other cohorts, such as the EOGRTS. In their comparison of the EOGRTS and the OECD TG 426, Makris and Voorhees[149] indicate that the EOGRTS does not routinely require a cognitive measure (i.e., learning and memory testing) of the offspring, which OECD eliminated on the basis of retrospective studies[43,179] that indicated that this measure in the offspring lacks sensitivity. Makris and Voorhees[149] question the omission of this end point, since they posit that OECD's decision was based on learning and memory data that were found to be problematic. Although learning and memory testing is not required by the EOGRTS, it does indicate that such an assessment by the investigator is optional as a follow-up study, should preliminary data indicate a deficit in learning and memory in treated animals. However, it is unlikely that the evaluation of associative learning will be triggered, since the EOGRTS only routinely evaluates habituation, a form of nonassociative learning, as part of the acoustic startle and open-field tests. Without including any measures of associative learning in the offspring, the sensitivity of the EOGRTS to detect DNT may be less than the OECD TG 426. In addition, the EOGRTS may also be less sensitive because it recommends half the number of animal test subjects (i.e., 10/sex/dose) as the OECD TG 426 prescribes for its behavioral testing and also assesses behavior in the offspring less frequently (i.e., behavioral testing is needed only at one time point during the postweaning period). In a separate review, the authors also highlight the advantages of the MOG over the EOGRTS, citing that that the MOG does not use internal triggers for follow-up testing, which they state has not been successful in the EOGRTS.[230] Instead, the MOG establishes how the animals will be utilized prior to the start of the study.

5.5.4.1.2 Behavioral Test Selection

Aside from selecting the appropriate guideline study, it may also be challenging to select adequate behavioral tests to characterize DNT effects. Currently, there tends to be a one size fits all approach in conducting DNT studies, which may result in false-negative outcomes.[230] However, not all tests are equally sensitive and detection of adverse effects will also largely reflect the nature of the substance being tested.[28] In an ideal setting, a battery of tests applied together would evaluate all the possible intracellular signaling events and integrated neural functions. However, at the present time, such a battery of tests has not been defined. As a result, most DNT assays allow flexibility for the sponsor to select appropriate behavioral tests. It remains the responsibility of the reviewing regulatory scientists to determine whether the selected behavioral tests adequately characterize the potential DNT hazard.

5.5.4.2 POSITIVE CONTROLS

An important component to any DNT study is a positive control that enables the performing lab to demonstrate that the sensitivity of the procedures being used is adequate to detect an adverse event. However, one positive control may not affect all end points.[42] Therefore, a single positive control may be insufficient for a DNT study and more than one positive control may be necessary. The selection of a positive control depends on the behavioral function that is under investigation. An ideal positive control produces a robust, reliable, and statistically significant change in task performance (i.e., assessed behavior) when properly conducted relative to the vehicle control. For example, amphetamine and chlorpromazine are frequently used as positive agents for motor activity assessments, whereas scopolamine is used in learning and memory testing.[43]

5.5.4.3   CONFOUNDING EFFECTS AND STUDY DESIGN CHALLENGES

There are several potential confounding effects that may limit the ability of a DNT study to reliably demonstrate that adverse effects observed in the study were due to test article exposure. These include:

1. Handling of rodents can induce a stress response that leads to adverse changes in behavior that are unrelated to the test chemical.[42,53]
2. Differences exist in the care required of young weanlings versus adult animals such that deprivation of food or water for a specific behavioral test (e.g., a reward-based cognitive test) can have a higher impact on the offspring.[71]
3. The housing condition of rodents can affect their response to a chemical and alter study outcomes. Variation in housing conditions includes whether the housing is socially enriched (i.e., animals are housed as a small group of the same sex versus being housed alone) or environmentally enriched (i.e., housed in larger versus smaller cages, housing contains only minimal essentials or is additionally supplemented with multisensory stimulating items such as toys and running wheels). Studies find that both social and environmental enrichment can significantly impact an animal's response to chemicals, and in some cases, reduce the severity of adverse effects in animals in enriched environments.[112,169,218,228] The optimal rodent housing condition is dependent on the age, sex, strain, and experimental objective of the study.[75] Currently, only the OECD TG 426, OECD TG 443, and the MOG guidelines stipulate the conditions of animal housing, which is not mentioned in the EPA's OPPTS 870.8600 and 870.6300 DNT guidelines. The OECD TG 426 guideline specifies that animals may be housed individually or be caged in small groups of the same sex and that mated females should be given defined nesting materials when parturition is near. The OECD TG

443 guideline specifies these same conditions, but adds that litters remain with their mothers until weaning, and are then subsequently housed in small groups of the same sex, but can be housed individually if scientifically justified. In addition, the levels of phytoestrogens contains in the bedding material should be minimal. Similarly, the MOG stipulates that pups remain with the dam until weaning, but then housed by litter and sex afterward.

4. Behavior can be influenced by the time of day that testing is performed, subtle variations in environmental conditions (e.g., light levels, humidity levels, visual cues), and by changes in the personnel who perform the tests. When confronted with the need to conduct DNT testing with a large number of animals per treatment group, it can be particularly challenging to maintain consistency in all testing parameters to minimize environmental confounding. In addition, there are logistical challenges to complete the behavioral testing for young rodents during the same developmental window. Comparing animals that are tested just a day apart in age may yield different behavioral responses even within untreated animals due to the rapid pace of neurodevelopment at juvenile stages.[71]

5. Experimental variability may be an additional source of confounding. Variability may be intrinsic or extrinsic depending on its source. Intrinsic variability, or biological variability, innately exists across test animals. Extrinsic variability includes any factors that exist beyond intrinsic variability, such as variations in methodology, approach to data analysis, or animal handling. Although researchers are limited in their ability to minimize biological variability, a framework has been developed to assist investigators to reduce extrinsic variability.[178]

6. The findings of DNT data can be misinterpreted if inappropriate approaches are undertaken in experimental design (e.g., number of animals per test group) and data analysis (e.g., using individual pup rather than litter as the statistical unit of analysis), which can result in false-positive or false-negative outcomes.[111] Accordingly, the Risk Science Institute expert working group from the International Life Science Institute (ILSI) Research Foundation established a framework to help researchers and reviewers of DNT studies properly interpret their data,[223] which includes considering the adequacy of study design, reliability of the conduct of study, and appropriate statistical analysis.

5.5.4.4 OPTIONAL END POINTS

The current DNT guidelines contain a select number of end points that are used to detect and characterize toxicity. However, collecting additional relevant information may be warranted in specific cases. These additional end points include the following:

1. *Pharmacokinetic (PK) or toxicokinetic data*: Currently, DNT studies do not typically require PK or TK data, such that test article exposures in pups are assumed to occur indirectly by lactational and placental transfer. The lack of exposure measurement in the pups introduces uncertainty regarding whether offspring are actually exposed to the test article prior to weaning.[3] Accordingly, adding PK or TK data collection may reduce uncertainty regarding pup exposure in DNT assays by confirming that chemicals are being transferred via those routes.

2. *Mating and social behavior*: Current DNT test guidelines do not routinely incorporate assessments of mating and social behavior, unless it is conducted as an added cohort to another study. Adding these end points may be instructive for specific chemicals, such as those that target the neuroendocrine axis.[40,77] Alternatively, neurotoxicants can also act directly on the brain (e.g., alter morphometry) and modify social behavior. For example, hamsters that are prenatally exposed to lead exhibit increased aggressive behavior[57] as cited in Cory-Slechta et al.[40] Social measures that are promising candidates for assessing social behavior in developmental neurotoxicology include maternal, aggressive, and sexual behaviors.

3. *Functional end points*: Measures of neural activity via electrophysiology or neurochemistry (i.e., neurotransmitter levels) are not included in the DNT guideline studies (Table 5.1). Despite their underrepresentation in routine safety assessments, experimental advances in the development of functional DNT testing have been progressing, which include optical and electrophysiological measurements of intra- and intercellular signaling in neural stem/progenitor cells, and evaluations of network activity in neuronal networks using multi-electrode arrays.[51] Although those additional functional parameters may offer more information concerning the mode of action of a developmental neurotoxic substance, their inclusion in routine DNT testing may be challenging experimentally and

introduce additional extrinsic variability. Therefore, while those neurocellular end points may provide useful mechanistic information for a subset of chemicals, they are not recommended for routine use at this point in time.

5.5.5   FUTURE METHODOLOGICAL DIRECTIONS

The traditional testing approaches recommended by current DNT guidelines are resource intensive. They require large numbers of animals (thousands) which incur large costs associated with their maintenance. In addition, these guideline studies are time- and effort-intensive to conduct and analyze. For those reasons, and because there are thousands of chemicals that have never been tested in a DNT study, researchers have dedicated a significant amount of effort to find new, reliable, and efficient means to screen and prioritize chemicals for DNT testing.[44] In this section, we will briefly summarize some of these new models as well as emerging regulatory approaches.

5.5.5.1   NEW EXPERIMENTAL MODELS

To help identify and prioritize chemicals for DNT testing, scientists have devised new high-throughput screening (HTS) approaches. Included among these approaches are those that are entirely in silico, relating the biological response of a set of molecules to their physicochemical properties. A simple example of this is when a new and untested chemical is compared to a structurally similar class of chemicals with known developmental neurotoxic potential, such as the organophosphorous insecticides. Additional approaches involve the application of in vitro assays in lieu of in vivo whole animal models. Some examples of these in vitro test systems include three-dimensional organotypic cultures, rat pheochromocytoma (PC12) cells,[44] multi-electrode array assays,[234] and primary human and rat neural progenitor cells that are grown as neurospheres to mimic basic processes of early fetal brain development.[13] Both in silico and in vitro approaches are currently used by the US EPA in the Toxcast Research Program that employs HTS of chemical libraries across a myriad of in vitro assays, including biochemical assays, human cells, and alternative models (e.g., mouse embryonic stem cells and zebrafish embryo development).[128,132] The Toxcast Research Program also uses computational toxicology to store and process these data to prioritize chemicals that may require further testing.

The translation of in vitro screening assays and toxicity studies to in vivo exposures is critical for risk assessors to identify an exposure level that could potentially be adverse to humans but is a difficult task due to the lack of human in vivo data. However, advances in in silico physiologically-based pharmacokinetic (PBPK) modeling, a mathematical tool that predicts the absorption, distribution, metabolism, and excretion (ADME) of chemicals in humans and other species, has made this in vitro to in vivo extrapolation (IVIVE) possible.[46] Therefore, PBPK modeling may be used to predict chemical exposures that are toxic to humans and indicate whether treatment-related effects in animal models are likely representative outcomes for humans. In addition, when evaluating whether a chemical is being transferred by lactational or placental routes, PBPK modeling can reduce the uncertainty of estimating pup exposure.

5.5.5.2 EMERGING METHODOLOGIES IN A REGULATORY CONTEXT

The US FDA CFSAN is currently updating the Redbook to incorporate recent advances in science and technology. A public meeting was held in December of 2014 to solicit feedback on this endeavor.[226] Since the US FDA CFSAN currently does not have its own formalized DNT study protocol, changes to Redbook may include DNT recommendations.

Traditional regulatory toxicology, as described earlier, involves decision making that relies primarily on observing apical effects in existing animal models. With the desire to reduce resource-intensive in vivo studies, there may be a shift toward a biological pathway-based approach, such as the Adverse Outcome Pathway (AOP). The AOP approach spearheaded by the OECD and the US EPA is conceptually designed to link a molecular initiating event to an adverse outcome of relevance to risk assessment.[5] Currently, there is an ongoing effort to develop specific AOPs for developmental and adult neurotoxicity.[10]

The US Federal Government has also taken the initiative to more efficiently test more chemicals and prioritize chemicals for extensive testing using traditional methods but reduce dependency on animal testing by establishing the Toxicology 21 (Tox21) Consortium. To accomplish this, the US FDA, NTP, the US EPA, and NIH (National Institutes of Health) established a collaboration in 2010 to test over 10,000 compounds of toxicological concern using quantitative HTS assays at the NIH Chemical Genomics Center (NCGC) and state-of-the art robotic technology. Projects include developing in vitro assays, establishing a library of compounds, and creating a publicly accessible database with chemicals and various health outcome data in experimental animals and humans (http://www.epa.gov/chemical-research/tox21-work-groups). Although high-throughput approaches such as Toxcast and Tox21 provide mechanistic information and prioritization for DNT testing, it remains a challenge to integrate the hazard identification information that emerges from these studies into the risk assessment paradigm in the absence of in vivo DNT studies.

KEYWORDS

- **developmental neurotoxicity**
- **DNT guidelines**
- **neural development**
- **behavioral milestones**
- **neurotoxicants**
- **behavioral tests**
- **histopathological assessment**

REFERENCES

1. Abkarian, G. G. Communication Effects of Prenatal Alcohol Exposure. *J. Commun Disord.* **1992,** *25* (4), 221–240.
2. Adams, J.; Barone, Jr., S.; LaMantia, A.; Philen, R.; Rice, D. C.; Spear, L.; Susser, E. Workshop to Identify Critical Windows of Exposure for Children's Health: Neurobehavioral Work Group Summary. *Environ. Health Perspect.* **2000,** *108* (Suppl. 3), 535–544.
3. Andersen, H. R.; Nielsen, J. B.; Grandjean, P. Toxicologic Evidence of Developmental Neurotoxicity of Environmental Chemicals. *Toxicology* **2000,** *144* (1–3), 121–127.
4. Andersen, S. L. Trajectories of Brain Development: Point of Vulnerability or Window of Opportunity? *Neurosci. Biobehav. Rev.* **2003,** *27* (1–2), 3–18.
5. Ankley, G. T.; Bennett, R. S.; Erickson, R. J.; Hoff, D. J.; Hornung, M. W.; Johnson, R. D.; Mount, D. R.; Nichols, J. W.; Russom, C. L.; Schmieder, P. K.; Serrrano, J. A.; Tietge, J. E.; Villeneuve, D. L. Adverse Outcome Pathways: A Conceptual Framework to Support Ecotoxicology Research and Risk Assessment. *Environ. Toxicol. Chem.* **2010,** *29* (3), 730–741.
6. Antunes, M.; Biala, G. The Novel Object Recognition Memory: Neurobiology, Test Procedure, and Its Modifications. *Cogn. Process* **2012,** *13* (2), 93–110.

7. Aoyama, H.; Naofumi, T.; Shutoh, Y.; Motomura, A.; Crofton, K. Developmental Neurotoxicology: History and Outline of Developmental Neurotoxicity Study. *Food Safety* **2015**, *3* (2), 48–61.

8. Bachevalier, J.; Vargha-Khadem, F. The Primate Hippocampus: Ontogeny, Early Insult and Memory. *Curr. Opin. Neurobiol.* **2005**, *15* (2), 168–174.

9. Bakker, J.; Honda, S.; Harada, N.; Balthazart, J. Sexual Partner Preference Requires a Functional Aromatase (cyp19) Gene in Male Mice. *Horm. Behav.* **2002**, *42* (2), 158–171.

10. Bal-Price, A.; Crofton, K. M.; Sachana, M.; Shafer, T. J.; Behl, M.; Forsby, A.; Hargreaves, A.; Landesmann, B.; Lein, P. J.; Louisse, J.; Monnet-Tschudi, F.; Paini, A.; Rolaki, A.; Schrattenholz, A.; Sunol, C.; van, T. C.; Whelan, M.; Fritsche, E. Putative Adverse Outcome Pathways Relevant to Neurotoxicity. *Crit. Rev. Toxicol.* **2015**, *45* (1), 83–91.

11. Barber, A. D.; Caffo, B. S.; Pekar, J. J.; Mostofsky, S. H. Developmental Changes in Within- and Between-Network Connectivity between Late Childhood and Adulthood. *Neuropsychologia* **2013**, *51* (1), 156–167.

12. Bauer, H. C.; Bauer, H.; Lametschwandtner, A.; Amberger, A.; Ruiz, P.; Steiner, M. Neovascularization and the Appearance of Morphological Characteristics of the Blood–Brain Barrier in the Embryonic Mouse Central Nervous System. *Brain Res. Dev. Brain Res.* **1993**, *75* (2), 269–278.

13. Baumann, J.; Gassmann, K.; Masjosthusmann, S.; DeBoer, D.; Bendt, F.; Giersiefer, S.; Fritsche, E. Comparative Human and Rat Neurospheres Reveal Species Differences in Chemical Effects on Neurodevelopmental Key Events. *Arch. Toxicol.* **2016**, *90* (6), 1415–1427.

14. Beck, A.; Schlagenhauf, F.; Wustenberg, T.; Hein, J.; Kienast, T.; Kahnt, T.; Schmack, K.; Hagele, C.; Knutson, B.; Heinz, A.; Wrase, J. Ventral Striatal Activation During Reward Anticipation Correlates with Impulsivity in Alcoholics. *Biol. Psychiatry* **2009**, *66* (8), 734–742.

15. Bellinger, D.; Leviton, A.; Waternaux, C.; Needleman, H.; Rabinowitz, M. Longitudinal Analyses of Prenatal and Postnatal Lead Exposure and Early Cognitive Development. *N. Engl. J. Med.* **1987**, *316* (17), 1037–1043.

16. Bellinger, D. C. A Strategy for Comparing the Contributions of Environmental Chemicals and Other Risk Factors to Neurodevelopment of Children. *Environ. Health Perspect.* **2012**, *120* (4), 501–507.

17. Berk, L. E. *Child Development*, 9th ed. Pearson, Upper Saddle River, NJ, 2012.

18. Bevins, R. A.; Besheer, J. Object Recognition in Rats and Mice: A One-Trial Nonmatching-to-Sample Learning Task to Study 'Recognition Memory'. *Nat. Protoc.* **2006**, *1* (3), 1306–1311.

19. Blanck, H. M.; Marcus, M.; Tolbert, P. E.; Rubin, C.; Henderson, A. K.; Hertzberg, V. S.; Zhang, R. H.; Cameron, L. Age at Menarche and Tanner Stage in Girls Exposed In Utero and Postnatally to Polybrominated Biphenyl. *Epidemiology* **2000**, *11* (6), 641–647.

20. Blood-Siegfried, J.; Rende, E. K. The Long-Term Effects of Prenatal Nicotine Exposure on Neurologic Development. *J. Midwifery Womens Health* **2010**, *55* (2), 143–152.

21. Boctor, S. Y.; Wang, C.; Ferguson, S. A. Neonatal PCP Is More Potent than Ketamine at Modifying Preweaning Behaviors of Sprague–Dawley Rats. *Toxicol. Sci.* **2008**, *106* (1), 172–179.

22. Bolon, B.; Garman, R.; Jensen, K.; Krinke, G.; Stuart, B.; Ad Hoc Working Group of the, S. T. P. S.; Regulatory Policy, C. A 'Best Practices' Approach to Neuropathologic

Assessment in Developmental Neurotoxicity Testing—For Today. *Toxicol. Pathol.* **2006,** *34* (3), 296–313.

23. Boothe, R. G.; Dobson, V.; Teller, D. Y. Postnatal Development of Vision in Human and Nonhuman Primates. *Annu. Rev. Neurosci.* **1985,** *8,* 495–545.

24. Brody, D. L.; Holtzman, D. M. Morris Water Maze Search Strategy Analysis in PDAPP Mice before and after Experimental Traumatic Brain Injury. *Exp. Neurol.* **2006,** *197* (2), 330–340.

25. Bryda, E. C. The Mighty Mouse: The Impact of Rodents on Advances in Biomedical Research. *Mol. Med.* **2013,** *110* (3), 207–211.

26. Budday, S.; Steinmann, P.; Kuhl, E. Physical Biology of Human Brain Development. *Front. Cell. Neurosci.* **2015,** *9,* 257.

27. Bunge, S. A.; Wright, S. B. Neurodevelopmental Changes in Working Memory and Cognitive Control. *Curr. Opin. Neurobiol.* **2007,** *17* (2), 243–250.

28. Bushnell, P. J. Testing for Cognitive Function in Animals in a Regulatory Context. *Neurotoxicol. Teratol.* **2015,** *52* (Pt. A), 68–77.

29. Buss, R. R.; Sun, W.; Oppenheim, R. W. Adaptive Roles of Programmed Cell Death During Nervous System Development. *Annu. Rev. Neurosci* **2006,** *29,* 1–35.

30. Byrnes, K. R.; Loane, D. J.; Stoica, B. A.; Zhang, J.; Faden, A. I. Delayed mGluR5 Activation Limits Neuroinflammation and Neurodegeneration after Traumatic Brain Injury. *J Neuroinflamm.* **2012,** *9,* 43.

31. Canfield, R. L.; Gendle, M. H.; Cory-Slechta, D. A. Impaired Neuropsychological Functioning in Lead-Exposed Children. *Dev. Neuropsychol.* **2004,** *26* (1), 513–540.

32. Carter, R. J.; Morton, J.; Dunnett, S. B. Motor Coordination and Balance in Rodents. *Curr. Protoc. Neurosci.* **2001,** *Chapter 8,* Unit 8.12.

33. Casey, B. J.; Duhoux, S.; Malter Cohen, M. Adolescence: What Do Transmission, Transition, and Translation Have to Do with It? *Neuron* **2010,** *67* (5), 749–760.

34. Caston, J.; Jones, N.; Stelz, T. Role of Preoperative and Postoperative Sensorimotor Training on Restoration of the Equilibrium Behavior in Adult Mice Following Cerebellectomy. *Neurobiol. Learn. Mem* **1995,** *64* (3), 195–202.

35. Cayre, M.; Canoll, P.; Goldman, J. E. Cell Migration in the Normal and Pathological Postnatal Mammalian Brain. *Prog. Neurobiol.* **2009,** *88* (1), 41–63.

36. Chan, W. Y.; Kohsaka, S.; Rezaie, P. The Origin and Cell Lineage of Microglia: New Concepts. *Brain. Res. Rev.* **2007,** *53* (2), 344–354.

37. Chen, S. Y.; Cheng, H. J. Functions of Axon Guidance Molecules in Synapse Formation. *Curr. Opin. Neurobiol.* **2009,** *19* (5), 471–478.

38. Choi, B. H. Methylmercury Poisoning of the Developing Nervous System: I. Pattern of Neuronal Migration in the Cerebral Cortex. *Neurotoxicology* **1986,** *7* (2), 591–600.

39. Conti-Ramsden, G.; Durkin, K. Language Development and Assessment in the Preschool Period. *Neuropsychol. Rev.* **2012,** *22* (4), 384–401.

40. Cory-Slechta, D. A.; Crofton, K. M.; Foran, J. A.; Ross, J. F.; Sheets, L. P.; Weiss, B.; Mileson, B. Methods to Identify and Characterize Developmental Neurotoxicity for Human Health Risk Assessment. I: behavioral Effects. *Environ. Health Perspect.* **2001,** *109 Suppl 1,* 79–91.

41. Cowan, N.; Nugent, L. D.; Elliott, E. M.; Ponomarev, I.; Saults, J. S. The Role of Attention in the Development of Short-Term Memory: Age Differences in the Verbal Span of Apprehension. *Child Dev.* **1999,** *70* (5), 1082–1097.

42. Crofton, K. M.; Foss, J. A.; Hass, U.; Jensen, K. F.; Levin, E. D.; Parker, S. P. Undertaking Positive Control Studies as Part of Developmental Neurotoxicity Testing: A

Report from the ILSI Research Foundation/Risk Science Institute Expert Working Group on Neurodevelopmental Endpoints. *Neurotoxicol. Teratol.* **2008,** *30* (4), 266–287.

43. Crofton, K. M.; Makris, S. L.; Sette, W. F.; Mendez, E.; Raffaele, K. C. A Qualitative Retrospective Analysis of Positive Control Data in Developmental Neurotoxicity Studies. *Neurotoxicol. Teratol.* **2004,** *26* (3), 345–352.

44. Crofton, K. M.; Mundy, W. R.; Lein, P. J.; Bal-Price, A.; Coecke, S.; Seiler, A. E.; Knaut, H.; Buzanska, L.; Goldberg, A. Developmental Neurotoxicity Testing: Recommendations for Developing Alternative Methods for the Screening and Prioritization of Chemicals. *ALTEX* **2011,** *28* (1), 9–15.

45. Crofton, K. M.; Peele, D. B.; Stanton, M. E. Developmental Neurotoxicity Following Neonatal Exposure to 3,3′-Iminodipropionitrile in the Rat. *Neurotoxicol. Teratol.* **1993,** *15* (2), 117–129.

46. Croom, E. L.; Shafer, T. J.; Evans, M. V.; Mundy, W. R.; Eklund, C. R.; Johnstone, A. F.; Mack, C. M.; Pegram, R. A. Improving In Vitro to In Vivo Extrapolation by Incorporating Toxicokinetic Measurements: A Case Study of Lindane-Induced Neurotoxicity. *Toxicol. Appl. Pharmacol.* **2015,** *283* (1), 9–19.

47. Curzon, P.; Zhang, M.; Radek, R. J.; Fox, G. B. The Behavioral Assessment of Sensorimotor Processes in the Mouse: Acoustic Startle, Sensory Gating, Locomotor Activity, Rotarod, and Beam Walking. In *Methods of Behavior Analysis in Neuroscience*, 2nd ed.; Buccafusco, J. J., Ed.; CRC Press/Taylor & Francis: Boca Raton, FL, 2009.

48. Davidson, A. J. Anesthesia and Neurotoxicity to the Developing Brain: The Clinical Relevance. *Paediatr. Anaesth.* **2011,** *21* (7), 716–721.

49. Davis, E. J.; Foster, T. D.; Thomas, W. E. Cellular Forms and Functions of Brain Microglia. *Brain Res. Bull.* **1994,** *34* (1), 73–78.

50. Davoust, N.; Vuaillat, C.; Androdias, G.; Nataf, S. From Bone Marrow to Microglia: Barriers and Avenues. *Trends Immunol.* **2008,** *29* (5), 227–234.

51. de Groot, M. W.; Westerink, R. H.; Dingemans, M. M. Don't Judge a Neuron Only by its Cover: Neuronal Function in In Vitro Developmental Neurotoxicity Testing. *Toxicol. Sci.* **2013,** *132* (1), 1–7.

52. Deacon, R. M. Appetitive Position Discrimination in the T-maze. *Nat. Protoc.* **2006,** *1* (1), 13–15.

53. Deacon, R. M. Housing, Husbandry and Handling of Rodents for Behavioral Experiments. *Nat. Protoc.* **2006,** *1* (2), 936–946.

54. Deacon, R. M.; Rawlins, J. N. T-Maze Alternation in the Rodent. *Nat. Protoc.* **2006,** *1* (1), 7–12.

55. Dehaene-Lambertz, G.; Spelke, E. S. The Infancy of the Human Brain. *Neuron* **2015,** *88* (1), 93–109.

56. Dekkers, M. P.; Barde, Y. A. Developmental Biology. Programmed Cell Death in Neuronal Development. *Science* **2013,** *340* (6128), 39–41.

57. Delville, Y. Exposure to Lead during Development Alters Aggressive Behavior in Golden Hamsters. *Neurotoxicol. Teratol.* **1999,** *21* (4), 445–449.

58. Dencker, L.; Eriksson, P. Susceptibility In Utero and upon Neonatal Exposure. *Food Addit. Contam.* **1998,** *15* Suppl., 37–43.

59. Denckla, M. B. Biological Correlates of Learning and Attention: What Is Relevant to Learning Disability and Attention-Deficit Hyperactivity Disorder? *J. Dev. Behav. Pediatr.* **1996,** *17* (2), 114–119.

60. Dheen, S. T.; Kaur, C.; Ling, E. A. Microglial Activation and its Implications in the Brain Diseases. *Curr. Med. Chem.* **2007,** *14* (11), 1189–1197.

61. Diamond, A. The Development and Neural Bases of Memory Functions as Indexed by the AB and Delayed Response Tasks in Human Infants and Infant Monkeys. *Ann. N.Y. Acad. Sci.* **1990,** *608,* 267–309.

62. Dix, S. L.; Aggleton, J. P. Extending the Spontaneous Preference Test of Recognition: Evidence of Object-Location and Object-Context Recognition. *Behav. Brain Res.* **1999,** *99* (2), 191–200.

63. Dobbing, J.; Sands, J. Quantitative Growth and Development of Human Brain. *Arch. Dis. Child.* **1973,** *48* (10), 757–767.

64. Doeppner, T. R.; Kaltwasser, B.; Bahr, M.; Hermann, D. M. Effects of Neural Progenitor Cells on Post-Stroke Neurological Impairment—A Detailed and Comprehensive Analysis of Behavioral Tests. *Front. Cell Neurosci.* **2014,** *8,* 338.

65. Doll, C. A.; Broadie, K. Impaired Activity-Dependent Neural Circuit Assembly and Refinement in Autism Spectrum Disorder Genetic Models. *Front. Cell. Neurosci.* **2014,** *8,* 30.

66. Doty, R. L. Olfactory Function in Neonates. In *The Human Sense of Smell*; Laing, D. G.; Doty, R. L.; Briepohl, W., Eds.; Springer-Verlag: Berlin-Heidelberg, 1991; pp 155–163.

67. Dubovicky, M.; Kovacovsky, P.; Ujhazy, E.; Navarova, J.; Brucknerova, I.; Mach, M. Evaluation of Developmental Neurotoxicity: Some Important Issues Focused on Neurobehavioral Development. *Interdiscip. Toxicol.* **2008,** *1* (3–4), 206–210.

68. Dudchenko, P. A. An Overview of the Tasks Used to Test Working Memory in Rodents. *Neurosci. Biobehav. Rev.* **2004,** *28* (7), 699–709.

69. EFSA, Scientific Opinion: Guidance for Submission for Food Additive Evaluations. *EFSA J.* **2012,** *10* (7), 2760.

70. Ehara, A.; Ueda, S. Application of Fluoro-Jade C in Acute and Chronic Neurodegeneration Models: Utilities and Staining Differences. *Acta Histochem. Cytochem.* **2009,** *42* (6), 171–179.

71. Ehman, K. D.; Moser, V. C. Evaluation of Cognitive Function in Weanling Rats: A Review of Methods Suitable for Chemical Screening. *Neurotoxicol. Teratol.* **2006,** *28* (1), 144–161.

72. Ek, C. J.; Dziegielewska, K. M.; Stolp, H.; Saunders, N. R. Functional Effectiveness of the Blood-Brain Barrier to Small Water-Soluble Molecules in Developing and Adult Opossum (*Monodelphis domestica*). *J. Comp. Neurol* **2006,** *496* (1), 13–26.

73. Emery, B. Regulation of Oligodendrocyte Differentiation and Myelination. *Science* **2010,** *330,* 779–782.

74. Engelhardt, B.; Liebner, S. Novel Insights into the Development and Maintenance of the Blood–Brain Barrier. *Cell Tissue Res.* **2014,** *355* (3), 687–699.

75. Everitt, J. I.; Foster, P. M. Laboratory Animal Science Issues in the Design and Conduct of Studies with Endocrine-Active Compounds. *ILAR J.* **2004,** *45* (4), 417–424.

76. Falluel-Morel, A.; Sokolowski, K.; Sisti, H. M.; Zhou, X.; Shors, T. J.; Dicicco-Bloom, E. Developmental Mercury Exposure Elicits Acute Hippocampal Cell Death, Reductions in Neurogenesis, and Severe Learning Deficits during Puberty. *J. Neurochem.* **2007,** *103* (5), 1968–1981.

77. Ferguson, S. A.; Law, C. D.; Kissling, G. E. Developmental Treatment with Ethinyl Estradiol, But not Bisphenol A, Causes Alterations in Sexually Dimorphic Behaviors in Male and Female Sprague–Dawley Rats. *Toxicol. Sci.* **2014,** *140* (2), 374–392.

78. Flaskos, J. The Developmental Neurotoxicity of Organophosphorus Insecticides: A Direct Role for the Oxon Metabolites. *Toxicol. Lett.* **2012,** *209* (1), 86–93.

79. Fleming, S. M.; Salcedo, J.; Fernagut, P. O.; Rockenstein, E.; Masliah, E.; Levine, M. S.; Chesselet, M. F. Early and Progressive Sensorimotor Anomalies in Mice Overexpressing Wild-Type Human Alpha-Synuclein. *J. Neurosci.* **2004**, *24* (42), 9434–9440.

80. Foster, P. M. Regulatory Forum Opinion Piece: New Testing Paradigms for Reproductive and Developmental Toxicity—The NTP Modified One Generation Study and OECD 443. *Toxicol. Pathol.* **2014**, *42* (8), 1165–1167.

81. Francis, E. Z.; Kimmel, C. A.; Rees, D. C. Workshop on the Qualitative and Quantitative Comparability of Human and Animal Developmental Neurotoxicity: Summary and Implications. *Neurotoxicol. Teratol.* **1990**, *12* (3), 285–292.

82. Fredriksson, K.; Kalimo, H.; Nordborg, C.; Olsson, Y.; Johansson, B. B. Cyst Formation and Glial Response in the Brain Lesions of Stroke-Prone Spontaneously Hypertensive Rats. *Acta Neuropathol.* **1988**, *76* (5), 441–450.

83. Futagi, Y.; Suzuki, Y. Neural Mechanism and Clinical Significance of the Plantar Grasp Reflex in Infants. *Pediatr. Neurol.* **2010**, *43* (2), 81–86.

84. Futagi, Y.; Suzuki, Y.; Goto, M. Clinical Significance of Plantar Grasp Response in Infants. *Pediatr. Neurol.* **1999**, *20* (2), 111–115.

85. Futagi, Y.; Toribe, Y.; Suzuki, Y. The Grasp Reflex and Moro Reflex in Infants: Hierarchy of Primitive Reflex Responses. *Int. J. Pediatr.* **2012**, *2012*, 191562.

86. Futagi, Y.; Yanagihara, K.; Mogami, Y.; Ikeda, T.; Suzuki, Y. The Babkin Reflex in Infants: Clinical Significance and Neural Mechanism. *Pediatr. Neurol.* **2013**, *49* (3), 149–155.

87. Garcia, S. J.; Seidler, F. J.; Slotkin, T. A. Developmental Neurotoxicity of Chlorpyrifos: Targeting Glial Cells. *Environ. Toxicol. Pharmacol.* **2005**, *19* (3), 455–461.

88. Gilbert, S. G.; Weiss, B. A Rationale for Lowering the Blood Lead Action Level from 10 to 2 microg/dL. *Neurotoxicology* **2006**, *27* (5), 693–701.

89. Ginhoux, F.; Lim, S.; Hoeffel, G.; Low, D.; Huber, T. Origin and differentiation of microglia. *Front. Cell. Neurosci.* **2013**, *7*, 45.

90. Giordano, G.; Costa, L. Developmental Neurotoxicity: Some Old and New Issues. *ISRN Toxicol.* **2012**, *2012*.

91. Go, K. G.; Hew, J. M.; Kamman, R. L.; Molenaar, W. M.; Pruim, J.; Blaauw, E. H. Cystic Lesions of the Brain. A Classification Based on Pathogenesis, with Consideration of Histological and Radiological Features. *Eur. J. Radiol.* **1993**, *17* (2), 69–84.

92. Gogtay, N.; Giedd, J. N.; Lusk, L.; Hayashi, K. M.; Greenstein, D.; Vaituzis, A. C.; Nugent, T. F., 3rd; Herman, D. H.; Clasen, L. S.; Toga, A. W.; Rapoport, J. L.; Thompson, P. M. Dynamic Mapping of Human Cortical Development during Childhood through Early Adulthood. *Proc. Natl. Acad. Sci. U.S.A.* **2004**, *101* (21), 8174–8179.

93. Goldberg, M. S.; Fleming, S. M.; Palacino, J. J.; Cepeda, C.; Lam, H. A.; Bhatnagar, A.; Meloni, E. G.; Wu, N.; Ackerson, L. C.; Klapstein, G. J.; Gajendiran, M.; Roth, B. L.; Chesselet, M. F.; Maidment, N. T.; Levine, M. S.; Shen, J. Parkin-Deficient Mice Exhibit Nigrostriatal Deficits But Not Loss of Dopaminergic Neurons. *J Biol. Chem.* **2003**, *278* (44), 43628–43635.

94. Goldstein, L. B. Model of Recovery of Locomotor Ability after Sensorimotor Cortex Injury in Rats. *ILAR J.* **2003**, *44* (2), 125–129.

95. Goldstein, L. B.; Davis, J. N. Beam-Walking in Rats: Studies towards Developing an Animal Model of Functional Recovery after Brain Injury. *J. Neurosci. Methods* **1990**, *31* (2), 101–107.

96. Gordon, G. R.; Mulligan, S. J.; MacVicar, B. A. Astrocyte Control of the Cerebrovasculature. *Glia* **2007**, *55* (12), 1214–1221.

97. Grandjean, P.; Landrigan, P. J. Developmental Neurotoxicity of Industrial Chemicals. *Lancet* **2006,** *368* (9553), 2167–2178.

98. Grandjean, P.; Weihe, P.; Debes, F.; Choi, A. L.; Budtz-Jorgensen, E. Neurotoxicity from Prenatal and Postnatal Exposure to Methylmercury. *Neurotoxicol. Teratol.* **2014,** *43*, 39–44.

99. Graven, S.; Browne, J. Auditory Development in the Fetus and Infant. *Newborn Infant Nurs. Rev.* **2008,** *8* (4), 187–193.

100. Graven, S.; Browne, J. Visual Development in the Human Fetus, Infant and Young Child. *Newborn Infant Nurs. Rev.* **2008,** *8* (4), 194–201.

101. Greene, N. D.; Copp, A. J. Development of the Vertebrate Central Nervous System: Formation of the Neural Tube. *Prenat. Diagn.* **2009,** *29* (4), 303–311.

102. Greene, N. D.; Copp, A. J. Neural Tube Defects. *Annu. Rev. Neurosci.* **2014,** *37*, 221–242.

103. Grumbach, M. M. The Neuroendocrinology of Human Puberty Revisited. *Horm. Res* **2002,** *57* (Suppl. 2), 2–14.

104. Guo, F.; Lu, X. W.; Xu, Q. P. Diagnosis and Treatment of Organotin Poisoned Patients. *World J. Emerg. Med.* **2010,** *1* (2), 122–125.

105. Harada, M. Minamata Disease: Methylmercury Poisoning in Japan Caused by Environmental Pollution. *Crit. Rev. Toxicol.* **1995,** *25* (1), 1–24.

106. Hass, U. The Need for Developmental Neurotoxicity Studies in Risk Assessment for Developmental Toxicity. *Reproduct. Toxicol.* **2006,** *22* (2), 148–156.

107. Heath, C. J.; Picciotto, M. R. Nicotine-Induced Plasticity During Development: Modulation of the Cholinergic System and Long-Term Consequences for Circuits Involved in Attention and Sensory Processing. *Neuropharmacology* **2009,** *56* (Suppl. 1), 254–262.

108. Herschkowitz, N. Neurological Bases of Behavioral Development in Infancy. *Brain Dev.* **2000,** *22* (7), 411–416.

109. Hirano, A.; Hirano, M. Benign Cysts in the Central Nervous System: Neuropathological Observations of the Cyst Walls. *Neuropathology* **2004,** *24* (1), 1–7.

110. Holland, A.; Haas, D. K.; Norman, D.; Brant-Zawadzki, M.; Newton, T. H. MRI of Normal Brain Maturation. *Am. J. Neuroradiol.* **1985,** *7*, 201–208.

111. Holson, R. R.; Freshwater, L.; Maurissen, J. P.; Moser, V. C.; Phang, W. Statistical Issues and Techniques Appropriate for Developmental Neurotoxicity Testing: A Report from the ILSI Research Foundation/Risk Science Institute Expert Working Group on Neurodevelopmental Endpoints. *Neurotoxicol. Teratol.* **2008,** *30* (4), 326–348.

112. Horvath, G.; Reglodi, D.; Vadasz, G.; Farkas, J.; Kiss, P. Exposure to Enriched Environment Decreases Neurobehavioral Deficits Induced by Neonatal Glutamate Toxicity. *Int. J. Mol. Sci.* **2013,** *14* (9), 19054–19066.

113. Huang, H.; Zhang, J.; Wakana, S.; Zhang, W.; Ren, T.; Richards, L. J.; Yarowsky, P.; Donohue, P.; Graham, E.; van Zijl, P. C.; Mori, S. White and Gray Matter Development in Human Fetal, Newborn and Pediatric Brains. *Neuroimage* **2006,** *33* (1), 27–38.

114. Huttenlocher, P. R.; Dabholkar, A. S. Regional differences in synaptogenesis in human cerebral cortex. *J. Comp. Neurol.* **1997,** *387* (2), 167–178.

115. Hutter-Saunders, J. A.; Gendelman, H. E.; Mosley, R. L. Murine Motor and Behavior Functional Evaluations for Acute 1-methyl-4-phenyl-1,2,3,6-Tetrahydropyridine (MPTP) Intoxication. *J. Neuroimmune Pharmacol.* **2012,** *7* (1), 279–288.

116. Ijomone, O.M.; Olaibi, O. K.; Biose, I. J.; Mba, C.; Umoren, K. E.; Nwoha, P. U. Performance of Motor Associated Behavioural Tests Following Chronic Nicotine Administration. *Ann. Neurosci.* **2014,** *21* (2), 42–46.

117. Ikonomidou, C.; Bosch, F.; Miksa, M.; Bittigau, P.; Vockler, J.; Dikranian, K.; Tenkova, T. I.; Stefovska, V.; Turski, L.; Olney, J. W. Blockade of NMDA Receptors and Apoptotic Neurodegeneration in the Developing Brain. *Science* **1999,** *283* (5398), 70–74.

118. Jakovcevski, I.; Filipovic, R.; Mo, Z.; Rakic, S.; Zecevic, N. Oligodendrocyte Development and the Onset of Myelination in the Human Fetal Brain. *Front. Neuroanat.* **2009,** *3*, 5.

119. Jevtovic-Todorovic, V.; Hartman, R. E.; Izumi, Y.; Benshoff, N. D.; Dikranian, K.; Zorumski, C. F.; Olney, J. W.; Wozniak, D. F. Early Exposure to Common Anesthetic Agents Causes Widespread Neurodegeneration in the Developing Rat Brain and Persistent Learning Deficits. *J. Neurosci.* **2003,** *23* (3), 876–882.

120. Johansson, P. A.; Dziegielewska, K. M.; Ek, C. J.; Habgood, M. D.; Liddelow, S. A.; Potter, A. M.; Stolp, H. B.; Saunders, N. R. Blood-CSF Barrier Function in the Rat Embryo. *Eur. J. Neurosci.* **2006,** *24* (1), 65–76.

121. Johansson, P. A.; Dziegielewska, K. M.; Liddelow, S. A.; Saunders, N. R. The blood–CSF Barrier Explained: When Development is Not Immaturity. *Bioessays* **2008,** *30* (3), 237–248.

122. Kabadi, S. V.; Stoica, B. A.; Hanscom, M.; Loane, D. J.; Kharebava, G.; Murray Ii, M. G.; Cabatbat, R. M.; Faden, A. I. CR8, a Selective and Potent CDK Inhibitor, Provides Neuroprotection in Experimental Traumatic Brain Injury. *Neurotherapeutics* **2012,** *9* (2), 405–421.

123. Kabadi, S. V.; Stoica, B. A.; Loane, D. J.; Luo, T.; Faden, A. I. CR8, a Novel Inhibitor of CDK, Limits Microglial Activation, Astrocytosis, Neuronal Loss, and Neurologic Dysfunction after Experimental Traumatic Brain Injury. *J. Cereb. Blood Flow. Metab.* **2014,** *34* (3), 502–513.

124. Kabadi, S. V.; Stoica, B. A.; Zimmer, D. B.; Afanador, L.; Duffy, K. B.; Loane, D. J.; Faden, A. I. S100B Inhibition Reduces Behavioral and Pathologic Changes in Experimental Traumatic Brain Injury. *J. Cereb. Blood Flow Metab.* **2015,** *35* (12), 2010–2020.

125. Kaufmann, W. Current Status of Developmental Neurotoxicity: An Industry Perspective. *Toxicol. Lett.* **2003,** *140–141*, 161–169.

126. Kaufmann, W.; Bolon, B.; Bradley, A.; Butt, M.; Czasch, S.; Garman, R. H.; George, C.; Groters, S.; Krinke, G.; Little, P.; McKay, J.; Narama, I.; Rao, D.; Shibutani, M.; Sills, R. Proliferative and Nonproliferative Lesions of the Rat and Mouse Central and Peripheral Nervous Systems. *Toxicol. Pathol.* **2012,** *40* (4 Suppl), 87S–157S.

127. Kaufmann, W.; Groters, S. Developmental Neuropathology in DNT-Studies—A Sensitive Tool for the Detection and Characterization of Developmental Neurotoxicants. *Reprod. Toxicol.* **2006,** *22* (2), 196–213.

128. Kavlock, R.; Dix, D. Computational Toxicology as Implemented by the U.S. EPA: Providing High Throughput Decision Support Tools for Screening and Assessing Chemical Exposure, Hazard and Risk. *J. Toxicol. Environ. Health B: Crit. Rev.* **2010,** *13* (2–4), 197–217.

129. Kimmel, C. A. Current Approaches to Risk Assessment for Developmental Neurotoxicity. In *Handbook of Developmental Neurotoxicology*; Slikker Jr., W., Chang, L., Eds.; Academic Press: San Diego, CA, 1998; pp 675–685.

130. Kinnavane, L.; Albasser, M. M.; Aggleton, J. P. Advances in the Behavioural Testing and Network Imaging of Rodent Recognition Memory. *Behav. Brain Res.* **2015,** *285*, 67–78.

131. Klejbor, I.; Ludkiewicz, B.; Turlejski, K. Effect of Light–Dark Changes on the Locomotor Activity in Open Field in Adult Rats and Opossums. *Folia Morphol. (Warsz.)* **2013,** *72* (4), 300–305.

132. Knudsen, T.; Martin, M.; Chandler, K.; Kleinstreuer, N.; Judson, R.; Sipes, N. Predictive Models and Computational Toxicology. *Methods Mol. Biol.* **2013**, *947*, 343–374.

133. Kobesova, A.; Kolar, P. Developmental Kinesiology: Three Levels of Motor Control in the Assessment and Treatment of the Motor System. *J. Bodyw. Mov. Ther.* **2014**, *18* (1), 23–33.

134. Konrad, K.; Firk, C.; Uhlhaas, P. J. Brain Development during Adolescence: Neuroscientific Insights into This Developmental Period. *Dtsch. Arztebl. Int.* **2013**, *110* (25), 425–431.

135. Kriegstein, A.; Alvarez-Buylla, A. The Glial Nature of Embryonic and Adult Neural Stem Cells. *Annu. Rev. Neurosci* **2009**, *32*, 149–184.

136. Kriegstein, A. R.; Noctor, S. C. Patterns of Neuronal Migration in the Embryonic Cortex. *Trends Neurosci.* **2004**, *27* (7), 392–399.

137. Kuan, C. Y.; Roth, K. A.; Flavell, R. A.; Rakic, P. Mechanisms of Programmed Cell Death in the Developing Brain. *Trends Neurosci.* **2000**, *23* (7), 291–297.

138. Kuhl, P.; Rivera-Gaxiola, M. Neural Substrates of Language Acquisition. *Annu. Rev. Neurosci.* **2008**, *31*, 511–534.

139. Kumar, J.; Hapidin, H.; Bee, Y. T.; Ismail, Z. Effects of the mGluR5 Antagonist MPEP on Ethanol Withdrawal Induced Anxiety-Like Syndrome in Rats. *Behav. Brain Funct.* **2013**, *9*, 43.

140. Labat-Moleur, F.; Guillermet, C.; Lorimier, P.; Robert, C.; Lantuejoul, S.; Brambilla, E.; Negoescu, A. TUNEL Apoptotic Cell Detection in Tissue Sections: Critical Evaluation and Improvement. *J. Histochem. Cytochem.* **1998**, *46* (3), 327–334.

141. Lalonde, R.; Bensoula, A. N.; Filali, M. Rotorod Sensorimotor Learning in Cerebellar Mutant Mice. *Neurosci. Res.* **1995**, *22* (4), 423–426.

142. Landrigan, P. J.; Garg, A. Chronic Effects of Toxic Environmental Exposures on Children's Health. *J. Toxicol. Clin. Toxicol.* **2002**, *40* (4), 449–456.

143. Landrigan, P. J.; Goldman, L. R. Children's Vulnerability to Toxic Chemicals: A Challenge and Opportunity to Strengthen Health and Environmental Policy. *Health Aff. (Millwood)* **2011**, *30* (5), 842–850.

144. Landrigan, P. J.; Sonawane, B.; Butler, R. N.; Trasande, L.; Callan, R.; Droller, D. Early Environmental Origins of Neurodegenerative Disease in Later Life. *Environ. Health Perspect.* **2005**, *113* (9), 1230–1233.

145. Lee, I.; Eriksson, P.; Fredriksson, A.; Buratovic, S.; Viberg, H. Developmental Neurotoxic Effects of Two Pesticides: Behavior and Biomolecular Studies on Chlorpyrifos and Carbaryl. *Toxicol. Appl. Pharmacol.* **2015**, *288* (3), 429–438.

146. Lenroot, R. K.; Giedd, J. N. Brain Development in Children and Adolescents: Insights from Anatomical Magnetic Resonance Imaging. *Neurosci. Biobehav. Rev.* **2006**, *30* (6), 718–729.

147. Loane, D. J.; Kumar, A. Microglia in the TBI Brain: The Good, the Bad, and the Dysregulated. *Exp. Neurol.* **2016**, *275* (Pt. 3), 316–327.

148. Makris, S. L.; Raffaele, K.; Allen, S.; Bowers, W.; Hass, U.; Alleva, E.; Calamandrei, G.; Sheets, L.; Amcoff, P.; Delrue, N.; Crofton, K. M. A Retrospective Performance Assessment of the Developmental Neurotoxicity Study in Support of OECD Test Guideline 426. *Environ. Health Perspect.* **2009**, *117* (1), 17–25.

149. Makris, S. L.; Vorhees, C. V. Assessment of Learning, Memory and Attention in Developmental Neurotoxicity Regulatory Studies: Introduction. *Neurotoxicol. Teratol.* **2015**, *52* (Pt. A), 62–67.

150. Malkova, N. V.; Yu, C. Z.; Hsiao, E. Y.; Moore, M. J.; Patterson, P. H. Maternal Immune Activation Yields Offspring Displaying Mouse Versions of the Three Core Symptoms of Autism. *Brain Behav. Immunol.* **2012**, *26* (4), 607–616.

151. Mandella, R. Neurotoxicology. In *Handbook of Toxicology*, 3rd ed.; Derelanko, M., Auletta, C., Eds.; CRC Press: Boca Raton, FL, 2014; pp 303–322.

152. Maren, S. Neurobiology of Pavlovian Fear Conditioning. *Annu. Rev. Neurosci.* **2001**, *24*, 897–931.

153. Markham, J. A.; Greenough, W. T. Experience-Driven Brain Plasticity: Beyond the Synapse. *Neuron Glia Biol.* **2004**, *1* (4), 351–363.

154. McGraw, L. A.; Young, L. J. The Prairie Vole: An Emerging Model Organism for Understanding the Social Brain. *Trends Neurosci.* **2010**, *33* (2), 103–109.

155. Miller, M. W. Effects of Alcohol on the Generation and Migration of Cerebral Cortical Neurons. *Science* **1986**, *233* (4770), 1308–1311.

156. Mishra, V.; Cheng, H.; Gong, G.; He, Y.; Dong, Q.; Huang, H. Differences of Inter-Tract Correlations between Neonates and Children around Puberty: A Study Based on Microstructural Measurements with DTI. *Front. Hum. Neurosci.* **2013**, *7*, 721.

157. Mitchell, R. G. The Moro reflex. *Cereb. Palsy Bull.* **1960**, *2*, 135–141.

158. Mollgard, K.; Saunders, N. R. The Development of the Human Blood–Brain and Blood–CSF Barriers. *Neuropathol. Appl. Neurobiol.* **1986**, *12* (4), 337–358.

159. Mollgoard, K.; Saunders, N. R. Complex Tight Junctions of Epithelial and of Endothelial Cells in Early Foetal Brain. *J. Neurocytol.* **1975**, *4* (4), 453–468.

160. Moretti, P.; Bouwknecht, J. A.; Teague, R.; Paylor, R.; Zoghbi, H. Y. Abnormalities of Social Interactions and Home-Cage Behavior in a Mouse Model of Rett Syndrome. *Hum. Mol. Genet* **2005**, *14* (2), 205–220.

161. Morris, R. G.; Garrud, P.; Rawlins, J. N.; O'Keefe, J. Place Navigation Impaired in Rats with Hippocampal Lesions. *Nature* **1982**, *297* (5868), 681–683.

162. Morrow, C. E.; Bandstra, E. S.; Anthony, J. C.; Ofir, A. Y.; Xue, L.; Reyes, M. B. Influence of Prenatal Cocaine Exposure on Early Language Development: Longitudinal Findings from Four Months to Three Years of Age. *J. Dev. Behav. Pediatr.* **2003**, *24* (1), 39–50.

163. Moser, V. C. Functional Assays for Neurotoxicity Testing. *Toxicol. Pathol.* **2011**, *39* (1), 36–45.

164. Murphy, M. P.; Rick, J. T.; Milgram, N. W.; Ivy, G. O. A simple and Rapid Test of Sensorimotor Function in the Aged Rat. *Neurobiol. Learn. Mem.* **1995**, *64* (2), 181–186.

165. Myers, K.; Goulet, M.; Rusche, J.; Boismenu, R.; Davis, M. Inhibition of Fear Potentiated Startle in Rats Following Peripheral Administration of Secretin. *Psychopharmacology (Berl.)* **2004**, *172* (1), 94–99.

166. Myers, K. M.; Davis, M. AX+, BX− Discrimination Learning in the Fear-Potentiated Startle Paradigm: Possible Relevance to Inhibitory Fear Learning in Extinction. *Learn. Mem.* **2004**, *11* (4), 464–475.

167. Neal, A. P.; Guilarte, T. R. Mechanisms of Lead and Manganese Neurotoxicity. *Toxicol. Res.* **2013**, *2*, 99–114.

168. Nelson, C. A. The Ontogeny of Human Memory: A Cognitive Neuroscience Perspective. *Dev. Psychol.* **1995**, *31* (5), 723–738.

169. Neugebauer, N. M.; Cunningham, S. T.; Zhu, J.; Bryant, R. I.; Middleton, L. S.; Dwoskin, L. P. Effects of Environmental Enrichment on Behavior and Dopamine Transporter Function in Medial Prefrontal Cortex in Adult Rats Prenatally Treated with Cocaine. *Brain Res. Dev. Brain Res.* **2004**, *153* (2), 213–223.

170. NRC. *Risk Assessment in the Federal Government: Managing the Process.* National Academy Press: Washington, DC, 1983.

171. NTP. Draft Protocol outline for the Modified One-generation Study (MOG#) of Test Article (CAS#, Test Article#) in Harlan Spargue Dawley Rats Exposed via Dosed feed, 2015.

172. OECD. Test No. 426: Developmental Neurotoxicity Study. *OECD Guidelines for the Testing of Chemicals, Section 4,* 2007.

173. OECD. Test No. 443: Extended One-Generation Reproductive Toxicity Study. *OECD Guidelines for the Testing of Chemicals, Section 4,* 2011.

174. Pedroso, F. S.; Rotta, N. T. Babkin Reflex and Other Motor Responses to Appendicular Compression Stimulus of the Newborn. *J. Child Neurol.* **2004,** *19* (8), 592–596.

175. Peper, J. S.; Hulshoff Pol, H. E.; Crone, E. A.; van Honk, J. Sex Steroids and Brain Structure in Pubertal Boys and Girls: A Mini-review of Neuroimaging Studies. *Neuroscience* **2011,** *191,* 28–37.

176. Peper, J. S.; van den Heuvel, M. P.; Mandl, R. C.; Hulshoff Pol, H. E.; van Honk, J. Sex Steroids and Connectivity in the Human Brain: A Review of Neuroimaging Studies. *Psychoneuroendocrinology* **2011,** *36* (8), 1101–1113.

177. Perry, B. D. Childhood Experience and the Expression of Genetic Potential: What Childhood Neglect Tells Us About Nature and Nurture. *Brain Mind* **2002,** *3,* 79–100.

178. Raffaele, K. C.; Fisher, J. E., Jr.; Hancock, S.; Hazelden, K.; Sobrian, S. K. Determining Normal Variability in a Developmental Neurotoxicity Test: A Report from the ILSI Research Foundation/Risk Science Institute Expert Working Group on Neurodevelopmental Endpoints. *Neurotoxicol. Teratol.* **2008,** *30* (4), 288–325.

179. Raffaele, K. C.; Rowland, J.; May, B.; Makris, S. L.; Schumacher, K.; Scarano, L. J. The Use of Developmental Neurotoxicity Data in Pesticide Risk Assessments. *Neurotoxicol. Teratol.* **2010,** *32* (5), 563–572.

180. Ramani, M.; van, G. T.; Kadish, I.; Bulger, A.; Ambalavanan, N. Neurodevelopmental Impairment Following Neonatal Hyperoxia in the Mouse. *Neurobiol. Dis.* **2013,** *50,* 69–75.

181. Rao, D. B.; Little, P. B.; Malarkey, D. E.; Herbert, R. A.; Sills, R. C. Histopathological Evaluation of the Nervous System in National Toxicology Program Rodent Studies: A Modified Approach. *Toxicol. Pathol.* **2011,** *39* (3), 463–470.

182. Rao, D. B.; Little, P. B.; Sills, R. C. Subsite Awareness in Neuropathology Evaluation of National Toxicology Program (NTP) Studies: A Review of Select Neuroanatomical Structures with their Functional Significance in Rodents. *Toxicol. Pathol.* **2014,** *42* (3), 487–509.

183. Rescorla, R. A. Conditioned Inhibition of Fear Resulting from Negative CS–US Contingencies. *J. Comp. Physiol. Psychol.* **1969,** *67* (4), 504–509.

184. Rice, D.; Barone, S., Jr. Critical Periods of Vulnerability for the Developing Nervous System: Evidence from Humans and Animal Models. *Environ. Health Perspect.* **2000,** *108* (Suppl. 3), 511–533.

185. Rose, S. A.; Feldman, J. F.; Jankowski, J. J. A Cognitive Approach to the Development of Early Language. *Child Dev.* **2009,** *80* (1), 134–150.

186. Rowitch, D. H. Glial Specification in the Vertebrate Neural Tube. *Nat. Rev. Neurosci.* **2004,** *5* (5), 409–419.

187. Ruben, R. J. The Ontogeny of Human Hearing. *Int. J. Pediatr. Otorhinolaryngol.* **1995,** *32 Suppl.,* S199–204.

188. Salvo, H.; Butt, M. Regulatory Guide to the Histopathological Assessment of Neurotox-
     icity Studies. In *Fundamental Neuropathology for Pathologists and Toxicologists: Prin-
     ciples and Techniques*; Bolon, B., Butt, M., Eds.; Johns Wiley & Sons, Inc.: Hoboken,
     NJ, 2011; pp 519–535.

189. Saunders, N. R.; Daneman, R.; Dziegielewska, K. M.; Liddelow, S. A. Transporters
     of the Blood–Brain and Blood-CSF Interfaces in Development and in the Adult. *Mol
     Aspects Med.* **2013**, *34* (2–3), 742–752.

190. Scallet, A. C.; Schmued, L. C.; Slikker, W., Jr.; Grunberg, N.; Faustino, P. J.; Davis,
     H.; Lester, D.; Pine, P. S.; Sistare, F.; Hanig, J. P. Developmental Neurotoxicity of
     Ketamine: Morphometric Confirmation, Exposure Parameters, and Multiple Fluores-
     cent Labeling of Apoptotic Neurons. *Toxicol. Sci.* **2004**, *81* (2), 364–370.

191. Scattoni, M. L.; Gandhy, S. U.; Ricceri, L.; Crawley, J. N. Unusual Repertoire of Vocal-
     izations in the BTBR T+tf/J Mouse Model of Autism. *PLoS One* **2008**, *3* (8), e3067.

192. Schaar, K. L.; Brenneman, M. M.; Savitz, S. I. Functional Assessments in the Rodent
     Stroke Model. *Exp. Transl. Stroke Med.* **2010**, *2* (1), 13.

193. Scheiffele, P. Cell–Cell Signaling during Synapse Formation in the CNS. *Annu. Rev.
     Neurosci.* **2003**, *26*, 485–508.

194. Schneider, S.; Peters, J.; Bromberg, U.; Brassen, S.; Miedl, S. F.; Banaschewski, T.;
     Barker, G. J.; Conrod, P.; Flor, H.; Garavan, H.; Heinz, A.; Ittermann, B.; Lathrop,
     M.; Loth, E.; Mann, K.; Martinot, J. L.; Nees, F.; Paus, T.; Rietschel, M.; Robbins, T.
     W.; Smolka, M. N.; Spanagel, R.; Strohle, A.; Struve, M.; Schumann, G.; Buchel, C.;
     Consortium, I. Risk Taking and the Adolescent Reward System: A Potential Common
     Link to Substance Abuse. *Am. J. Psychiatry* **2012**, *169* (1), 39–46.

195. Schoeters, G.; Den Hond, E.; Dhooge, W.; van Larebeke, N.; Leijs, M. Endocrine
     Disruptors and Abnormalities of Pubertal Development. *Basic Clin. Pharmacol.
     Toxicol.* **2008**, *102* (2), 168–175.

196. Selevan, S. G.; Rice, D. C.; Hogan, K. A.; Euling, S. Y.; Pfahles-Hutchens, A.; Bethel,
     J. Blood Lead Concentration and Delayed Puberty in Girls. *N. Engl. J. Med.* **2003**, *348*
     (16), 1527–1536.

197. Selwyn, R.; Hockenbury, N.; Jaiswal, S.; Mathur, S.; Armstrong, R. C.; Byrnes, K. R.
     Mild Traumatic Brain Injury Results in Depressed Cerebral Glucose Uptake: An (18)
     FDG PET Study. *J. Neurotrauma* **2013**, *30* (23), 1943–1953.

198. Shafer, T. J.; Meyer, D. A.; Crofton, K. M. Developmental Neurotoxicity of Pyrethroid
     Insecticides: Critical Review and Future Research Needs. *Environ. Health Perspect.*
     2005, *113* (2), 123–136.

199. Shiotsuki, H.; Yoshimi, K.; Shimo, Y.; Funayama, M.; Takamatsu, Y.; Ikeda, K.;
     Takahashi, R.; Kitazawa, S.; Hattori, N. A Rotarod Test for Evaluation of Motor Skill
     Learning. *J Neurosci. Methods* **2010**, *189* (2), 180–185.

200. Shoji, H.; Hagihara, H.; Takao, K.; Hattori, S.; Miyakawa, T. T-Maze Forced Alternation
     and Left-Right Discrimination Tasks for Assessing Working and Reference Memory in
     Mice. *J. Vis. Exp.* **2012**, *60*. doi:10.3791/3300.

201. Singleton, R. H.; Yan, H. Q.; Fellows-Mayle, W.; Dixon, C. E. Resveratrol Attenu-
     ates Behavioral Impairments and Reduces Cortical and Hippocampal Loss in a Rat
     Controlled Cortical Impact Model of Traumatic Brain Injury. *J. Neurotrauma* **2010**, *27*
     (6), 1091–1099.

202. Sisk, C. L.; Foster, D. L. The Neural Basis of Puberty and Adolescence. *Nat. Neurosci.*
     2004, *7* (10), 1040–1047.

203. Sisk, C. L.; Zehr, J. L. Pubertal Hormones Organize the Adolescent Brain and Behavior. *Front. Neuroendocrinol.* **2005,** *26* (3–4), 163–174.

204. Slattery, D. A.; Cryan, J. F. Using the Rat Forced Swim Test to Assess Antidepressant-Like Activity in Rodents. *Nat. Protoc.* **2012,** *7* (6), 1009–1014.

205. Slotkin, T. A.; Skavicus, S.; Card, J.; Stadler, A.; Levin, E. D.; Seidler, F. J. Developmental Neurotoxicity of Tobacco Smoke Directed Toward Cholinergic and Serotonergic Systems: More Than Just Nicotine. *Toxicol. Sci.* **2015,** *147* (1), 178–189.

206. Smirnova, L.; Hogberg, H. T.; Leist, M.; Hartung, T. Developmental Neurotoxicity— Challenges in the 21st Century and In Vitro Opportunities. *ALTEX* **2014,** *31* (2), 129–156.

207. Sobotka, T. J.; Ekelman, K. B.; Slikker, W., Jr.; Raffaele, K.; Hattan, D. G. Food and Drug Administration Proposed Guidelines for Neurotoxicological Testing of Food Chemicals. *Neurotoxicology* **1996,** *17* (3–4), 825–836.

208. Sofroniew, M. V. Molecular Dissection of Reactive Astrogliosis and Glial Scar Formation. *Trends Neurosci.* **2009,** *32* (12), 638–647.

209. Sofroniew, M. V.; Vinters, H. V. Astrocytes: Biology and Pathology. *Acta Neuropathol.* **2010,** *119* (1), 7–35.

210. Soltys, Z.; Ziaja, M.; Pawlinski, R.; Setkowicz, Z.; Janeczko, K. Morphology of Reactive Microglia in the Injured Cerebral Cortex. Fractal Analysis and Complementary Quantitative Methods. *J. Neurosci. Res.* **2001,** *63* (1), 90–97.

211. Spear, L. P. The Adolescent Brain and Age-Related Behavioral Manifestations. *Neurosci. Biobehav. Rev.* **2000,** *24* (4), 417–463.

212. Stahnke, T.; Richter-Landsberg, C. Triethyltin-Induced Stress Responses and Apoptotic Cell Death in Cultured Oligodendrocytes. *Glia* **2004,** *46* (3), 334–344.

213. Stanton, M. E.; Spear, L. P. Workshop on the Qualitative and Quantitative Comparability of Human and Animal Developmental Neurotoxicity, Work Group I Report: Comparability of Measures of Developmental Neurotoxicity in Humans and Laboratory Animals. *Neurotoxicol. Teratol.* **1990,** *12* (3), 261–267.

214. Stefovska, V. G.; Uckermann, O.; Czuczwar, M.; Smitka, M.; Czuczwar, P.; Kis, J.; Kaindl, A. M.; Turski, L.; Turski, W. A.; Ikonomidou, C. Sedative and Anticonvulsant Drugs Suppress Postnatal Neurogenesis. *Ann. Neurol.* **2008,** *64* (4), 434–445.

215. Steru, L.; Chermat, R.; Thierry, B.; Simon, P. The Tail Suspension Test: A New Method for Screening Antidepressants in Mice. *Psychopharmacology (Berl.)* **1985,** *85* (3), 367–370.

216. Stevens, C.; Sanders, L.; Neville, H. Neurophysiological Evidence for Selective Auditory Attention Deficits in Children with Specific Language Impairment. *Brain Res.* **2006,** *1111* (1), 143–152.

217. Svirsky, M. A.; Robbins, A. M.; Kirk, K. I.; Pisoni, D. B.; Miyamoto, R. T. Language Development in Profoundly Deaf Children with Cochlear Implants. *Psychol. Sci.* **2000,** *11* (2), 153–158.

218. Szabadfi, K.; Atlasz, T.; Horvath, G.; Kiss, P.; Hamza, L.; Farkas, J.; Tamas, A.; Lubics, A.; Gabriel, R.; Reglodi, D. Early Postnatal Enriched Environment Decreases Retinal Degeneration Induced by Monosodium Glutamate Treatment in Rats. *Brain Res.* **2009,** *1259*, 107–112.

219. Tilson, H. A. Behavioral Indices of Neurotoxicity: What Can Be Measured? *Neurotoxicol. Teratol.* **1987,** *9* (6), 427–443.

220. Tilson, H. A. Neurobehavioral Methods Used in Neurotoxicological Research. *Toxicol. Lett.* **1993,** *68* (1–2), 231–240.

221. Treit, S.; Lebel, C.; Baugh, L.; Rasmussen, C.; Andrew, G.; Beaulieu, C. Longitudinal MRI Reveals Altered Trajectory of Brain Development During Childhood and Adolescence in Fetal Alcohol Spectrum Disorders. *J. Neurosci.* **2013**, *33* (24), 10098–10109.

222. Trotman, H. D.; Holtzman, C. W.; Ryan, A. T.; Shapiro, D. I.; MacDonald, A. N.; Goulding, S. M.; Brasfield, J. L.; Walker, E. F. The Development of Psychotic Disorders in Adolescence: A Potential Role for Hormones. *Horm. Behav.* **2013**, *64* (2), 411–419.

223. Tyl, R. W.; Crofton, K.; Moretto, A.; Moser, V.; Sheets, L. P.; Sobotka, T. J. Identification and Interpretation of Developmental Neurotoxicity Effects: A Report from the ILSI Research Foundation/Risk Science Institute Expert Working Group on Neurodevelopmental Endpoints. *Neurotoxicol. Teratol.* **2008**, *30* (4), 349–381.

224. USEPA. OPPTS 870.6300 Developmental Neurotoxicity Study. *Health Effects Test Guidelines*, 1998 (EPA 712-C-98-239).

225. USEPA. Revised OP (Organophosphate) Cumulative Risk Assessment. US Environmental Protection Agency: Washington, DC, 2002.

226. USFDA. Toxicological Principles for the Safety Assessment of Food Ingredients: Public Meeting on Updates and Safety Risk Assessment Considerations: Request for Comments. Federal Register 79, 2014; Vol. 79; pp 64603–64604.

227. Vandamme, T. F. Use of Rodents as Models of Human Diseases. *J. Pharm. Bioallied Sci.* **2014**, *6* (1), 2–9.

228. Verwer, C. M.; van der Ven, L. T.; van den Bos, R.; Hendriksen, C. F. Effects of Housing Condition on Experimental Outcome in a Reproduction Toxicity Study. *Regul. Toxicol. Pharmacol.* **2007**, *48* (2), 184–193.

229. von Stackelberg, K.; Guzy, E.; Chu, T.; Henn, B. C. Exposure to Mixtures of Metals and Neurodevelopmental Outcomes: A Multidisciplinary Review Using an Adverse Outcome Pathway Framework. *Risk Anal.* **2015**, *35* (6), 971–1016.

230. Vorhees, C. V.; Makris, S. L. Assessment of Learning, Memory, and Attention in Developmental Neurotoxicity Regulatory Studies: Synthesis, Commentary, and Recommendations. *Neurotoxicol. Teratol.* **2015**, *52* (Pt. A), 109–115.

231. Vorhees, C. V.; Williams, M. T. Morris Water Maze: Procedures for Assessing Spatial and Related Forms of Learning and Memory. *Nat. Protoc.* **2006**, *1* (2), 848–858.

232. Vorhees, C. V.; Williams, M. T. Value of Water Mazes for Assessing Spatial and Egocentric Learning and Memory in Rodent Basic Research and Regulatory Studies. *Neurotoxicol. Teratol.* **2014**, *45*, 75–90.

233. Walf, A. A.; Frye, C. A. The Use of the Elevated Plus Maze as an Assay of Anxiety-Related Behavior in Rodents. *Nat. Protoc.* **2007**, *2* (2), 322–328.

234. Wallace, K.; Strickland, J. D.; Valdivia, P.; Mundy, W. R.; Shafer, T. J. A Multiplexed Assay for Determination of Neurotoxicant Effects on Spontaneous Network Activity and Viability from Microelectrode Arrays. *Neurotoxicology* **2015**, *49*, 79–85.

235. Wallingford, J. B.; Niswander, L. A.; Shaw, G. M.; Finnell, R. H. The Continuing Challenge of Understanding, Preventing, and Treating Neural Tube Defects. *Science* **2013**, *339* (6123), 1222002.

236. Wood, S. L.; Beyer, B. K.; Cappon, G. D. Species Comparison of Postnatal CNS Development: Functional Measures. *Birth Defects Res. B: Dev. Reprod. Toxicol.* **2003**, *68* (5), 391–407.

237. Workman, A. D.; Charvet, C. J.; Clancy, B.; Darlington, R. B.; Finlay, B. L. Modeling Transformations of Neurodevelopmental Sequences across Mammalian Species. *J. Neurosci.* **2013**, *33* (17), 7368–7383.

238. Wu, C. S.; Jew, C. P.; Lu, H. C. Lasting Impacts of Prenatal Cannabis Exposure and the Role of Endogenous Cannabinoids in the Developing Brain. *Fut. Neurol.* **2011,** *6* (4), 459–480.

239. Wu, T.; Buck, G. M.; Mendola, P. Blood Lead Levels and Sexual Maturation in U.S. Girls: the Third National Health and Nutrition Examination Survey, 1988–1994. *Environ. Health Perspect.* **2003,** *111* (5), 737–741.

240. Yamaguchi, Y.; Miura, M. How to Form and Close the Brain: Insight into the Mechanism of Cranial Neural Tube Closure in Mammals. *Cell Mol. Life Sci.* **2013,** *70* (17), 3171–3186.

241. Yamaguchi, Y.; Shinotsuka, N.; Nonomura, K.; Takemoto, K.; Kuida, K.; Yosida, H.; Miura, M. Live Imaging of Apoptosis in a Novel Transgenic Mouse Highlights its Role in Neural Tube Closure. *J. Cell Biol.* **2011,** *195* (6), 1047–1060.

242. Yonemori, F.; Yamaguchi, T.; Nakayama, H.; Narita, K.; Hojo, S.; Tamura, A. Effect of JTP-2942, a Novel Thyrotropin-Releasing Hormone Analog, on Motor Deficits after Chronic Focal Cerebral Ischemia in Rats. *J. Cereb. Blood Flow Metab.* **2000,** *20* (1), 74–81.

243. Zafeiriou, D. I. Primitive Reflexes and Postural Reactions in the Neurodevelopmental Examination. *Pediatr. Neurol.* **2004,** *31* (1), 1–8.

244. Zafeiriou, D. I.; Tsikoulas, I. G.; Kremenopoulos, G. M.; Kontopoulos, E. E. Plantar Response Profile of High-Risk Infants at One Year of Life. *J. Child Neurol.* **1999,** *14* (8), 514–517.

245. Zeliger, H. I. Exposure to Lipophilic Chemicals as a Cause of Neurological Impairments, Neurodevelopmental Disorders and Neurodegenerative Diseases. *Interdiscip. Toxicol.* **2013,** *6* (3), 103–110.

246. Zhan, X.; Kim, C.; Sharp, F. R. Very Brief Focal Ischemia Simulating Transient Ischemic Attacks (TIAs) Can Injure Brain and Induce Hsp70 Protein. *Brain Res.* **2008,** *1234,* 183–197.

247. Zheng, W. Neurotoxicology of the Brain Barrier System: New Implications. *J. Toxicol. Clin. Toxicol.* **2001,** *39* (7), 711–719.

248. Zovkic, I. B.; Sweatt, J. D. Epigenetic Mechanisms in Learned Fear: Implications for PTSD. *Neuropsychopharmacology* **2013,** *38* (1), 77–93.

CHAPTER 6

PROTECTIVE EFFECT OF FOOD-GRADE LACTIC ACID BACTERIA AGAINST OXIDATIVE STRESS

J. E. AGUILAR-TOALÁ[1], B. VALLEJO-CORDOBA[1],
A. F. GONZÁLEZ-CÓRDOVA[1], R. GARCÍA-VARELA[2],
H. S. GARCÍA[3], and A. HERNÁNDEZ-MENDOZA[1*]

[1]Centro de Investigación en Alimentación y Desarrollo A. C., Carretera a la Victoria Km. 0.6, Hermosillo, Sonora 83304, México

[2]CIATEJ, Autopista Monterrey-Aeropuerto, Km 10, Parque PIIT, Via de Innovacion 404, Apodaca, NL 66629, México

[3]Unidad de Investigación y Desarrollo de Alimentos, Instituto Tecnológico de Veracruz, M.A. de Quevedo 2779, Col. Formando Hogar, Veracruz, Veracruz 91897, México

*Corresponding author. E-mail: ahernandez@ciad.mx

CONTENTS

ABSTRACT

Oxidative stress is a physiological condition characterized by a disrupted balance between oxidants such as reactive oxygen and nitrogen species (ROS/RNS) and antioxidants (e.g., glutathione, peroxidase, catalase) in favor of the former. Such condition has been associated to different human pathologies (e.g., cancer, neurogenerative, and cardiovascular diseases) by causing damage to basic component for cell function and survival. Therefore, several strategies have been explored in order to regulate the redox homeostasis including the use of phytochemicals (polyphenols, carotenoids), vitamins (E and C), minerals (Zn), and proteins (bioactive peptides). Recent in vitro and in vivo studies have shown that ingestion of food-grade lactic acid bacteria (LAB) may reinforce antioxidant defense systems of the host, thereby alleviate oxidative stress related-pathologies. Nevertheless, the protective mechanisms of LAB are not fully elucidated. It has been suggested that the possible effect may be due to (1) bacteria ROS- and RNS-scavenging properties, (2) chelating metal ions (iron and copper) involved in ROS/RNS formation, and (3) bacteria capacity to produce antioxidant compounds such as bioactive peptides, polysaccharides, and fatty acids. Therefore, the aim of this chapter is to provide an overview of potential protective role of food-grade lactic acid bacteria against oxidative stress as well as to describe the possible mechanism involved.

6.1   INTRODUCTION

Under normal metabolism, the continuous formation of free radicals and radical-derived compounds are important for natural physiological functions. For example, reactive oxygen species (ROS) play an important role in receptor-mediated signaling pathways.[28] Many of the ROS-mediated responses protect cells against oxidative stress and reestablish redox homeostasis.[32,94,97] ROS are generated as a response to inflammation and act as a defense mechanism.[13] They can also induce cellular senescence and apoptosis. These characteristics provide ROS with antitumorigenic functions.[79,83] ROS produced in the mitochondria during oxidative phosphorylation, serve as sensors of oxygen tension in the control of ventilation and erythropoietin production. Another major contributor of ROS is the metabolism of nitric oxide.[22,27,101]

Nitric oxide has been recognized as a signaling molecule for vasodilation and neurotransmission; however, when reacting with ROS can form the highly toxic reactive nitrogen species (RNS).[2,95] Furthermore, when

ROS are found within the cell, they can act as a secondary messenger for intracellular signaling cascades, which induces and maintains the oncogenic phenotype of cancer cells.[96] The excessive production and accumulation of ROS inside the cell may occur by an increase in the mitochondrial electron-transport chain, excessive stimulation of NAD(P)H and/or by deficiency of enzymatic and nonenzymatic antioxidants,[22] which converts the RNS into reactive species hazardous for living organisms, resulting in oxidative stress, which is a physiological condition characterized by a disruption in the balance between highly reactive oxidant molecules and antioxidants.[62,63] This physiological condition has been implicated in the process of aging and age-associated disorders (e.g., neurodegenerative diseases) as well as to different chronic and degenerative pathologies such as cancer, cardiovascular diseases, diabetes mellitus, rheumatoid arthritis, ischemia/reperfusion injury, and obstructive sleep apnea, among others.[22,109]

The human body possesses specific defense mechanisms against oxidative stress. However, it may not be sufficient to prevent damage caused by oxidative stress; therefore, food-containing antioxidants may be used to aid our bodies in preventing or reducing oxidative stress damage.[30,51] In this manner, several strategies to regulate the redox homeostasis including the use of phytochemicals (polyphenols, carotenoids), vitamins (E and C), minerals (Zn), and proteins (bioactive peptides) have been previously reported.[14,82,84]

Recent scientific evidence has proven the importance of food-grade lactic acid bacteria (LAB) as potential antioxidants.[5] Both in vivo as well as in vitro studies have revealed that these bacteria, mainly the species of *Lactobacillus* and *Bifidobacterium*, and the food products that contain them exhibit antioxidant properties. However, the mechanisms implicated in their protective effect are not fully elucidated. It has been suggested that the protective effect may be due to (1) bacterial ROS- and RNS-scavenging properties, (2) chelation of metal ions, mainly iron and copper, involved in ROS/RNS formation, and (3) the capacity of bacteria to produce antioxidant compounds such as bioactive peptides, polysaccharides, and fatty acids. Thus, this chapter provides an overview of the protective role of food-grade LAB against oxidative stress, as well as a description of the possible mechanism involved.

6.2 OXIDATIVE AND NITROSATIVE STRESS

Oxidative stress occurs when cells have an overload of ROS and/or RNS, which means that their endogenous antioxidant capacity is overrun.[59]

ROS and RNS are the most significant biological free radicals which include hydroxyl (OH·), superoxide ($O_2^{·-}$), lipid peroxyl (LOO·), nitric oxide (NO·), and peroxynitrite (OONO⁻); and nonradical molecules: hydrogen peroxide (H_2O_2), ozone (O_3), and lipid peroxide (LOOH). Oxidative stress is mainly triggered by the excessive production of ROS and RNS, insufficient antioxidants defense mechanism, and the incorporation of ROS and RNS from the surrounding environment.[104]

Research has provided evidence proving that free radicals are responsible for a variety of degenerative diseases[40,45,68] by damaging lipids, proteins and DNA, as depicted in Figure 6.1.[19] Such damage has proven to be the cause of a number of human pathologies such as cancer, cardiovascular, and neurodegenerative diseases.[62] Additionally, ROS and RNS participate in the normal aging process, as well as contribute to age-related diseases such as type 2 diabetes mellitus and rheumatoid arthritis.[26,27]

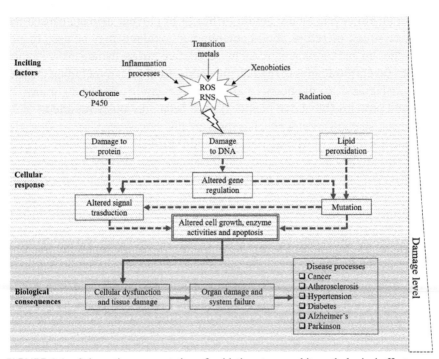

FIGURE 6.1   Schematic representation of oxidative stress and its pathological effects.

Oxidative stress may occur in various situations: (1) increased levels of endogenous and exogenous compounds entering autoxidation coupled with ROS and RNS production (Fig. 6.2); (2) a decrease of antioxidant reserves;

(3) inactivation of antioxidant enzymes; (4) a decrease in the production of antioxidant enzymes and low molecular mass antioxidants; and (5) combinations of the above factors.[63] ROS are produced endogenously by the mitochondria, cytochrome P450 metabolism, peroxisomes, inflammation processes, ischemia/reperfusion injury and exogenous sources that include cigarette smoke, environmental pollutants, radiation, pesticides, and ozone.[62]

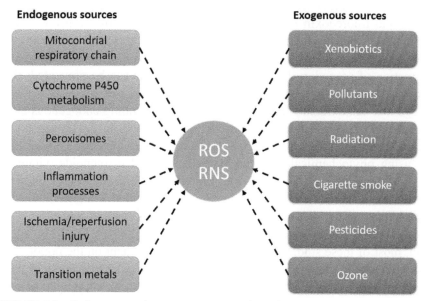

FIGURE 6.2   Endogenous and exogenous sources of reactive oxygen and nitrogen species.

6.2.1 ANTIOXIDANTS AND THE HUMAN BODY

The human body has adapted defense mechanisms against oxidative stress; these involve a highly complex system of antioxidant comprised by enzymatic and nonenzymatic components. These components work synergistically to protect cells from oxidative stress.[75,76] Under normal conditions, cells are capable of preventing ROS/RNS-induced oxidative damage, by generating endogenous antioxidants, which include glutathione (GSH), superoxide dismutase (SOD), catalase (CAT) and glutathione reductase (GR) and peroxidase (GPx) enzymes, and nonenzymatic antioxidants such as GSH, ubiquinol, and cysteine.[7,62,65] However, endogenous antioxidants may not be sufficient to maintain their protective effect under promoted oxidative stress. In these cases, dietary antioxidants may be required for optimal cellular functions.[76]

6.2.2 DIETARY ANTIOXIDANTS

The main source of exogenous antioxidants is found in our daily food consumption.[14] Examples of the predominant dietary antioxidants are β-carotene, lycopene, and vitamins A, C, and E (α-tocopherol). Furthermore, the essential trace element selenium (Se) is often thought to be a dietary antioxidant originated by the formation of selenocysteine, a fundamental component of the active site or second redox-active site of GSH peroxidases and thioredoxin reductase, respectively,[20,67] and by Se itself. These antioxidants play a vital role in both food systems as well as in the human body to reduce oxidative processes. In food systems, lipid peroxidation and formation of secondary lipid peroxidation products can be prevented; this will help to maintain flavor, texture, and color during storage.[90] Conversely, exogenous antioxidants play an important role by aiding endogenous antioxidants in the regulation of the excessive ROS/RNS in the human body. In this manner, studies have reported the regulation of redox homeostasis by using phytochemicals such as polyphenols and carotenoids, vitamins E and C, minerals as zinc (Zn), and proteins and bioactive peptides.[14,82,84] Furthermore, epidemiological studies have shown that a higher intake of antioxidants in the diet may be associated to lower risks of cancer, as well as cardiovascular and neurodegenerative diseases.[4,10,21,24,29,36,37,99] In addition, most of these studies are mostly in good agreement on the notion that antioxidants are more effective in the prevention of pathologies, rather than in their treatment. For example, antioxidant therapy using a preparation of green tea, containing epigallocatechin-3-gallate [51.9%], epigallocathechin [5.5%], epicatechin [12.2%], and epicatechin-3-gallate [6.1%]), was effective against precancerous prostate lesions in human volunteers;[12] however, the antioxidant therapy with a preparation of green tea-containing epigallocatechin-3-gallate [22.5%] and other nonspecified polyphenols, was not efficient on human volunteers afflicted with metastatic prostate cancer.[18,42]

6.3 PROTECTIVE EFFECT OF SOME FOOD-GRADE LACTIC ACID BACTERIA AGAINST OXIDATIVE STRESS

Food-grade LAB are generally regarded as safe microorganisms which have a wide use in industry. Since they have been proven safe for humans, they are ideal microbial biopreservatives. They also do not adversely alter nutritional properties in foods, are effective at low concentrations, as well as active under refrigerated storage, and the specific strains are considered to own health-promoting effects.[53,77,105] Specific food-grade LAB, when ingested by humans, have been associated with various essential biological functions, including manipulation of intestinal microbiota, suppression of pathogens, immunomodulation, stimulation of epithelial cell proliferation, fortification of intestinal barrier, prevention of diarrhea, and reduction of lactose intolerance.[73,87,93,106]

Furthermore, it has been suggested that some LAB, especially *Lactobacillus* strains might show antioxidant properties.[5] Hence, these strains of bacteria are commonly found in human intestines,[86] they are able to interact with their host and may exhibit their antioxidant effect as either whole cells or by the release of different metabolites such as bioactive peptides and polysaccharides. In addition, these microorganisms may also release their intracellular content into the gut, therefore, contributing as an antioxidant in the intestines. On the other hand, these food-grade LAB produce metabolites with an antioxidant activity can be used to prepare fermented products.[100]

There are several approaches reported to measure antioxidative capacity. However, the application of more than two different methods is recommended due to variables such as free radical initiator, oxidizing substrates, antioxidant mechanism evaluated hydrogen atoms or electron transfer, metals chelation, biological relevance, and the interpretation and expression of results.[8]

6.3.1 PRODUCTION OF ANTIOXIDANT COMPOUNDS BY LACTIC ACID BACTERIA

Specific LAB produce bioactive compounds may exhibit antioxidant activities. These compounds include conjugated linoleic acid (CLA), exopolysaccharides (EPS), and bioactive peptides.

Ha et al.[33] reported that CLA is an effective antioxidant, with a higher potency than α-tocopherol and as effective as butylated hydroxytoluene. The main mechanism reported for this fatty acid involves its scavenging capacity of free radicals by donating of hydrogen atoms.[112] CLA is synthesized mainly by ruminal bacteria, raising the interrogative if there may be other microorganisms capable of producing it.[89] This supposition has promoted extensive research of LAB with the potential ability to produce CLA in growth medium and milk, via linoleate isomerase activity.[41,89] It is possible that the addition of these cultures to milk can produce fermented milks with ca. 0.7–1.5-fold CLA content.[34,98] In this regard, several LAB strains have been identified as potential CLA producers, of which the main genus are *Bifidobacterium*

(*Bb.*), *Enterococcus* (*E.*), *Lactobacillus* (*Lb.*), *Lactococcus* (*Lc.*), *Propionibacterium*, and *Streptococcus* (*Strep*).[6,89]

Previous research has described the synthesis of EPS by *Lc. lactis*, *Strep. thermophilus*, *Lb. acidophilus*, *Lb. delbrueckii*, *Lb. helveticus*, *Lb. paracasei*, and *Lb. rhamnosus*.[31,50] These mentioned strains are commonly used in commercial products involving diverse milk fermentations. It was reported that EPS may exhibit antioxidant activity through radical scavenging capacity, inhibit lipid oxidation and chelate metals,[61,72,107] as well as to increase antioxidant enzyme activities in an experimental rat colitis,[85] and in a D-galactose-induced aged mice model.[107] Studies suggest that antioxidant properties may be attributed, at least partly, to the free carboxylic groups and their relatively homogenous distribution in the polysaccharide, with contributions by the content of uronic acid.[103] Therefore, some authors[55,107,114] found that uronic acid plays an important role in the antioxidant properties of polysaccharides from *Sargassum fusiforme*, *Bb. animalis* RH and *Lb. helveticus* MB2-1, respectively. Similarly, Li et al.[55] reported an exopolysaccharide from *Lb. helveticus* MB2-1, with a greater proportion of uronic acid that was negatively charged, thereby resulting in a higher ferrous ion-chelating capacity. Ferrous ions are involved in the formation of free radicals by means of the Fenton and Haber–Weiss reaction. This generates the highly reactive hydroxyl radical. In a related research, a direct relationship between uronic acid content and the radical scavenging capacity of tea polysaccharide has been reported.[16,17]

In parallel, Virtanen et al.[100] evaluated the antioxidant effect of fermented milk by 25 LAB strains. The strains with higher activity were *Leuconostoc mesenteroides* ssp. *cremoris*, *Lb. jensenii* (ATCC 25258), and *Lb. acidophilus* (ATCC 4356). The authors observed a positive relationship between the antioxidant capacity and fermentation time (0–24 h or 48 h). In addition, Ramesh et al.[78] evaluated 19 *Lactobacillus* strains for their ability to produce antioxidants in milk. The authors observed that strains with the highest capacity in fermented milks were *Lb. rhamnosus* NCDC24, and *Lb. delbrueckii* subsp. *bulgaricus* NCDC08 after 24 h fermentation time. The antioxidant capacity in both studies correlated positively with the degree of proteolysis, suggesting that peptides were responsible for the antioxidative effect. In addition, Liu et al.[58] reported that *Lb. helveticus*, isolated from kefir grains, possessed highly proteolytic capacities with an important antioxidant effect in fermented milk and soymilk. Such activity increased significantly nearing the end of the fermentation period. Osuntoki and Korie[69] evaluated the antioxidant effect of fermented milk with various *Lactobacillus* strains, isolated from Nigerian fermented foods. Results revealed that four of the five tested strains increased their antioxidant effect as fermentation time went from 0 to 24 h. In addition, the antioxidant effect remained unchanged during the fermentation period for the remaining strain. These results indicated even though all LAB belong to the same genus, their proteolytic system varied. These studies suggest that the antioxidant capacity developed by fermentation with different LAB is strain-specific and does not necessarily depend on the fermentation time or their proteolytic capacity.

6.3.2 ANTIOXIDANT CAPACITY OF LACTIC ACID BACTERIA

The antioxidant capacity exhibited by LAB has been tested as whole cells and cell-free extract of various LAB. However, the components responsible for their whole cells antioxidant capacity are limited. It has been suggested that the cell wall polysaccharides, peptidoglycan, lipoteichoic acid, and proteins could be responsible for their established antioxidant effect.[54,60,110,113] However, further research is needed to accurately assess the component(s) responsible for the antioxidant capacity of LAB whole cells from a molecular perspective.

Several studies suggested that the antioxidant effect of LAB intracellular content is based on enzyme activity of SOD, GPx, NADH-oxidase, NADH-peroxide, and the molecule of GSH. SOD is responsible for the degradation of toxic superoxides, such as superoxide radical (O_2^-) to oxygen (O_2) and hydrogen peroxide (H_2O_2). There are four types of SOD enzymes classified in terms of their metal cofactors: iron SOD, manganese SOD, copper–Zn SOD, and nickel SOD.[1] Yoon and Byun[111] demonstrated the antioxidant capacity of cell-free extracts of 11 strains of *Lactobacillus* spp. The authors reported a positive correlation between inhibitory activity on linoleic acid peroxidation and intracellular GSH levels, which suggested that strains having high levels of GSH had high level of hydroxyl radical scavenging activity. The authors also reported that differences in media composition and the type of nitrogen source could cause variation in the cellular GSH level. Research by Kim et al.[44] determined the antioxidant activity of four strains of *Lactobacillus*. They found that *Lb. brevis* KCTC 3498 showed higher GPx activity, a potent scavenger that protects cells from oxidative damage, when compared to *Lb. acidophilus*. However, they could not find a positive correlation between an antioxidant effect and GPx activity in other strains and suggested that other compounds may be involved in determining the antioxidant effect. LAB have been tested and determined as CAT-negative; it has been reported that *Bb. infantis, Bb. breve, Bb. adolescentis,* and *Bb. longum* are able to degrade hydrogen peroxide by producing NADH peroxidase.[88] Han et al.[35] found that the strain NZ9800 (wild-type) of *Lc. lactis* that produce SOD spontaneously had a positive significant effect on the reduction of oxidative damage induced by trinitrobenzenesulfonic acid in an inflammatory murine model. Moreover, Rochat et al.[80] reported the preventive effect of the antioxidative *Lb. casei* BL23 containing the CAT-carrier plasmid pLEM415mnkat (MnKat[+]) on dextran sodium sulfate-induced colitis in mice. Their results showed that mice treated with probiotic *Lb. casei* BL23 MnKat[+] had a significant reduction in damage induced by dextran sodium sulfate by producing manganese-dependent CAT, when compared to control mice treated with *Lb. casei* BL23 MnKat[−].

6.3.3 IN VITRO STUDIES OF ANTIOXIDANT PROPERTIES OF FOOD-GRADE LACTIC ACID BACTERIA

Most studies analyze the antioxidant capacity of whole cells and cell-free extract of various LAB strains, including some probiotic bacteria (Table 6.1).

Lin and Yen[57] analyzed the in vitro antioxidant potential of cell-free extracts of 19 LAB strains by inhibition of ascorbate autoxidation. The authors found that all strains exhibited antioxidant properties and that the possible antioxidative mechanisms included metal-ion-chelating ability, scavenging of ROS, and reducing activity of intracellular cell-free extract of LAB. Lin and Chang[56] studied the antioxidant properties of *Bb. longum* ATCC 15708 and *Lb. acidophilus* ATCC 4356. Overall, cell-free extracts showed a significantly ($P < 0.05$) higher inhibition mean percentage of linoleic acid peroxidation activity (ranging from 45.3% to 48.0%) than the whole cells (ranging from 28.1% to 33.2%). However, the 2,2-diphenyl-1-picrylhydrazyl (DPPH) scavenging activity showed an inverse behavior, that is, whole cells showed significantly ($P < 0.05$) higher activity, ranging from 43.2% to 52.1%, than cell-free extracts, which ranged from 20.8 to 41.6%. Kullisaar et al.[46] found similar results in the inhibition of linoleic acid peroxidation activity using two *Lb. fermentum* strains, E-3 and E-18. Cell-free extracts showed a significantly ($P < 0.05$) higher mean percentage of inhibition of linoleic acid peroxidation activity (ranging from 51% to 59%) than whole cells (ranging from 21% to 29%). In addition, authors found that both strains contained significant levels of GSH and expressed Mn-SOD, which are associated with the prevention of lipid peroxidation and degradation of hydrogen peroxide. These results were consistent with those reported by Ahotupa et al.[3], who found that *Lb. rhamnosus* GG inhibited lipid peroxidation by iron chelation and superoxide anion scavenging activity.

TABLE 6.1 *In Vitro* Studies of Antioxidant Properties of Some Food Grade LAB.

Strains	Samples analyzed WC	Samples analyzed C-FE	Applied antioxidant method	Possible antioxidant mechanism	Reference
Six *Lb. acidophilus*, six *Lb. bulgaricus*, five *Strep. thermophilus*, and two *Bb. longum*		+	Inhibition of ascorbate autoxidation, metal ion chelating assay, scavenging of reactive oxygen species, reducing activity	Scavenge of reactive oxygen species, metal ion chelating ability, reducing activity	Lin and Yen[57]
Bb. longum ATCC 15708, *Lb. acidophilus* ATCC 4356	+	+a	Inhibition of linoleic acid peroxidation, DPPH scavenging activity	Scavenge of reactive DPPH and oxygen species	Lin and Chang[56]
Lb. fermentum E-3 and E-18	+	+*	Inhibition of linoleic acid peroxidation, ABTS scavenging activity	Scavenge of reactive ABTS and oxygen species	Kullisaar et al.[46]
Lb. acidophilus LA5 and LA100, *Lb. casei* 01, *Lb. bulgaricus* LB207 and *Lb. rhamnosus* GG744		+	Inhibition of linoleic acid peroxidation, metal ion chelating assay, scavenging of reactive oxygen species, reducing activity	Scavenge of reactive oxygen species, metal ion chelating ability, reducing activity	Kim et al.[43]
Thirteen probiotic *Lactobacillus* strains		+	Inhibition of linoleic acid peroxidation	Scavenge of reactive oxygen species, glutathione activity	Yoon and Byun[111]
Lb. rhamnosus GG and *Lb. casei* KCTC 3260, 01, and KCTK 3109	+	+a	Inhibition of linoleic acid peroxidation	Scavenge of reactive oxygen species	Lee et al.[52]
Lb. delbrueckii ssp. *lactis* (RM2-5, RM6-5, RM5-4), *Lb. delbrueckii* ssp. *bulgaricus* (O16, NCFM, L-1), *Lb. acidophilus* (18, 10442, Y-23) and *Lb. casei* (9018, E5, E10)	+	+a	Oxygen radical absorbance capacity	Scavenge of reactive oxygen species	Saide and Gilliland[81]

TABLE 6.1 (Continued)

Strains	Samples analyzed		Applied antioxidant method	Possible antioxidant mechanism	Reference
	WC	C-FE			
Strep. salivarius ssp. thermophilus ATCC 19258 and Lb. delbrueckii spp. bulgaricus ATCC 11842		+	Inhibition of linoleic acid and plasma lipid peroxidation	Scavenge of reactive oxygen species	Ou et al.[70]
Lb. acidophilus KCTC 3111, Lb. jonsonnii KCTC 3141, Lb. acidophilus KCTC 3151, Lb. brevis KCTC 3498	+	+[a]	Inhibition of linoleic acid peroxidation, scavenging of reactive oxygen species	Scavenge of reactive oxygen species	Kim et al.[44]
Lb. fermentum	+		DPPH scavenging activity	Scavenge of reactive DPPH	Wang et al.[102]
Lb. casei ssp. casei SY13 and Lb. delbrueckii ssp. bulgaricus LJJ	+		Inhibition of linoleic acid peroxidation, DPPH scavenging activity	Scavenge of reactive DPPH and oxygen species	Zhang et al.[113]
7 Bifidobacterium, 11 Lactobacillus, 6 Lactococcus, and 10 Streptococcus thermophilus strains	+		Inhibition of linoleic acid peroxidation and ascorbic acid oxidation	Scavenge of reactive DPPH and oxygen species	Amaretti et al.[5]

WC, whole cells; C-FE, cell-free extracts; +, type of food-grade LAB sample analyzed.
[a]Significantly higher activity obtained compared to other samples analyzed in the same study.

Other studies regarding the evaluation of antioxidant activity of several commercial starter cultures have been reported. Kim et al.[43] evaluated the antioxidant effect of cell-free extracts of *Lb. acidophilus* LA5 and LA100, *Lb. casei* 01, *Lb. bulgaricus* LB207 and *Lb. rhamnosus* GG744. Authors found that their antioxidant capacity varied, ranging from 38% to 81% of lipid peroxidation inhibition among all assessed strains. In a related research, Yoon and Byun[111] reported similar results for LAB antioxidant capacities by inhibition of lipid peroxidation, using cell-free extracts of 13 probiotic *Lactobacillus* strains. Their findings demonstrated that inhibitory activity ranged from 5% to 70% and also found a positive correlation (correlation coefficient = 0.65, $n = 11$) among cellular GSH content in the cell-free extract and antioxidant activity, which suggested that this molecule may partly be responsible for the inhibitory activity on linoleic acid peroxidation. Ou et al.[70] assessed the antioxidant capacity of cell-free extract of starter cultures of yogurt (*Strep. salivarius* ssp. *thermophilus* ATCC 19258 and *Lb. delbrueckii* spp. *bulgaricus* ATCC 11842), using two-model system based on the inhibition of lipid peroxidation, namely free fatty acid, linoleic acid, biological lipid, and plasma lipid. The results showed that both systems demonstrated potentially higher antioxidant capacity, with values of inhibition percentage ranging from 56.9% to 61.4% and 40.6% to 57.2% for linoleic acid and plasma lipid systems, respectively.

In another study, Zhang et al.[113] assessed the antioxidant capacity of starter cultures isolated from traditional yogurt, specifically *Lb. casei* subsp. *casei* SY13 and *Lb. delbrueckii* subsp. *bulgaricus* LJJ. Their results showed that whole cells and cell-free extract exhibited DPPH scavenging activity and inhibition of linoleic acid peroxidation, in which whole cells showed a significantly ($P < 0.05$) higher mean regarding DPPH scavenging activity that ranged from 10% to 27.5%), compared to cell-free extract that ranged from 3% to 18%. Results showed that there were no statistical differences in linoleic acid peroxidation inhibition percentages among whole cells and cell-free extracts. Lee et al.[52] evaluated the antioxidant capacity of whole cells and cell-free extracts of *Lb. rhamnosus* GG, *Lb. casei* KCTC 3260, *Lb. casei* KCTC 3109, and *Lb. casei* 01. Cell-free extracts showed significantly ($P < 0.05$) higher mean percentage of linoleic acid peroxidation activity inhibition that ranged from 37.4% to 57.1%; than whole cells, which ranged from 9.8% to 46.2%. Similarly, Saide and Gilliland[81] also assessed the antioxidant capacity of whole cells and cell-free extract of some species of *Lactobacillus* to protect a protein from free radicals by using the oxygen radical absorbance capacity method. Results showed that cell-free extracts had significantly ($P < 0.05$) higher mean antioxidant capacity, which ranged from 212 to 6070 µM Trolox equivalents/10^9 cells; than whole cells, that ranged from 127 to 3152 µM Trolox equivalents/10^9 cells. This behavior on higher antioxidant capacity of cell-free extracts compared to whole cells was consistent with previous reports by Lin and Chang,[56] Kullisaar et al.,[46] and Lee et al.[52]

Wang et al.[102] demonstrated the antioxidant capacity of *Lb. fermentum* as whole cells adjusted to different bacterial concentrations (10^6–10^9 CFU/ml). The results showed that antioxidant capacity increased in a concentration-dependent manner, for example, at 10^6 CFU/ml *Lb. fermentum* showed 64.26% scavenging capacity while at 10^9 *Lb. fermentum* showed 87.89% scavenging capacity. Recently, Amaretti et al.[5] evaluated the antioxidant capacity of 34 strains of LAB (7 *Bifidocaterium*, 11 *Lactobacillus*, 6 *Lactococcus*, and 10 *Strep. thermophilus*). The authors found a wide dispersion in values of ascorbic and linoleic acid oxidation, suggesting that antioxidative properties were strain-specific.

In vitro antioxidative studies described above have shown that antioxidant properties of LAB are strain-specific. Moreover, most of the in vitro antioxidant assays suggest that the mechanism of antioxidant action may occur mainly by scavenging ROS and/or by chelating transition metal ions, such as iron and copper, which are involved in ROS/RNS formation. On the other hand, most studies proved that the cell-free extracts of various LAB strains exhibit higher antioxidant capacity than whole cells, which suggest that the main compounds responsible for antioxidant properties might be SOD, GPx, NADH-oxidase, NADH-peroxide, and GSH; however, further studies are needed to identify the precise antioxidant compounds involved, as well to elucidate the molecular mechanism underlying their antioxidant activity. Furthermore, despite the fact that in vitro methods are popular given their swiftness and sensitivity, these methods do not consider certain parameters under complex cell environments, such as bioavailability and membrane permeability. Therefore, the traditional indexes of chemical antioxidant capacity might poorly reflect the antioxidant effects in vivo.

6.3.4 *IN VIVO STUDIES OF THE PROTECTIVE EFFECT OF FOOD-GRADE LACTIC ACID BACTERIA AGAINST OXIDATIVE STRESS*

6.3.4.1 *STUDIES USING ANIMAL MODELS*

Yadah et al.[108] evaluated the effect of Dahi, an Indian-style yogurt-containing *Lb. acidophilus* NCDC14 and *Lb. casei* NCDC19, on streptozotocin-induced oxidative damaged rats. The authors reported that Dahi significantly suppressed the oxidative damage in pancreatic tissue by inhibition of lipid peroxidation and formation of nitric oxide, and by preserving an antioxidant pool (activities of SOD, CAT, GPx, and GSH content). In a similar study, Arvind Sinha et al.[9] studied the effect of probiotic Dahi curd containing *Lb. acidophilus* and *Lb. casei*; compared to normal curd containing *Lc. lactis* spp. *lactis* biovar. *diacetylactis*, on lipid peroxidation and intestinal carcinogenesis induced by 1,2-dimethylhydrazine in rats. Results showed higher effectiveness of the probiotic Dahi curd that inhibited lipid peroxidation in both liver and colon in comparison with the normal curd.

Kumar et al.[48] evaluated the effect of probiotic fermented milk, containing *Lb. rhamnosus* GG and *Lb. casei* Shirota, on aflatoxin B_1-induced hepatocellular carcinoma in rats. A significant reduction of lipid peroxidation and enhanced activate antioxidant enzymes such as GPx, SOD, CAT, and GSH-S-transferase were observed in the group with probiotic fermented milk when compared with the aflatoxin B_1 group. Moreover, Ishii et al.[38] studied the effect of oral administration of *Bb. breve* on UV-induced skin damage in a hairless mouse model. The results showed that *Bb. breve* significantly suppressed hydrogen peroxide levels, proteins and lipid oxidation, and xanthine oxidase activity in the skin. The authors suggested that *Bb. breve* attenuated UV-induced oxidative stress in the skin due to the prevention of ROS generation.

In a related study, Sun et al.[92] evaluated the protective effect of different *Lactobacillus* strains, *Lb. rhamnosus* LGG, *Lb. paracasei* Fn032 and *Lb. plantarum* Fn001, against Fe-induced colon oxidative stress in mice. The authors observed that *Lb. rhamnosus* LGG significantly inhibited the increase of colonic mucosal free radicals and malondialdehyde content, a lipid peroxidation indicator, while *Lb. paracasei* Fn032 only inhibited malondialdehyde content. These data suggest that different *Lactobacillus* strains have specific free-radical scavenging activities. In another study, Sengül et al.[85] evaluated the protective effect of EPS produced by the probiotic strains *Lb. delbrueckii* subsp. *bulgaricus* B3 or A13, against gut oxidative damage, present during colitis induced by of acetic acid dispensed into rat colon using a polypropylene catheter placed 8 cm from the anus. The results evidenced a higher antioxidant enzyme activity (SOD, CAT) as well as higher total GSH, reduced GSH and GSH disulfide level probiotic-treated groups, compared to the control group. Conversely, a lower lipid peroxidation was observed. These findings suggest that probiotic bacteria may significantly attenuate oxidative stress in a rat colitis model.

In vivo studies using animal models described above have shown that some LAB have antioxidant properties by increasing the activity of antioxidant enzymes and decreasing lipid peroxidation in tissues (e.g., liver, pancreas, and colon). These results suggested that specific LAB are able to interact with their host as either whole cells or by the releasing different metabolites and/or their intracellular content, thereby improving the total antioxidant status (TAS) of the animal host, which may be relevant to a wide range of human pathologies that involve oxidative damage. Despite the efficacy found in animal studies, these might not necessarily be translated to human trials. Animal models can serve as an important source of in vivo information but alternative translational approaches are needed to establish links between animal models assays and clinical applications.

6.3.4.2 CLINICAL TRIALS

Human studies have been carried out in order to confirm the suitability and efficacy of the microorganism (Table 6.2). This may allow their use in food products as well as to determine the antioxidant effect at a physiological level.[47] Some food-grade LAB have the ability to resist the transit through the gastrointestinal tract and adhere to the intestinal epithelium cells, thereby lining and colonizing the lumen of the tract. These LAB produce different metabolites, released compounds during cellular lysis such as enzymes or themselves, whole cells or intracellular component as antioxidants. Thus, food-grade LAB function as antioxidants to their host by supplying potential antioxidants during the cell life cycle in the gastrointestinal tract.[5]

Mohammadi et al.[66] studied the effect of probiotic yoghurt, containing *Lb. acidophilus* and *Bb. lactis* and multispecies probiotic capsule such as *Lb. casei, Lb. acidophilus, Lb. rhamnosus, Lb. bulgaricus, Bb. breve, Bb. longum,* and *Strep. thermophilus.* Yoghurt supplementation (100 g/day) was studied on biomarkers of oxidative stress of petrochemicals workers (probiotic yoghurt group, n = 12; conventional yoghurt group, $n = 10$), who were exposed to benzene, toluene, ethylbenzene, xylenes, and diesel. After 6 weeks of supplementation, a significant decrease ($P < 0.05$) of iso-prostaglandin levels in plasma was observed in both groups when compared to baseline levels. In addition, the group (n = 13) receiving multispecies probiotic capsule showed a significant decrease ($P < 0.05$) of protein carbonyl levels in plasma. In another study, Marterelli et al.[64] evaluated the effect of two probiotic strains (*Lb. rhamnosus* IMC 501 and *Lb. paracasei* IMC 502) administered (~10^9 cells/day) to athletes (probiotic group, $n = 12$; control group, $n = 12$) during intense training. Because of their intense physical activity oxygen consumption increased, thereby an excess of circulating ROS were produced. After 4 weeks of treatment, a significant decrease ($P < 0.05$)

TABLE 6.2 Studies Using Animal and Human Models.

Strains	Model	Results	Reference
Lb. casei, Lb. acidophilus, Lb. rhamnosus, Lb. bulgarius, Bb. longum, Strep. thermophilus	Petrochemical workers	In blood: ↓ Protein carbonyl ↓ Isoprostaglandin	Mohammadi et al.[66]
Lb. rhamnosus IMC 501, Lb. paracasei IMC 502	Athletes during intense exercise training	In blood: ↓ Reactive oxygen metabolites ↑ Biological antioxidant potential	Marterelli et al.[64]
Bb. bifidum W23, Bb. lactis W51, E. faecium W54, Lb. acidophilus W22, Lb. brevis W63, Lc. lactis W58	Athletes during intense exercise training	In blood: ↓ Protein carbonyl	Lamprecht et al.[49]
Bb. breve	Mice with skin damage	Skin: ↓ Protein oxidation ↓ Lipid peroxidation	Ishii et al.[38]
Lb. rhamnosus GG, Lb. casei Shirota	Rats with hepatocellular carcinoma	Liver: ↓ Lipid peroxidation ↑ Antioxidant enzymes activity	Kumar et al.[48]
Lb. rhamnosus LGG, Lb. paracasei Fn032 and Lb. plantarum Fn001	Rats with intestinal carcinogenesis	Colon: ↓ Lipid peroxidation	Sun et al.[92]
Lb. acidophilus and Lb. casei	Rats with intestinal carcinogenesis	Liver and colon: ↓ Lipid peroxidation	Arvind Sinha et al.[9]
Lb. acidophilus NCDC14, Lb. casei NCDC19	Rats with diabetes	Pancreas: ↓ Lipid peroxidation ↑ Antioxidant enzymes activity	Yadah et al.[108]

TABLE 6.2 (*Continued*)

Strains	Model	Results		Reference
Bacillus polyfermenticus	Rats with intestinal carcinogenesis	Blood:	↓ Lipid peroxidation	Park et al.[74]
			↑ Total antioxidant level	
Bacillus polyfermenticus SCD	High-fat and cholesterol fed rat	Blood:	↓ Conjugated dienes	Paik et al.[71]
			↑ Total antioxidant level	
Strep. thermophilus YIT 2001	Mice with an iron overloaded colon	Colon:	↓ Lipid peroxidation	Ito et al.[39]

of reactive oxygen metabolites levels was observed. In contrast, an increase of biological antioxidant in plasma was noted. In a similar a study, a reduced level of protein carbonyl in plasma was measured in athletes supplied with a mixture of probiotic strains, after 14 weeks of intense exercise.[49]

Consumption of fermented goat milk with Lb. fermentum ME-3 (150 g/day) by healthy volunteers ($n = 16$) showed an increase of total antioxidative status and a decrease in oxidative stress markers.[47] Similarly, Ejtahed et al.[23] described that probiotic yoghurt (300 g/day) containing Lb. acidophilus La5 and Bb. lactis Bb12 consumed by people with type 2 diabetes (probiotic group, $n = 30$; control group, $n = 30$) for 6 weeks significantly decreased fasting blood glucose and increased SOD and GPx activities, when compared to conventional yogurt consumption. Furthermore, the peroxidation lipid concentration significantly decreases in both groups. However, CAT activity remained unchanged in the treatment group, while the TAS increased significantly compared with conventional yogurt consumption. Previous studies by Chamari et al.[15] and Songisepp et al.[91] also reported a significant increment in TAS levels in healthy subjects by the consumption of probiotic yoghurt (300 g/day, $n = 60$) during 6 weeks and probiotic fermented milk (150 ml/day, $n = 21$) or capsules (9 log CFU/day, $n = 23$) during 3 weeks, respectively. However, these results differ from those reported by Asemi et al.[11] and Fabian and Elmadfa,[25] who did not find a significant increase in TAS in healthy pregnant women (probiotic group, $n = 37$; control group, $n = 33$) and young healthy women (probiotic group, $n = 17$; control group, $n = 16$) by the consumption of 200 and 100 g/day of probiotic yoghurt, respectively. A possible explanation for these findings could be attributed to differences on the experimental period and also to the amount of fermented milk consumed by the subjects.

All the evidence described above strongly suggests that food-grade LAB may reduce oxidative damage, ROS-free radicals and modulate cellular defense antioxidant enzymes, which may contribute to prevent and control different oxidative stress-associated diseases. However, potential protective effects should be first established using antioxidant in vitro methods and subsequently in vivo studies using animal models; thus, the gathered information may eventually lead to human clinical trials.

6.4 CONCLUDING REMARKS

Recent scientific evidence suggests the importance of food-grade LAB as potential antioxidants, which entails that these bacteria, as part of the daily diet could play an important role in the prevention of chronic degenerative diseases associated with oxidative stress. However, detailed mechanistic studies must be carried out to better understand the antioxidant involvement of bacterial cell-surface and/or intracellular components in their protective effect, such that more food-grade LAB can be uncovered. Furthermore, it is still a huge challenge to identify the fate of antioxidant components of LAB and their precise biological activity once entering the digestion process and crossing to the circulatory system. Prior to the advancement of functional products against oxidative stress-related diseases onto the market, the technofunctional properties of food-grade lactic bacteria as new antioxidant food products must be assessed, since unknown changes in cell structure, organization, and composition of bacteria may occur during food formulation, processing, and storage. Such changes may significantly influence the possible antioxidant benefits of the bacteria.

KEYWORDS

- lactic acid bacteria
- free radicals
- oxidative stress
- bioactive peptides
- glutathione

REFERENCES

1. Abreu, I. A.; Cabelli, D. E. Superoxide Dismutases—A Review of the Metal-Associated Mechanistic Variations. *Biochim. Biophys. Acta* **2010**, *1804*, 263–274.
2. Adams, L.; Franco, M. C.; Estevez, A. G. Reactive Nitrogen Species in Cellular Signaling. *Exp. Biol. Med. (Maywood)* **2015**, *240*, 711–717.
3. Ahotupa, M.; Saxelin, M.; Korpela, R. Antioxidative Properties of *Lactobacillus* GG. *Nutr. Today* **1996**, *31*, 51–52.
4. Amara, F.; Berbenni, M.; Fragni, M.; Leoni, G.; Viggiani, S.; Ippolito, V. M.; Larocca, M.; Rossano, R.; Alberghina, L.; Riccio, P.; Colangelo, A. M. Neuroprotection by Cocktails of Dietary Antioxidants under Conditions of Nerve Growth Factor Deprivation. *Oxid. Med. Cell Longv.* **2015**. doi:10.1155/2015/217258.
5. Amaretti, A.; di Nunzio, M.; Pompei, A.; Raimondi, S.; Rossi, M.; Bordoni, A. Antioxidant Properties of Potentially Probiotic Bacteria: *In Vitro* and *In Vivo* Activities. *Appl. Microbiol. Biotechnol.* **2013**, *97*, 809–817.

6. Andrade, J. C.; Ascencao, K.; Gullón, P.; Henriques, S. M. S.; Pinto, J. M. S.; Rocha-Santos, T. A.; Freitas, A. C.; Gomes, A. M. Production of Conjugated Linoleic Acid by Food-Grade Bacteria: A Review. *Int. J. Dairy Technol.* **2012,** *65,* 467–481.

7. André, C. M.; Larondelle, Y.; Evers, D. Dietary Antioxidants and Oxidative Stress from a Human and Plan Perspective: A Review. *Curr. Nutr. Food Sci.* **2010,** *6,* 2–12.

8. Antolovich, M.; Prenzler, P. D.; Patsalides, E.; McDonald, S.; Robards, K. Methods for Testing Antioxidant Activity. *Analyst* **2002,** *127,* 183–198.

9. Arvind Sinha, P. R.; Singh, N. K.; Kumar, R. Effects of *Acidophilus casei* Dahi (Probiotic Curd) on Lipids in 1,2-Dimethylhydrazine Induced Intestinal Cancer in Rats. *Int. J. Probiotics Prebiotics* **2009,** *4,* 195–200.

10. Ascherio, A.; Weisskopf, M. G.; O'Reilly, E. J; Jacobs, E. J.; McCullough, M. L.; Calle, E. E.; Cudkowicz, M. Thun, M. J. Vitamin E Intake and Risk of Amyotrophic Lateral Sclerosis. *Ann. Neurol.* **2005,** *57,* 104–110.

11. Asemi, Z.; Jazayeri, S.; Najafi, M.; Samimi, M.; Mofid, V.; Shidfar, F.; Shakeri, H.; Esmaillzadeh, A. Effect of Daily Consumption of Probiotic Yogurt on Oxidative Stress in Pregnant Women: A Randomized Controlled Clinical Trial. *Ann. Nutr. Metab.* **2012,** *60,* 60–62.

12. Bettuzzi, S.; Brausi, M.; Rizzi, F.; Castagnetti, G.; Peracchia, G.; Corti, A. Chemoprevention of Human Prostate Cancer by Oral Administration of Green Tea Catechins in Volunteers with High-Grade Prostate Intraepithelial Neoplasia: A Preliminary Report from a One-Year Proof-of-Principle Study. *Cancer Res.* **2006,** *66,* 1234–1240.

13. Bodamyali, T.; Stevens, C. R.; Blake, D. R.; Winyard, P. G. Reactive Oxygen/Nitrogen Species and Acute Inflammation: A Physiological Process. In *Free Radicals and Inflammation*; Winyard, P. G., Blake, D. R., Evans, C. H., Eds.; Springer Science and Business Media: *Birkhauser*-Verlag Basel, 2000; 259 p.

14. Bouayed, J.; Bohn, T. Exogenous Antioxidants-Double-Edged Swords in Cellular Redox State. *Oxid. Med. Cell. Longev.* **2010,** *3,* 228–237.

15. Chamari, M.; Djazayery, A.; Jalali, M.; Yeganeh, H. S.; Hosseini, S.; Heshmat, R.; Haeri, B. B. The Effect of Daily Consumption of Probiotic and Conventional Yogurt on Some Oxidative Stress Factors in Plasma of Young Healthy Women. *ARYA Atheroscler.* **2008,** *4,* 175–179.

16. Chen, H.; Zhang, M.; Xie, B. Quantification of Uronic Acids in Tea Polysaccharide Conjugates and their Antioxidant Properties. *J. Agric. Food Chem.* **2004,** *52,* 3333–3336.

17. Chen, H.; Zhang, M.; Qu, Z.; Xie, B. Antioxidant Activities of Different Fractions of Polysaccharide Conjugates from Green Tea (*Camelia sinensis*). *Food Chem.* **2008,** *106,* 559–563.

18. Choan, E.; Segal, R.; Jonker, D.; Malone, S.; Reaume, N.; Eapen, L.; Gallant, V. A Prospective Clinical Trial of Green Tea for Hormone Refractory Prostate Cancer: An Evaluation of the Complementary/Alternative Therapy Approach. *Urol. Oncol.* **2005,** *23,* 108–113.

19. Dalle-Donne, I.; Rosii, R.; Colombo, R.; Giustarini, D.; Milzani, A. Biomarkers of Oxidative Damage in Human Disease. *Clin. Chem.* **2006,** *52,* 601–623.

20. Davis, C. D.; Tsuji, P. A.; Milner, J. A. Selenoproteins and Cancer Prevention. *Ann. Rev. Nutr.* **2012,** *32,* 73–95.

21. de Lau, L. M.; Koudstaal, P. J.; Witterman, J. C.; Hofman, A.; Breteler, M. M. Dietary Folate, Vitamin B12, and Vitamin B6 and the Risk of Parkinson Disease. *Neurology* **2006,** *67,* 315–318.

22. Drögue, W. Free Radicals in the Physiological Control of Cell Function. *Physiol. Rev.* **2002**, *82*, 47–95.

23. Ejtahed, H. S.; Mohtadi-Nia, J.; Homayouni-Rad, A.; Niafar, M.; Asghari-Jafarabadi, M.; Mofid, V.; Akbarian-Moghari, A. Effect of Probiotic Yogurt Containing *Lactobacillus acidophilus* and *Bifidobacterium lactis* on Lipid Profile in Individuals with Type 2 Diabetes Mellitus. *J. Dairy Sci.* **2011**, *94*, 3288–3294.

24. Engelhart, M. J.; Geerlings, M. I.; Ruitenberg, A.; van Swieten, J. C.; Hofman, A.; Witterman, J. C.; Breteler, M. M. Dietary Intake of Antioxidants and Risk of Alzheimer Disease. *JAMA* **2002**, *287*, 3223–3229.

25. Fabian, E.; Elmadfa, I. The Effect of Daily Consumption of Probiotic and Conventional Yoghurt on Oxidant and Antioxidant Parameters in Plasma of Young Healthy Women. *Int. J. Vitam. Nutr. Res.* **2007**, *77*, 79–88.

26. Finkel, T.; Holbrook, N. J. Oxidants, Oxidative Stress and the Biology of Ageing. *Nature* **2000**, *408*, 239–247.

27. Finkel, T. Oxidant Signals and Oxidative Stress. *Curr. Opin. Cell Biol.* **2003**, *15*, 247–254.

28. Finkel, T. Signal Transduction by Reactive Oxygen Species. *J. Cell Biol.* **2011**, *194*, 7–15.

29. Geleijnse, J. M.; Launer, L. J.; Van der Kuip, D. A.; Holfman, A.; Witterman, J. C. Inverse Association of Tea and Flavonoid Intakes with Incident Myocardial Infraction: The Rotterdam Study. *Am. J. Clin. Nutr.* **2002**, *75*, 880–886.

30. Ginter, E.; Simko, V.; Panakova, V. Antioxidants in Health and Disease. *Bratisl. Lek. Listy.* **2013**, *115*, 603–606.

31. Girard, M.; Schaffer-Lequart, C. Gelation and Resistance to Shearing of Fermented Milk: Role of Exopolysaccharides. *Int. Dairy J.* **2007**, *17*, 666–673.

32. Gutowski, M.; Kowalczyk, S. A Study of Free Radical Chemistry: Their Role and Pathophysiological Significance. *Acta Biochim. Pol.* **2013**, *60*, 1–16.

33. Ha, Y. L.; Storkson, J.; Pariza, M. W. Inhibition of Benzo(*a*)pyrene-Induced Mouse for Stomach Neoplasia by Conjugated Dienoic Derivatives of Linoleic Acid. *Cancer Res.* **1990**, *50*, 1097–1101.

34. Ham, J. S.; In, Y. M.; Jeong, S. G.; Kim, J. G.; Lee, E. H.; Kim, H. S.; Yoon, S. K.; Lee, B. H. Screening of Conjugated Linoleic Acid Producing Lactic Acid Bacteria from Fecal Samples of Healthy Babies. *Asian Australas. J. Anim. Sci.* **2002**, *15*, 1031–1035.

35. Han, W.; Merceneir, A.; Ait-Belgnaoui, A.; Pavan, S.; Lamine, F.; van Swam, I. I.; Kleerebezem, M.; Salvador-Cartier, C.; Hisbergues, M.; Bueno, L.; Theodorou, V.; Fioramonti, J. Improvement of an Experimental Colitis in Rats by Lactic Acid Bacteria Producing Superoxide Dismutase. *Inflamm. Bowel Dis.* **2006**, *12*, 1044–1052.

36. Henríquez-Sánchez, P.; Sánchez-Villegas, A.; Ruano-Rodríguez, C.; Gea, A.; Lamuela-Raventós, R. M.; Estruch, R.; Salas-Salvadó, J.; Covas, M. I.; Corella, D.; Schröder, H.; Gutiérrez-Bedmar, M.; Santos-Lozano, J. M.; Pintó, X.; Arós, F.; Fiol, M.; Tresserra-Rimbau, A.; Ros, E.; Martínez-González, M. A.; Serra-Majem, L. Dietary Total Antioxidant Capacity and Mortality in the PREDIMED Study. *Eur. J. Nutr.* **2016**, *55*, 227–236.

37. Huxley, R. R.; Neil, H. A. The Relation between Dietary Flavonol Intake and Coronary Heart Disease Mortality: A Meta-Analysis of Prospective Cohort Studies. *Eur. J. Clin. Nutr.* **2003**, *57*, 904–908.

38. Ishii, Y.; Sugimoto, S.; Izawa, N.; Sone, T.; Chiba, K.; Miyazaki, K. Oral Administration of *Bifidobacterium breve* Attenuates UV-Induced Barrier Perturbation and Oxidative Stress in Hairless Mice Skin. *Arch. Dermatol. Res.* **2014**, *306*, 467–473.

39. Ito, M.; Ohishi, K.; Yoshida, Y.; Yokoi, W.; Sawada, H. Antioxidative Effects of Lactic Acid Bacteria on the Colonic Mucosa of Iron-Overloaded Mice. *J. Agric. Food Chem.* **2003,** *51,* 4456–4460.

40. Jay, D.; Hitomi, H.; Griendling, K. K. Oxidative Stress and Diabetic Cardiovascular Complications. *Free Radic. Biol. Med.* **2006,** *40,* 183–192.

41. Jiang, J.; Björck, L.; Fondén, R. Production of Conjugated Linoleic Acid by Dairy Starter Cultures. *J. Appl. Microbiol.* **1998,** *85,* 95–102.

42. Khan, N.; Adhami, V. M.; Mukhtar, H. Review: Green Tea Polyphenols in Chemoprotection of Prostate Cancer: Preclinical and Clinical Studies. *Nutr. Cancer* **2009,** *61,* 836–841.

43. Kim, H. S.; Chae, H. S.; Jeong, S. G.; Ham, J. S.; Im, S. K.; Ahn, C. N.; Lee, J. M. Antioxidant Activity of Some Yogurt Starter Cultures. *Asian-Australas. J. Anim. Sci.* **2005,** *18,* 255–258.

44. Kim, H. S.; Chae, H. S.; Jeong, S. G.; Ham, J. S.; Im, S. K.; Ahn, C. N.; Lee, J. M. *In Vitro* Antioxidative Properties of Lactobacilli. *Asian-Australas. J. Anim. Sci.* **2006,** *19,* 262–265.

45. Klauning, J. E.; Kamendulis, L. M. The Role of Oxidative Stress in Carcinogenesis. *Annu. Rev. Pharmacol. Toxicol.* **2004,** *44,* 239–267.

46. Kullisaar, T.; Zilmer, M.; Mikelsaar, M.; Vihalemm, T.; Annuk, H.; Kairane, C.; Kilk, A. Two Antioxidative Lactobacilli Strains as Promising Probiotics. *Int. J. Food Microbiol.* **2002,** *72,* 215–224.

47. Kullisaar, T.; Songisepp, E.; Milkesarr, M.; Zilmer, K.; Vihalemm, T.; Zilmer, M. Antioxidative Probiotic Fermented Goats' Milk Decreases Oxidative Stress-Mediated Atherogenicity in Human Subjects. *Br. J. Nutr.* **2003,** *90,* 449–456.

48. Kumar, M.; Verma, V.; Nagpal, R.; Kumar, A.; Behare, P. V.; Singh, B.; Aggarwal, P. K. Anticarcinogenic Effect of Probiotic Fermented Milk and Chlorophyllin on Aflatoxin-B-Induced Liver Carcinogenesis in Rats. *Br. J. Nutr.* **2012,** *107,* 1006–1016.

49. Lamprecht, M.; Bogner, S.; Schippinger, G.; Steinbauer, K.; Fankhauser, F.; Hallstroem, S.; Schuettz, B. Probiotic Supplementation Affects Markers of Intestinal Barrier, Oxidation, and Inflammation in Trained Men: A Randomized, Double-Blinded, Placebo-Controlled Trial. *J. Int. Soc. Sports Nutr.* **2012,** *9,* 45.

50. Laws, A.; Gu, Y.; Marshall, V. Biosynthesis, Characterization, and Design of Bacterial Exopolysaccharides from Lactic Acid Bacteria. *Biotechnol. Adv.* **2001,** *19,* 597–625.

51. Lee, J.; Koo, N.; Min, D. B. Reactive Oxygen Species, Aging, and Antioxidative Nutraceuticals. *Compr. Rev. Food Sci.* **2004,** *3,* 21–33.

52. Lee, J.; Hwang, K. Y.; Chung, M. Y.; Cho, D. H.; Park, C. S. Resistance of *Lactobacillus casei* KCTC 3260 to Reactive Oxygen Species (ROS): Role for a Metal Ion Chelating Effect. *J. Food Sci.* **2005,** *70,* 388–391.

53. Leroy, F.; De Vuyst, L. Lactic Acid Bacteria as Functional Starter Cultures for the Food Fermentation Industry. *Trends Food Sci. Technol.* **2004,** *15,* 67–78.

54. Li, S.; Zhao, Y.; Zhang, L.; Zhang, X.; Huang, L.; Li, D.; Niu, C.; Yang, Z.; Wang, Q. Antioxidant Activity of *Lactobacillus plantarum* Strains Isolated from Traditional Chinese Fermented Foods. *Food Chem.* **2012,** *135,* 1914–1919.

55. Li, W.; Ji, J.; Rui, X.; Yu, J.; Tang, W.; Chen, X.; Jiang, M.; Dong, M. Production of Exopolysaccharides by *Lactobacillus helveticus* MB2-1 and its Functional Characteristics *In Vitro*. *LWT-Food Sci. Technol.* **2014,** *59,* 732–739.

56. Lin, M. Y.; Chang, F. J. Antioxidative Effect of Intestinal Bacteria *Bifidobacterium longum* ATCC 15708 and *Lactobacillus acidophilus* ATCC 4356. *Dig. Dis. Sci.* **2000,** *45,* 1617–1622.

57. Lin, M. Y.; Yen, C. L. Antioxidative Ability of Lactic Acid Bacteria. *J. Agric. Food Chem.* **1999**, *47*, 1460–1466.

58. Liu, J.-R.; Chen, M.-J.; Lin, C.-W. Antimutagenic and Antioxidant Properties of Milk-Kefir and Soymilk-Kefir. *J. Agric. Food Chem.* **2005**, *53*, 2467–2474.

59. Liu, C. S; Nam, T. G.; Han, M. W.; Ahn, S. M.; Choi, H. S.; Kim, T. Y.; Chun, O. K.; Koo, S. I.; Kim, D. O. Protective Effect of Detoxified *Rhus verniciflua* Stokes on Human Keratinocytes and Dermal Fibroblasts against Oxidative Stress and Identification of the Bioactive Phenolics. *Biosci. Biotechnol. Biochem.* **2013**, *77*, 1682–1688.

60. Liu, C. F.; Pan, T. M. *In Vitro* Effects of Lactic Acid Bacteria on Cancer Cell Viability and Antioxidant Activity. *J. Food Drug Anal.* **2010**, *18*, 77–86.

61. Liu, C. F.; Tseng, K. C.; Chiang, S. S.; Lee, B. H.; Hsu, W. H.; Pan, T. M. Immunomodulatory and Antioxidant Potential of *Lactobacillus* Exopolysaccharides. *J. Sci. Food Agric.* **2011**, *91*, 2284–2291.

62. Lobo, V.; Patil, A.; Phatak, A.; Chandra, N. Free Radicals, Antioxidants and Functional Foods: Impact on Human Health. *Pharmacogn. Rev.* **2010**, *4*, 118–126.

63. Lushchak, V. I. Free Radicals, Reactive Oxygen Species, Oxidative Stress and Its Classification. *Chem. Biol. Interact.* **2014**, *224*, 164–175.

64. Marterelli, D.; Verdenelli, M. C.; Scuri, S.; Cocchioni, M.; Silvi, S.; Cecchini, C.; Pompei, P. Effect of a Probiotic Intake on Oxidant and Antioxidant Parameters in Plasma of Athletes during Intense Exercise Training. *Curr. Microbiol.* **2011**, *62*, 1689–1696.

65. Matés, J. M.; Segura, J. A.; Alonso, F. J.; Márquez, J. Intracellular Redox Status and Oxidative Stress: Implications for Cell Proliferation, Apoptosis, and Carcinogenesis. *Arch. Toxicol.* **2008**, *82*, 273–299.

66. Mohammadi, A. A.; Jazayeri, S.; Khosravi-Darani, K.; Solati, Z.; Mohanmmadpour, N.; Asemi, Z.; Adab, Z.; Djalali, M.; Tehrani-Doost, M.; Hosseini, M.; Eghtesadi, S. Effects of Probiotics on Biomarkers of Oxidative Stress and Inflammatory Factors in Petrochemical Workers: A Randomized, Double-Blind, Placebo-Controlled Trail. *Int. J. Prev. Med.* **2015**, *4*, 83–88.

67. Mustacich, D.; Powis, G. Thioredoxin Reductase. *Biochem. J.* **2000**, *346*, 1–8.

68. Oigo, K.; Wang, D. H. Biomarkers of Oxidative/Nitrosative Stress: An Approach to Disease Prevention. *Acta Med. Okayama* **2007**, *61*, 181–189.

69. Osuntoki, A.; Korie, I. Antioxidant Activity of Whey from Milk Fermented with *Lactobacillus* Species Isolated from Nigerian Fermented Foods. *Food Technol. Biotech.* **2010**, *48*, 505–511.

70. Ou, C. C.; Ko, J. L.; Lin, M. Y. Antioxidative Effects of Intracellular Extracts of Yogurt Bacteria on Lipid Peroxidation and Intestine 407 Cells. *J. Food Drug Anal.* **2006**, *3*, 304–310.

71. Paik, H. D.; Park, J. S.; Park, E. Effects of *Bacillus polyfermenticus* SCD on Lipid and Antioxidant Metabolisms in Rats Fed a High-Fat and High-Cholesterol Diet. *Biol. Pharm. Bull.* **2005**, *28*, 1270–1274.

72. Pan, D.; Mei, X. Antioxidant Activity of an Exopolysaccharide Purified from *Lactococcus lactis* subsp. *lactis* 12. *Carbohydr. Polym.* **2010**, *80*, 908–914.

73. Papadimitriou, K.; Zoumpopoulou, G.; Foligné, B.; Alexandraki, V.; Kazou, M.; Pou, B.; Tsakalidou, E. Discovering Probiotic Microorganisms: *In Vitro*, *In Vivo*, Genetic and Omics Approaches. *Front. Microbiol.* **2015**, *6*, 58.

74. Park, E.; Jeon, G. I.; Park, J. S.; Paik, H. D. A Probiotic Strain of *Bacillus polyfermenticus* Reduces DMH Induced Precancerous Lesions in F344 Male Rat. *Biol. Pharm. Bull.* **2007**, *30*, 569–574.

75. Pisoschi, A. M.; Pop, A. The Role of Antioxidants in the Chemistry of Oxidative Stress: A Review. *Eur. J. Med. Chem.* **2015**, *97*, 55–74.

76. Rahman, K. Studies on Free Radicals, Antioxidants, and Co-factors. *Clin. Interv. Aging* **2007**, *2*, 219–236.

77. Rahman, M. M.; Hossain, K. M.; Rahman, S. M. M. Isolation, Characterization, and Properties Study of Probiotic Lactic Acid Bacteria of Selected Yoghurt from Bangladesh. *Afr. J. Microbiol. Res.* **2016**, *10*, 23–31.

78. Ramesh, V.; Kumar, R.; Singh, R. R. B.; Kaushik, J. K.; Mann, B. Comparative Evaluation of Selected Strains of Lactobacilli for the Development of Antioxidant Activity in Milk. *Dairy Sci. Technol.* **2012**, *92*, 179–188.

79. Reuter, S.; Gupta, S. C.; Chaturvedi, M. M.; Aggarwal, B. B. Oxidative Stress, Inflammation, and Cancer: How Are They Linked?. *Free Radic. Biol. Med.* **2010**, *49*, 1603–1616.

80. Rochat, T.; Bermúdez-Humarán, L.; Gratadoux, J. J.; Fourage, C.; Hoebler, C.; Corthier, G.; Langella, P. Anti-inflammatory Effects of *Lactobacillus casei* BL23 Producing or Not a Manganese-Dependent Catalase on DSS-Induced Colitis in Mice. *Microb. Cell Fact.* **2007**, *6*, 1–10.

81. Saide, J. A. O.; Gilliland, S. E. Antioxidative Activity of Lactobacilli Measured by Oxygen Radical Absorbance Capacity. *J. Dairy Sci.* **2005**, *88*, 1352–1357.

82. Sarmadi, B. H.; Ismail, A. Antioxidative Peptides from Food Proteins: A Review. *Peptides* **2010**, *31*, 1949–1956.

83. Sakuma, S.; Abe, M.; Kohda, T.; Fujimoto, Y. Hydrogen Peroxide Generated by Xanthine/Xanthine Oxidase System Represses the Proliferation of Colorectal Cancer Cell Line Caco-2. *J. Clin. Biochem. Nutr.* **2015**, *56*, 15–19.

84. Seifried, H. E.; Anderson, D. E.; Fisher, E. I.; Milner, J. A. A Review of the Interaction among Dietary Antioxidants and Reactive Oxygen Species. *J. Nutr. Biochem.* **2007**, *18*, 567–579.

85. Sengül, N.; Isik, S.; Aslim, B.; Ucar, G.; Demirbag, A. The Effect of Exopolysaccharide-Producing Probiotic Strains on Gut Oxidative Damage in Experimental Colitis. *Dig. Dis. Sci.* **2011**, *56*, 707–714.

86. Shah, N. P. Functional Cultures and Health Benefits. *Int. Dairy J.* **2007**, *17*, 1262–1277.

87. Shafi, A.; Farooq, U.; Akram, K.; Hayat, Z.; Murtaza, A. Prevention and Control of Diseases by Use of Pro- and Prebiotics (Synbiotics). *Food Rev. Int.* **2014**, *30*, 291–316.

88. Shimamura, S.; Abe, F.; Ishibashi, N.; Miyakawa, H.; Yaeshima, T.; Araya, T.; Tomita, M. Relationship between Oxygen Sensitivity and Oxygen Metabolism of *Bifidobacterium* Species. *J. Dairy Sci.* **1992**, *75*, 3296–3306.

89. Sieber, R.; Collomb, M.; Aeschlimann, A.; Jelen, P.; Eyer, H. Impact of Microbial Cultures on Conjugated Linoleic Acid in Dairy Products—A Review. *Int. Dairy J.* **2004**, *14*, 1–15.

90. Sindhi, V.; Gupta, V.; Sharma, K.; Bhatnagar, S.; Kumari, R.; Dhaka, N. Potential Application of Antioxidants—A Review. *J. Pharm Sci.* **2013**, *7*, 828–835.

91. Songisepp, E.; Kals, J.; Kullisaar, T.; Mändar, R.; Hütt, P.; Zilmer, M.; Mikelsaar, M. Evaluation of the Functional Efficacy of an Antioxidative Probiotic in Healthy Volunteers. *Nutr. J.* **2005**, *4*, 22–32.

92. Sun, J.; Hu, X. L.; Le, G. W.; Shi, Y. H. Inhibition of Fe-Induced Colon Oxidative Stress by Lactobacilli in Mice. *World J. Microbiol. Biotechnol.* **2013**, *29*, 209–216.

93. Thomas, C.; Versalovic, J. Probiotics–Host Communication: Modulation of Signaling Pathways in the Intestine. *Gut Microbes* **2010**, *1*, 148–163.

94. Trachootham, D.; Lu, W.; Ogasawara, M. A.; Valle, R. D. V.; Huang, P. Redox Regulation of Cell Survival. *Antioxid. Redox Signal.* **2008,** *10,* 1343–1374.
95. Tsang, A. H.; Chung, K. K. Oxidative and Nitrosative Stress in Parkinson's Disease. *Biochim. Biophys. Acta* **2009,** *1792,* 643–650.
96. Valko, M.; Leibfritz, D.; Moncol, J.; Cronin, M. T.; Mazur, M.; Telser, J. Free Radicals and Antioxidants in Normal Physiological Functions and Human Disease. *Int. J. Biochem. Cell Biol.* **2007,** *39,* 44–84.
97. Valavanidis, A.; Vlachogianni, T.; Fiotakis, K.; Loridas, S. Pulmonary Oxidative Stress, Inflammation and Cancer: Respirable Particulate Matter, Fibrous Dust and Ozone as Major Causes of Lung Carcinogenesis through Reactive Oxygen Species Mechanisms. *Int. J. Environ. Res. Public Health* **2013,** *10,* 3886–3907.
98. Van Nieuwenhove, C. P.; Oliszewski, R.; González, S. N.; Pérez Chaia, A. B. Conjugated Linoleic Acid Conversion by Dairy Bacteria Cultured in MRS Broth and Buffalo Milk. *Lett. Appl. Microbiol.* **2007,** *44,* 467–474.
99. Vance, T. M.; Azabdaftari, G.; Pop, E. A.; Lee, S. G.; Su, L. J.; Fontham, E. T.; Bensen, J. T.; Steck, S. E.; Arab, L.; Mohler, J. L.; Chen, M. H.; Koo, S. I.; Chun, O. K. Intake of Dietary Antioxidants is Inversely Associated with Biomarkers of Oxidative Stress among Men with Prostate Cancer. *Br. J. Nutr.* **2016,** *115,* 68–74.
100. Virtanen, T.; Pihlanto, A.; Akkanen, S.; Korhonen, H. Development of Antioxidant Activity in Milk Whey during Fermentation with Lactic Acid Bacteria. *J. Appl. Microbiol.* **2007,** *102,* 106–115.
101. Waltz, P.; Escobar, D.; Botero, A. M.; Zuckerbraun, B. S. Nitrate/Nitrite as Critical Mediators to Limit Oxidative Injury and Inflammation. *Antioxid. Redox Signal.* **2015,** *23,* 328–329.
102. Wang, A. N.; Yi, X. W.; Yu, H. F.; Dong, B.; Qiao, S. Y. Free Radical Scavenging Activity of *Lactobacillus fermentum In Vitro* and its Antioxidative Effect on Growing-Finishing Pigs. *J. Appl. Microbiol.* **2009,** *107,* 1140–1148.
103. Wang, J.; Hu, S.; Nie, S.; Yu, Q.; Xie, M. Reviews on Mechanisms of *In Vitro* Antioxidant Activity of Polysaccharides. *Oxid. Med. Cell. Longev.* **2015.** doi:10.1155/2016/5692852.
104. Weidinger, A.; Kozlov, A. V. Biological Activities of Reactive Oxygen and Nitrogen Species: Oxidative Stress *versus* Signal Transduction. *Biomolecules* **2015,** *5,* 472–484.
105. Wessels, S.; Axelsson, L.; Bech Hansen, E.; De Vuyst, L.; Laulund, S.; Lähteenmäki, L.; Lindgren, S.; Mollet, B.; Salminen, S.; Wright, A. The Lactic Acid Bacteria, the Food Chain, and their Regulation. *Trends Food Sci. Technol.* **2004,** *15,* 498–505.
106. Wolvers, D.; Antoine, J. M.; Myllyluoma, E.; Schrezenmeir, J.; Szajewska, H.; Riijkers, T. Guidance for Substantiating the Evidence for Beneficial Effects of Probiotics: Prevention and Management of Infections by Probiotics. *J. Nutr.* **2010,** *140,* 698–712.
107. Xu, R.; Shang, N.; Li, P. *In Vitro* and *In Vivo* Antioxidant Activity of Exopolysaccharide Fractions from *Bifidobacterium animalis* RH. *Anaerobe* **2011,** *17,* 226–231.
108. Yadah, H.; Jain, S.; Sinha, P. R. Oral Administration of Dahi Containing *Lactobacillus acidophilus* and *Lactobacillus casei* Delay the Progression of Streptozoocin-Induced Diabetes in Rats. *J. Dairy Sci.* **2008.** 75, 189–195.
109. Yang, H. Y.; Lee, T. H. Antioxidant Enzymes as Redox-Based Biomarkers: A Brief Review. *BMB Rep.* **2015,** *48,* 200–208.
110. Yi, Z.; Fu, Y.; Li, M.; Gao, K.; Zhang, X. Effect of LTA Isolated from Bifidobacteria on D-Galactose-Induced Aging. *Exp. Gerontol.* **2009,** *44,* 760–765.

111. Yoon, Y. H.; Byun, J. R. Occurrence of Glutathione Sulphydryl (GSH) and Antioxidant Activities in Probiotic *Lactobacillus* spp. *Asian-Austral. J. Anim. Sci.* **2004,** *17,* 1582–1585.

112. Yu, L.; Adams, D.; Gabel, M. Conjugated Linoleic Acid Isomers Differ in their Free Radical Scavenging Properties. *J. Agric. Food Chem.* **2002,** *50,* 4135–4140.

113. Zhang, S.; Liu, L.; Su, Y.; Li, H.; Sun, Q.; Liang, X.; Lv, J. Antioxidative Activity of Lactic Acid Bacteria in Yogurt. *Afr. J. Microbiol. Res.* **2011,** *5,* 5194–5201.

114. Zhou, J.; Hu, N.; Wu, Y. L.; Pan, Y. J.; Sun, C. R. Preliminary Studies on the Chemical Characterization and Antioxidant Properties of Acidic Polysaccharides from *Sargassum fusiforme. J. Zhejiang Univ. Sci. B* **2008,** *9,* 721–727.

CHAPTER 7

MYCOTOXINS

SUZANNE HENDRICH[*]

Food Science and Human Nutrition, Iowa State University, Ames IA 50011-4149, United States

[*]*E-mail: shendric@iastate.edu*

CONTENTS

ABSTRACT

Research activity on mycotoxins of major concern to human health (afla-toxins (AFB), deoxynivalenol (DON), fumonisins, ochratoxins (OTA)) in recent years has emphasized mechanisms of action, mitigation strategies, and novel detection methods. Risk assessment of these mycotoxins is not well studied. DON exposure may be of somewhat more concern than the other three mycotoxins, although some world regions have much greater concern for AFB exposure than other regions. Metabolism of AFB, DON, and OTA may be modifiable in humans, and for DON, in plants. Micro-bial detoxification of mycotoxins holds some promise for toxin mitigation. Development of mycotoxin-resistant food crop species is ongoing, as well as work on "green" technologies emphasizing biocontrol of mycotoxins. Food system resource managers should carefully consider enhancing efforts to monitor and control mycotoxins.

7.1 INTRODUCTION

Mycotoxins, toxins produced by fungi that commonly contaminate food crops, remain an important global food safety concern. Aflatoxins and fumonisins mainly pose a cancer risk, whereas deoxynivalenol (DON) poses a risk to gastrointestinal and immune function. Ochratoxin A (OTA) poses a risk for kidney disease. Grains and some legumes are the predominant sources of these toxins, but they vary in the range of foods that they contaminate. For example, fumonisins occur mainly in corn, whereas DON is mainly found in wheat, barley, and corn. Aflatoxins are mainly found in peanuts and corn. The nature of the fungi that produce each toxin seems to be the main determinant of which crop species will be the main sources of the mycotoxins.

Aflatoxins, most importantly aflatoxin B1 (AFB1), are produced, and named for *Aspergillus flavus*, but other *Aspergillus* species also produce aflatoxins, especially *Aspergillus parasiticus*. Crosses of these two fungi produce greater amounts of aflatoxins than do either parent species, but the two species are typically isolated from each other, with *A. flavus* infecting peanuts, corn, cottonseed, and tree nuts and *A. parasiticus* infecting mainly peanuts.[1] Aflatoxin B1 is a human liver carcinogen and is also involved in impairing growth, development, and immune function of children in regions with significant aflatoxin contamination of staple foods.[2]

Fumonisins are produced by at least 15 *Fusarium* species, especially *Fusarium verticillioides*, *Fusarium proliferatum*, and *F. subglutinans*. These fungi are corn pathogens, causing stalk rot as well as potentially harmful levels of the predominant fumonisin, B1 in corn kernels.[3] Fumonisin B1 has been associated with human esophageal cancer and neural tube defects,[3] especially in regions where corn is a staple food and where contamination of corn by this toxin is not well recognized and managed. In vivo studies of *Fusarium* mycotoxins have been reviewed recently, showing a broad array of effects across many species.[4]

DON is mainly produced by *Fusarium graminearum* and also by *Fusarium culmorum*.[5] These fungi cause *Fusarium* head blight in wheat, a main source of this toxin. Other cereals, such as barley and corn, can also be significant DON sources. DON is linked with immune dysfunction and gastroenteritis, hence its prior common name, vomitoxin.[6]

Ochratoxins are produced by *Aspergillus ochraceus*, *Aspergillus carbonarius*, and *Penicillium verrucosum*. *A. ochraceus* grows and produces ochratoxin mainly in stored grains under dry conditions and in moderate temperatures. *A. carbonarius* grows in grapes, so ochratoxins may be found in wines and other grape-derived foods. *P. verrucosum* grows well in cooler climates, so northern European and North American grains, especially wheat, experience ochratoxin contamination mainly from this source.[7] OTA is the main ochratoxin important to human health and is associated with nephritic syndrome, but only in regions with very high exposure to OTA, such as in parts of Egypt and Sierra Leone.[6]

A recent casual survey of scientific literature through PubMed indicates significant research activity, especially focused on mechanisms and mitigation of toxicity of aflatoxins and DON, fungal biology associated with fumonisin production, and novel detection methods for ochratoxins (Table 7.1). How this pattern of research activity aligns with public health needs associated with these toxins will be discussed in summarizing key recent studies related to mycotoxin risk assessment, metabolism, and mitigation. Mycotoxigenic fungi will continue to evolve, so continual improvement of techniques to identify and assess health risks of emerging mycotoxins is needed but is seemingly not being addressed systematically at this time.

7.2   MYCOTOXIN RISK ASSESSMENT

Connecting human health risks with dietary exposure to mycotoxins poses severe challenges. Outbreaks of acute illness are associated with aflatoxin B1 (aflatoxicosis causing hepatic toxicity), DON (gastroenteritis), and OTA

TABLE 7.1 Survey of Recent Scientific Papers Published in English on Mycotoxins Catalogued by PubMed from January to May 2016.

Mycotoxin	Total papers	Quantitation in foods	Novel detection methods	Exposure and risk assessment	Mechanisms of action and mitigation of toxicity	Detoxification in foods	Fungal biology
Aflatoxins	161	26 (15%)	35 (20%)	15 (9%)	49 (28%)	17 (10%)	19 (11%)
Deoxynivalenol	91	11 (12%)	10 (11%)	4 (4%)	35 (37%)	5 (5%)	28 (30%)
Fumonisins	64	7 (11%)	13 (20%)	4 (6%)	16 (25%)	4 (6%)	20 (31%)
Ochratoxin	110	12 (11%)	40 (36%)	7 (6%)	26 (24%)	9 (8%)	16 (15%)

(nephritic syndrome). Verification of mycotoxin outbreaks requires myco-toxin analysis of grain samples verified to be of the same lot or source as ingested immediately prior to the onset of illness. Blood or urinary myco-toxin analysis and assessment of disease symptoms are also required, concomitantly. Although numerous methods are available for mycotoxin analysis, most such methods require expensive instrumentation such as LC/MS. Medical personnel with appropriate diagnostic expertise are also required. Public health systems coordinating such efforts are largely lacking worldwide. To assess cancer risk from aflatoxins and fumonisins, much longer term exposure surveillance is required. For a genotoxic agent such as aflatoxin, exposure in early life causing genetic damage may result in much later development of cancer. For fumonisins, chronic exposure seems to be required for its carcinogenic effects. There is yet an incomplete under-standing of human dietary exposure patterns for mycotoxins in regions where mycotoxin-related health concerns exist. It may be that OTA causes kidney impairment at lower doses than seen in nephritic syndrome, but establishing this as a solid connection requires multivariate analysis coordi-nated across human populations. Likewise, DON may impair immune and intestinal function in important but relatively subtle ways that are difficult to discern. Increased global scientific cooperation and coordination are crucial to address these needs. It is unfortunate that disease presence rather than disease prevention seems to drive investment in such endeavors. Myco-toxin prevention systems that are sustainable will need to include permanent investment in agricultural practices, health surveillance, and basic and trans-lational research. Mitigation of human health risks from mycotoxins is ethi-cally and practically important. Global burden from mycotoxin-associated diseases was estimated recently at ~200,000 excess liver cancer cases per year attributable to aflatoxin. Disease burdens from fumonisin, DON, and OTA remain uncertain, however likely, especially for fumonisins.[6] Effects of ingestion of combinations of mycotoxins also needs greater attention.

Recent studies of dietary exposure to aflatoxin modeling intake of three maize foods based on aflatoxin analysis of these foods in regions of Kenya. Eating whole kernel maize would result in a 5–10-fold greater exposure to aflatoxin than eating maize meal or muthokoi (dehulled and processed maize), about 300 ng aflatoxin/kg body weight. This exposure is 1000-fold greater than noted in the United States.[8]

In a study in Lebanon, mean aflatoxin B1 exposure was 0.63 ng/kg/day, extrapolating to an increased risk of cancer of ~0.05 cases/100,000 indi-viduals,[9] a relatively low additional risk. In a survey of aflatoxin intake from foods in Malaysia, mean aflatoxin intake was much greater, about 30 ng/

kg/day, contributing ~0.7 liver cancer cases/100,000 individuals. With the current maximum limit for aflatoxin of 15 ppb in Malaysia, this finding indicates some need for continued vigilance in limiting intake of foods contaminated with aflatoxin above that maximum.[10]

From the most recent French Total Diet Study, only DON exposure and not exposure to aflatoxin, fumonisin, or OTA exceeded the health-based guidance value (HBGV) of estimated intake of 1000 ng DON/kg/day. Only 0.5% of adults and 5% of children exceeded this estimated DON intake. Mean DON exposures were estimated at ~400 ng/kg/day for adults and ~550 ng/kg/day for children from this study.[11] This study should be seen as a model for other countries to better assess health risks from mycotoxins.

A total diet study of urban Lebanese showed that mean DON intake exceeded the European Food Safety Authority's (EFSA) HBGV (1000 ng/kg/day) at 1560 ng/kg/day, whereas mean OTA intake of 4.3 ng/kg/day was 80% of EFSA's HBGV.[9] When exposure to DON was combined with exposures to 3- and 15-acetyl DON in a case study of 1269 individuals in Shanghai, China, mean DON exposure from these three forms slightly exceeded the HBGV at 1085 ng/kg/day.[12] More work on public health effects of such findings related to DON are needed.

A Tunisian case control study of 69 women with breast cancer and 41 controls showed significantly greater urinary α-zearalenol in women with breast cancer, with mean concentration of 4.6 ng/mL, threefold greater than in controls.[13] This estrogenic metabolite of zearalenone might enhance growth of estrogen-responsive breast cancer cells. This study suggests that it would be worth studying the extent to which urinary α-zearalenol might predict breast cancer risk. However, a recent study biomonitoring mycotoxins in Belgium showed that only 1 adult out of 239 studied had any urinary content of α-zearalenol, and this metabolite was not detected in children ($n = 155$).[14] The study in Belgium implies that in countries with more highly developed food safety systems, zearalenone would not pose a human breast cancer risk. A study associating zearalenone exposure with reproductive development in one hundred and sixty-three 9–10-year-old girls in New Jersey showed mean urinary α-zearalenol 10-fold less than seen in Tunisian women, and lesser breast development in girls with greater zearalenone exposure,[15] suggesting an antiestrogenic effect of the mycotoxin at these exposure levels. It is intriguing to consider further work to investigate possible breast cancer protective effects of zearalenone at exposures similar to those noted above in Belgium or the United States.

7.3   MYCOTOXIN METABOLISM

The metabolism of mycotoxins, by animals, bacteria associated with the gut, and by plants may be a significant factor in mitigating health risks of these compounds, but this aspect of mycotoxins has not been incorporated directly into risk assessment or mitigation strategies. It may be that dietary and other health habits of populations either enhance or inhibit mycotoxin detoxification. Such possibilities will be explored in this chapter section.

Among the four major mycotoxins, aflatoxin is known to undergo significant mammalian metabolism, both in activation to its proximate carcinogenic (mutagenic) form, aflatoxin 8,9-epoxide, by cytochromes P-450 (P450),[16] and its detoxification, especially by glutathione-S-transferases (GSTs) to transform the epoxide site's to a hydroxyl and a glutathione adduct.[17] P450s are inducible by dietary components including flavonoids[18] and cruciferous vegetables such as broccoli and cabbage.[19] Chronic food restriction also may induce P450s,[20] suggesting enhanced susceptibility to AFB1 toxicity in regions where food shortages and undernutrition are more common. Paradoxically, P450s may also be inhibited by flavonoids.[21] Some flavonoids such as apigenin, a flavonoid in parsley, inhibited AFB1 mutagenicity in vitro mediated by the human P450 enzyme thought to be important for AFB1 activation, hCYP1A2.[22] The significance of this finding for prevention of AFB1-associated human cancers remains to be determined. Several studies have shown in animal models the possible mitigation of aflatoxin toxicity and carcinogenesis by dietary alterations of its metabolism. Marked inhibition of AFB1 carcinogenesis in rainbow trout, the most sensitive species to AFB1, was shown for beta-naphthoflavone (BNF) and indole-3-carbinole (a component of cruciferous vegetables) but only BNF induced P450 in this model.[23] This early study illustrated the complexity of attempting to prevent AFB1 carcinogenicity by dietary components, as mediated by modulation of P450. Chickens fed 100 ppb AFB1 showed induction of P450, which was prevented by supplementation with 0.5 mg selenium (Se)/kg diet compared with 0.2 mg Se/kg, suggesting that diets containing this moderately greater amount of Se might mitigate the activation of AFB1. This approach may be feasible to investigate as a human intervention in regions where AFB1 contamination is common. Dietary induction of GSTs as a strategy to mitigate AFB1 carcinogenicity has been shown using a model antioxidant, oltipraz, in rats.[24] Oltipraz increased production of AFB-glutathione metabolites in a human clinical trial, demonstrating the feasibility of this approach.[25] The identification of effective dietary inducers of GSTs that can mitigation AFB1 toxicity in humans remains to be accomplished. It has been recently proposed that strategies not involving AFB1 metabolism, such as increasing dietary chlorophyllin content, which binds to and inhibits absorption of AFB1, may be more useful to consider because altering P450s and GSTs is likely to alter metabolism of many drugs, thus making public policy recommendations about such dietary constituents highly problematic.[26]

DON is also metabolized by inducible biotransformation in animals. In particular, the hydroxyls of DON are sites for addition of sulfate (by sulfotransferases [STs]) or glucuronide. Glucuronidation by UDP-glucuronosyltransferases (UGTs) is favored in species possessing both types of biotransformation enzymes, based on limited data.[27] Human UGTs have less capability to form DON glucuronides than do rat UGTs in vitro, but such metabolites are the predominant urinary excretion products across species.[28] DON-3-glucuronide was shown to have negligible toxicity compared with DON in human K562 cells, consistent with the general idea in toxicology that such metabolites are detoxification products.[29] Because UGTs are highly inducible by some dietary constituents, such induction may mitigate DON toxicity in humans. Neither the extent of UGT induction nor the effect of this phenomenon on DON toxicity has been established yet in humans.

The biotransformation of DON in plants to DON glucosides, especially DON-3-glucoside (D3G) has been observed.[30] Many hydroxylated secondary plant metabolites are also stored in plants in glucoside form. This conversion of DON initially was shown to "mask" DON to its detection. Since then, D3G has been recognized as a minor but not insignificant form of DON in DON-contaminated grains, constituting as much as 25% of total DON in wheat and maize.[31] D3G can be readily converted back to DON by bacterial β-glucosidases in the mammalian gut. D3G per se is practically unabsorbed, so the absorption of DON from dietary D3G would occur mainly in the ileum and colon which contain most of the bacteria in the intestine. Enhanced presence of D3G in the diet could alter the site of intestinal toxicity. The development of grains that have increased ability to convert DON to D3G would not be advisable unless DON de-epoxidation capacity, and hence DON detoxification was also commonly occurring. This has been demonstrated in rats, in which the urinary excretion of D3G was fivefold less than that seen for DON; most D3G was excreted as DON or de-epoxy DON (DOM-1) in rat feces.[32] When DON was fed to pigs as D3G, its apparent bioavailability was about twofold less.[33] DON de-epoxidation in the rumen is the main fate of DON in cattle,[34] which seems to be why ruminants are relatively protected from DON toxicity. De-epoxidation in the lower intestine of pigs is also common, but this has no protective benefit from DON toxicity because DON seems to be absorbed in the small intestine before the

DON de-epoxidating bacteria can be effective.[35] Likewise, in one study of French farmers, about 30% of the humans tested had DON de-epoxidating activity in fecal bacteria.[36] A lesser extent of this metabolism was observed in individuals from the United Kingdom.[37] De-epoxidation of DON in humans would not be expected to mitigate DON toxicity appreciably unless DON were present mainly as D3G which is not currently the case. If we presume that DON can be rapidly and extensively converted to DON glucuronides, the DON glucuronides would be expected to be eliminated mainly in bile. These metabolites would not be reabsorbed until they were converted back to DON by bacterial glucuronidases in the lower intestines. At that point in DON metabolism DON might be detoxified by gut bacterial DON de-epoxidases. It may be worth exploring the feasibility of modifying the human-gut microbiome to include DON de-epoxidating bacteria in individuals who do not naturally carry such bacterial species. Some such species have been identified and might be seen as a new class of probiotics, potentially beneficial bacteria that might be introduced into the food supply. The need for such alteration of human-gut bacteria remains to be established and would not be a trivial process. But if such a need were confirmed (in humans who do not already have this metabolic capability in their gut microbiomes), standards exist for assuring the efficacy and safety of probiotic bacteria.[38]

OTA is the main ochratoxin of concern to human health. It can be hydrolyzed by proteases to form an apparently nontoxic form, ochratoxin-alpha, and phenylalanine. The lack of toxicity of OTA-alpha has been demonstrated in zebrafish recently.[39] The percentage of ochratoxin absorbed in humans has not been directly determined, presumably due to ethical concerns about deliberately exposing humans to this presumed carcinogen, but across several species, uptake has been estimated at ~50% of ingested dose.[40] In limited human studies, OTA-alpha seemed to be the predominant urinary form of OTA. In one study, human urinary contents of OTA and OTA-alpha were about equal.[41] Pregnant women showed about 10-fold greater urinary OTA-alpha than OTA,[42] indicating seemingly greater OTA detoxification ability of pregnant women than of nonpregnant women. Hydrolysis of OTA by microbial enzymes may be a strategy for mitigation of the mycotoxin,[43] but the capability of enhancing such hydrolysis in vivo in humans remains to be determined.

The metabolism of fumonisins has been shown to be virtually nil in vivo, as might be expected for these highly water-soluble, relatively large, and, therefore, poorly absorbable toxins. Seemingly their water-solubility facilitates their rapid excretion and poor retention in body tissues, as has been demonstrated by studies of the fate of radio-labeled FB1 in a rodent model.[44]

These compounds may be altered during some food processing reactions (e.g., addition of reducing sugars to the primary amine of FB1) and by microbial enzymes (e.g., carboxylesterase FumD) that are potentially useful in mitigating the toxins.[45]

In summary, it may be worth considering incorporating alteration of human/gut microbial metabolism of some mycotoxins in future mitigation strategies. Insofar, as the future may hold the ability for genetic characterization of individual biotransformation enzyme genetics and polymorphisms, and thus the prospect of tailoring diet to contain the right mix of biotransformation enzyme inducers or inhibitors depending on dietary circumstances, metabolism modification may need to be part of the defense arsenal against these toxins.

7.4   MYCOTOXIN MITIGATION

Mitigating the presence of mycotoxins in the human food supply is mostly about reducing mycotoxin burden in grain crops used for human food. Significant attention is also directed toward reducing mycotoxins in the feed of livestock, for example, in the case of aflatoxin, due to carry over of the aflatoxin metabolite AFM1 into dairy milk. Mitigation strategies involve improving mycotoxin detection and regulation, which currently means removing from the food supply foods that exceed action, advisory, or guidance levels developed by the US Food and Drug Administration (FDA). Analogous standards that may be somewhat more or less stringent exist in many countries, including the European Union (EU). An action level is mandated by FDA only for aflatoxins in the United States, currently set at 20 ppb for foods for human consumption. FDA advisory levels are set at 1 ppm for DON in foods for human consumption, and FDA has also provided guidance levels for fumonisins of 2–4 ppm for foods for human consumption.[46]

A strong research publication focus has been on innovative detection methods, but this research does not seem to be well-aligned with the needs, especially in low income countries, for rapid, accurate, and inexpensive mycotoxin detection. Mycotoxin analytical methods have been recently reviewed[47] and major constraints in this field were noted, including the varied chemistries of the mycotoxins, the need to assess multiple mycotoxins in food samples and to assure that samples are appropriately representative of the scope of possible contamination, and the need for speed and economy. QuEChERs (quick, easy, cheap, effective, rugged, and safe) technologies were also noted to be especially important in the realm of mycotoxin analysis. Portability of analytical methods is improving, with a number of promising advances coming from the realm of nanotechnology coupled with alternatives to antibody-based mycotoxin detection. Such alternatives include aptamers (RNA-binding) and molecular imprint polymers. Nanomaterial sensors for mycotoxins including AFB1, DON, FB1, and OTA have been developed.[48] Because many nanomaterials do not occur in nature, particular caution in assessing safety related to disposal, environmental persistence, and potential health effects on humans and ecosystems is warranted. Spectroscopic detection coupled with chromatographic separation methods of varying types and expense remain the state-of-the-art in terms of reliability, but portability of spectroscopy is also improving.[47] A few recent studies on mycotoxin detection show promise. An aptamer-based dipstick for AFB1 was shown to have comparable detectability compared with a standard ELISA method in the ppb range, as needed for food samples. The method took 30 min to complete, with simple solvent extraction (20% methanol) of grain samples including maize.[49] An antibody-based microarray system for simultaneous detection of AFB1 and FB1 was shown to be feasible and comparable to standard ELISAs in detection levels. This method will require further validation for food samples.[50] A portable evanescent wave optical aptasensor with a reversible ligand-grafted biosensing surface was demonstrated for OTA, with detection limit of 0.4 ppb, and OTA recoveries from powdered wheat of 89–106%, with ~15% CV. This detection limit is sufficient to meet current regulatory policies,[51] but this seemingly relatively cost-effective and reusable method will need further validation across food sources of OTA. DON-specific nanobodies (single-domain antibodies) that can mimic DON have been recently discovered and might be useful in further optimizing DON detection.[52] The adoption of the Food Safety Modernization Act (FSMA) in the United States in 2011 emphasizes prevention of food contamination. It remains to be seen how FSMA will affect mycotoxin detection, but rapid, reliable, and inexpensive methods available to farmers are likely to be needed, thus promising emerging technologies such as these will be crucial.

Preventing mycotoxins in the field is a burden for grain producers that currently rely on their ability to identify fungal contamination and insofar as feasible and permissible to apply appropriate fungicides. Fungicide resistance is an ongoing concern, as well as general environmental and human health concern about use of synthetic chemical fungicides, so potentially toxigenic fungi-inhibiting plants and their extracts are under investigation as alternatives.[53] Commercialization remains to be achieved; significant technical and economic barriers exist in this field of "green chemicals."

Identification and development of mycotoxin-resistant crop varieties has shown particular promise in maize, a species for which at least a few naturally occurring variants are resistant to AFB1.[54] AFB1 resistance associated proteins have been identified, and current genomic technologies may permit engineering of such proteins into other crop species. But no commercially available AFB1-resistant maize lines are yet available.[54] Several cross bred maize lines were recently identified as resisting both AFB1 and FB1 contamination in field trials in South Africa, in which at least a few crosses were developed that did not accumulate AFB1 above 5 ppb or FB1 above 4 ppm (current regulatory levels).[55] For DON in wheat, the quantitative trait locus Fhb1 permitted conjugation of DON with glucose and several glucose derivatives as well as glutathione conjugates, significantly increasing D3G/DON ratio.[56] As noted in the above section on DON metabolism, this conversion would not be expected to significantly detoxify DON unless DON-de-epoxidating capability was also present in individuals ingesting this grain. It might be presumed that any DON conjugate, whether with glucose, glutathione, or other glucose derivatives would likely be deconjugated by gut bacteria but that remains to be proven. Barley is another major source of DON; it has been discovered recently that black barley showed about half the DON contamination of yellow barley, so switching to this barley type might be a feasible mitigation approach.[57] A yeast species, *Kluyveromyces thermotolerans* was capable of decreasing OTA in grapes,[58] which may be a significant source of this toxin, so some types of biocontrol may be feasible, but will certainly need to be developed on a crop-specific basis. No work on OTA resistance in grain crop species was uncovered for this review. But progress is being made, seemingly especially for AFB1 and FB1 resistance in maize.

A number of potential strategies to decrease mycotoxins during food processing have been demonstrated in the literature. Current US regulatory policies do not permit blending of a crop contaminated above key limits with noncontaminated crop; exceptions may be made when a severe mycotoxin contamination epidemic occurs. Diversion of mycotoxin contaminated crops into animal feeds may occur where regulatory levels permit.[46] Regulators, scientists, and citizens should engage in effective global discourse about the problem of mycotoxin contamination of crops used to feed humans. It is important to determine a rational future for feeding a world in which mycotoxin contamination is likely to be a problem of increasing severity due to increasingly extreme weather and climate conditions that have been occurring and are predicted. More attention to development of low cost and effective means of decreasing mycotoxins in human foods as a part of food processing is warranted.

Aflatoxin decontamination methods have been developed. Screening grain kernels under UV light which can recognize grain grossly contaminated with aflatoxins and mechanical sorting to cull contaminated kernels is permitted in the United States to achieve grain batches compliant with the action level for aflatoxin. Ammoniation of cottonseed is permitted by FDA. Although this method has been established to effectively detoxify maize containing aflatoxins,[59] with several trials across livestock and laboratory animals showing a reduction in aflatoxin content to 1% or less than in the starting contaminated grain,[59] this method is not approved for grains in the United States. As summarized in a recent review focused on an African perspective on mycotoxin remediation,[60] sorting and cleaning before storage, and keeping stored grain dry may be quite effective in reducing aflatoxin contamination of grains and peanuts. In a study using visibly moldy maize in Malawi, hand sorting to remove obviously damaged or shriveled seeds and seed fragments removed ~95% of AFB1 or FB1. Floating the kernels in water before sorting only removed about 60% of either type of contamination; adding flotation to hand sorting showed no additional benefit.[61] Thus, simple but labor-intensive methods may be beneficial where farmers and consumers are educated about the health benefits of removing aflatoxins from foods. A novel method of treating hazelnuts with cold atmospheric plasma in a controlled pressure chamber using power of 1000 W decreased AFB1 content of the nuts by two-thirds after 12 min.[62] This technique should not interfere with food quality and might be useful for many other AFB-containing foods. The wider feasibility, mainly cost-effectiveness, of such technologies remains for future work.

For DON, as might be expected from its hydrophilicity, processing foods in water such that the water is removed from the final product can remove significant amounts of DON. This is pertinent but probably not practical for pasta. Boiling of 310 g pasta from 0 to 10 min showed progressive loss of DON from 0.62 ppm to 0.16 ppm (75% loss of DON),[63] but as this was "fresh" pasta, eating quality would not be acceptable to many consumers after the longer boiling times that were more effective. Wheat flour bread making and baking did not diminish DON concentrations.[63] Treatment of DON-contaminated dried distillers grain solids for nursery swine feed with 5% sodium metabisulfite,(SMB) autoclaving and drying decreased DON concentration in this feed by more than 80%. Heating DON with SMB causes formation of a DON-10-sulfonate, which was nontoxic to the pigs. Average daily gain was restored to control levels by this treatment.[64] Practically, such treatments of grain flours for human intake might be feasible, but prevention of toxic effects to workers from sulfur dioxide gas release during processing would be important. Also, some individuals may have allergic-like reactions to sulfites as food additives, and warning labels would be needed. Heat processing per se, such as during extrusion of corn flour[65] may remove DON by as much as 98%, but results from another lab with wheat flour did not show this ability of heat processing.[66] It seems prudent to conduct additional studies on SMB using human foods because this may be a cost-effective approach that could be necessary depending on the extent of DON contamination that may emerge in some regions. Additional investigations as to the potential of SMB treatment of DON contaminated grain flours to adversely affect sulfite sensitive individuals, and appropriate additional food labeling may be needed as well.

Regulatory limits for ochratoxins in foods range from 2 to 10 ppb in the EU. Pre-harvest control by good agricultural practices, careful use of fungicides and biocontrol agents (e.g., yeasts, natamycin) are thought to be most effective against OTA,[67] as well as low-moisture storage. Adsorption of OTA from beverages may be feasible but must be evaluated carefully for effects on nutritional quality and taste; modified zeolites may be particularly useful.[67] Quaternary ammonium beta-cyclodextrin was shown *in vitro* to have 200-fold stronger affinity for OTA than beta-cyclodextrin, as measured by fluorescent spectroscopic changes; this cyclodextrin derivative may be a good candidate for pass-through adsorption of OTA from beverages.[68] More practical conditions will need to be investigated for such an adsorbent, as well as determination of any significant adverse effects on beverage quality from use of the adsorbent.

The prospect that human metabolic capabilities may also mitigate health risks from mycotoxins also deserves greater attention, based on the theoretical framework developed above (see Section 7.3).

7.5   RECOMMENDATIONS

Human risk assessment of mycotoxins that includes better recognition of disease and cost burdens of these food-borne toxins is a primary need. Such risk assessment should move toward incorporating the assessment of dietary and other health habits in addition to mycotoxin exposure assessment. A number of dietary constituents, as discussed previously, might mitigate adverse effects of mycotoxins. Exercise is increasingly recognized as a strong factor in mitigation of cancer risk,[69] but taking a global perspective, does intensive physical activity in the case of subsistence farmers confer benefits or add health stress?

Discovery and development of mycotoxin resistant crop species is progressing. This work will need to continue permanently as it is reasonable to consider that mycotoxigenic species will continue to evolve. Integrated pest management systems that employ "green" technologies of biocontrol against mycotoxins must become feasible and affordable in the future.

The recognition of the potential of microbes to degrade and detoxify mycotoxins may extend from the field to the fork, in that antimycotoxin microbes might be developed into a new generation of probiotics that could be incorporated into an array of ready-to-eat food products. Such a recommendation should be approached with great caution and respect for the many unknowns that need careful testing as fundamental discoveries move into product development. Engineering or manipulation of the human-gut microbiome to contain microbes beyond what are naturally present across diverse human populations seems unwise without a great deal more understanding of gut microbial populations and interactions between these microbes and complex food constituents.

A focus on extending adequate resources for mycotoxin management and mitigation to low income world regions must be the greatest priority. Advances in the ability of human populations to effectively govern themselves and abide by fair rules of law will be needed to accomplish the needed eradication of excess liver cancer due to aflatoxin. Humanity deserves better assurance of food safety; mycotoxins remain an important global consideration in that regard.

KEYWORDS

- aflatoxin
- deoxynivalenol
- fumonisin
- ochratoxin
- remediation
- risk assessment

REFERENCES

1. Olarte, R. A.; Worthington, C. J.; Horn, B. W.; Moore, G. G.; Singh, R.; Monacell, J. T.; Dorner, J. W.; Stone, E. A.; Xie, D. Y.; Carbone, I. Enhanced Diversity and Aflatoxigenicity in Interspecific Hybrids of *Aspergillus flavus* and *Aspergillus parasiticus*. *Mol. Ecol.* **2015,** *24* (8), 1889–1909.

2. Wild, C. P.; Miller, J. D.; Groopman, J. D. In *Mycotoxin Control in Low- and Middle-Income Countries*; Wild, C. P., Miller, J. D., Groopman, J. D., Eds.; International Agency for Research on Cancer : Lyon, France, 2015. For more information contact publications@iarc.fr.

3. Alberts, J. F.; van Zyl, W. H.; Gelderblom, W. C. Biologically Based Methods for Control of Fumonisin-Producing Fusarium Species and Reduction of the Fumonisins. *Front. Microbiol.* **2016,** *7,* 548.

4. Escriva, L.; Font, G.; Manyes, L. *In Vivo* Toxicity Studies of Fusarium Mycotoxins in the Last Decade: A Review. *Food Chem. Toxicol.* **2015,** *78,* 185–206.

5. Hellin, P.; Dedeurwaerder, G.; Duvivier, M.; Scauflaire, J.; Huybrechts, B.; Callebaut, A.; Munaut, F.; Legreve, A. Relationship between *Fusarium* spp. Diversity and Mycotoxin Contents of Mature Grains in Southern Belgium. *Food Addit. Contam., A Chem. Anal. Control Expo Risk Assess.* **2016,** *33* (7), 1–13.

6. Wu, F.; Groopman, J. D.; Pestka, J. J. Public Health Impacts of Foodborne Mycotoxins. *Annu. Rev. Food Sci. Technol.* **2014,** *5,* 351–372.

7. Ostry, V.; Malir, F.; Ruprich, J. Producers and Important Dietary Sources of Ochratoxin A and Citrinin. *Toxins (Basel)* **2013,** *5* (9), 1574–1586.

8. Kilonzo, R. M.; Imungi, J. K.; Muiru, W. M.; Lamuka, P. O.; Njage, P. M. Household Dietary Exposure to Aflatoxins from Maize and Maize Products in Kenya. *Food Addit. Contam., A Chem. Anal. Control Expo Risk Assess.* **2014,** *31* (12), 2055–2062.

9. Raad, F.; Nasreddine, L.; Hilan, C.; Bartosik, M.; Parent-Massin, D. Dietary Exposure to Aflatoxins, Ochratoxin A and Deoxynivalenol from a Total Diet Study in an Adult Urban Lebanese Population. *Food Chem. Toxicol.* **2014,** *73,* 35–43.

10. Chin, C. K.; Abdullah, A.; Sugita-Konishi, Y. Dietary Intake of Aflatoxins in the Adult Malaysian Population—An Assessment of risk. *Food Addit. Contam., B: Surveill.* **2012,** *5* (4), 286–294.

11. Sirot, V.; Fremy, J. M.; Leblanc, J. C. Dietary Exposure to Mycotoxins and Health Risk Assessment in the Second French Total Diet Study. *Food Chem. Toxicol.* **2013,** *52,* 1–11.

12. Han, Z.; Nie, D.; Ediage, E. N.; Yang, X.; Wang, J.; Chen, B.; Li, S.; On, S. L.; De Saeger, S.; Wu, A. Cumulative Health Risk Assessment of Co-occurring Mycotoxins of Deoxynivalenol and its Acetyl Derivatives in Wheat and Maize: Case Study, Shanghai, China. *Food Chem. Toxicol.* **2014,** *74,* 334–342.

13. Belhassen, H.; Jimenez-Diaz, I.; Arrebola, J. P.; Ghali, R.; Ghorbel, H.; Olea, N.; Hedili, A. Zearalenone and its Metabolites in Urine and Breast Cancer Risk: A Case-Control Study in Tunisia. *Chemosphere* **2015,** *128,* 1–6.

14. Heyndrickx, E.; Sioen, I.; Huybrechts, B.; Callebaut, A.; De Henauw, S.; De Saeger, S. Human Biomonitoring of Multiple Mycotoxins in the Belgian Population: Results of the BIOMYCO Study. *Environ. Int.* **2015,** *84,* 82–89.

15. Bandera, E. V.; Chandran, U.; Buckley, B.; Lin, Y.; Isukapalli, S.; Marshall, I.; King, M.; Zarbl, H. Urinary Mycoestrogens, Body Size and Breast Development in New Jersey Girls. *Sci. Total Environ.* **2011,** *409* (24), 5221–5227.

16. Croy, R. G.; Essigmann, J. M.; Reinhold, V. N.; Wogan, G. N. Identification of the Principal Aflatoxin B1-DNA Adduct Formed *In Vivo* in Rat Liver. *Proc. Natl. Acad. Sci. U.S.A.* **1978,** *75* (4), 1745–1749.

17. Degen, G. H.; Neumann, H. G. The Major Metabolite of Aflatoxin B1 in the Rat Is a Glutathione Conjugate. *Chem. Biol. Interact.* **1978,** *22* (2–3), 239–255.

18. Siess, M. H.; Guillermic, M.; Le Bon, A. M.; Suschetet, M. Induction of Monooxygenase and Transferase Activities in Rat by Dietary Administration of Flavonoids. *Xenobiotica* **1989,** *19* (12), 1379–1386.

19. Prochaska, H. J.; Santamaria, A. B.; Talalay, P. Rapid Detection of Inducers of Enzymes that Protect against Carcinogens. *Proc. Natl. Acad. Sci. U.S.A.* **1992,** *89* (6), 2394–2398.

20. Sohn, H. O.; Lim, H. B.; Lee, Y. G.; Lee, D. W.; Lee, K. B. Modulation of Cytochrome P-450 Induction by Long-Term Food Restriction in Male Rats. *Biochem. Mol. Biol. Int.* **1994,** *32* (5), 889–896.

21. Moon, Y. J.; Wang, X.; Morris, M. E. Dietary Flavonoids: Effects on Xenobiotic and Carcinogen Metabolism. *Toxicol. In Vitro* **2006,** *20* (2), 187–210.

22. Peterson, S.; Lampe, J. W.; Bammler, T. K.; Gross-Steinmeyer, K.; Eaton, D. L. Apiaceous Vegetable Constituents Inhibit Human Cytochrome P-450 1A2 (hCYP1A2) Activity and hCYP1A2-Mediated Mutagenicity of Aflatoxin B1. *Food Chem. Toxicol.* **2006,** *44* (9), 1474–1484.

23. Nixon, J. E.; Hendricks, J. D.; Pawlowski, N. E.; Pereira, C. B.; Sinnhuber, R. O.; Bailey, G. S. Inhibition of Aflatoxin B1 Carcinogenesis in Rainbow Trout by Flavone and Indole Compounds. *Carcinogenesis* **1984,** *5* (5), 615–619.

24. Kensler, T. W.; Egner, P. A.; Dolan, P. M.; Groopman, J. D.; Roebuck, B. D. Mechanism of Protection against Aflatoxin Tumorigenicity in Rats Fed 5-(2-Pyrazinyl)-4-methyl-1,2-dithiol-3-thione (Oltipraz) and Related 1,2-Dithiol-3-thiones and 1,2-Dithiol-3-ones. *Cancer Res.* **1987,** *47* (16), 4271–4277.

25. Kensler, T. W.; Curphey, T. J.; Maxiutenko, Y.; Roebuck, B. D. Chemoprotection by Organosulfur Inducers of Phase 2 Enzymes: Dithiolethiones and Dithiins. *Drug Metabol. Drug Interact.* **2000,** *17* (1–4), 3–22.

26. Gross-Steinmeyer, K.; Eaton, D. L. Dietary Modulation of the Biotransformation and Genotoxicity of Aflatoxin B(1). *Toxicology* **2012,** *299* (2–3), 69–79.

27. Chen, L.; Yu, M.; Wu, Q.; Peng, Z.; Wang, D.; Kuca, K.; Yao, P.; Yan, H.; Nussler, A. K.; Liu, L.; Yang, W. Gender and Geographical Variability in the Exposure Pattern and Metabolism of Deoxynivalenol in Humans: A Review. *J. Appl. Toxicol.* **2016,** *37* (1), 60–70.

28. Maul, R.; Warth, B.; Schebb, N. H.; Krska, R.; Koch, M.; Sulyok, M. In Vitro Glucuronidation Kinetics of Deoxynivalenol by Human and Animal Microsomes and Recombinant Human UGT Enzymes. *Arch. Toxicol.* **2015,** *89* (6), 949–960.

29. Wu, X.; Murphy, P.; Cunnick, J.; Hendrich, S. Synthesis and Characterization of Deoxynivalenol Glucuronide: Its Comparative Immunotoxicity with Deoxynivalenol. *Food Chem. Toxicol.* **2007,** *45* (10), 1846–1855.

30. Berthiller, F.; Dall'Asta, C.; Schuhmacher, R.; Lemmens, M.; Adam, G.; Krska, R. Masked Mycotoxins: Determination of a Deoxynivalenol Glucoside in Artificially and Naturally Contaminated Wheat by Liquid Chromatography-Tandem Mass Spectrometry. *J. Agric. Food Chem.* **2005,** *53* (9), 3421–3425.

31. Berthiller, F.; Dall'asta, C.; Corradini, R.; Marchelli, R.; Sulyok, M.; Krska, R.; Adam, G.; Schuhmacher, R. Occurrence of Deoxynivalenol and its 3-Beta-D-Glucoside in

Wheat and Maize. *Food Addit. Contam., A Chem. Anal. Control Expo Risk Assess.* **2009,** *26* (4), 507–511.

32. Nagl, V.; Schwartz, H.; Krska, R.; Moll, W. D.; Knasmuller, S.; Ritzmann, M.; Adam, G.; Berthiller, F. Metabolism of the Masked Mycotoxin Deoxynivalenol-3-Glucoside in Rats. *Toxicol. Lett.* **2012,** *213* (3), 367–373.

33. Nagl, V.; Woechtl, B.; Schwartz-Zimmermann, H. E.; Hennig-Pauka, I.; Moll, W. D.; Adam, G.; Berthiller, F. Metabolism of the Masked Mycotoxin Deoxynivalenol-3-Glucoside in Pigs. *Toxicol. Lett.* **2014,** *229* (1), 190–197.

34. Seeling, K.; Danicke, S.; Valenta, H.; Van Egmond, H. P.; Schothorst, R. C.; Jekel, A. A.; Lebzien, P.; Schollenberger, M.; Razzazi-Fazeli, E.; Flachowsky, G. Effects of *Fusarium toxin*-Contaminated Wheat and Feed Intake Level on the Biotransformation and Carry-Over of Deoxynivalenol in Dairy Cows. *Food Addit. Contam.* **2006,** *23* (10), 1008–1020.

35. Danicke, S.; Valenta, H.; Doll, S. On the Toxicokinetics and the Metabolism of Deoxynivalenol (DON) in the Pig. *Arch. Anim. Nutr.* **2004,** *58* (2), 169–180.

36. Turner, P. C.; Hopton, R. P.; Lecluse, Y.; White, K. L.; Fisher, J.; Lebailly, P. Determinants of Urinary Deoxynivalenol and De-epoxy Deoxynivalenol in Male Farmers from Normandy, France. *J. Agric. Food Chem.* **2010,** *58* (8), 5206–5212.

37. Turner, P. C.; Hopton, R. P.; White, K. L.; Fisher, J.; Cade, J. E.; Wild, C. P. Assessment of Deoxynivalenol Metabolite Profiles in UK Adults. *Food Chem. Toxicol.* **2011,** *49* (1), 132–135.

38. Tuomola, E.; Crittenden, R.; Playne, M.; Isolauri, E.; Salminen, S. Quality Assurance Criteria for Probiotic Bacteria. *Am. J. Clin. Nutr.* **2001,** *73* (2 Suppl.), 393s–398s.

39. Haq, M.; Gonzalez, N.; Mintz, K.; Jaja-Chimedza, A.; De Jesus, C. L.; Lydon, C.; Welch, A.; Berry, J. P. Teratogenicity of Ochratoxin A and the Degradation Product, *Ochratoxin alpha*, in the Zebrafish (*Danio rerio*) Embryo Model of Vertebrate Development. *Toxins (Basel)* **2016,** *8* (2), 40.

40. Galtier, P.; Alvinerie, M.; Charpenteau, J. L. The Pharmacokinetic Profiles of Ochratoxin A in Pigs, Rabbits and Chickens. *Food Cosmet. Toxicol.* **1981,** *19* (6), 735–738.

41. Coronel, M. B.; Marin, S.; Tarrago, M.; Cano-Sancho, G.; Ramos, A. J.; Sanchis, V. Ochratoxin A and its Metabolite Ochratoxin Alpha in Urine and Assessment of the Exposure of Inhabitants of Lleida, Spain. *Food Chem. Toxicol.* **2011,** *49* (6), 1436–1442.

42. Klapec, T.; Sarkanj, B.; Banjari, I.; Strelec, I. Urinary Ochratoxin A and Ochratoxin Alpha in Pregnant Women. *Food Chem. Toxicol.* **2012,** *50* (12), 4487–4492.

43. Dobritzsch, D.; Wang, H.; Schneider, G.; Yu, S. Structural and Functional Characterization of Ochratoxinase, a Novel Mycotoxin-Degrading Enzyme. *Biochem. J.* **2014,** *462* (3), 441–452.

44. Dantzer, W. R.; Hopper, J.; Mullin, K.; Hendrich, S.; Murphy, P. A. Excretion of (14) C-Fumonisin B(1), (14)C-Hydrolyzed Fumonisin B(1), and (14)C-Fumonisin B(1)- Fructose in Rats. *J. Agric. Food Chem.* **1999,** *47* (10), 4291–4296.

45. Masching, S.; Naehrer, K.; Schwartz-Zimmermann, H. E.; Sarandan, M.; Schaumberger, S.; Dohnal, I.; Nagl, V.; Schatzmayr, D. Gastrointestinal Degradation of Fumonisin B(1) by Carboxylesterase FumD Prevents Fumonisin Induced Alteration of Sphingolipid Metabolism in Turkey and Swine. *Toxins (Basel)* **2016,** *8* (3). doi:10.3390/toxins8030084.

46. N. G. a. F. Association. FDA Mycotoxin Regulatory Guidance, 2011. https://www.ngfa.org/wp-content/uploads/NGFAComplianceGuide-FDARegulatoryGuidanceforMycotoxins8-2011.pdf.

47. Turner, N. W.; Bramhmbhatt, H.; Szabo-Vezse, M.; Poma, A.; Coker, R.; Piletsky, S. A. Analytical Methods for Determination of Mycotoxins: An Update (2009–2014). *Anal. Chim. Acta* **2015,** *901,* 12–33.

48. Rai, M.; Jogee, P. S.; Ingle, A. P. Emerging Nanotechnology for Detection of Mycotoxins in Food and Feed. *Int. J. Food Sci. Nutr.* **2015,** *66* (4), 363–370.

49. Shim, W. B.; Kim, M. J.; Mun, H.; Kim, M. G. An Aptamer-Based Dipstick Assay for the Rapid and Simple Detection of Aflatoxin B1. *Biosens. Bioelectron.* **2014,** *62,* 288–294.

50. Lamberti, I.; Tanzarella, C.; Solinas, I.; Padula, C.; Mosiello, L. An Antibody-Based Microarray Assay for the Simultaneous Detection of Aflatoxin B1 and Fumonisin B 1. *Mycotoxin Res.* **2009,** *25* (4), 193–200.

51. Liu, L. H.; Zhou, X. H.; Shi, H. C. Portable Optical Aptasensor for Rapid Detection of Mycotoxin with a Reversible Ligand-Grafted Biosensing Surface. *Biosens. Bioelectron.* **2015,** *72,* 300–305.

52. Qiu, Y. L.; He, Q. H.; Xu, Y.; Bhunia, A. K.; Tu, Z.; Chen, B.; Liu, Y. Y. Deoxynivalenol-Mimic Nanobody Isolated from a Naive Phage Display Nanobody Library and its Application in Immunoassay. *Anal. Chim. Acta* **2015,** *887,* 201–208.

53. Santino, A.; Poltronieri, P.; Mita, G. Advances on Plant Products with Potential to Control Toxigenic Fungi: A Review. *Food Addit. Contam.* **2005,** *22* (4), 389–395.

54. Brown, R. L.; Menkir, A.; Chen, Z.-Y.; Bhatnagar, D.; Yu, J.; Yao, H.; Cleveland, T. E. Breeding Aflatoxin-Resistant Maize Lines Using Recent Advances in Technologies—A Review. *Food Addit. Contam., A: Chem., Anal., Ctrl., Expos. Risk Assess.* **2013,** *30* (8), 1382–1391.

55. Chiuraise, N.; Derera, J.; Yobo, K.; Magorokosho, C.; Nunkumar, A.; Qwabe, N. Progress in Stacking Aflatoxin and Fumonisin Contamination Resistance Genes in Maize Hybrids. *Euphytica* **2016,** *207* (1), 49–67.

56. Kluger, B.; Bueschl, C.; Lemmens, M.; Michlmayr, H.; Malachova, A.; Koutnik, A.; Maloku, I.; Berthiller, F.; Adam, G.; Krska, R.; Schuhmacher, R. Biotransformation of the Mycotoxin Deoxynivalenol in Fusarium Resistant and Susceptible Near Isogenic Wheat Lines. *PLoS ONE* **2015,** *10* (3), 1–19.

57. Choo, T. M.; Vigier, B.; Savard, M. E.; Blackwell, B.; Martin, R.; Junmei, W.; Jianming, Y.; Abdel-Aal, E.-S. M. Black Barley as a Means of Mitigating Deoxynivalenol Contamination. *Crop Sci.* **2015,** *55* (3), 1096–1103.

58. Chulze, S. N.; Palazzini, J. M.; Torres, A. M.; Barros, G.; Ponsone, M. L.; Geisen, R.; Schmidt-Heydt, M.; Köhl, J. Biological Control as a Strategy to Reduce the Impact of Mycotoxins in Peanuts, Grapes and Cereals in Argentina. *Food Addit. Contam., A: Chem., Anal., Ctrl., Expos. Risk Assess.* **2015,** *32* (4), 471–479.

59. Park, D. L. Perspectives on Mycotoxin Decontamination Procedures. *Food Addit. Contam.* **1993,** *10* (1), 49–60.

60. Gnonlonfin, G. J. B.; Hell, K.; Adjovi, Y.; Fandohan, P.; Koudande, D. O.; Mensah, G. A.; Sanni, A.; Brimer, L. A Review on Aflatoxin Contamination and Its Implications in the Developing World: A Sub-Saharan African Perspective. *Crit. Rev. Food Sci. Nutr.* **2013,** *53* (4), 349–365.

61. Matumba, L.; Van Poucke, C.; Njumbe Ediage, E.; Jacobs, B.; De Saeger, S. Effectiveness of Hand Sorting, Flotation/Washing, Dehulling and Combinations Thereof on the Decontamination of Mycotoxin-Contaminated White Maize. *Food Addit. Contam., A: Chem., Anal., Ctrl., Expos. Risk Assess.* **2015,** *32* (6), 960–969.

62. Siciliano, I.; Spadaro, D.; Prelle, A.; Vallauri, D.; Cavallero, M. C.; Garibaldi, A.; Gullino, M. L. Use of Cold Atmospheric Plasma to Detoxify Hazelnuts from Aflatoxins. *Toxins (Basel)* **2016,** *8* (5). doi:10.3390/toxins8050125.

63. Cano-Sancho, G.; Sanchis, V.; Ramos, A. J.; Marín, S. Effect of Food Processing on Exposure Assessment Studies with Mycotoxins. *Food Addit. Contam., A: Chem., Anal., Ctrl., Expos. Risk Assess.* **2013,** *30* (5), 867–875.

64. Frobose, H. L.; Fruge, E. D.; Tokach, M. D.; Hansen, E. L.; DeRouchey, J. M.; Dritz, S. S.; Goodband, R. D.; Nelssen, J. L. The Influence of Pelleting and Supplementing Sodium Metabisulfite ($Na_2S_2O_5$) on Nursery Pigs Fed Diets Contaminated with Deoxynivalenol. *Anim. Feed Sci. Technol.* **2015,** *210*, 152–164.

65. Cazzaniga, D.; Basilico, J. C.; Gonzalez, R. J.; Torres, R. L.; de Greef, D. M. Mycotoxins Inactivation by Extrusion Cooking of Corn Flour. *Lett. Appl. Microbiol.* **2001,** *33* (2), 144–147.

66. Dänicke, S.; Valenta, H.; Gareis, M.; Lucht, H. W.; Reichenbach, H. V. On the Effects of a Hydrothermal Treatment of Deoxynivalenol (DON)-Contaminated Wheat in the Presence of Sodium Metabisulphite ($Na_2S_2O_5$) on DON Reduction and on Piglet Performance. *Anim. Feed Sci. Technol.* **2005,** *118* (1/2), 93–108.

67. Amézqueta, S.; González-Peñas, E.; Murillo-Arbizu, M.; López de Cerain, A. Ochratoxin A Decontamination: A Review. *Food Control* **2009,** *20* (4), 326–333.

68. Poór, M.; Kunsági-Máté, S.; Szente, L.; Matisz, G.; Secenji, G.; Czibulya, Z.; Kőszegi, T. Interaction of Ochratoxin A with Quaternary Ammonium Beta-cyclodextrin. *Food Chem.* **2015,** *172*, 143–149.

69. Printz, C. A 'Field in Motion.' *Cancer (0008543X)* **2013,** *119* (6), 1117–1118.

CHAPTER 8

ENDOCRINE-DISRUPTING CHEMICALS IN FOOD AND THEIR TOXICOLOGICAL IMPLICATIONS

S. RAISUDDIN* and SHIKHA SHARMA

Department of Medical Elementology & Toxicology, Jamia Hamdard (Hamdard University), New Delhi 110062, India

Corresponding author. E-mail: sraisuddin@jamiahamdard.ac.in

CONTENTS

ABSTRACT

Endocrine-disrupting chemicals (EDCs) are exogenous agents which interfere with or disrupt the normal functions of natural hormones. Most of the EDCs are synthetic chemicals; some are natural food compounds such as phytoestrogens. Humans are exposed to complex mixtures of EDCs throughout their lives, as they are emitted into the atmosphere during anthropogenic activities and physicochemical reactions in nature. Inhalation of EDCs as particulates and gaseous vapors triggers their interaction with endocrine glands and exerts several adverse effects on activities of hormone receptors. EDCs can affect endocrine function at multiple sites and through numerous specific modes of action. The EDC-induced alteration in synthesis, secretion, transportation, binding, and metabolism of natural hormones has been documented. Eventually, these alterations lead to dysregulation of homeostatic mechanisms, reproduction and development. Foods are the major sources of EDCs. Exposure to EDCs through the food web brings these toxic compounds into human exposure. The current research on EDCs is focused on study of their effects at molecular levels and to identify novel targets. Studies are also underway to test natural compounds, especially compounds with antioxidant potential for their efficacy to minimize the toxic effects of EDCs. In this contribution, we have highlighted the various toxic consequences of EDCs exposed through foods. We will also discuss various strategies which could be used to minimize health hazards to EDCs.

8.1 INTRODUCTION

Endocrine-disrupting chemicals (EDCs) are generally defined as exogenous chemical agents which interfere with the production, release, transport, metabolism, binding, action, or elimination of the natural hormones within the body responsible for the maintenance of physiological conditions and therefore the regulation of endocrine functions.[1,2] EDCs when being ingested by humans cause unusual reproductive, developmental, neurological, cardiovascular, metabolic effects, and disrupts immune system,[3] resulting from their interaction with various receptors such as estrogen receptors (ERs), retinoid X receptors (RXRs), peroxisome proliferator-activated receptors (PPARs), aryl hydrocarbon receptors (AhRs), and thyroid receptors (TRs).[4-7]

Mechanisms of action of EDCs are exceptionally diverse and exposure responses could involve many tissues, organs, or system functions. For instance, EDCs may adversely affect puberty and reproduction by causing disruption in hypothalamic–pituitary level and target tissue like breasts. Similarly, energy balance may also be disturbed by EDCs through hypothalamic centers and fat tissue. Moreover, factors such as sex steroids and leptin can also be disturbed by the interference of EDCs which may lead to the disruption of physiological feedback systems.[2,8] Disorders of endocrine system function resulting from over stimulation or depression may lead to an excessive or deficient secretion of hormones (Fig. 8.1). Various chemical substances such as phytoestrogens, pesticides, and plasticizers may have an effect on the endocrine system. The majority of environmental EDCs are both produced and introduced into the environment by human activities.[9,10] These environmentally persistent compounds do biomagnify through the food web and bioaccumulate in human beings and animals. Bonefeld-Jørgensen et al.[11] reviewed the effect of persistent organic pollutants (POPs) including polychlorinated dibenzodioxin, polychlorinated dibenzofurans, polychlorinated biphenyls, organochlorine pesticides (OCPs), and the amphiphillic perfluoroalkyl acids and concluded that these compounds are potential endocrine disrupters. Authors observed that POPs have carcinogenic potentials and can play an important role in health risk. Human beings are exposed to complex mixtures of chemicals, having individually very different biological potentials and effects. Endocrine disruption in human is multifactorial. Some EDCs have been shown to act additively. Waring et al.[12] reviewed the effect of a range of dietary flavonoids and environmental phenolic contaminants on inhibition of sulfotransferase isoforms which inactivate estrogens by sulfonation. They stated that the environmental pollutants like chlorinated phenolic disinfectants and plasticisers, if combined with dietary flavonoids might cause long-term deleterious effects.

For more than two decades, it has been accepted that food contact materials contribute to human xenobiotic exposure.[13,14] Owing to the worldwide debate on substances like bisphenol A (BPA) and their impact on the endocrine system, there is interest in EDCs and publications are rising on some attention-grabbing actions of EDCs.[15–18] The high migration of phthalates discovered from food-packaging material into foods needed the European Food Safety Authority (EFSA) to undertake a new risk assessment of EDCs.[19]

Although the materials and articles meant to contact food, including food packaging, are not usually perceived to be a chemical hazard compared to pesticides, veterinary drugs, or heavy metals, the environment or improper food storage can facilitate release of toxicants together with EDCs with potential hazard into the foods.[20] This underpins risk management decisions for putting limits on human exposure to these compounds, particularly when directed toward vulnerable consumer groups like infants, young children, and adults.

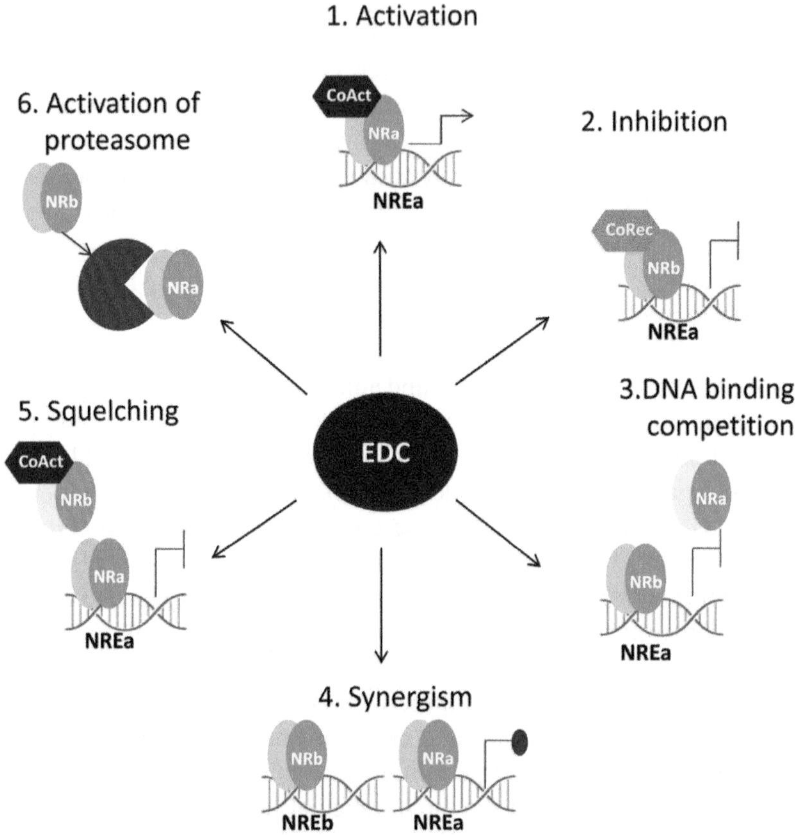

FIGURE 8.1 *Various interactions of EDCs with cellular receptors.* Endocrine-disrupting chemicals (EDCs) interact with nuclear receptor (NR) through multiple mechanisms. In this figure, NRs represented as NRa and NRb to show interactions of EDCs with NRs in general, whereas NREa and NREb are nuclear receptor response elements. (1) EDCs can replace the endogenous ligand and can act as direct agonists and antagonists of NRa. They can regulate the recruitment of coactivators (CoAct) by the NR and initiate target-gene transcription, although some EDCs act as modulators rather than as full agonists by inducing the recruitment of only some of the coactivators. (2) EDCs can also act as antagonists by recruiting corepressors (CoRe) and DNA binding or target-gene expression. Moreover, EDCs can interfere by indirect mechanisms, (3) EDC binding to NRb leads to disturbances of NRa signaling via molecular cross-talk such as competition or for DNA-binding sites, (5) or for coactivators, (4) other indirect mechanisms are the binding of NRa and NRb to neighboring sequences, which may lead to either synergism or inhibition of the regulatory activity, (6) or to the EDC-mediated activation of NRb, which results in NRa degradation through proteasome activation.

Several studies have investigated and reviewed the impact of EDCs on humans and other species.[21–24] When the relevance of these exposures is evaluated, serious difficulties arise because of many variables that include assessing the complete spectrum of possible endocrine responses.[24,25] Overall, the biological response to EDCs are complex because of the various relationships between organs and glands that serve to preserve the organism's physiological conditions.[22]

8.2   COMMON ENDOCRINE-DISRUPTING CHEMICALS IN FOOD

Numerous chemical substances, which have been recognized as safe, have now been found to adversely affect hormonal balance in humans and other organisms. They include BPA, phthalates, benzophenone, and its derivatives and organic compounds of tin.[26,27] These compounds migrate from food contact materials into foods. The migration of potential EDCs into foods has stimulated efforts for developing a uniform approach, specially tailored for their identification and evaluation. About 50 chemical substances authorized in food contact materials are known for being potentially endocrine disruptors.[27] For example, nanotechnology-enabled packaging materials have become a major area for innovation in the food sector. Nanomaterials contained in food contact materials are shown to migrate into the food at a detectable level.[28] No generic threshold for nanomaterials has been defined till now. In Europe, specific legislation on plastic-based food contact materials limits the use of nanomaterials to those explicitly mentioned in the positive list, although the number of authorized nanomaterials is limited to be used with very few of the compounds. Currently, the regulation require that packaging materials be tested for their suitability for use in food contact materials.[29] To describe the potential of nanomaterials in food contact applications in a more accurate and conclusive manner, all applications have to undergo a specific risk assessment, even those applications where the nonmaterial is used behind a functional barrier. This practice may partially explain the slow authorization rate for new nanomaterials. These tests for the risk assessment are generally carried out using food simulants that mimic the migration effects in foods. Similar approach is accepted to be appropriate for assessing the migration of nanoparticles in the food.[30–33]

The endocrine system regulates many biological functions such as reproduction, development, homeostasis, etc. There are a number of receptor sites on the tissues where hormones act. For this reason, testing for EDCs is very tough.[34] In case of pesticide residues in foods that are of special concern, a plan and temporary criteria for undertaking procedures preceding risk management have been suggested by Max-Stoelting.[35] This takes into consideration consumer safety and proposed two approaches for classifying chemical substances as EDCs. The primary approach describes exposure assessment through quantifying EDCs that enter the organism via food as the amount or dose of any compound to which an individual has exposed, is one of the determining factor of its potentiality of assessment of risk hazard, while the second approach evaluates and reports the endocrine-disrupting potential of the given substance.[35] However, Ruden[36] stressed that the planned approaches should be treated as temporary, until scientific criteria enabling risk assessment are developed in step with EU laws. It may be anticipated that such criteria would also be relevant for the chemical substances that migrate from food-contacting materials into foods.

The range of materials intended to contact with foods results from the various functions they are designed for. By implication, this additionally results in diverse issues for evaluating under which circumstances a given material may safely be used for making contact with foods.[37,38] Food packaging is intended to protect food against external factors like chemical and biological contaminants, and prevent oxidation by atmospheric oxygen, light, loss of gas from beverages, loss or absorption of humidity and aroma, etc.

Presently used food packaging is made from different plastic materials and laminates. Moreover, the packed food may come into contact with the inner walls of cans, gaskets, and coatings utilized in lids usually containing fully or partly plastic material. These materials may be sources of harmful or inadequately tested substances.[19,39] Such concerns about EDCs have been thoroughly elaborated by Muncke,[14] who demonstrated that food packaging may indeed contain various substances suspected of acting as EDCs. Since foods can interact with the inner surface of packaging, migration of its constituents may be expected. Migrating substances may be expected to include monomers, polymerization initiators, catalyzers, and various other chemical ingredients. Chemical substances such as polymer degradation products including other substances that are intentionally added at the time of production and food processing showed estrogenic potency.[40–42]

There are no internationally recognized classifications or analysis criteria for EDCs in foods so far that can be used as a standard for risk assessment. Although the categorization of EDCs on the basis of most known adverse effects relevant for human health is possible, especially in terms of risk assessment approaches. These include the main adverse human health effects such as carcinogenesis, reprotoxicity, and mutagenesis.[43]

In the case of pesticides, for instance, the EU anticipates new criteria for permitting EDC identification, proposing that those substances classified as carcinogenic and/or toxic for reproduction should also be treated as endocrine disruptors.[43] Some examples of substances such as BPA, di-(2-eth-ylhexyl) phthalate (DEHP), di-*n*-butyl phthalate (DBP) in food contact materials together with their toxicological evaluations that resulted in their use becoming limited or restricted owing to legal decisions arising from risk management, are delineated in EU strategy,[43] while describing these substances their endocrinal effects were not mentioned. Therefore, in this contribution, we describe the toxic effects of commonly occurring chemicals in foods with endocrine-disrupting effects with specific relation to situations of food processing and storage and their likely implications in human health in Tables 8.1 and 8.2.

TABLE 8.1   Endocrine Disrupting Chemicals Used as Food Additives for Various Purposes.

Category	Compound	Effects	References
Antioxidants	Butylated hydroxytoluene Butylated hydroxyanisole Propyl gallate Tocopherols	Estrogenic, androgenic, and other toxic effects	Sonnenschein and Soto[44]
Taste enhancers	MSG Disodium inosinate	Endocrine and metabolic disruption and other toxic effects	Skultetyova et al.[45]
Synergists	Citric acid Tricalcium phosphate Other phosphates Ascorbic acid	Cytotoxicity and other toxic effects	Wang et al.[46]
Antimicrobials	Sodium benzoate Calcium propionate	Endocrine disruption and other toxic effects	Khasnavis and Pahan[47]

MSG, monosodium glutamate.

TABLE 8.2　List of EDCs that Migrate into Food Stuffs from Food Packaging.

Category	Compound	Effects	References
Monomer	Bisphenol A	Potent estrogenic and other endocrine disrupting effects	Acconcia et al.[48] Rodríguez et al.[49]
Plasticizers	Dibutyl phthalate Bis(2-ethylhexyl) phthalate	Estrogenic effects, reproductive toxicity	Ahmadivand et al.[50] Howdeshell et al.[51]
Printing ink-photo initiator	Benzophenone	Weak estrogenic toxicity	Bradley et al.[52]
Colorants (for monomers and plasticizers)	Pthalo blues Lead chromate Lead molybdate Chromium oxide	Thyroid disruption Highly toxic but endocrine disrupting capacity not confirmed	EU strategy[53]
Coating materials (food containers)	PFOA Semicarbazide	Thyroid disruption Highly toxic and endocrine disrupting	EFSA[54]

PFOA, perflurooctanoic acid.

8.2.1　ALKYLPHENOLS

Alkylphenols (APs) are one of the high production volume man-made compounds, primarily used to manufacture alkylphenol ethoxylates (APEs). APs and APEs are in use for over 50 years and are necessary for manufacture of textiles, pulp and paper, agricultural pesticides, coatings, fill oils and fuels, metals, and plastics.[55] Nonylphenol (NP) and octylphenol (OP) are the two most important APs due to their high commercial use. They exist in various forms or "isomers" and are used to build nonylphenol ethoxylates (NPEs) and octylphenol ethoxylates. NPEs were used primarily as surfactant detergents. NPs are also used to manufacture resins and plastics and as stabilizers and antioxidants in plastics.[56]

Due to the contamination of the environment and food chain, exposure to APs can occur via food and, to a lesser extent, drinking water. APs have been found in edible fish and shellfish as well in other packaged foods.[57] Inputs of APs and APEs into the aquatic environment leads to their presence in fish and shellfish, and application of sewage sludge to agricultural land introduces APs and APEs to soils used for growing crops and grazing livestock. This may become a potential source of their entry into food. NPs may also migrate from food contact materials, as are used as additives (stabilizers, antioxidants) in plastic packaging.[58,59] The discharge of wastewaters, waste-matter effluents, and waste from industries using APs or APEs, results in the widespread presence of APs and APEs in the surface waters and aquatic environment (rivers, estuaries, coastal marine environments)[60–62] creating a possibility of their entry into the food.

A study was conducted on Italian women to detect contamination of NP and OP in breast milk.[63] The purpose of study was to calculate the maximum daily intake of NP and OP by the infants with maximum body weight of 5 kg on the basis of the concentrations found in the breast milk samples. Maximum NP daily intake by the infants through breastfeed was 3.94 µg/kg/day, which was found to be close to the tolerable daily intake (TDI) of 5 µg/kg/bw proposed by the Danish Institute of Safety and Toxicology. Although the maximum OP daily intake could not be calculated as no TDI is available, the authors observed that the intake of OP was at least six orders of magnitude lower than the No Observed Adverse Effect Level (NOAEL) of 10 mg/kg/day. Authors also observed a significant correlation with the consumption of fish and the concentration of NPs in the milk samples.[63] This study raised an important point for the risk assessment of breastfeed population exposed to NPs and other chemicals many of which probably endowed with endocrine-disrupting properties.

Another study was conducted in Sweden markets to check the concentrations of NP and BPA in food based on Sweden per capita food consumption.[64] Fruits, vegetables, and cereals were the foods with the highest levels of NP in this study. Even then, the estimated intake of NP in the present study was found 10 times lower than the TDI (5 µg/kg/bw). Authors observed that more than 30% of the nursing women population had levels of free or total NP in blood serum at or above Limit of detection (LOD) (0.5–0.8 ng/g serum). Women with blood NP levels at or above LOD consumed more fruits and vegetables than women with levels below LOD. On the basis of their observation, authors concluded that these food groups are important sources of NP exposure from food in Sweden.[64]

There is a large body of proof on the endocrine-disrupting properties of APs. The oestrogenic properties of APs were demonstrated as early as 1938 and that they have since been shown to be estrogenic in animals.[65–67] Recent attention has focused on the results on fish and wild fish populations (including roach, rainbow trout, gudgeon, eel, and flounder) of estrogenic chemicals (including APs) in sewage effluents discharged into rivers and estuaries. [68–71] In some cases, where industries using APE surfactants (such as the textile industry), APs are suspected of contributing to majority of the estrogenic activity of the effluent.[72] Estrogenic elements of sewage effluents are implicated within the widespread "feminization" of male fish observed in rivers and estuaries in the Great Britain and Europe.[65,70,73] Effects determined in male fish include the expression of the female egg yolk macromolecule vitellogenin, the presence of intersex characteristics like "ovotestes" (the development of eggs [oocytes] within the testis), feminization of the reproductive ducts, reduced sperm production, and reduced fertilization success.[70,74–76] A study on the Pacific oyster (*Crassostrea gigas*) has shown that NP may induce long-term and transgenerational effects.[77] The study showed that a single 48-h exposure of larvae at a key stage in their development to environmentally relevant concentrations of NP induced long-run, sexual biological process effects (altered sex quantitative relation toward females and an increase within the incidence of hermaphroditism). Germ-cell viability was also affected leading to poor embryonic and larval development of the following generation showing up to 100% mortality in some cases.

8.2.2 BISPHENOL A

BPA is one of the most reported EDC. It has been reported to be found in food-contacting materials made up of polycarbonate and internal lacquer coatings in metal cans. For instance, food cans receive high heat exposure during the process of sterilization. This high heat may cause leaching of BPA present at the internal coatings of the cans. In a study conducted by Grumetto et al.[78] on 42 cans of tomatoes, it was observed that BPA was detected up to a very high concentration (115.3 µg/g/kg of tomato) in 22 cans. This value was extremely higher than that of the TDI value of BPA (4 µg/kg bw/day). Ćwiek-Ludwicka.[79] reviewed the exposure of BPA through dietary and nondietary sources in different age groups of human populations with a special reference to food contact materials. He observed that the highest assessed dietary exposure from food contact materials in infants and toddlers was 0.086 µg/kg bw/day. Moreover, the highest level of dietary exposure in adult population including the women of childbearing age ranged from 0.335 to 0.388 µg/kg bw/day. The assessed values were quite lower than that of the TDI value (4 µg/kg bw/day) of BPA. However, the authors suggested that available data on the exposure of BPA need to be updated further for the human health risk assessment.

From the toxicological analysis report on BPA, the EFSA concluded that exposure to comparatively high doses (above TDI), can also be associated with some estrogenic effects.[77,80] However, additional analysis by EFSA became necessary due to new, but not yet evaluated results suggesting that BPA affects neurodevelopment when in utero exposure of experimental animals was followed by exposure throughout infancy through the milk of mothers.[81] Furthermore, new toxicokinetics studies together with placental transport have demonstrated need for renewing risk assessment concerning perinatal exposure to BPA.[82] In a review by Roger et al.[83] the effect of BPA on immune cells of humans and other species has been reported. They also described that the immunological alteration caused by BPA may be mediated through ERs, AhR, PPARs.

There are numerous in vitro and in vivo studies demonstrating of adverse effects of BPA at doses higher than the current TDI (4 µg/kg bw/ day) and lower than the previous TDI (50 µg/kg bw/day). Exposure to BPA at or below 50 µg/kg bw/day may cause alteration in the time of puberty, alteration in estrous cycle, changes in prostate, alteration in mammary gland development with evidence of intraductal hyperplasias and preneo-plastic mammary gland, changes in the uterus and ovary, alterations in brain sexual dimorphisms, changes in brain steroid receptor levels, behavioral changes, alteration in glucose homeostasis, and body weight.[48,49,84] Some important mechanisms and effects of BPA are shown in Figure 8.2. BPA has apparently become one amongst the most researched compounds in recent past. EFSA has demonstrated that food materials constitute a significant source of BPA exposure.[85] In the previous assessment report of EFSA, high exposure in toddlers was up to 300 ng/kg bw and in 3-month-old infants which reached to 11,000 ng/kg bw. As per the current assessments (2013), the exposure in toddlers was 857 and 495 ng/kg bw in infants 3–5 days after birth. EFSA additionally drew attention to the fact that thermal paper employed in printers and cash registers can also be considered as second highest source of BPA. It had been further concluded that biomonitoring of BPA in body excretion fluids provides a reliable estimation of the overall exposure from all sources, opening promising views for large-scale obser-vation programs.[86,87]

8.2.3 PHTHALATES

Phthalates are primarily used as plasticizers in food-contacting materials, generally used to prepare plastic containers to extend their flexibility, trans-parency, and durability. Butyl benzyl phthalate is used as a plasticizer for polyvinyl and cellulose resins and organic intermediates.[88] DBP is used in paper coatings, printing inks and elastomers. DEHP is used as a plasticizer

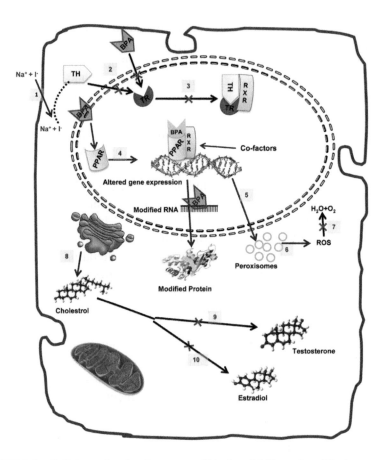

FIGURE 8.2 *Cellular and molecular targets of bisphenol A.* Examples of the known effects and mechanisms of action of bisphenol A (BPA) (1) BPA increase the activity of the sodium/ iodide symporter, leading to increased iodide uptake. Although iodide is necessary for thyroid hormone (TH) production, the effects of BPA on TH synthesis are unclear. (2) BPA impede with the binding of TH to thyroid hormone receptors (TR) by binding to TR themselves. (3) The gene expression of retinoid X receptor (RXR) is reduced by BPA, thus reducing the capacity to form heterodimers with TR. By doing so, the regulation of TH-related genes is disrupted, leading to effects at the individual level. (4) Peroxisome proliferator-activated receptors (PPARs) gene and protein expressions are increased. BPA can also activate PPARs (5). After forming heterodimers and binding to response elements, peroxisome proliferation is increased. (6) Fatty acid oxidation then leads to the formation of reactive oxygen species (ROS). (7) Transcription, translation, and activity of scavenger are decreased, leading to detrimental effects at the organ and individual levels. (8) BPA reduces cholesterol transport by reducing scavenger receptor class B-1. (9) The gene expression of a series of enzymes necessary for steroidogenesis is repressed, leading to lower testosterone levels. (10) The gene expression, protein expression, and activity of aromatase are also decreased, resulting in lessened estradiol synthesis when animals are exposed to BPA. These effects then induce changes at the organ, individual, and population levels.

for polyvinyl chloride, particularly in preparing medical devices and a plasticizer for resins and elastomers. Bicalho[89] estimated TDI values for DEHP, di-*n*-butylpthalate (DnBP), di-isobutylpthalate (DiDP), and di-isononylpthalate (DiNP) are 0.05, 0.01, 0.15, and 0.15 mg/kg bw/day, respectively.[90]

Food is the major source of exposure of pthalates in human beings. Number of studies are available which provide the information about the migration and potentiallity of pthalates as endocrine disruptors. A study conducted on US population to examines associations between recent fast-food intake of BPA and urinary metabolites of DEHP and DiNP. Total participants of 8877 from the National Health and Nutrition Examination Survey (NHANES 2003–2010) were used. Authors quantified the levels of DEHP and DiNP using 24-h dietary recall data and observed that the levels of these two pthalates were higher in the urine samples of participants with high consumption of fast food than nonconsumers. Authors also observed the evidence of a positive, dose–response relationship between fast-food intake and phthalates exposure.[91]

Ge et al.[92] investigated the residual profile and levels of 15 phthalates in 90 goat milk-based infant formulas from 15 commercial brands of 10 dairy enterprises located in Shaanxi Province, China. According to their findings DBP, bis-DEHP, DiDP, and dimethyl phthalate were the most detectable pthalates with their geometric mean concentrations in the formulas were 38.1, 24.2, 16.6, and 8.7 μg/kg, respectively. On the basis of the observations made, authors demonstrated that goat milk-based infant formulas may be one of major source of exposure to phthalates in infants.

Another recent study reported by Kim et al.[93] on the exposure of the most vulnerable group (infants) to the pthalates through breast fed was conducted in South Korea in 2012. Samples were collected from 62 lactating mothers at 1 month postpartum from 4 cities of Korea and evaluated for 6 phthalate metabolites (mono-isobutyl phthalate (MiBP), mono-*n*-butyl phthalate (MnBP), mono(2-ethyl-hexyl) phthalate (MEHP), mono-(2-ethyl-5-hydroxyhexyl) phthalate, mono-(2-ethyl-5-oxohexyl) phthalate, and monoethyl phthalate (MEP). MEP, MiBP, MnBP, and MEHP were the most detected pthalates in all breast milk samples with a median concentration of 0.37, 1.10, 1.70, and 2.08 μg/L, respectively. As per the study, the median daily intake estimates of phthalates including both monoester and diester forms through breast milk consumption ranged between 0.91 and 6.52 μg/kg/bw for DEHP and between 0.38 and 1.43 μg/kg bw for DnBP. Based on the estimated daily intake up to 8% of infants exceeded the reference dose of anti-androgenicity for DEHP and 6% of infants exceeded the TDI for DnBP. Breast milk MiBP and MnBP concentrations showed significant positive associations with maternal consumption of whipped cream and purified water.

Several in vitro, in silico, and preclinical studies demonstrate the association of pthalate esters with endocrine disruption, reproductive developmental damage, neurodevelopmental toxicity, and growth-related problems in humans.[94–96] Phthalates are reported to target the women and adversely affect the two essential processes of folliculogenesis and steroidogenesis in ovary.[95] According to an epidemiological study published in 2012, pthalate esters MEHP and DEHP can inhibit testosterone production in adult human testis.[96] Moreover, in line with the EFSA's scientific opinion, even if the DEHP dietary intake is below the TDI (0.05 mg/kg bw), there are other sources contributing toward the overall exposure and need to be taken into consideration.[82,85]

8.2.4 BENZOPHENONE

Benzophenone is used as an additive (photo-initiator UV) for printing inks and can be transferred from cardboard food packaging into the packed foods.[14] Several studies demonstrating the migration of photoinitiators (benzophenone and its derivatives) into the food are available for example a recent study has been published on migration of photoinitiators used on the inks of food-packaging materials. Conventional migration testing for long-term storage was done for 10 days at 60°C. Authors followed the test conditions according to the EU Regulation (No. 10/2011). The results obtained were then compared with the migration in cereals after the storage of 6 months at room temperature. The simulation with Tenax at 60°C overestimated actual migration in cereals up to a maximum of 92%. In addition, the effect of a lower contact temperature and the impact of the Tenax pore size were investigated. Analogous simulation performed with rice instead of Tenax resulted in insufficiently low migration rates showing Tenax is a much stronger adsorbent than rice and cereals.[97]

Toxicological studies aimed at the hormone-mimetic potential of benzophenone and its derivatives are equivocal.[14] A recent epidemiological study conducted on 501 couples who were discontinuing use of contraceptives to become pregnant for the Longitudinal Investigation of Fertility and the Environment Study (Michigan and Texas, 2005–2009). Concentration of the benzophenone filters in the urine samples was checked. On the basis of their findings, authors suggested that male exposure to benzophenone UV filters may diminish couples' fecundity, which may lead to a longer time to pregnancy.[98] Another epidemiological study was published in 2015. The purpose of the study was to check the concentration of benzophenone type UV filters in semen and urine samples of 413 men (collected during 2005–2009). The findings suggest that specific UV filters may be associated with some aspects of semen end points but await future corroboration.[99] Both the studies are clearly supporting the endocrine disruption potential of benzophenone. In an analysis of potential health threats arising from benzophenone in some foods, Muncke[14] stressed that the presence of this compound was confirmed in food packed in multilayer cardboard packaging and that the present TDI of benzophenone, as determined by the EU, is 0.01 mg/kg bw. The migration limit into the foods through food contact materials was set at 0.6 mg/kg of the food assuming that the consumption of packaged food by an average adult weighting 60 kg will not exceed 1 kg/day.[39]

8.2.5 ORGANOTINS

Organotins have been used for several years in plastic food packaging as light and heat-stabilizing agents or polymerization initiators.[100] They protect polyvinylchloride (PVC) from degradation by binding chloride leading to chlorinated organotin species.[101] Major source of human exposure to organotins is considered to be food, especially sea food as it is more prone to be contaminated with the organotins used as antifouling agents on ships. However, there are some other food sources of organotin exposure like margarine, mayonnaise, wine and beer, etc.[14] Endocrine disruption potentiality of organotins is well documented but data in context of human exposure is limited.[102–104]

Monobutylin chloride is used as a hot-end coating in the manufacturing of glass packaging.[105] It is applied to the hot glass surface that leads to instant pyrolization of the organic moiety. Butyltin has been found in wine and liquors usually together with dibutyltin.[106] Dibutyltin degrades to monobutyltin under acidic conditions.[105] Both the compounds in these samples are thought to have been originated from PVC plastic bulk storage containers. Many organotins have been tested in vitro for their effect on 17β-estradiol (E2) synthesis. Dibutyltin and monobutyltin were found to upregulate 17β-hydroxysteroid dehydrogenase type I (an enzyme which highly activated in human placenta cells and catalyses the conversion of estrone to E2).[26] Tributyltin (TBT) triphenyltin (TPT) are well established EDCs. The TDI value of TBT is 0.25 μg/kg bw/day derived from a toxicity end point based on the immune response of rats, and a safety factor of 100 has been applied to extrapolate the result from rats to humans.[107,108] Both TBT and TPT are not approved for use in food contact material in the United States and EU.[85,109] However, due to historic use of TBT in shipping industry as an antifouling agent, it has entered into the marine food chain, and it could still exist there.

A study was conducted to detect the level of organotins in the sea food of Portuguese market for the assessment of risk hazard of the consumers. Levels of butylins in edible parts of fish, crustaceans, and cephalopods were found in the lower range than that of reported for these animal groups from other locations (i.e., below 30 ng/g wet weight). Although the concentration for bivalves (up to 275 ng/g wet weight) was reported to be moderate, most of the samples showed TBT and DBT levels below the tolerable average residue levels (93 ng/g wet weight) of the Portuguese consumers, which may indicate low risk. However, the authors suggested these results to be integrated with further findings on the molecular targets of organotins in mammals.[110]

LOD and limit of quantification of organotins in beers and wines ranged from 1 to 40 and 3 to 80 ng(Sn)/L, respectively, owing to acidity-enhanced migration from PVC food-contacting material.[111,112] This situation highlights potential migration of compounds with endocrine-disrupting properties in food and drinks.

8.2.6   ARSENIC

Arsenic represents the 20th most prevalent metal in the earth crust.[113] It is widely distributed in the environment in the form of both pure arsenic and complexes with oxygen, chlorine, sulfur, carbon, and hydrogen. Drinking water is the major source of supply of environmental exposure. It had been reported that arsenic may accumulate in ground water wells and other water bodies. Also, arsenic is a major component of many pesticides widely used in agriculture. Consumption of arsenic by humans can occur not only through drinking water but also through the food chain. Crops receiving arsenic through the pesticides and contaminated irrigation water (e.g., rice) take up and accumulate it in numerous degrees depending on the species and variety.[114,115]

Cereal and cereal products, and particularly rice and rice based products, have reported to have arsenic concentrations with 0.1–0.4 mg As/kg dry weight, by the absorption through their roots which either directly uptake arsenic from soil or through the airborne arsenic deposited on the leaves.[116] Brown rice appears to possess a higher arsenic concentration than white rice indicating that the arsenic is connected to bran elements.[117] Fish and other seafood are major source of organic species of arsenic (organoarsenicals). Metabolites of organoarsenicals produced after their biotransformation is found to be highly toxic alike the metabolites of extremely noxious inorganic arsenic.[117]

Arsenic-mediated endocrine disruption has been reported to be potential both in animals and humans. Mechanisms proposed for endocrinal effects of arsenic include alterations in cell signaling, cell cycle control, DNA repair, oxidative stress, and others.[118–120] Although effects of arsenic vary at different doses/concentration, time of exposure, and there may be tissue-specific differences of the effects. Few studies reported endocrine-disrupting potential of arsenic by altering steroid hormone receptors, that is, glucocorticoid, estrogen, progesterone, mineralocorticoid, and androgen receptors.[120–122] Arsenic is also reported for its immunosuppressive effects.[123] Furthermore, studies were published on the effect of arsenic on other subfamilies of nuclear receptors. Davey et al.[124] studied the RXR and TR-mediated effect of sodium arsenate on human embryonic NT2 and rat pituitary GH3 cells. NT2 and GH3 cells were treated with 0.01–5 μM of sodium arsenite for 24 h. Authors observed significant alteration in RAR-dependent gene transcription of a transfected RAR response element–luciferase construct and the native RA-inducible cytochrome P450 (CYP26A) gene in NT2 cells. For thyroid hormone (TH)-mediated mechanism, an amphibian ex vivo tail metamorphosis assay was used to examine whether endocrine disruption by low-dose arsenic could have specific pathophysiologic consequences because tail metamorphosis is tightly controlled by TH through TR. Inhibition in TH-dependent tail shrinkage was observed in dose-dependent manner. Nuclear receptors and their subclasses play important roles in normal human development and adult function and their dysregulation is associated with many disease processes. Therefore, disruption of these hormone receptor dependent processes by arsenic is also potentially relevant to human developmental problems and disease risk.

8.2.7 CADMIUM

Cadmium is found as an environmental pollutant, both through natural occurrence and from industrial and agricultural sources.[125] Food is one of the main sources of cadmium exposure for the nonsmoking general population.[125,126]

Cadmium absorption after dietary exposure in humans is comparatively low (3–5%). However, it is efficiently retained in the kidney and liver in the human body with longer biological half-life ranging from 10 to 10–30 years.[127]

The EFSA's concise European Food Consumption database was used to determine cadmium dietary exposure and the consumption data as reported by Sand et al.[128] and Ferrari et. al.[129] National food consumption survey data were used to evaluate the consumption pattern of specific subgroups like children and vegetarians. The food groups that contributed the most to dietary cadmium exposure were cereals and cereal products, pulses, vegetables, nuts, starchy roots or potatoes, meat, and meat products. The mean dietary exposure across European countries was calculated to be 2.3 µg/kg bw/week (range from 1.9 to 3.0 µg/kg bw/week) and the high exposure was calculated to be 3.0 µg/kg bw/week (range from 2.5 to 3.9 µg/kg bw/week). Due to the high consumption of pulses, cereals, nuts, and oilseeds, vegetarian population have a higher dietary exposure to up to 5.4 µg/kg bw/week. Regular consumers of bivalve molluscs and wild mushrooms were also found to have higher dietary exposures of 4.6 and 4.3 µg/kg bw/week, respectively.[128–130]

A report has been published on the hormonal effects of cadmium with special focus on the risk of breast, endometrial, and prostate cancer in humans in 2015.[131] The report was based on a collaborative meeting of experimental toxicologists and epidemiologists conducted by National Food Agency of Sweden and Swedish Chemicals Agency. As per the report, dietary exposure of cadmium was statistically significantly found associated with increased risk of endometrial cancer. The multivariable adjusted relative risk was found 1.39 (i.e., 95% confidence interval (CI), 1.04–1.086, $P_{trend} = 0.019$), comparing highest tertile of exposure with lowest. On the other hand, women with low exogenous and endogenous estrogen a fold increased risk was found 2.9 (95% CI, 1.05–7.79). Statistically significant but a small risk of breast and prostate cancers (21% and 13%, respectively) was observed.

Cadmium disrupts steroidogenesis by interfering with the biosynthesis of androgens, estrogens, and progesterone in vivo and in vitro, thus, leading to disturbed sex differentiation and altered gametogenesis.[132–135] On the other hand, it may bind both the estrogen and androgen receptors. The experimental data suggest that cadmium-induced estrogen like effects are mediated via membrane associated signaling and do not involve classical ER signaling.[136,137]

8.2.8   LEAD

Lead is an environmental pollutant that is present naturally and, to a great extent, from anthropogenic activities like mining, smelting, and battery manufacturing.[138,139] Lead occurs in organic and inorganic forms; the latter predominates in the environment. Measures have been taken to regulate lead in food cans, paint, petrol, and pipes in Europe since the 1970s.[140] Human exposure to lead may occur via food, water, air, soil, and dust. Food is one of the major sources of exposure to lead.[141] Presence of lead in food and water should take into consideration for lead contamination of the domestic environment. Old lead paints on walls and woodwork, and paints on toys can be the important sources of excessive lead intake in children. Lead glazes are used on ceramic kitchen utensil, earthenware, and stoneware vessels because they permit more flexibility in the kiln temperatures for firing pottery.[142,143] Lead may also occur in decorative glazes on some forms of pottery. The leaching of lead from inadequately fired glazes has been investigated, and it was observed that lead may cause serious health hazards from vessels used as containers for acidic foods and beverages.[142,143] Lead glazes for decorative purposes must not be placed in contact with food. Tinware, in which the use of impure tin was a frequent source of lead, has been replaced by aluminum and stainless steel containers. Tinplate cans with soldered seams have been investigated as possible sources of lead contamination for a range of foods. In a survey carried out in the United Kingdom, the mean lead concentration for canned baby food was estimated about 0.24 mg/kg compared with a level of 0.04 mg/kg for baby food in jars.

A study focused on the effect of lead on glucose homeostasis subjected to metabolic function of pancreas and liver was reported by Mostafalou et al.[144] Experiments were conducted in two parts. In the first part, islets of Langerhans were isolated from the pancreas and exposed to lead acetate at 0.05% and 0.2% for 24 h. In the second part, lead acetate at 0.05% and 0.2% in drinking water was administered to rats for 32 days. It was observed that lead disrupted insulin secretion function of islets through activating GSK-3β and endoplasmic reticulum stress and increased activity of gluconeogenic enzymes in the liver featured by glucose intolerance. Authors concluded that chronic exposure to lead can disrupt glucose homeostasis by affecting pancreas and liver mainly through induction of insulin resistance.

In summary, lead contamination can alter many processes regulated by endocrine system, such as longevity, development, sexual receptivity, fertility, and locomotion.[145,146]

8.2.9   MERCURY

Mercury concentration can be detected in air, water, and soil, all around the world. Humans have extracted and used mercury for hundreds of years. Mercury is used in certain medical devices. Consequently, the level of concentration of mercury, mobilized, and released because of human activities has highly increased, leading to elevated concentrations in water, soil, air, and living organisms.[147] Major mercury bioaccumulation in the environment occurs through the food chain.[148,149] In water, plants and small organisms like phytoplankton take up mercury through passive surface absorption or through food intake (Fig. 8.3).

A cross-sectional study conducted by Gump et al.[152] on children aged 9–11 years showing children who consumed fish had a significantly more atheroprotective lipid profile but higher levels of mercury in their blood relative to children that did not consume fish. Although the levels of blood mercury were very low in these children ($M = 0.77$ µg/L, all but one participant had levels below 3.27 µg/L). Increasing blood mercury was significantly associated with blunted diurnal cortisol levels. Blood mercury was also significantly associated with acute phase proteins suggesting systemic inflammation, and several of these proteins were found to significantly reduce the association between mercury and diminished cortisol when included in the model.

Reports indicated that both organic and inorganic mercury compounds highly accumulate in major endocrine glands such as thyroid, hypothalamus, testes, ovaries, and adrenal cortex.[153,154] Mercury-based compounds disrupt steroidogenesis including sex hormone synthesis, male and female fertility as well, and also the hypothalamic pituitary thyroid axis and hypothalamic pituitary adrenal axis.[155,156] Most data available indicate the fact that mercury can act as a potent endocrine disrupter.[157]

Various effects and mechanisms of action of heavy metals on endocrine functions are summarized in Table 8.3.

8.2.10   PHYTOESTROGENS

The relation between health and diet has been accepted by the nutrition community within the past many years. Interest of research on the physiological roles of bioactive compounds found in plants has increased. The phytoestrogens are present in widely consumed fruits and vegetables as bioactive molecules.[165,166] They are called as "phytoestrogens" because they

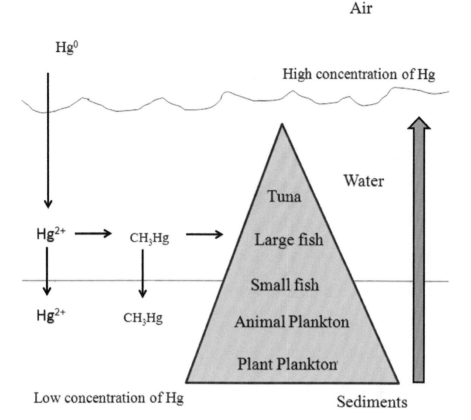

FIGURE 8.3 *Environmental biogeochemical cycling of mercury.* Mercury released from power plants and other pollution enters the aquatic environment, where it undergoes methylation and gets convert into methyl mercury. Mercury and methyl mercury are present in only very small concentrations in seawater. However, they are absorbed, usually as methyl mercury, by algae at the start of the food chain. This alga is then eaten by fish and other organisms higher in the food chain. Fish efficiently absorb methyl mercury but only very slowly excrete it. Methyl mercury is not soluble and therefore is not apt to be excreted. Instead, it accumulates, primarily in the viscera, although also in the muscle tissue. This results in the bioaccumulation of mercury in a build-up in the adipose tissue of successive trophic levels: phytoplankton, zooplankton, small fish, larger fish, etc.One severe case of mercury poisoning was highlighted in 1950s, when Chisso Corporation in Minamata, Japan, disposed off untreated effluent containing methylmercury chloride into Minamata Bay.[150,151] Once in the bay's sediments, the mercury was readily absorbed by marine species, contaminating the whole ecosystem. Fish consumed by local residents resulted in the deaths of more than 1000 people and severely impacted the developing fetuses of pregnant women.[150,151]

TABLE 8.3 Effect and Mechanism of Action of Heavy Metals on Endocrine System.

Heavy metal	Effects	Mechanism	References
Arsenic	Increase in GR, MR, PR, AR, and TR-mediated transcription (low doses)	Stimulation or inhibition of nuclear transcription activity mediated by several hormone receptors bind with ERs	Davey et al.[157] Kaltreider et al.[158] Jana et al.[159]
	Inhibition of GR, MR, PR, AR, and TR-mediated transcription (high doses)		
	Inhibition of ER-mediated transcription estrogenic effect		
Mercury	Stimulation of progesterone synthesis	Induction of 3-beta hydroxysteroid dehydrogenase	Mondal et al.[160] Barregard et al.[161]
	Reduction in plasma levels of testosterone and 17-beta-estradiol	Inhibition of the type I iodothyronine deiodinase	
	Reduction in sperm motility and sperm count		
Manganese	Increase in serum levels of LH, FSH, and testosterone	Activation of the soluble guanylyl cyclase (sGC)	Prestifilippo et al.[162] Lee et al.[163]
	Stimulation of spermatogenesis		
	Stimulation in the secretion of LH		
	Early onset of puberty		
Zinc	Stimulation of spermatogenesis	Membrane-stabilizing activity	Aitken and Clarkson[164]

GR, Glucocorticoid receptor; MR, mineralocorticoid receptor; PR, progesterone receptor; AR, androgen receptor; TR, thyroid receptor; ER, estrogen receptor; LH, luteinizing hormone; FSH, follicle stimulating hormone; LDL-R, low-density lipoprotein receptor.

have the capability to bind with ERs and induce an estrogenic or antiestrogenic response in target tissues like pituitary gland, uterus, breast, and hormone-responsive tumors.[167,168] Phytoestrogens include a wide range of structurally different compounds like isoflavones mainly found in soy, lignans found in grains, and stilbenes found in the skin of grapes. Other less investigated phytoestrogens are flavones, flavans, isoflavones, and coumestans. Both estrogenic and antiestrogenic activities of these compounds are widely reported in literature.[169,170] Classification and examples of commonly found phytoesrogens with representative structures are shown in Table 8.4.

TABLE 8.4 Classification of Phytoestrogens.

Class of phytoestrogen	Examples	Structure of representative compound
Isoflavones	Biochenin A Daidzein Daidzin Formononetin Genistein Genistin	 Genistein
Flavones	Apigenin Baicalein Chrysin Norwogonin	 Apigenin
Flavonols	Galangin Keampferol Myricetin Quercetin	 Quercetin
Flavanones	Naringenin Pinocembrin Taxifolin	 Naringenin

TABLE 8.4 *(Continued)*

Class of phytoestrogen	Examples	Structure of representative compound
Coumestans	Coumestrol 4-Methyl coumestrol	 Coumestrol
Lignans	Dibenzylbutan(diol)e lignans Enterodiol Secoisolariciresinol	 Secoisolariciresinol
	Tetrahydrofuran lignans Enterolactone Matairesinol	 Matairesinol
Stilbenes	Resveratrol Pterostilbene	 Resveratrol

Source: Gencel et al.[170] (modified).

8.2.10.1 PHYTOESTROGEN-RICH FOODS

Phytoestrogens are present in many food including vegetables, fruits, wine, and tea. Phytoestrogens have also been identified in a variety of herbal dietary supplements.[171,172] Plants vary in concentrations of phytoestrogens due to variation in soil, climate, and the age of the plant. Chemically, phytoestrogens are polyphenolic phytochemicals. These constitute the largest class of phytochemicals and are most widely distributed in the plants.[173] Polyphenols show multiple functions in plants, and they exert a protective action on insects, fungi, viruses, and bacteria. Polyphenols also act as antioxidants plant hormone controllers.[173]

8.2.10.2 FLAVONOIDS

Flavonoids are a group of plant phenols including over 4000 different compounds of which are amongst the most studied phytochemicals. The structure allows a large number of variations giving rise to isoflavonoids (glycitein, genistein, daidzein), flavonols (quercetin, kaempferol, myricetin), flavones (apigenin, luteolin), anthocyanidins, and flavanones (catechin, epicatechin).[174] An important effect of flavonoids is the scavenging of oxygen-derived free radicals. The main source of isoflavonoids in the diet is soy-based food.[175]

The isoflavonoids from soy beans, genistein, and daidzein are the most studied phytoestrogens.[176] They either exist as aglycones or glucosides. The aglycones are normally transported across intestinal epithelial cells, whereas the glucosides are hydrolyzed in the gut to their aglycones. Genistein shows one-third the potency of estradiol when it interacts with ERα, and one-thousandth of the potency of estradiol when it interacts with ERβ (determined by expression of luciferase reporter gene construct in kidney cells that have been cotransfected with ERα and ERβ).[177] Genistein can show similar effects to estradiol in tissues including ovary, endometria, prostate, vascular, breast, bone, and cell lines.[178–181] Moreover, genistein induces responses that are not related to the ER due to inhibition of tyrosine kinase and DNA topoisomerase. Results shown even in the presence of antiestrogens revealing a nongenomic action that might explain the difference between genistein and estradiol.[182] Estrogenic effects of dietary genistein are also documented. Hsieh et al.[183] conducted in vitro and in vivo studies using MCF-7 cell tumors reporting that dietary genistein (750 ppm) act as an estrogen agonist resulting in the stimulation of mammary gland growth and enhance the growth of MCF-7 cell tumors in athymic mice. Another study on different concentrations of genistein (15, 150, or 300 ppm) in soy isolates was conducted showing varying amounts of genistein increased estrogen-dependent tumor growth in a dose-dependent manner.[184]

Isoflavonoids have potential to prevent cardiovascular disease. Setchell and Cassidy[185] discussed the cardioprotective effect of isoflavonoids (soy food) during estrogen deficiency. The mechanism of cardioprotective effect of isoflavonoids in estrogen replacement therapy includes decrease in low-density lipoprotein cholesterol and increase in high-density lipoprotein cholesterol. In addition, in vitro experiments showed that flavonoids possess anti-inflammatory, anti-allergic, antiviral, anticarcinogenic properties, and various molecules, notably isoflavonoids, are identified as phytoestrogens having the ability to bind estrogens receptors and possess estrogenic or anti-estrogenic activities.[186–188]

8.2.10.3 LIGNANS

Lignans are constituents of higher plants, such as whole grains, legumes, vegetables, and seeds with exceptionally high concentrations of lignans found in flaxseed.[189,190] Although earlier thought to be present only in higher plants, lignans have also been detected in the biological fluids of humans and animals. Lignans act as either estrogen agonists or antagonists and have generated interest due to their potential use in hormone replacement and prevention of hormone-dependent cancers of uterus and prostate and breast.[191] Apart from health benefits, lignans are reported to cause infertility and hyperestrogenization in many species.[192] Dietary substances capable of altering secretion levels of estrogen during pregnancy are of major concern because pregnancy is a hormone-sensitive period for both the mother and offspring. During pregnancy, high estrogen levels are needed by the mother to maintain pregnancy by stimulating uterine changes to allow implantation, enhancing uterine growth to accommodate a growing fetus and by acting as a trigger for parturition. On contrary, administration of estrogens to pregnant animals has been found to cause detrimental effects on pregnancy.[193]

8.2.10.4 COUMESTANS

Foods like alfalfa and clover sprouts are the major sources of coumestans.[194,195] Overall, consumption of coumestans is not as high as that of isoflavones and lignans. Despite its relatively reduced prevalence in foods compared to lignans and isoflavones, many biochemical properties of coumestans have been elucidated. Once digested, coumestans are broken down into many compounds including coumestrol, which has strong affinity for the ER. Coumestrol has shown preferential binding to ERβ than to ERα.[194] Amongst number of phytoestrogen metabolites studied (including daidzein and genistein), coumestrol have shown the strongest binding affinity for the ER. It has been shown that coumestrol binds to the ER with an affinity only 5–10-fold lesser than that of 17β-estradiol.[196] This comparatively strong binding affinity for the ER has led some researchers to postulate that coumestans are the most potent phytoestrogen. However, this has yet to be tested clinically.

Much information is not available concerning the safety profile of isolated coumestan use. Their effects seem to be like those of isoflavones and lignans. Coumestrol intake was not associated with changes in endometrial cancer risk.[197] Similarly, the study conducted by Bandera et al.[198] reported no association between coumestan-containing foods and increased cancer risk. Moreover, in models of ovariectomized rats, coumestrol did not cause endometrial hyperplasia.[199] The limited information on isolated coumestan supplementation makes risk assessment difficult. Further studies in future studies may elucidate their full side-effect profile.

8.2.10.5   STILBENES

Resveratrol is a natural phenol which is found in grape skin and a wide range of medicinal plants. Due its high concentration in grape skin, good amount of resveratrol is present in red wine.[200] The effects of resveratrol have been proposed as a partial explanation for the apparent ability of moderate consumption of red wine to lower down the risk of cardiovascular disease.[201] Platelets from human volunteers who consumed 2 mg of resveratrol per day showed diminished thromboxane B2 synthesis and reduced thrombin-induced aggregation compared with controls. Daily consumption of some red wines might produce pharmacologically significant serum concentrations of resveratrol.[202] In addition to cardioprotective effects, many in vivo studies have reported that resveratrol also carries anti-inflammatory, antiviral, antidiabetic, and neuroprotective effects.[203–205] Furthermore, animal studies have also discussed its effects as a cancer preventative agent.[206]

Resveratrol has been shown to bind to the human ER.[201] As a phytoestrogen, resveratrol is a selective estrogen receptor modulator and its relative agonistic and antagonistic effects have yet to be fully elucidated. Some studies show that in the absence of 17β-estradiol, resveratrol demonstrates mixed agonistic and antagonistic effects in breast; however, in the presence of 17β-estradiol, it functions as an antagonist. Bhat[207] showed that in animal models, carcinogen-induced preneoplastic lesions and mammary tumors are inhibited by resveratrol showing its chemopreventive effect. It has also been suggested that resveratrol exerts its cancer protective properties via its combination of interaction with the ER and as its role in tyrosine-kinase signaling. In a study, resveratrol reported to be a competitive inhibitor of 17β-estradiol of rat and human with IC_{60} values of 3.87 ± 0.06 and 8.48 ± 0.04 µM, respectively. However, it did not inhibit hydroxysteroid dehydrogenase 3, leading to specific inhibition of 3β-17β-estradiol (100 µM) tested. Therefore, it was concluded that resveratrol had the potential to interfere with androgen biosynthesis of rat Leydig cells.[153,208]

8.3 SOURCES OF EXPOSURE AND EXPOSURE ASSESSMENT

8.3.1 TRANSFER OF FOOD CONTACT MATERIALS INTO FOODS

Materials and articles intended to come in food contact such as packaging, containers, food-processing equipments and kitchen wares are the potential source of migrating chemicals into the food. Safety of these food contact materials should be evaluated to prevent the chemical migration and contamination in food. Food-packaging material has gained much attention for food safety assessment.[209] Foods may also be highly contaminated with packaging-related xenobiotic chemicals from printing inks and recycled paperboard. As discussed in previous sections of this chapter, many of the food contact materials such as BPA, phthalate esters, alkyl phenols reported to have the potential of endocrine disruption. DEHP, DBP, and benzophenone were found to transfer from recycled carton food packaging into breadcrumbs and rice.[210] In a statement by EFSA in 2009, it was found that the concentration of 4-methylbenzophenone in cereal samples packaged in paperboard was up to the level of 3.7 ppm in the food.[211] Benzophenone was reported in the vary of 232.7–580.9 mg/m² in freshly done carton food packaging. In this study, sufficient inner bag material was found to be multilayer plastic foils, whereas single layer polypropylene could not prevent migration of benzophenone into the food.[212] Highest migration levels of 3–5 ppm were found mainly in finely ground foods like flour or icing sugar.

The risk assessment for intensity of migration and potential of toxicity of the food-contacting materials, two approaches are to be considered: (1) exposure assessment and (2) hazard characterization.[14] Both the approaches have their advantages and limitations, and there appears to be no consensus among the scientific community for a harmonized approach.

8.3.2 MIGRATION THROUGH LIQUID FOODS FROM PLASTIC BOTTLES AND REGULATORY SCENE: THE EXAMPLE OF BPA

Human exposure to BPA is described as "continuous and widespread," and that it is "primarily through food."[213] Babies are supposed to be the most sensitive group exposed to BPA.[24,214] Babies that are not breast fed will drink powdered or liquid milk formula through bottles. Additionally, cans of milk formula may be coated with epoxy resin and may release BPA into the milk. Powdered milk formulations are normally not packed in epoxy resin-lined cans, whereas the liquid milks are for the required heat sterilization. Milk formula may be served in a polycarbonate bottle adding to the exposure to BPA. Breast-fed infants may also receive additional feedings of bottled milk.[215] Babies are also likely to drink other liquids such as water and juices, apart from milk in plastics bottles. Babies spend much time on the floor and put many plastic items in their mouths. These scenarios lead to higher exposure levels of BPA in them than adults.[216]

Several risk assessments of BPA have reported that exposure levels are far below the NOAEL.[217] Because of several limitations in risk assessment methodologies of toxicological studies, a strong scientific discussion arose. These limitations were lesser number of animals, nonoral administration routes, fewer or single-dose groups, etc. Following the precautionary principle, Denmark decided in March 2010 to provisionally ban all BPA-containing food containers made for children less than 3 years of age. This decision followed a similar ban imposed in Canada and in some US states. In May 2010, the National Assembly of France voted in favor of a ban of BPA in plastic baby feeding bottles.[215,217]

8.3.2.1 PREVIOUS STUDIES AND CONTROVERSIES

BPA can leach from polycarbonate into liquid foods through diffusion and hydrolysis of polycarbonate catalyzed by hydroxide.[217] Release of BPA from polycarbonate containers into food also depends on the contact time, temperature, and type of food.[217,218] Some scientists studied the difference between release of BPA by new and used baby bottles and came to contradictory conclusions.[215,219,220] Furthermore, the researchers did not report whether the used bottles were of the same production lot as the new ones. Therefore, their experiments cannot be taken as conclusive on whether the rate of release of BPA would decrease or increase with age of the bottle. Washing the bottles also induces chemical ageing of the internal surface. However, experimental results on this topic are not consistent.

Mercea[220] and Biedermann-Brem et al.[221] reported that the bottles released lower amounts of BPA into aqueous food simulants after repeated washing. On the contrary, Brede et al.[222] and Nam et al.[223] reported an increase, whereas Kubwabo et al.[224] observed no significant change in the releasing concentration of BPA in washed bottles.

Few studies reported other influencing factors that can affect the releasing concentrations of BPA in washed bottles. A higher pH clearly increases migration of BPA, independent from other physical factors like exposure time, temperature, and heating mode (microwave/thermal oven).[216,221,225] This has strong relevance, if one considers that some food preparation processes can cause an increase of the pH above 8, which is generally the highest value for food. For example, boiling tap water in a pan or microwave for several minutes during which carbon dioxide may be released, consequently increases the pH.[214] Mineral content can also influence release. However, there is no clear evidence that the effect was actually caused by the mineral composition and not by the pH.

Maia et al.[226] observed the aminolysis of polycarbonate into BPA by some amines. It should be a relevant method for generally present biogenic amines in milk. Solutions of 1,3-cyclohexane-bis(methylamine) and 1,4-diaminobutane (600 mg/L) showed clear aminolysis at 121°C for 1 h, whereas trimethylamine and 1,3-xylylenediamine were not much effective. The pH result (10.4–11.4) was excluded by comparing with relevant sodium hydroxide solutions. During contact at 25°C for 5 days, aminolysis was not detectable. Aminolysis of polycarbonate can occur with 50 mM solutions of methionine and adjusted to pH 11 by sodium hydroxide at lower temperatures of 37°C.[227-229] The migration rates range from 0.38 to 0.49 mg/L/day and corrected for the realistic surface to volume ratio of baby bottles of about 8/dm, 0.11–0.14 mg/L/day. These migration rates are much lesser than those that can be calculated from the work of Maia et al.[226] that range from 48 to 12,000 mg/L/day or corrected for S/V, 19–5000 mg/L/day. Observation of Sajiki and Yonekubo[228] state that amino acids can result in about 0.01 mg/L of BPA at pH 11 and 37°C for 30 min contact. Few values on migration of BPA into milk that exist were below the LOD of 0.03 mg/L. It cannot be concluded from present information whether or not aminolysis contributes to migration of BPA into milk or not. However, it is clear that only a few substances might have an effect and the question is whether or not these substances are present in milk at relevant concentrations.

Another relevant depolymerization process that may influence migration of BPA from polycarbonate during migration testing with 500th grain alcohol is ethanolysis.[232] Methanolysis is an important process to recycle BPA only under mild conditions because at critical conditions (243°C and half-dozen 6.4 MPa) polycarbonate is converted by ethanol into BPA and diethyl carbonate.[230–232]

8.3.2.2 INITIATIVES TO REDUCE THE EXPOSURE OF BPA TO BABIES

Canada has taken the initiative to protect young children from BPA.[225] Since the exposure to BPA during development poses high risk for adverse effects, the fetus and baby are believed to be more prone to the estrogenic effects of BPA.[233,234] Due to their small body size and limited capacity to metabolize BPA there is increased risk of toxicity in children. Early exposure might have lifelong adverse consequences. BPA may alter the programing of genes during critical periods of cell differentiation throughout fetal and neonatal development.[234,235] This method, referred to as "epigenetics' programing," may result in the expression of metabolic disruption and cancers during later life.[234,235] Concern regarding BPA stretches far beyond the nursery, although that is a fine place to start a retreat from EDCs in general and BPA especially.

8.3.2.3 SUGGESTIONS TO REDUCE THE BPA EXPOSURE IN BABIES

Studies recommend minimizing baby and infant exposure to BPA. Suggestions include breast feed as much as and as long as possible, if using an infant formula, select a powder and avoid liquid, avoid polycarbonate containers, use glass bottles, and use as few cans as possible. Whereas the problems are complicated, and BPA contamination is multifaceted and multisource, individual action should be prioritized, based on what is already understood: "of particular concern is that the use of BPA in food and beverage plastic storage and heating containers and to line metal cans." Pregnant women, and others, could decrease their BPA exposure by reducing their intake of canned food (where the plastic lining will leach BPA), and by avoiding BPA-containing food containers like polycarbonate drink bottles. For infants, the use of glass bottles is highly recommended. If using processed baby food, glass jars may be preferred instead of cans. Citizens may be encouraged to particpate in actions for reducing risk of BPA. Of late, the organic and biodynamic food sectors have aimed to provide the market with food free from xenobiotics including BPA, and we can hope that in future its hazard will be minimized.

8.4   EFFECT ASSESSMENT

Effects are generally assessed for additives and monomers, however, not for complete leacheate that can contaminate foods. Further, the final packaging with printing inks, adhesives, and labels may possibly leach additional compounds into the foods. At present, there is no systematic assessment of toxicity of complete-packaging leacheates. Presently, effect assessment for new substance authorizations is based on exposure levels. For a low exposure to a deliberately added substance, toxicity testing is needed focusing on genotoxicity and mutagenicity. Studies assessing reproductive toxicity are only needed, if the leaching substance will be present in food simulants at 5 ppm or higher in the EU and 1 ppm (1 mg/kg food) or higher in the United States.[236,237]

EDCs act not only via receptor binding but can also act via enzymatic pathways involved in steroid biosynthesis and metabolism, and numerous other mechanisms that converge upon endocrine and reproductive systems.[6,7,238] Originally, there was concern for developmental and reproductive toxicities. These chemicals have recently been linked to various effects like cardiovascular disease, diabetes, cancers, obesity, and neurological effects and are of a major concern to public health.[2,235] In case of chronic exposure to BPA effects are seen in animal studies particularly if exposure occurs during the perinatal period. The ongoing debate regarding low-dose effects of BPA is controversial. EU and US regulations are the most widely used international regulations for food contact materials.[239,240] However, countries like China, Japan, and Canada follow their own regulations. EFSA also plays an important role in regulation. It conducts conventional risk assessment and prescribes the way in which an application for a listing is to be submitted. EFSA gives opinion on recommendation or restriction of any substance based on the results of safety assessments. Safety assessment is further followed by risk management decision for the final approval. However, many changes and modifications for the real-time health-hazard assessment need to be done. A study published by Dionisi and Oldring[241] suggested to involve per capita exposure to assess health hazard more accurately. Authors observed that default regulation of EU based on the assumption of individuals eating 1 kg of food/day in a cube of surface area 6 dm^2, packaged with the same material, is inappropriate for canned foods. They observed that the surface of canned foods and canned beverages was only one-sixth (1 dm^2) of EU assumption. Another study reported by de Fátima Pocas et al.[239] different approaches of assessments were described including US and EU regulations. The major concern of the study was those compounds with uncertain effect or about which prolonged exposure data are not available. Authors suggested probabilistic approaches of health-risk assessment for additives and other contaminants. Considering the laboratory and sustained epidemiological data, several EDCs have been banned or their daily intake regulated to minimize the exposure. Although the ban of organotins had its origin in maritime usgae, Canada became the first country to ban BPA in baby feeding bottles (Table 8.5).

8.5 CONCLUSIONS AND FUTURE RESEARCH DIRECTION

Much evidence confirms the potential effects of EDCs on human health even at low concentrations, as they are generally encountered in foods. Chronic exposure to EDCs is of toxicological concern and this concern increases when humans are exposed to mixtures of similarly acting EDCs and during sensitive windows of development. The widespread use of chemicals with endocrine-disrupting properties in foods and food packaging thus might present a risk, and it requires dedicated assessment. EDCs have not been much researched with respect to presence in food, migration, human body burdens, and their impact on human health. Not all substances migrating from food and food packaging have been characterized for their endocrine disruptive effects, which is sometimes discovered only by chance long after they are widely in use. For example, black nylon utensils were notified as nonconforming food contact material in the markets. These products exhibited excessive migration of aromatic amines which are considered to be harmful to human health. However, risk from the use of these utensils was extremely low. The current situation implies chronic exposure to EDCs with significant unknown effects that present a health risk that also might affect future generations considering effect in infants.

TABLE 8.5 Epidemiological Effects, Levels in Human Samples, and Legal Status of EDCs.

EDCs	Human epidemiological studies	Levels in human	Legal status	References
Bisphenol A	Associated with diabetes and liver abnormalities	TDI: range from <50 µg/kg/bw/day (US-EPA)	Canada became the first country to ban BPA in baby bottles; WHO begins assessing BPA safety	Lang et al.[242] Maiolini[243] Rubin[244]
Pthalates (e.g., DEHP, DBP)	Associated with obesity and insulin resistance	DBP's TDI: 10 mg/kg/bw/day (EU-EFSA)	Restricted in toys in EU, 1999 and United States, 2009	Stahlhut et al.[245]
Alkylphenols (e.g., octylphenol)		In urine: 0.4–13.9 ng/mL; In adipose tissue: 57 ng/g (Spain)	Restricted for some uses in EU	Lee et al.[246]
Organochlorine (e.g., DDT)	Associated with diabetes and MetS	DDE range from <5 to 15,000 mg/kg/bw	Banned in most developed countries since 1970; restricted by Stockholm Convention in 2000s	Smink et al.[247] Park et al.[248] Turyk et al.[249]
Organotins (e.g., TBT, TPTO)		In serum: 27 nM; In human tissue: 3–100 nM	Banned worldwide by International Maritime Organization	Grun et al.[250] Kanayama et al.[251] Le Maire et al.[252]
Dioxins (e.g., PCB, TCDD)	Associated with obesity, diabetes, and MetS	In adipose tissue: 3.6 pg/(g lipid); In blood: 2.2 ppt	PCB banned and other dioxins restricted by Stockholm Convention	Cimafranca et al.[253]

MetS, metabolic syndrome; *BW*, body weight; *TDI*, tolerable daily intake; *ppt*, parts per trillion; *WHO*, World Health Organization; *EU*, European Union; *US-EPA-US*, Environmental Protection Agency, *EU-EFSA*, European Food Safety Authority.

Taken together, policies governing the use of EDCs in foods and food contact materials should be revised as needed to reflect contemporary scientific understanding. Substances that have been in use for many years should be reassessed for their endocrine disruptive potential using the latest toxicology principles. Several aspects of risk assessment are important when drawing conclusions from toxicological studies on EDCs released from food contact materials into foods. Above all, consumer safety must be taken into account and concerns more than just a single substance having endocrine-disrupting potential. In the cases in which some of EDCs induce similar effects, the possibility of competing toxicity should be considered. Furthermore, recommendations on acceptable intake of particular EDCs refer to the average consumer and may not adequately secure those individuals who consume food products containing abnormally high amounts of these substances and have high vulnerability index such as children.

With today's toxicological knowledge threshold concepts for unidentified EDC migrants from foods and food-packaging materials need thorough reassessment and validation according to modern scientific developments. An additional screening for endocrine-disrupting effect might prove useful for exposure and risk assessment; it could determine the complete migrants from the finished packaging into foods, for existing as well as new authorizations. This requires establishment of the relevance of such bioassays for human and environmental health. Strategies as to how to minimize EDC exposure need to address packaged food and food contact material as a potentially large source of contamination and exposure. Detailed information on human exposure to EDCs from foods is required from academics. In terms of fetal exposure to EDCs and chronic health effects later in life, longitudinal studies are needed to assess the contribution of foods to EDCs exposure. All these efforts would certainly benefit from a larger awareness within the scientific and social communities for different types of foods and food contact material as a major food contaminant source of EDCs. However, due to their low level of exposure and several uncertainly and confounding factors, appropriate risk assessment of EDCs still remains elusive.[254] There is also a need to have some kind harmonization of methodological approaches for exposure and risk assessment of EDCs, especially for the ones which have substantial exposure such as BPA.

KEYWORDS

- exogenous agents
- anthropogenic activities
- physicochemical reactions
- hormone receptors
- toxic effects

REFERENCES

1. EPA. *Endocrine*. US Environmental Protection Agency, 2012.
2. Gore, A. C.; Chappell, V. A.; Fenton, S. E.; Flaws, J. A.; Nadal, A.; Prins, G. S.; Toppari, J.; Zoeller, R. T. EDC-2: The Endocrine Society's Second Scientific Statement on Endocrine-Disrupting Chemicals. *Endocr. Rev.* **2015,** *36*, E1–E150.
3. Schug, T. T.; Janesick, A.; Blumberg, B.; Heindel, J. J. Endocrine Disrupting Chemicals and Disease Susceptibility. *J. Steroid Biochem. Mol. Biol.* **2011,** *127*, 204–215.
4. Wetheril, Y. B.; Akingbemi, B.; T.; Kanno, J.; McLachlan, J. A.; Nadal, A.; Sonnenschein, C.; Watson, C. S.; Zoeller, R. T.; Belcher, S. M. *In Vitro* Molecular Mechanisms of Bisphenol A Action. *Reprod. Toxicol.* **2007,** *24*, 178–198.
5. Riu, A.; le, Maire, A.; Grimaldi, M.; Audebert, M.; Hillenweck, A.; Bourguet, W.; Balaguer, P.; Zalko, D. Characterization of Novel Ligands of ERα, Erβ, and PPARγ: The Case of Halogenated Bisphenol A and their Conjugated Metabolites. *Toxicol. Sci.* **2011,** *122*, 372–382.
6. Delfosse, V.; Maire, A. L.; Balaguer, P.; Bourguet, W. A Structural Perspective on Nuclear Receptors as Targets of Environmental Compounds. *Acta Pharmacol. Sin.* **2015,** *36*, 88–101.
7. Delfosse, V., Grimaldi, M., le Maire, A., Bourguet, W., Balaguer, P. Nuclear Receptor Profiling of Bisphenol-A and its Halogenated Analogues. *Vitam. Horm.* **2014,** *94*, 229–251.
8. Fudvoye, J.; Bourguignon, J. P.; Parent, A. S. Endocrine-Disrupting Chemicals and Human Growth and Maturation: A Focus on Early Critical Windows of Exposure. *Vitam. Horm.* **2014,** *94*, 1–25.
9. Pivnenko, K.; Eriksen, M. K.; Martín-Fernández, J. A.; Eriksson, E.; Astrup, T. F. Recycling of Plastic Waste: Presence of Phthalates in Plastics from Households and Industry. *Waste Manage.* **2016,** *54*, 44–52.
10. Toro-Velez, A. F.; Madera-Parra. C. A.; Peña-Varón, M. R.; Lee, W. Y.; Bezares-Cruz, J. C.; Walker, W. S.; Cárdenas-Henao, H.; Quesada-Calderón, S.; García-Hernández, H.; Lens, P. N. BPA and NP Removal from Municipal Wastewater by Tropical Horizontal Subsurface Constructed Wetlands. *Sci. Total Environ.* **2016,** *542*, 93–101.
11. Bonefeld-Jørgensen, E. C.; Ghisari, M.; Wielsoe, M.; Bjerregaard-Olesen, C.; Kjeldsen, L. S.; Long, M. Biomonitoring and Hormone-Disrupting Effect Biomarkers of Persistent

Organic Pollutants In Vitro and *Ex Vivo. Basic Clin. Pharmacol. Toxicol.* **2014,** *115,* 118–128.

12. Waring, R. H.; Ayers, S.; Gescher, A. J.; Glatt, H. R.; Meinl, W.; Jarratt, P.; Kirk, C. J.; Pettitt, T.; Rea, D.; Harris, R. M. Phytoestrogens and Xenoestrogens: The Contribution of Diet and Environment to Endocrine Disruption. *J. Steroid Biochem. Mol. Biol.* **2008,** *108,* 213–220.

13. Haighton, L. A.; Hlywka, J. J.; Doull, J.; Kroes, R.; Lynch, B. S.; Munro, I. C. An Evaluation of the Possible Carcinogenicity of Bisphenol A to Humans. *Regul. Toxicol. Pharmacol.* **2002,** *35,* 238–254.

14. Muncke, J. Endocrine Disrupting Chemicals and Other Substances of Concern in Food Contact Materials: An Updated Review of Exposure, Effect and Risk Assessment. *J. Steroid Biochem. Mol. Biol.* **2011,** *127,* 118–127.

15. Alonso-Magdalena, P.; Quesada, I.; Nadal, Á. Prenatal Exposure to BPA and Offspring Outcomes: The Diabesogenic Behaviour of BPA. *Dose Respon.* **2015,** *13,* 1559325815590395.

16. Stel, J.; Legler, J. The role of Epigenetics in the Latent Effects of Early Life Exposure to Obesogenic Endocrine Disrupting Chemicals. *Endocrinology* **2015,** *156,* 3466–3472.

17. Vaiserman, A. Early-Life Exposure to Endocrine Disrupting Chemicals and Later-Life Health Outcomes: An Epigenetic Bridge. *Aging Dis.* **2014,** *5,* 419–429.

18. Chen, J. Q.; Brown, T. R.; Russo, J. Regulation of Energy Metabolism Pathways by Estrogens and Estrogenic Chemicals and Potential Implications in Obesity Associated with Increased Exposure to Endocrine Disruptors. *BBA—Mol. Cell. Res.* **2009,** *1793,* 1128–1143.

19. Petersen, J. H.; Jensen, L. K. Phthalates and Food-Contact Materials: Enforcing the 2008 European Union Plastics Legislation. *Food Addit. Contam., A: Chem. Anal. Control Expo. Risk Assess.* **2010,** *27,* 1608–1616.

20. US-EPA. *Endocrine Disruptor Screening Program Comprehensive Management Plan.* February 14, 2014.

21. Lee, M. J.; Lin, H.; Liu, C. W.; Wu, M. H.; Liao, W. J.; Chang, H. H.; Ku, H. C.; Chien, Y. S.; Ding, W. H.; Kao, Y. H. Octylphenol Stimulates Resistin Gene Expression in 3T3-L1 Adipocytes via the Estrogen Receptor and Extracellular Signal-Regulated Kinase Pathways. *Am. J. Physiol. Cell Physiol.* **2008,** *294,* 1542–1551.

22. Diamanti-Kandarakis, F. E.; Bourguignon, J. P.; Giudice, L. C.; Hauser, R.; Prins, G. S.; Soto, A. M.; Zoeller, R. T.; Gore, A. C. Endocrine-Disrupting Chemicals: An Endocrine Society Scientific Statement. *Endocr. Rev.* **2009,** *30,* 293–342.

23. Anjum, S.; Rahman, S.; Kaur, M.; Ahmad, F.; Rashid, H.; Ansari, R. A.; Raisuddin, S. Melatonin Ameliorates Bisphenol A-Induced Biochemical Toxicity in Testicular Mito-chondria of Mouse. *Food Chem. Toxicol.* **2011,** *49,* 2849–2854.

24. Vandenberg, L. N.; Colborn, T.; Hayes, T. B.; Heindel, J. J.; Jacobs, D. R.; Lee, D. H.; Shioda, T.; Soto, A. M.; vom Saal, F. S.; Welshons, W. V.; Zoeller, R. T.; Myers, J. P. Hormones and Endocrine-Disrupting Chemicals: Low-Dose Effects and Nonmonotonic Dose Response. *Endocr. Rev.* **2012,** *33,* 378–455.

25. Zlatnik, M. G. Endocrine-Disrupting Chemicals and Reproductive Health. *J. Midwifery Womens Health* **2016,** *61,* 442–455.

26. Nakamishi, T.; Hiromori, Y.; Yokoyama, H.; Koyanagi, M.; Itoh, N.; Nishikawa, J.; Tanaka, K. Organotin Compounds Enhance 17-Beta-Hydroxysteroid Dehydroge-nase Type I Activity in Human Choriocarcinoma JAr Cells: Potential Promotion of

17Beta-Estradiol Biosynthesis in Human Placenta. *Biochem Pharmacol.* **2006,** *71,* 1349–1357.

27. Muncke, J. Exposure to Endocrine Disrupting Compounds via the Food Chain: Is Packaging a Relevant Source? *Sci. Total Environ.* **2009,** *407,* 4549–4559.

28. Iavicoli, I.; Fontana, L.; Leso, V.; Bergamaschi, A. The Effects of Nanomaterials as Endocrine Disruptors. *Int. J. Mol. Sci.* **2013,** *14,* 16732–16801.

29. Wyser, Y.; Adams, M.; Avella, M.; Carlander, D.; Garcia, L.; Pieper, G.; Rennen, G.; Schuermans, J.; Weiss, J. Outlook and Challenges of Nanotechnologies for Food Packaging. *Packag. Technol. Sci.* **2016.** DOI:10.1002/pts.2221.

30. EFSA CEP Panel. Scientific Opinion on the Safety Evaluation of the Active Substance Iron(II) Modified Bentonite as Oxygen Absorber for Use in Active Food Contact Materials. *EFSA J.* **2012,** *10,* 2906.

31. EFSA CEP Panel. Scientific Opinion on the Safety Evaluation of the substance, Titanium Nitride, Nanoparticles, For Use in Food Contact Materials. *EFSA J.* **2012,** *10,* 641.

32. EFSA CEP Panel. Scientific Opinion on the Safety of the Substances, Kaolin and Polyacrylic Acid, Sodium Salt, for Use in Food Contact Materials. *EFSA J.* **2014,** *12,* 3637.

33. EFSA CEP Panel. Scientific Opinion on the Safety Evaluation of the Substance Zinc Oxide, Nanoparticles, Uncoated or Coated with [3-(Methacryloxy)Propyl] Trimethoxisilane, for Use in Food Contact Materials. *EFSA J.* **2015,** *13,* 4063.

34. Harvey, P. W.; Everett, D. J. Regulation of Endocrine Disrupting Chemicals: Critical Overview and Deficiencies in Toxicology and Risk Assessment for Human Health. *Best Pract. Res. Clin. Endocrinol. Metab.* **2006,** *20,* 145–165.

35. Max-Stoelting, P.; Pfeil, R.; Solecki, R.; Ulbrich, B.; Grote, K.; Ritz, V.; Banasiak, U.; Heinrich-Hirsch, B.; Moeller, T.; Chahoud, I.; Hirsch-Ernst, K. I. Assessment Strategies and Decision Criteria for Pesticides with Endocrine Disrupting Properties to Humans. *Reprod. Toxicol.* **2011,** *31,* 574–584.

36. Rudén, Ch. Principles and Practices of Health Risk Assessment under Current EU 37 Regulations. *Reg. Toxicol. Pharmacol.* **2006,** *44,* 14–23.

37. Meeker, J. D. Exposure to Environmental Endocrine Disruptors and Child Development. *Arch. Pediatr. Adolesc. Med.* **2012,** *166,* 952–958.

38. Hotchkiss, A. K.; Rider, C. V.; Blystone, C. R.; Wilson, V. S.; Hartig, P. C.; Ankley, G. T.; Foster, P. M.; Gray, C. L.; Gray, L. E. Fifteen Years after "Wingspread"—Environmental Endocrine Disrupters and Human and Wildlife Health: Where We Are Today and Where We Need to Go. *Toxicol. Sci.* **2008,** *105,* 235–259.

39. Cwiek-Ludwicka, K. Hazards for Health Related to the Migration of Chemical Substances from Packaging into Food. *Rocz. Panstw. Zakl. Hig.* **2010,** *61,* 341–347.

40. Grob, K.; Biedermann, M.; Scherbaum, E.; Roth, M.; Rieger, K. Food Contamination with Organic Materials in Perspective: Packaging Materials as the Largest and Least Controlled Source? A View Focusing on the European Situation. *Crit. Rev. Food Sci. Nutr.* **2006,** *46,* 529–535.

41. Bradley, E.; Coulier, L. *An Investigation into the Reaction and Breakdown Products from Starting Substances Used to Produce Food Contact Plastics.* Central Science Laboratory: London, 2007, FD 07/01.

42. Ter Veld, M. G.; Schouten, B.; Louisse, J.; van Es, D. S.; van der Saag, P. T.; Rietjens, I. M.; Murk, A. J. Estrogenic Potency of Food-Packaging-Associated Plasticizers and Antioxidants as Detected in ERα and ERβ Reporter Gene Cell Lines. *J. Agric. Food Chem.* **2006,** *54,* 4407–4416.

43. European Commission. *Defining Criteria for Identifying Endocrine Disruptors in the Context of the Implementation of the Plant Protection Product Regulation and Biocidal Products Regulation*, 2014, DG ENV.A.3, DG.SANCO.E.3.

44. Sonnenschein, C.; Soto, A. M. An Updated Review of Environmental Estrogen and Androgen Mimics and Antagonists. *J. Steroid Biochem. Mol. Biol.* **1998**, *65*, 143–150.

45. Skultétyová, I.; Tokarev, D.; Jezová, D. Stress-Induced Increase in Blood–Brain Barrier Permeability in Control and Monosodium Glutamate-Treated Rats. *Brain Res. Bull.* **1998**, *45*, 175–178.

46. Wang, C. F.; Tian, Y. Reproductive Endocrine-Disrupting Effects of Triclosan: Population Exposure, Present Evidence and Potential Mechanisms. *Environ. Pollut.* **2015**, *206*, 195–201.

47. Khasnavis, S.; Pahan, K. Sodium Benzoate, a Metabolite of Cinnamon and a Food Additive, Upregulates Neuroprotective Parkinson Disease Protein DJ-1 in Astrocytes and Neurons. *J. Neuroimmune Pharmacol.* **2012**, *7*, 424–435.

48. Acconcia, F.; Pallottini, V.; Marino, M. Molecular Mechanisms of Action of BPA. *Dose Respon.* **2015**, *13*, 1559325815610582.

49. Rodríguez, D. A. O.; de Lima, R. F.; Campos, M. S.; Costa, J. R.; Biancardi, M. F.; Marques, M. R.; Taboga, S. R.; Santos, F. C. Intrauterine Exposure to Bisphenol A Promotes Different Effects in Both Neonatal and Adult Prostate of Male and Female Gerbils (*Meriones unguiculatus*). *Environ. Toxicol.* **2015**. DOI:10.1002/tox.22176.

50. Ahmadivand, S.; Farahmand, H.; Mirvaghefi. A.; Eagderi, S.; Zargar, A. Effects of (Anti)Androgenic Endocrine Disruptors (Dehp and Butachlor) on Immunoglobulin m(IgM) and Leukocytes Counts of Male Rainbow Trout (*Oncorhynchus mykiss*). *Bull. Environ. Contam. Toxicol.* **2015**, *94*, 695–700.

51. Howdeshell, K. L.; Rider, C. V.; Wilson, V. S.; Furr, J. R.; Lambright, C. R.; Gray, L. E. Dose Addition Models Based on Biologically Relevant Reductions in Fetal Testosterone Accurately Predict Postnatal Reproductive Tract Alterations by a Phthalate Mixture in Rats. *Toxicol. Sci.* **2015**, *148*, 488–502.

52. Bradley, E. L.; Stratton, J. S.; Leak, J.; Lister, L.; Castle, L. Printing Ink Compounds in Foods: UK Survey Results. *Food Addit. Contam., B: Surveill.* **2013**, *6*, 73–83.

53. EU. *EU Strategy for EDC*, 2010. http://eng.mst.dk/topics/chemicals/endocrine-disruptors/the-eu-list-of-potential-endocrine-disruptors/.

54. EFSA. Scientific Opinions of the Panel on Food Additives, Flavourings, Processing Aids and Materials in Contact with Food (AFC) on a Request from the Commission on the Toxicokinetics of Bisphenol A. *EFSA J.* **2008**, *759*, 1–10.

55. APERC (APE Research Council). *Product Information*, 2014. http://www.aperc.org/productinfo.htm.

56. European Chemicals Bureau, European Commission. *European Union Risk Assessment Report. 4-Nonylphenol (Branched) and Nonylphenol*. Institute for Health and Consumer Protection, European Communities, Luxemburg, 2002.

57. Tsuda, T.; Suga, K.; Kaneda, E.; Ohsuga, M. Determination of 4-Nonylphenol, Nonylphenol Monoethoxylate, Nonylphenol Diethoxylate and Other Alkylphenols in Fish and Shellfish by High-Performance Liquid Chromatography with Fluorescence Detection. *J. Chromatogr., B: Biomed. Sci. Appl.* **2000**, *746*, 305–309.

58. Greenpeace. *The Determination of Additives in Food Products*. Greenpeace, 2003.

59. Guenther, K.; Heinke, V.; Thiele, B.; Kleist, E.; Prast, H.; Raecker, T. Endocrine Disrupting Nonylphenols Are Ubiquitous in Food. *Environ. Sci. Technol.* **2002**, *36*, 1676–1680.

60. Kannan, K.; Kober, J. L.; Kang, Y. S.; Masunaga, S.; Nakanishi, J.; Ostaszewski, A.; Giesy, J. P. Polychlorinated Naphthalenes, Biphenyls, Dibenzo-*p*-dioxins, and Dibenzofurans as well as Polycyclic Aromatic Hydrocarbons and Alkylphenols in Sediment from the Detroit and Rouge Rivers, Michigan, USA. *Environ. Toxicol. Chem.* **2002**, *20*, 1878–1889.

61. Cespedes, R.; Lacorte, S.; Ginebreda, A.; Barcelo, D. Chemical Monitoring and Occurrence of Alkylphenols, Alkylphenol Ethoxylates, Alcohol Ethoxylates, Phthalates and Benzothiazoles in Sewage Treatment Plants and Receiving Waters along the Ter River Basin (Catalonia, N.E. Spain). *Anal. Bioanal. Chem.* **2006**, *385*, 992–1000.

62. Fenet, H.; Gomez, E.; Pillon, A.; Rosain, D.; Nicolas, J. C.; Casellas, C.; Balaguer, P. Estrogenic Activity in Water and Sediments of a French River: Contribution of Alkylphenols. *Arch. Environ. Contam. Toxicol.* **2003**, *44*, 1–6.

63. Ademollo, N.; Ferrara, F.; Delise, M.; Fabietti, F.; Funari, E. Nonylphenol and Octylphenol in Human Breast Milk. *Environ. Int.* **2008**, *34*, 984–987.

64. Gyllenhammar, I.; Glynn, A.; Darnerud, P, O., Lignell, S.; van Delft, R.; Aune, M. 4-Nonylphenol and Bisphenol A in Swedish Food and Exposure in Swedish Nursing Women. *Environ. Int.* **2012**, *43*, 21–28.

65. Nimrod, A. C.; Benson, W. H. Environmental Estrogenic Effects of Alkylphenol Ethoxylates. *Crit. Rev Toxicol.* **1996**, *26*, 335–364.

66. Soto, A. M.; Sonnenschein, C.; Chung, K. L.; Fernandez, M. F.; Olea, N.; Serrano, F. O. The E-SCREEN Assay as a Tool to Identify Estrogens: An Update on Estrogenic Environmental Pollutants. *Environ. Health Perspect.* **1995**, *103*, 113–122.

67. White, R.; Jobling, S.; Hoare, S. A.; Sumpter, J. P.; Parker, M. G. Environmentally Persistent Alkylphenolic Compounds Are Estrogenic. *Endocrinology* **1994**, *135*, 175–182.

68. Lange, A.; Paull, G. C.; Hamilton, P. B.; Iguchi, T.; Tyler, C. R. Implications of Persistent Exposure to Treated Wastewater Effluent for Breeding in Wild Roach (*Rutilus rutilus*) Populations. *Environ. Sci. Technol.* **2011**, *45*, 1673–1679.

69. Kelly, M. A.; Reid, A. M.; Quinn-Hosey, K. M.; Fogarty, A. M.; Roche, J. J.; Brougham, C. A. Investigation of the Estrogenic Risk to Feral Male Brown Trout (*Salmo trutta*) in the Shannon International River Basin District of Ireland. *Ecotoxicol. Environ. Saf.* **2010**, *73*, 1658–1665.

70. Jobling, S.; Williams, R.; Johnson, A.; Taylor, A.; Gross-Sorokin, M.; Nolan, M.; Tyler, C. R.; van Aerle, R.; Santos, E.; Brighty, G. Predicted Exposures to Steroid Estrogens in UK rivers Correlate with Widespread Sexual Disruption in Wild Fish Populations. *Environ. Health Perspect.* **2006**, *114*, 32–39.

71. Chokwe, T. B.; Okonkwo, J. O.; Sibali, L. L.; Ncube, E. J. Alkylphenol Ethoxylates and Brominated Flame Retardants in Water, Fish (Carp) and Sediment Samples from the Vaal River, South Africa. *Environ. Sci. Pollut. Res. Int.* **2015**, *22*, 11922–1192.

72. Sheahan, D. A.; Brighty, G. C.; Daniel, M.; Kirby, S. J.; Hurst, M. R.; Kennedy, J.; Morris, S.; Routledge, E. J.; Sumpter, J. P., Waldock, M. J. Reduction in the Estrogenic Activity of a Treated Sewage Effluent Discharge to an English River as a Result of a Decrease in the Concentration of Industrially Derived Surfactants. *Environ. Toxicol. Chem. 21* (3), **2002**, 515–519.

73. Rodgers-Gray, T. P.; Jobling, S.; Kelly, C.; Morris, S.; Brighty, G.; Waldock, M. J.; Sumpter, J. P.; Tyler, C. R. Exposure of Juvenile Roach (*Rutilus rutilus*) to Treated Sewage Effluent Induces Dose-Dependent and Persistent Disruption in Gonadal Duct Development. *Environ. Sci. Technol.* **2001**, *35*, 462–470.

74. Heemken, O. P.; Reincke, H.; Stachel, B.; Theobald, N. The Occurrence of Xenoestrogens in the Elbe River and the North Sea. *Chemosphere* **2001**, *45*, 245–259.

75. Allen, Y.; Matthiessen, P.; Scott, A. P.; Haworth, S.; Feist, S.; Thain, J. E. The Extent of Oestrogenic Contamination in the UK Estuarine and Marine Environments—Further Surveys of Flounder. *Sci. Total Environ.* **1999**, *233*, 5–20.

76. Bjerregaard, L. B.; Korsgaard, B.; Bjerregaard, P. Intersex in Wild Roach (*Rutilus rutilus*) from Danish Sewage Effluent-Receiving Streams. *Ecotoxicol. Environ. Saf.* **2006**, *64*, 321–328.

77. Nice, H. E.; Morritt, D.; Crane, M.; Thorndyke, M. Long-Term and Transgenerational Effects of Nonylphenol Exposure at a Key Stage in the Development of *Crassostrea gigas*. Possible Endocrine Disruption? *Mar. Ecol. Prog. Ser.* **2003**, *256*, 293–300.

78. Grumetto, L.; Montesano, D.; Seccia, S.; Albrizio, S.; Barbato, F. Determination of Bisphenol A and Bisphenol B Residues in Canned Peeled Tomatoes by Reversed-Phase Liquid Chromatography. *J. Agric. Food Chem.* **2008**, *56*, 10633–10637.

79. Ćwiek-Ludwicka, K. Bisphenol A (BPA) in Food Contact Materials—New Scientific Opinion from EFSA Regarding Public Health Risk. *Rocz. Panstw. Zakl. Hig.* **2015**, *66*, 299–307.

80. EFSA. Opinion of the Scientific Panel on Food Additives, Flavourings, Processing Aids and Materials in Contact with Food on a Request from the Commission Related to 2,2-Bis-(4-hydroxyphenyl)propane (Bisphenol A). *EFSA J.* **2006**, *428*, 1–75.

81. Stump, D. G.; Beck, M. J.; Radowsky, A.; Garman, R. H.; Freshwater, L. L.; Sheets, L. P.; Marty, M. S.; Waechter, J. M.; Dimond, S. S.; Van Miller, J. P.; Shiotsuka, R. N.; Beyer, D.; Chapelle, A. H.; Hentges, S. G. Developmental Neurotoxicity Study of Dietary Bisphenol A in Sprague–Dawley Rats. *Toxicol. Sci.* **2009**, *115*, 167–182.

82. Cwiek-Ludwicka, K.; Ludwicki, J. K.; Endocrine Disruptors in Food Contact Materials: Is there a Health Threat? *Rocz. Panstw. Zakl. Hig.* **2014**, *65*, 169–177.

83. Rogers, J. A.; Metz, L.; Yong, V. W. Review: Endocrine Disrupting Chemicals and Immune Responses: A Focus on Bisphenol-A and its Potential Mechanisms. *Mol. Immunol.* **2013**, *53*, 421–430.

84. Rubin, B. S. Bisphenol A: An Endocrine Disruptor with Widespread Exposure and Multiple Effects. *J. Steroid Biochem. Mol. Biol.* **2011**, *127*, 27–34.

85. EFSA Draft. *Scientific Opinion on the Risks to Public Health Related to the 4 Presence of Bisphenol A (BPA) in Foodfoods—Part: Exposure Assessment.* EFSA Panel on Food Contact Materials, Enzymes, Flavourings and Processing Aids. European Food Safety Authority, 2013. http://www.efsa.europa.eu/en/consultations/call/130725.htm.

86. Gerona, R. R.; Pan, J.; Zota, A. R.; Schwartz, J. M.; Friesen, M.; Taylor, J. A.; Hunt, P. A.; Woodruff, T. J. Direct Measurement of Bisphenol A (BPA), BPA Glucuronide BPA Sulfate in a Diverse and Low-Income Population of Pregnant Women Reveals High Exposure, with Potential Implications for Previous Exposure Estimates: A Cross-Sectional Study. *Environ. Health* **2016**, *15*, 50.

87. Teeguarden, J. G.; Twaddle, N. C.; Churchwell, M. I.; Doerge, D. R. Urine and Serum Biomonitoring of Exposure to Environmental Estrogens I: Bisphenol A in Pregnant Women. *Food Chem. Toxicol.* **2016**, *92*, 129–142.

88. Alam, M. S.; Kurohmaru, M. Butyl Benzyl Phthalate Induces Spermatogenic Cell Apoptosis in Prepubertal Rats. *Tissue Cell* **2016**, *48*, 35–42.

89. Bicalho, B.; Serrano, K.; Dos, Santos, Pereira, A.; Devine, D. V.; Acker, J. P. Blood Bag Plasticizers Influence Red Blood Cell Vesiculation Rate without Altering the Lipid Composition of the Vesicles. *Transfus. Med. Hemother.* **2016**, *43*, 19–26.

90. EFSA. Opinion of the Scientific Panel on Food Additives, Flavourings, Processing Aids and Materials in Contact with Food (AFC) on a Request from the Commission Related to Bis(2-ethylhexyl) Phthalate (DEHP) for Use in Food Contact Materials. *EFSA J.* **2005**, *243*, 1–20.

91. Zota, A. R.; Phillips, C. A.; Mitro, S. D. Recent Fast Food Consumption and Bisphenol A and Phthalates Exposures among the US Population in NHANES, 2003–2010. *Environ Health Perspect.* **2016**, *124*, 1521–1528.

92. Ge, W. P.; Yang, X. J.; Wu, X. Y.; Wang, Z.; Geng, W.; Guo, C. F. Phthalate Residue in Goat Milk-Based Infant Formulas Manufactured in China. *J. Dairy Sci.* **2016**, *99*, 7776–7781.

93. Kim, S.; Lee, J.; Park, J.; Kim, H. J.; Cho, G.; Kim, G. H.; Eun, S. H.; Lee, J. J.; Choi, G.; Suh, E.; Choi, S, Kim, S., Kim, Y. D., Kim, S. K.; Kim, S. Y.; Kim, S.; Eom, S.; Moon, H.; B.; Kim, S.; Choi, K. Concentrations of Phthalate Metabolites in Breast Milk in Korea: Estimating Exposure to Phthalates and Potential Risks among Breast-Fed Infants. *Sci. Total Environ.* **2015**, *8*, 13–19.

94. Benjamin, S.; Pradeep, S.; Josh, M. S.; Kumar, S.; Masai, E. A Monograph on the Remediation of Hazardous Phthalates. *J. Hazard. Mater.* **2015**, *298*, 258–272.

95. Hannon, P. R.; Flaws, J. A. The Effects of Phthalates on the Ovary. *Front. Endocrinol. (Lausanne)* **2015**, *2*, 6–8.

96. Desdoits-Lethimonier, C.; Albert, O.; Le Bizec, B.; Perdu, E.; Zalko, D.; Courant, F.; Lesné, L.; Guillé, F.; Dejucq-Rainsford, N.; Jégou, B. Human Testis Steroidogenesis Is Inhibited by Phthalates. *Hum. Reprod.* **2012**, *27*, 1451–1459.

97. Van Den Houwe, K.; Evrard, C.; Van Loco, J.; Lynen, F.; Van Hoeck, E. Migration of Photoinitiators from Cardboard into Dry Food: Evaluation of Tenax® as a Food Simulant. *Food Addit. Contam., A: Chem. Anal. Control Expo. Risk Assess.* **2016**, *33*, 913–920.

98. Buck Louis, G. M.; Kannan, K.; Sapra, K. J.; Maisog, J.; Sundaram, R. Urinary Concentrations of Benzophenone-Type Ultraviolet Radiation Filters and Couples' Fecundity. *Am. J. Epidemiol.* **2014,** *180*, 1168–1175.

99. Louis, G. M., Chen, Z., Kim, S., Sapra, K. J., Bae, J.; Kannan, K. Urinary Concentrations of Benzophenone-Type Ultraviolet Light Filters and Semen Quality. *Fertil. Steril.* **2015**, *104* (4), 989–996.

100. Yingxia, L.; Yaqian, M.; Yiqun, W.; Lan, G.; Xiaofen, W. Fast and Effective Low-Temperature Freezing Extraction Technique to Determine Organotin Compounds in Edible Vegetable Oil. *J. Sep. Sci.* **2016**, *39*, 2380–2387.

101. Davies, A. G.; Smith, P. J.; Emeléus, H. J.; Sharpe, A. G. Recent Advances in Organotin Chemistry. *Adv. Inorg. Chem. Radiochem.* **1980**, *23*, 1–77.

102. Horiguchi, T. Masculinization of Female Gastropod Mollusks Induced by Organotin Compounds, Focusing on Mechanism of Actions of Tributyltin and Triphenyltin for Development of Imposex. *Environ. Sci.* **2006**, *13*, 77–87.

103. Penza, M.; Jeremic, M.; Marrazzo, E.; Maggi, A.; Ciana, P.; Rando, G.; Grigolato, P. G.; Di, Lorenzo, D. The environmental Chemical Tributyltin Chloride (TBT) Shows both Estrogenic and Adipogenic Activities in Mice which Might Depend on the Exposure Dose. *Toxicol. Appl. Pharmacol.* **2011**, *255*, 65–75.

104. Cheshenko, K.; Pakdel, F.; Segner, H.; Kah, O.; Eggen, R. I. Interference of Endocrine Disrupting Chemicals with Aromatase CYP19 Expression or Activity, and Consequences for Reproduction of Teleost Fish. *Gen. Comp. Endocrinol.* **2008**, *155*, 31–62.

105. Hoch, M. Organotin Compounds in the Environment—An Overview. *Appl. Geochem.* **2001**, *16*, 719–743.

106. Liu, J. Y.; Jiang, G. B. Survey on the Presence of Butyltin Compounds in Chinese Alcoholic Beverages, Determined by Using Headspace Solid-Phase Microextraction Coupled with Gas Chromatography-Flame Photometric Detection. *J. Agric. Food Chem.* **2002**, *50*, 6683–6687.

107. Ho, K. K.; Leung, K. M. Organotin Contamination in Seafood and Its Implication for Human Health Risk in Hong Kong. *Mar. Pollut. Bull.* **2014**, *85*, 634–640.

108. Bettin, C.; Oehlmann, J.; Stroben, E. TBT-Induced Imposex in Marine Neogastropods Is Mediated by an Increasing Androgen Level. *Helgolander. Meeresun.* **1996**, *50*, 299–317.

109. Dybing, E.; Doe, J.; Grotten, J.; Kleiner, J.; O'Birien, J.; Renwick, A. G.; Schlatter, J.; Steinberg, P.; Tritscher, A.; Walker, R.; Younes, M. Hazard Characterisation of Chemicals in Food, and Diet, Dose Response, Mechanisms and Exytrapolation Issues. *Food. Chem. Toxicol.* **2002**, *40*, 237–282.

110. Santos, M. M; Enes, P.; Reis-Henriques, M. A.; Kuballa, J.; Castro, L. F; Vieira, M. N. Organotin Levels in Seafood from Portuguese Markets and the Risk for Consumers. *Chemosphere* **2009**, *75*, 661–666.

111. Papaspyrou, S. D.; Thomaidis, N. S.; Lampi, E. N.; Lioupis, A. Determination of Migration of *n*-Butyltins and *n*-Octyltins to Food Simulants by Gas Chromatography–Mass Spectrometry. *Appl. Organomet. Chem.* **2007**, *21*, 412–424.

112. Heroult, J.; Bueno, M.; Potin-Gautier, M.; Lespes, G. Organotin Speciation in French Brandies and Wines by Solid-Phase Microextraction and Gas Chromatography—Pulsed Flame Photometric Detection. *J. Chromatogr. A* **2008**, *1180*, 122–130.

113. Georgescu, B.; Georgescu, C.; Daraban, S.; Bouaru, A.; Pascalau, S. Heavy Metals Acting as Endocrine Disruptors. *Sci. Papers: Anim. Sci. Biotechnol.* **2011**, *44*, 89–93.

114. Sinha, B.; Bhattacharyya, K. Arsenic Toxicity in Rice with Special Reference to Speciation in Indian Grain and Its Implication on Human Health. *J. Sci. Food. Agric.* **2015**, *95*, 1435–1444.

115. Xu, X. Y.; McGrath, S. P.; Meharg, A. A.; Zhao, F. J. Growing Rice Aerobically Markedly Decreases Arsenic Accumulation. *Environ. Sci Technol.* **2008**, *42*, 5574–5579.

116. Smith, E.; Kempson, I.; Juhasz, A. L.; Weber, J.; Skinner, W. M.; Gräfe, M. Localization and Speciation of Arsenic and Trace Elements in Rice Tissues. *Chemosphere* **2009**, *76*, 529–535.

117. Dopp, E.; Kligerman, A. D.; Diaz-Bone, R. A. Organoarsenicals. Uptake, Metabolism, and Toxicity. *Met. Ions Life Sci.* **2010**, *7*, 231–265.

118. Sun, H. J.; Xiang, P.; Luo, J.; Hong, H.; Lin, H.; Li, H. B.; Ma, L. Q. Mechanisms of Arsenic Disruption on Gonadal, Adrenal and Thyroid Endocrine Systems in Humans: A Review. *Environ. Int.* **2016**, *95*, 61–68.

119. Watson, W. H.; Yager, J. D. Arsenic: Extension of its Endocrine Disruption Potential to Interference with Estrogen Receptor-Mediated Signaling. *Toxicol. Sci.* **2007**, *98*, 1–4.

120. Aposhian, H. V.; Aposhian, M. M. Arsenic Toxicology: Five Questions. *Chem. Res. Toxicol.* **2006**, *19*, 1–15.

121. Bodwell, J. E; Kingsley, L. A.; Hamilton, J. W. Arsenic at Very Low Concentrations Alters Glucocorticoid Receptor (GR)-Mediated Gene Activation But Not GR-Mediated Gene Repression: Complex Dose–Response Effects Are Closely Correlated with Levels of Activated GR and Require a Functional GR DNA Binding Domain. *Chem. Res. Toxicol.* **2004**, *17*, 1064–1076.

122. Bodwell, J. E.; Gosse, J. A.; Nomikos, A. P.; Hamilton, J. W. Arsenic Disruption of Steroid Receptor Gene Activation: Complex Dose–Response Effects Are Shared by Several Steroid Receptors. *Chem. Res. Toxicol.* **2006**, *19*, 1619–1629.

123. Kozul, C. D.; Hampton, T. H.; Davey, J. C.; Gosse.; J.; A.; Nomikos, A. P. Eisenhauer, P. L.; Weiss, D. J.; Thorpe, J. E.; Ihnat, M. A.; Hamilton, J. W. Chronic Exposure to Arsenic in the Drinking Water Alters the Expression of Immune Response Genes in Mouse Lung. *Environ. Health. Perspect.* **2009**, *117*, 1108–1115.

124. Davey, J. C.; Nomikos, A. P.; Wungjiranirun, M.; Sherman, J. R.; Ingram, L.; Batki, C.; Lariviere, J. P.; Hamilton, J. W. Arsenic as an Endocrine Disruptor: Arsenic Disrupts Retinoic Acid Receptor- and Thyroid Hormone Receptor-Mediated Gene Regulation and Thyroid Hormone-Mediated Amphibian Tail Metamorphosis. *Environ. Health Perspect.* **2008**, *116*, 165–172.

125. Garner, R.; Levallois, P. Cadmium Levels and Sources of Exposure among Canadian Adults. *Health Rep.* **2016**, *27*, 10–18.

126. Jean, J; Sirot, V.; Vasseur, P.; Narbonne, J. F.; Leblanc, J. C.; Volatier, J. L.; Rivière, G. Impact of a Modification of Food Regulation on Cadmium Exposure. *Regul. Toxicol. Pharmacol.* **2015**, *73*, 478–483.

127. Chen, M.; Yin, H.; Bai, P.; Miao, P.; Deng, X.; Xu, Y.; Hu, J.; Yin, J. ABC Transporters Affect the Elimination and Toxicity of CdTe Quantum Dots in Liver and Kidney Cells. *Toxicol. Appl. Pharmacol.* **2016**, *303*, 11–20.

128. Sand, S.; Heraud, F.; Arcella, D. The Use of Chemical Occurrence Data at European vs. National Level in Dietary Exposure Assessments: A Methodological Study. *Food Chem. Toxicol.* **2013**, *62*, 7–15.

129. Ferrari, P.; Arcella, D.; Heraud, F.; Cappé, S.; Fabiansson, S. Impact of Refining the Assessment of Dietary Exposure to Cadmium in the European Adult Population. *Food Addit. Contam., A: Chem. Anal. Control Expo. Risk Assess.* **2013**, *30*, 687–697.

130. Schwarz, M. A.; Lindtner, O.; Blume, K.; Heinemeyer, G.; Schneider, K. Cadmium Exposure from Food: The German LExUKon Project. *Food Addit. Contam., A: Chem. Anal. Control Expo. Risk Assess.* **2014**, *31*, 1038–1051.

131. National Food Agency. Endocrine Active Substances in the Food—What is the Problem? *Livsmedelsverkets Rapportserie Nr. 8*, 2016.

132. Pillai, P.; Pandya, C.; Gupta, S.; Gupta, S. Biochemical and Molecular Effects of Gestational and Lactational Coexposure to Lead and Cadmium on Ovarian Steroidogenesis Are Associated with Oxidative Stress in F1 Generation Rats. *J. Biochem. Mol. Toxicol.* **2010**, *24*, 384–394.

133. Prajapati, A.; Rao, A.; Patel, J.; Gupta, S.; Gupta. S. A. Single Low Dose of Cadmium Exposure Induces Benign Prostate Hyperplasia like Condition in Rat: A Novel Benign Prostate Hyperplasia Rodent Model. *Exp. Biol. Med. (Maywood).* **2014**, *239*, 829–884.

134. Yu, X.; Hong, S.; Faustman, E. M. Cadmium-Induced Activation of Stress Signaling Pathways, Disruption of Ubiquitin-Dependent Protein Degradation and Apoptosis in Primary Rat Sertoli Cell–Gonocyte Cocultures. *Toxicol. Sci.* **2008**, *104*, 385–396.

135. Ketata, I.; Smaoui-Damak, W.; Guermazi, F.; Rebai, T.; Hamza-Chaffai, A. *In Situ* Endocrine Disrupting Effects of Cadmium on the Reproduction of *Ruditapes decussatus*. *Comp. Biochem. Physiol., C: Toxicol. Pharmacol.* **2007**, *146*, 415–330.

136. Byrne, C.; Divekar, S. D.; Storchan, G. B.; Parodi, D. A.; Martin, M. B. Cadmium—A Metallohormone? *Toxicol. Appl. Pharmacol.* **2009**, *238*, 26–71.

137. Benbrahim-Tallaa, L.; Liu, J.; Webber, M. M.; Waalkes, M. P. Estrogen Signaling and Disruption of Androgen Metabolism in Acquired Androgen-Independence during

Cadmium Carcinogenesis in Human Prostate Epithelial Cells. *Prostate* **2007**, *67*, 135–145.

138. Leech, T. G.; Adams, E. A.; Weathers, T. D.; Staten, L. K.; Filippelli, G. M. Inequitable Chronic Lead Exposure: A Dual Legacy of Social and Environmental Injustice. *Fam Community Health* **2016**, *39*, 151–159.

139. Li, Y.; Xie, C.; Murphy, S. K.; Skaar, D.; Nye, M.; Vidal, A. C.; Cecil, K. M.; Dietrich, K. N.; Puga, A.; Jirtle, R. L.; Hoyo, C. Lead Exposure during Early Human Development and DNA Methylation of Imprinted Gene Regulatory Elements in Adulthood. *Environ. Health Perspect.* **2016**, *124*, 666–673.

140. Bierkens, J.; Smolders, R.; Van, Holderbeke, M.; Cornelis, C. Predicting Blood Lead Levels from Current and Past Environmental Data in Europe. *Sci. Total Environ.* **2011**, *409*, 5101–5110.

141. Rebeniak, M.; Wojciechowska-Mazurek, M.; Mania, M.; Szynal, T.; Strzelecka, A.; Starska, K. Exposure to Lead and Cadmium Released from Ceramics and Glassware Intended to Come into Contact with Food. *Rocz. Panstw. Zakl. Hig.* **2014**, *65*, 3091–3099.

142. Dorevitch, S.; Babin, A. Health Hazards of Ceramic Artists. *Occup. Med.* **2001**, *16*, 563–575.

143. Anderson, D. L.; Cunningham, W. C.; Lindstrom, T. R.; Olmez, I. Identification of Lead and Other Elements in Ceramic Glazes and Housewares by ^{109}Cd-Induced X-ray Fluorescence Emission Spectrometry. *J. AOAC. Int.* **1995**, *78*, 407–412.

144. Mostafalou, S.; Baeeri, M.; Bahadar, H.; Soltany-Rezaee-Rad, M.; Gholami., M.; Abdollahi, M. Molecular Mechanisms Involved in Lead Induced Disruption of Hepatic and Pancreatic Glucose Metabolism. *Environ. Toxicol. Pharmacol.* **2015**, *39*, 16–26.

145. Hirsch, H. V.; Possidente, D.; Possidente, B. Pb^{2+}: An Endocrine Disruptor in *Drosophila? Physiol. Behav.* **2010**, *99*, 254–259.

146. Taupeau, C.; Poupon, J.; Treton. D.; Brosse, A.; Richard, Y.; Machelon, V. Lead Reduces Messenger RNA and Protein Levels of Cytochrome p450 Aromatase and Estrogen Receptor Beta in Human Ovarian Granulosa Cells. *Biol. Reprod.* **2003**, *68*, 1982–1988.

147. Finster, M. E.; Raymond, M. R.; Scofield, M. A.; Smith, K. P. Mercury-Impacted Scrap Metal: Source and Nature of the Mercury. *J. Environ. Manage.* **2015**, *161*, 303–308.

148. Marshall, B. G.; Forsberg, B. R.; Thomé-Souza, M.; Peleja, R., Moreira, M. Z.; Freitas, C. E. Evidence of Mercury Biomagnification in the Food Chain of the Cardinal Tetra *Paracheirodon axelrodi* (Osteichthyes: Characidae) in the Rio Negro, Central Amazon, Brazil. *J. Fish Biol.* **2016**, *89*, 220–240.

149. Lavoie, R. A.; Jardine, T. D.; Chumchal, M. M.; Kidd, K. A.; Campbell, L. M. Biomagnification of Mercury in Aquatic Food Webs: A Worldwide Meta-analysis. *Environ. Sci. Technol.* **2013**, *47*, 13385–13394.

150. Ishihara, N. History of Ignorance of Methylmercury Toxicity and Intoxication in Japan in Relation to Minamata Disease. *Nihon Eiseigaku Zasshi* **2014**, *69*, 75–79.

151. Normile, D. Mercury Pollution. In Minamata, Mercury Still Divides. *Science* **2013**, *341*, 1446–1447.

152. Gump, B. B.; MacKenzie, J. A.; Dumas, A. K.; Palmer, C. D.; Parsons, P. J.; Segu, Z. M.; Mechref, Y. S.; Bendinskas, K. G. Fish Consumption, Low-Level Mercury, Lipids, and Inflammatory Markers in Children. *Environ. Res.* **2012**, *112*, 204–211.

153. Li, L., Chen, X.; Zhu, Q; Chen, D.; Guo, J.; Yao, W.; Dong, Y.; Wei, J.; Lian, Q.; Ge, R. S.; Yuan, B. Disrupting Androgen Production of Leydig Cells by Resveratrol via Direct

Inhibition of Human and Rat 3β-Hydroxysteroid Dehydrogenase. *Toxicol Lett.* **2014,** *226,* 14–19.

154. Zhu, X.; Kusaka, Y.; Sato, K.; Zhang, Q. The Endocrine Disruptive Effects of Mercury. *Environ. Health. Prev. Med.* **2000,** *4,* 174–183.

155. Richter, C. A.; Martyniuk, C. J.; Annis, M. L.; Brumbaugh, W. G.; Chasar, L. C.; Denslow, N. D.; Tillitt, D. E. Methylmercury-Induced Changes in Gene Transcription Associated with Neuroendocrine Disruption in Largemouth Bass (*Micropterus salmoides*). *Gen. Comp. Endocrinol.* **2014,** *203,* 215–224.

156. Tan, S. W.; Meiller, J. C.; Mahaffey, K. R. The Endocrine Effects of Mercury in Humans and Wildlife. *Crit. Rev. Toxicol.* **2009,** *39,* 228–269.

157. Davey, J. C.; Bodwell, J. E.; Gosse, J. A.; Hamilton, J. W. Arsenic as an Endocrine Disruptor: Effects of Arsenic on Estrogen Receptor-Mediated Gene Expression In Vivo and in Cell Culture. *Toxicol. Sci.* **2007,** *98,* 75–86.

158. Kaltreider, R. C.; Davis, A. M.; Lariviere, J. P.; Hamilton, J. W. Arsenic alters the Function of the Glucocorticoid Receptor as a Transcription Factor. *Environ. Health Perspect.* **2001,** *109,* 245–251.

159. Jana, K.; Jana, S.; Samanta, P. K. Effects of Chronic Exposure to Sodium Arsenite on Hypothalamo–Pituitary–Testicular Activities in Adult Rats: Possible an Estrogenic Mode of Action. *Reprod. Biol. Endocrinol.* **2006,** *16,* 4–9.

160. Mondal, S.; Mukhopadhyay, B.; Bhattacharya, S. Inorganic Mercury Binding to Fish Oocyte Plasma Membrane Induces Steroidogenesis and Translatable Messenger RNA Synthesis. *Biometals* **1997,** *10,* 285–290.

161. Barregard, L.; Lindstedt, G.; Schütz, A.; Sällsten, G. Endocrine Function in Mercury Exposed Chloralkali Workers. *Occup. Environ. Med.* **1994,** *51,* 536–540.

162. Prestifilippo, J. P.; Fernández-Solari, J.; Mohn, C.; De Laurentiis, A.; McCann, S. M.; Dees, W.; Rettori, V. Effect of Manganese on Luteinizing Hormone-Releasing Hormone Secretion in Adult Male Rats. *Toxicol. Sci.* **2007,** *97,* 75–80.

163. Lee, D. H.; Jacobs, Jr., D. R. Methodological Issues in Human Studies of Endocrine Disrupting Chemicals. *Rev. Endocr. Metab. Disord.* **2015,** *16,* 289–297.

164. Aitken, R. J.; Clarkson, J. S. Cellular Basis of Defective Sperm Function and its Association with Genesis of Reactive Oxygen Species by Human Spermatozoa. *J. Reprod. Fertil.* **1987,** *81,* 459–469.

165. Bacciottini, L.; Falchetti, A.; Pampaloni, B.; Bartolini, E.; Carossino, A. M.; Brandi, M. L. Phytoestrogens: Food or Drug? *Clin. Cases Miner. Bone Metab.* **2007,** *4,* 123–130.

166. Ganai, A. A.; Farooqi, H. Bioactivity of Genistein: A Review of In Vitro and In Vivo Studies. *Biomed. Pharmacother.* **2015,** *76,* 30–38.

167. Hwang, Y. P.; Jeong, H. G. Mechanism of Phytoestrogen Puerarin-Mediated Cytoprotection Following Oxidative Injury: Estrogen Receptor-Dependent Up-regulation of PI3K/Akt and HO-1. *Toxicol. Appl. Pharmacol.* **2008,** *233,* 371–381.

168. Andres, S.; Hansen, U.; Niemann, B.; Palavinskas, R.; Lampen, A. Determination of the Isoflavone Composition and Estrogenic Activity of Commercial Dietary Supplements Based on Soy or Red Clover. *Food Funct.* **2015,** *6,* 2017–2025.

169. Islam, M. A.; Bekele, R.; Vanden Berg, J. H.; Kuswanti, Y.; Thapa, O.; Soltani, S.; van Leeuwen, F. X.; Rietjens, I. M.; Murk, A. J. Deconjugation of Soy Isoflavone Glucuronides Needed for Estrogenic Activity. *Toxicol. In Vitro* **2015,** *29,* 706–715.

170. Gencel, V. B.; Benjamin, M. M.; Bahou, S. N., Khalil, R. A. Vascular Effects of Phytoestrogens and Alternative Menopausal Hormone Therapy in Cardiovascular Disease. *Mini Rev. Med Chem.* **2012,** *12,* 149–174.

171. Dalais, F. S.; Rice, G. E.; Wahlqvist, M. L.; Grehar, M.; Murkies, A. L.; Medley, G.; Ayton, R.; Strauss, B. J. Effects of Dietary Phytoestrogens in Postmenopausal Women. *Climacteric* **1998**, *1*, 124–129.

172. Murkies, A. L.; Wilcox, G.; Davis, S. R. Phytoestrogens 1. *J. Clin. Endocrinol. Metab.* **1998**, *83*, 297–303.

173. Stark, A.; Madar, Z. Phytoestrogens: A Review of Recent Findings. *J. Pediatr. Endocrinol. Metab.* **2002**, *15*, 561–572.

174. Yao, L. H.; Jiang, Y. M.; Shi, J.; Tomás-Barberán, F. A.; Datta, N.; Singanusong, R.; Chen, S. S. Flavonoids in Food and their Health Benefits. *Plant Foods Hum. Nutr.* **2004**, *59*, 113–122.

175. Ross, J. A.; Kasum, C. M. Dietary Flavonoids: Bioavailability, Metabolic Effects, and Safety. *Annu. Rev. Nutr.* **2002**, *22*, 19–34.

176. Terahara, N. Flavonoids in Foods: A Review. *Nat. Prod. Commun.* **2015**, *10*, 521–528.

177. McCarty, M. F. Isoflavones Made Simple—Genistein's Agonist Activity for the Beta-type Estrogen Receptor Mediates their Health Benefits. *Med. Hypoth.* **2006**, *66*, 1093–1014.

178. Losa, S. M.; Todd, K. L.; Sullivan, A. W.; Cao, J.; Mickens, J. A.; Patisaul, H. B. Neonatal Exposure to Genistein Adversely Impacts the Ontogeny of Hypothalamic Kisspeptin Signalling Pathways and Ovarian Development in the Peripubertal Female Rat. *Reprod. Toxicol.* **2011**, *31*, 280–289.

179. Bolca, S.; Urpi-Sarda, M.; Blondeel, P.; Roche, N.; Vanhaecke, L.; Possemiers, S.; Al-Maharik, N.; Botting, N.; De Keukeleire, D.; Bracke, M.; Heyerick, A.; Manach, C.; Depypere, H. Disposition of Soy Isoflavones in Normal Human Breast Tissue. *Am. J. Clin. Nutr.* **2010**, *91*, 976–984.

180. Molzberger, A. F.; Vollmer, G.; Hertrampf, T.; Möller, F. J.; Kulling, S.; Diel, P. *In Utero* and Postnatal Exposure to Isoflavones Results in a Reduced Responsivity of the Mammary Gland towards Estradiol. *Mol. Nutr. Food Res.* **2012**, *56*, 399–409.

181. Cai, L. Q.; Cai, J.; Wu, W.; Zhu, Y. S. 17α-Estradiol and Genistein Inhibit High Fat Diet Induced Prostate Gene Expression and Prostate Growth in the Rat. *J. Urol.* **2011**, *186*, 1489–1496.

182. Santell, R. C.; Kieu, N.; Helferich, W. G. Genistein Inhibits Growth of Estrogen-Independent Human Breast Cancer Cells in Culture but Not in Athymic Mice. *J. Nutr.* **2000**, *130*, 1665–1669.

183. Hsieh, C. Y; Santell, R. C.; Haslam, S. Z.; Helferich, W. G. Estrogenic Effects of Genistein on the Growth of Estrogen Receptor-Positive Human Breast Cancer (MCF-7) Cells *In Vitro* and *In Vivo*. *Cancer Res.* **1998**, *58*, 3833–3838.

184. Allred, C. D.; Allred, K. F.; Ju, Y. H.; Virant, S. M.; Helferich,W. G. Soy Diets Containing Varying Amounts of Genistein Stimulate Growth of Estrogen-Dependent (MCF-7) Tumors in a Dose-Dependent Manner. *Cancer Res.* **2001**, *61*, 5045–5050.

185. Setchell, K. D.; Cassidy, A. Dietary Isoflavones: Biological Effects and Relevance to Human Health. *J. Nutr.* **1999**, *129*, 758S–767S.

186. Kim, H. P.; Son, K. H.; Chang, H. W.; Kang, S. S. Anti-inflammatory Plant Flavonoids and Cellular Action Mechanisms. *J. Pharmacol. Sci.* **2004**, *96*, 229–245.

187. Hämäläinen, M.; Nieminen, R.; Vuorela, P.; Heinonen, M.; Moilanen, E. Anti-inflammatory Effects of Flavonoids: Genistein, Kaempferol, Quercetin, and Daidzein Inhibit STAT-1 and NF-κB Activations, Whereas Flavone, Isorhamnetin, Naringenin, and Pelargonidin Inhibit only NF-κB Activation along with their Inhibitory Effect on iNOS

Expression and NO Production in Activated Macrophages. *Mediat. Inflamm.* **2007,** Article ID 45673. DOI:10.1155/2007/45673.

188. Kuo, S. C.; Chen, S. C.; Chen, L. H.; Wu, J. B.; Wang, J. P.; Teng, C. M. Potent Anti-platelet, Anti-inflammatory and Antiallergic Isoflavanquinones from the Roots of *Abrus precatorius*. *Planta Med.* **1995,** *61,* 307–312.

189. Hu, Y.; Song, Y.; Franke, A. A.; Hu, F. B.; van Dam, R. M.; Sun, Q. A Prospective Inves-tigation of the Association between Urinary Excretion of Dietary Lignan Metabolites and Weight Change in US Women. *Am. J. Epidemiol.* **2015,** *182,* 503–511.

190. Oomah, B. D. Flaxseed as a Functional Food Source. *J. Sci. Food Agric.* **2001,** *81,* 889–894.

191. Martinchik, A. N.; Zubtsov, V. V. Phytoestrogenis Properties of Flaxseed Lignans. *Vopr. Pitan.* **2012,** *81,* 61–66 [in Russian].

192. Orcheson, L. J.; Rickard, S. E.; Seidl, M. M.; Thompson, L. U. Flaxseed and Its Mammalian Lignan Precursor Cause a Lengthening or Cessation of Estrous Cycling in Rats. *Cancer Lett.* **1998,** *125,* 69–76.

193. Tou, J. C.; Chen, J.; Thompson, L. U. Flaxseed and Its Lignan Precursor, Secoisolar-iciresinol Diglycoside, Affect Pregnancy Outcome and Reproductive Development in Rats. *J. Nutr.* **1998,** *128,* 1861–1888.

194. Nehybova, T.; Smarda, J.; Benes, P. Plant Coumestans: Recent Advances and Future Perspectives in Cancer Therapy. *Anticancer Agents Med. Chem.* **2014,** *14,* 1351–1362.

195. Bedell, S.; Nachtigall, M.; Naftolin, F. The Pros and Cons of Plant Estrogens for Meno-pause. *J. Steroid Biochem. Mol. Biol.* **2014,** *139,* 225–236.

196. Scarlata, S.; Miksicek, R. Binding Properties of Coumestrol to Expressed Human Estrogen Receptor. *Mol. Cell. Endocrinol.* **1995,** *115,* 65–72.

197. Horn-Ross, P. L.; John, E. M.; Canchola, A. J.; Stewart, S. L.; Lee, M. M. Phytoestrogen Intake and Endometrial Cancer Risk. *J. Natl. Cancer Inst.* **2003,** *95,* 1158–1164.

198. Bandera, E. V.; Williams, M. G.; Sima, C.; Bayuga, S.; Pulick, K.; Wilcox, H.; Soslow, R..; Zauber, A. G.; Olson, S. H. Phytoestrogen Consumption and Endometrial Cancer Risk: A Population-Based Case-Control Study in New Jersey. *Cancer Causes Control* **2009,** *20,* 1117–1127.

199. Markaverich, B. M.; Webb, B.; Densmore, C. L.; Gregory, R. R. Effects of Coumestrol on Estrogen Receptor Function and Uterine Growth Inovariectomized Rats. *Environ. Health Perspect.* **1995,** *103,* 574–581.

200. Lippi, G.; Franchini, M.; Favaloro, E. J.; Targher, G. Moderate Red Wine Consump-tion and Cardiovascular Disease Risk: Beyond the "French Paradox". *Semin. Thromb. Hemost.* **2010,** *31,* 59–70.

201. Gehm, B.; McAndrews, J.; Chien, P.; Jameson, J. Resveratrol, a Polyphenolic Compound Found in Grapes and Wine, Is an Agonist for the Estrogen Receptor. *Proc. Natl. Acad. Sci. U.S.A.* **1997,** *94,* 14138–14143.

202. Pace-Asciak, C. R.; Rounova, O.; Hahn, S. E.; Diamandis, E. P.; Goldberg, D. M. Wines and Grape Juices as Modulators of Platelet Aggregation in Healthy Human Subjects. *Clin. Chim. Acta* **1996,** *246,* 163–182.

203. Docherty, J. J.; Fu, M. M.; Stiffler, B. S.; Limperos, R. J.; Pokabla, A. L.; DeLucia, A. L. Resveratrol Inhibition of Herpes Simplex Virus Replication. *Antivir. Res.* **1999,** *43,* 145–155.

204. Gentilli, M.; Mazoit, J. X.; Bouaziz, H.; Fletcher, D.; Casper, R. F.; Benhamou, D.; Savouret, J. F. Resveratrol Decreases Hyperalgesia Induced by Carrageenan in the Rat Hind Paw. *Life Sci.* **2001,** *68,* 1317–1321.

205. Karuppagounder, S. S.; Pinto, J. T.; Xu, H.; Chen, H. L.; Beal, M. F.; Gibson, G. E. Dietary Supplementation with Resveratrol Reduces Plaque Pathology in a Transgenic Model of Alzheimer's Disease. *Neurochem Int.* **2009**, *54*, 111–118.

206. Wenzel, E.; Soldo, T.; Erbersdobler, H.; Somoza, V. Bioactivity and Metabolism of Trans-Resveratrol Orally Administered to Wistar Rats. *Mol. Nutr. Food Res.* **2005**, *49*, 482–494.

207. Bhat, K. P.; Lantvit, D.; Christov, K.; Mehta, R. G.; Moon, R. C.; Pezzuto, J. M. Estrogenic and Antiestrogenic Properties of Resveratrol in Mammary Tumor Models. *Cancer Res.* **2001**, *61*, 7456–7463.

208. Svechnikov, K.; Spatafora, C.; Svechnikova, I.; Tringali, C.; Söder, O. Effects of Resveratrol Analogs on Steroidogenesis and Mitochondrial Function in Rat Leydig Cells In Vitro. *J. Appl. Toxicol.* **2009**, *29*, 673–680.

209. Gartner, S.; Balski, M.; Koch, M.; Nehls, I. Analysis and Migration of Phthalates in Infant Food Packed in Recycled Paperboard. *J. Agric. Food Chem.* **2009**, *57*, 10675–10681.

210. Brauer, B.; Funke, T. Bestimmung von kontaminanten. Papier aus recycleten fasern und verpackte lebensmittel. [Detection of Contaminants. Paper from Recycled Fibre and Packaged Food.] *Dtsch. Lebensmitt. Rundsch.* **2008**, *104*, 330–335 [in German].

211. EFSA. In *EFSA Statement on the Presence of 4-Methylbenzophenone Found in Breakfast Cereals*; E. F. S. Authority, Ed.; EFSA: Parma, Italy, **2009**.

212. Yueh, M. F.; Tukey, R. H. Triclosan: A. Widespread Environmental Toxicant with Many Biological Effects. *Annu. Rev. Pharmacol. Toxicol.* **2016**, *56*, 251–272.

213. Baluka, S. A.; Rumbeiha, W. K. Bisphenol A and Food Safety: Lessons from Developed to Developing Countries. *Food Chem. Toxicol.* **2016**, *92*, 58–63.

214. Cirillo, T.; Latini, G.; Castaldi, M. A.; Dipaola, L.; Fasano, E.; Esposito, F.; Scognamiglio, G.; Francesco, F. D.; Cobellis, L. Exposure to Di-2-Ethylhexyl Phthalate, Di-*n*-Butyl Phthalate and Bisphenol A through Infant Formulas. *J. Agric. Food Chem.* **2015**, *63*, 3303–3310.

215. Hoekstra, E. J.; Simoneau, C. Release of Bisphenol A from Polycarbonate: A Review. *Crit. Rev. Food Sci. Nutr.* **2013**, *53*, 386–402.

216. Lorenzini, R.; Biedermann, M.; Grob, K.; Garbini, D.; Barbanera, M.; Braschi, I. Migration Kinetics of Mineral Oil Hydrocarbons from Recycled Paperboard to Dry Food: Monitoring of Two Real Cases. *Food Addit. Contam., A* **2013**, *30*, 760–770.

217. Aschberger, K.; Castello, P.; Hoekstra, E.; Karakitsios, S.; Munn, S.; Pakalin, S.; Sarigiannis, D. Bisphenol A and Baby Bottles: Challenges and Perspectives. *JRC Scientific and Technical Reports*, European Union, 2010.

218. Long, F.; Zhu, A.; Zhou, X.; Wang, H.; Zhao, Z.; Liu, L.; Shi, H. Highly Sensitive and Selective Optofluidics-Based Immunosensor for Rapid Assessment of Bisphenol A Leaching Risk. *Biosens. Bioelectron.* **2014**, *55*, 19–25.

219. Le, H. H.; Carlson, E. M.; Chua, J. P.; Belcher, S. M. Bisphenol A is Released from Polycarbonate Drinking Bottles and Mimics the Neurotoxic Actions of Estrogen in Developing Cerebellar Neurons. *Toxicol. Lett.* **2008**, *176*, 149–156.

220. Mercea, P. Physicochemical Processes Involved in Migration of Bisphenol A from Polycarbonate. *J. Appl. Polym. Sci.* **2009**, *112*, 579–593.

221. Biedermann, M.; Ingenhoff, J. E.; Zurfluh, M.; Richter, L.; Simat, T.; Harling, A.; Altkofer, W.; Helling, R.; Grob, K. Migration of Mineral Oil, Photoinitiators and Plasticisers from Recycled Paperboard into Dry Foods: A Study under Controlled Conditions. *Food Addit. Contam., A* **2013**, *30*, 885–898.

222. Brede, C.; Fjeldal, P.; Skjevrak, I.; Herikstad, H. Increased Migration Levels of Bisphenol A from Polycarbonate Baby Bottles after Dishwashing, Boiling and Brushing. *Food Addit. Contam.* **2003**, *20*, 684–689.

223. Nam, S. H.; Seo, Y. M.; Kim, M. G. Bisphenol A Migration from Polycarbonate Baby Bottle with Repeated Use. *Chemosphere* **2010**, *79*, 949–952.

224. Kubwabo, C; Kosarac, I.; Stewart, B.; Gauthier, B. R.; Lalonde, K.; Lalonde, P. J. Migration of Bisphenol A from Plastic Baby Bottles, Baby Bottle Liners and Reusable Polycarbonate Drinking Bottles. *Food Addit. Contam., A: Chem. Anal. Control Expo. Risk Assess.* **2009**, *26*, 928–937.

225. Johnson, S.; Saxena, P.; Sahu, R. Leaching of Bisphenol A from Baby Bottles. *Proc. Natl. Acad. Sci. India, Sect., B: Biol. Sci.* **2015**, *85*, 131–135.

226. Maia, J.; Cruz, J. M.; Sendón, R.,; Bustos, J.; Cirugeda, M. E; Sanchez, J. J.; Paseiro, P; Effect of Amines in the Release of Bisphenol A from Polycarbonate Baby Bottles. *Food Res. Intl.* **2010**, *43*, 1283–1288.

227. Zimmerer, C.; Nagel, J.; Steiner, G.; Heinrich, G. Nondestructive Molecular Characterization of Polycarbonate–Polyvinylamine Composites after Thermally Induced Aminolysis. *Macromol. Mater. Eng.* **2016**, *301*, 648–652.

228. Sajiki, J.; Yonekubo, J. Leaching of Bisphenol A (BPA) from Polycarbonate Plastic to Water Containing Amino Acids and Its Degradation by Radical Oxygen Species. *Chemosphere* **2004**, *55*, 861–867.

229. Geens, T.; Goeyens, L.; Covaci, A. Are Potential Sources for Human Exposure to Bisphenol-A Overlooked? *Int. J. Hyg. Environ. Health* **2011**, *214*, 339–347.

230. Liu, F.; Li, Z.; Yu, S.; Cui, X.; Ge, X. Environmentally Benign Methanolysis of Polycarbonate to Recover Bisphenol A and Dimethyl Carbonate in Ionic Liquids. *J. Hazard. Mater.* **2010**, *174*, 872–875.

231. Ji, G.; Zhang, H.; Huang, F.; Huang, X. Effects of Nonionic Surfactant Triton X-100 on the Laccase-Catalyzed Conversion of Bisphenol A. *J. Environ. Sci. (China)* **2009**, *21*, 1486–1490.

232. Jie, H.; Ke, H.; Qing, Z.; Lei, C.; Yongqiang, W.; Zibin, Z. Study on Depolymerization of Polycarbonate in Supercritical Ethanol. *Adv. Polym. Sci.* **2006**, *91*, 2307–2314.

233. Vafeiadi, M.; Roumeliotaki, T.; Myridakis, A.; Chalkiadaki, G.; Fthenou, E.; Dermitzaki, E.; Karachaliou, M.; Sarri, K.; Vassilaki, M.; Stephanou, E. G.; Kogevinas, M.; Chatzi, L. Association of Early Life Exposure to Bisphenol A with Obesity and Cardiometabolic Traits in Childhood. *Environ. Res.* **2016**, *146*, 379–387.

234. Kundakovic, M.; Champagne, F. A. Epigenetic Perspective on the Developmental Effects of Bisphenol A. *Brain Behav. Immun.* **2011**, *25*, 1084–1093.

235. Manikkam, M.; Tracey, R.; Guerrero-Bosagna, C.; Skinner, M. K. Plastics Derived Endocrine Disruptors (BPA, DEHP and DBP) Induce Epigenetic Transgenerational Inheritance of Obesity, Reproductive Disease and Sperm Epimutations. *PLoS ONE* **2013**, *8*, e55387.

236. Beronius, A.; Rudén, C.; Hanberg, A.; Håkansson, H. Health Risk Assessment Procedures for Endocrine Disrupting Compounds within Different Regulatory Frameworks in the European Union. *Regul. Toxicol. Pharmacol.* **2009**, *55*, 111–122.

237. Bars, R.; Fegert, I.; Gross, M.; Lewis, D.; Weltje, L.; Weyers, A.; Wheeler, J. R.; Galay-Burgos, M. Risk Assessment of Endocrine Active Chemicals: Identifying Chemicals of Regulatory Concern. *Regul. Toxicol. Pharmacol.* **2012**, *64*, 143–154.

238. Reinen, J.; Vermeulen, N. P. Biotransformation of Endocrine Disrupting Compounds by Selected Phase I and Phase II Enzymes—Formation of Estrogenic and Chemically Reactive Metabolites by Cytochromes P450 and Sulfotransferases. *Curr. Med. Chem.* **2015,** *22,* 500–527.

239. de Fátima Pocas, M.; Hogg; T. Exposure Assessment of Chemicals from Packaging Materials in Foods: A Review. *Trends Food Sci. Technol.* **2007,** *18,* 219e230.

240. Beronius, A.; Rudén, C.; Hanberg, A.; Håkansson, H. Health Risk Assessment Procedures for Endocrine Disrupting Compounds within Different Regulatory Frameworks in the European Union. *Regul. Toxicol. Pharmacol.* **2009,** *55,* 111–122.

241. Dionisi, G.1.; Oldring, P. K.; Estimates of Per Capita Exposure to Substances Migrating from Canned Foods and Beverages. *Food Addit. Contam.* **2002,** *19,* 891–903.

242. Lange, A.; Paull, G. C.; Hamilton, P. B.; Iguchi, T.; Tyler, C. R. Implications of Persistent Exposure to Treated Wastewater Effluent for Breeding in Wild Roach (*Rutilus rutilus*) Populations. *Environ. Sci. Technol.* **2011,** *45,* 1673–1679.

243. Maiolini, E.; Ferri, E.; Pitasi, A. L.; Montoya, A.; Di Giovanni, M.; Errani, E.; Girotti, S. Bisphenol A Determination in Baby Bottles by Chemiluminescence Enzyme-Linked Immunosorbent Assay, Lateral Flow Immunoassay and Liquid Chromatography Tandem Mass Spectrometry. *Analyst* **2014,** *139,* 318–324.

244. Rubin, B. S. Bisphenol A: An Endocrine Disruptor with Widespread Exposure and Multiple Effects. *Steroid Biochem. Mol. Biol.* **2011,** *127,* 27–34.

245. Stahlhut, R.; van Wijngaarden, E.; Dye, T.; Cook, S.; Swan, S. Concentrations of Urinary Phthalate Metabolites Are Associated with Increased Waist Circumference and Insulin Resistance in Adult US Males. *Environ. Health Perspect.* **2007,** *115,* 876–882.

246. Lee, M. J.; Lin, H.; Liu, C. W.; Wu, M. H.; Liao, W. J.; Chang, H. H.; Ku, H. C.; Chien, Y. S.; Ding, W. H.; Kao, Y. H. Octylphenol Stimulates Resistin Gene Expression in 3T3-L1 Adipocytes via the Estrogen Receptor and Extracellular Signal-Regulated Kinase Pathways. *Am. J. Physiol. Cell Physiol.* **2008,** *294,* 1542–1551.

247. Smink, A.; Ribas-Fito, N.; Garcia, R.; Torrent, M.; Mendez, M. A.; Grimalt, J. O.; Sunyer, J. Exposure to Hexachlorobenzene during Pregnancy Increases the Risk of Overweight in Children Aged 6 Years. *Acta Paediatr.* **2008,** *97,* 1465–1469.

248. Park, S. K.; Son, H. K.; Lee, S. K.; Kang, J. H.; Chang, Y. S.; Jacobs, D. R.; Lee, D. H. Relationship between Serum Concentrations of Organochlorine Pesticides and Metabolic Syndrome among Non-diabetic Adults. *J. Prev. Med. Public Health,* **2010,** *43,* 1–8.

249. Turyk, M.; Anderson, H.; Knobeloch, L.; Imm, P.; Persky, V. Organochlorine Exposure and Incidence of Diabetes in a Cohort of Great Lakes Sport Fish Consumers. *Environment* **2009,** *117,* 1076–1082.

250. Grun, F.; Watanabe, H.; Zamanian, Z.; Maeda, L.; Arima, K.; Cubacha, R.; Gardiner, D. M.; Kanno, J.; Iguchi, T.; Blumberg, B. Endocrine-Disrupting Organotin Compounds Are Potent Inducers of Adipogenesis in Vertebrates. *Mol. Endocrinol.* **2006,** *20,* 2141–2155.

251. Kanayama, T.; Kobayashi, N.; Mamiya, S.; Nakanishi, T.; Nishikawa, J. Organotin Compounds Promote Adipocyte Differentiation as Agonists of the Peroxisome Proliferator-Activated Receptor Gamma/Retinoid X Receptor Pathway. *Mol. Pharmacol.* **2005,** *67,* 766–774.

252. Le Maire, A.; Grimaldi, M.; Roecklin, D.; Dagnino, S.; Vivat-Hannah, V.; Balaguer, P.; Bourguet, W. Activation of RXR–PPAR Heterodimers by Organotin Environmental Endocrine Disruptors. *EMBO Rep.* **2009,** *10*, 367–373.

253. Cimafranca, M. A.; Hanlon, P. R.; Jefcoate, C. R. TCDD Administration after the Proad-ipogenic Differentiation Stimulus Inhibits PPARγ through a MEK-Dependent Process but Less Effectively Suppresses Adipogenesis. *Toxicol. Appl. Pharmacol.* **2004,** *196*, 156–168.

254. Lee, D. H.; Jacobs, D. R., Jr. Methodological Issues in Human Studies of Endocrine Disrupting Chemicals. *Rev. Endocr. Metab. Disord.* **2015,** *16*, 289–297.

CHAPTER 9

RADIONUCLIDES IN FOODS

MOHAMED ABDELRAZEK ABDELALEEM*

Nuclear Research Center, Atomic Energy Authority, Cairo 13759, Egypt

*E-mail: Abdelrazek_MD@yahoo.com

CONTENTS

ABSTRACT

Radionuclides, whether manmade or natural in origin, are present in the environment. These radioactive materials can contaminate food and can affect our health. Long-term consumption of radiological contaminated foods can damage DNA, resulting in the mutation of genes contained in the DNA and lead to cancer. When exposed, the degree of harm to human health depends on the type and the dose of radionuclides. It is also critical to know the radioactivity of the contaminated food. The common radionuclides found in food are the naturally occurring radionuclides such as Potassium-40, Uranium-238, Radium-226, and their associated progeny. The manmade radionuclides can also be discharged into the environment following a major nuclear incident. Iodine-131 is an immediate concern since it's distributed over a large area and rapidly transferred from contaminated feed into milk. Other radioisotopes that could be of long-term concern if released are Cesium-134, Cesium-137, Strontium-90, Strontium-89, and Plutonium. They can remain in the environment for a long time. Derived intervention levels as setting limits and the protective actions are used to determine the levels of radionuclides in human food and to reduce the amount of contamination.

9.1   INTRODUCTION

Radionuclides are chemical elements with unstable atomic structures called radioactive isotopes. The number after the element name is the identification of the mass number of the specific radioactive isotopes. The unstable structures break down to release or emit radiation energy from the nucleus or other parts of the atom. The types of radiation that can be released are alpha particles, beta particles, and gamma rays (photons). Radionuclides emitting more than one type of radiation are classified by the predominant type released. Most naturally occurring radionuclides are alpha-particle emitters (uranium and radium-226), but some beta-particle emitters also occur naturally (radium-228 and potassium-40). Artificial radionuclides (man-made) are mainly beta and gamma emitters. Tritium is a beta-particle emitter that may be formed naturally in the atmosphere or by human activities.[26]

Our bodies, food, and water include natural radionuclides from the environment. Humans are exposed to radiation (also known as background radiation) from these radionuclides on a daily basis. Radiation comes from space (i.e., cosmic rays) as well as from naturally occurring radioactive materials found in the soil, water, and air. The concentration of naturally occurring radionuclides, which can be detected in food and water, varies depending on several factors such as local geology, climate and agricultural practices.[114]

The chemical hazards of radioactive materials should never be neglected. As radioactive substances are also chemical substances, their handling may involve hazards of nonradiological origin. Furthermore, radioactive substances are often intimately mixed with chemical products having an aggressive action on the human body (mutagenic, toxic, and carcinogenic effects).[21,23]

However, radiotoxicity is the toxicity of both an incorporated nuclide and its daughter products due to ionizing radiation and/or chemical nature of nuclide element.

9.2   RADIONUCLIDES DESCRIPTION

Although several hundred radionuclides are either produced by nuclear explosions or present in irradiated reactor fuel, a limited number contribute significantly to human exposure. These would normally include fission products and activation products. Radioactive noble gases, such as ^{85}Kr or ^{133}Xe are not considered since they are unlikely to contribute significantly to internal exposure via the food chain. Table 9.1 showed radionuclides produced in fission and activation processes which may contribute significantly to human exposure in the event of an accident.

9.2.1   FISSION

Atmospheric testing of nuclear weapons is the primary source of radionuclides produced in the fission process and found in the environment. However, the major source of fission product radionuclides in recent years has been from nuclear accidents.[1] A nuclear reactor meltdown could release a spectrum of radionuclides similar to that of a nuclear bomb explosion, but the ratios of nuclides would greatly differ for the two cases. The reason for the differences in ratios of radionuclides is that during the reactor operation *the long-lived* radionuclides tend to build up progressively, whereas *the short-lived* radionuclides tend to reach an equilibrium state at which the rate of decay equals the rate of production. Operating time and fuel burned up are controlled in the proportion of various radionuclides in the operation of nuclear reactors.[49]

9.2.2 ACTIVATION PRODUCTS

Activation products are created in nuclear reactors and other nuclear devices by the reactions of neutrons with fuel and construction materials. Activation products include the isotopes of the transuranic elements and radioisotopes of cesium, cobalt, hydrogen, carbon, iron, manganese, zinc, and a host of other radionuclides, all of which should be recognized and considered in determining the environmental pathways of human exposure.[49]

TABLE 9.1   Fission and Activation Products of Certain Radionuclides.[29]

	Nuclide	Half-life	Major decay
Fission products	^{89}Sr	50.5 d	β^-
	^{90}Sr, ^{90}Y	28.7 y, 64.1 h	β^-, β^-
	^{95}Zr, ^{95}Nb	64.09 d, 35.0 d	$\beta^-\gamma$, $\beta^-\gamma$
	^{99}Mo, ^{99m}Tc	39.272 d, 6.006 h	$\beta^-\gamma$, $\beta^-\gamma$
	^{106}Ru, ^{106}Rh	372.6 d, 29.92 s	β^-, $\beta^-\gamma$
	^{129}Te	33.6 d	$\beta^-\gamma$
	^{131}I	8.021 d	$\beta^-\gamma$
	^{132}Te, ^{132}I	76.856 h, 2.3 h	$\beta^-\gamma$, $\beta^-\gamma$
	^{137}Cs, ^{137}Ba	30.0 y, 2.55 min	$\beta^-\gamma$
	^{140}Ba, ^{140}La	12.751 d, 1.6779 d	$\beta^-\gamma$, $\beta^-\gamma$
	^{144}Ce, ^{144}Pr	284.45 d, 17.28 d	$\beta^-\gamma$, $\beta^-\gamma$
Activation products	^{3}H	12.35 y	β^-
	^{14}C	5730 y	β^-
	^{55}Fe	2.75 y	EC
	^{59}Fe	44.53 d	$\beta^-\gamma$
	^{54}Mn	312.5 d	EC, γ
	^{60}Co	5.27 y	$\beta^-\gamma$
	^{65}Zn	243.9 d	EC, γ
	^{134}Cs	754.2 d	$\beta^-\gamma$
	^{239}Np	2.355 d	$\beta^-\gamma$
	^{241}Pu, ^{241}Am	14.35 y, 432.0 y	β^-, $\alpha\gamma$
	^{242}Cm	162.94 d	α
	^{238}Pu	87.7 y	α
	^{239}Pu	2.411×10^4 y	α
	^{240}Pu	6.563×10^3 y	α
	^{242}Pu	3.735×10^5 y	α

Half-life is given in minutes (min), hours (h), days (d) and years (y)—1 year = 365.25 days.

Source: Adapted from IAEA (1989. *Measurement of Radionuclides in Food and the Environment (A Guidebook)*, Vienna; Technical Reports Series No. 295. http://www-pub.iaea.org/books/IAEABooks/1398/Measurement-of-Radionuclides-in-Food-and-the-Environment).

9.2.3 NATURAL RADIOACTIVITY

Uranium and thorium isotopes plus their daughter products and ^{40}K are considered the major naturally occurring radionuclides. These natural activities influence on the backgrounds of counting instruments.[103]

9.3 HOW FOOD ARE CONTAMINATED WITH RADIONUCLIDES

Within 2 years of a radiation accident, the direct deposition of radionuclides on the surface of plant leaves is a more relevant pathway for human intake than root uptake. Milk and meat livestock, fed on radiocontaminated pasture, are considered an important pathway for transfer of radionuclides to humans. Although rain is known to wash out radionuclides efficiently from the atmosphere and to boost the contamination level of soil, dry deposition causes a higher relative deposition on the plants and hence a higher surface activity on the plant surface.[59] In this respect, the date of the Fukushima accident (March 11, 2011) may have lessened impacts on food contamination accident because it happened prior to the main agricultural season. At the time of the Chernobyl accident (April 26, 1986), the season had already begun and radionuclides were directly deposited on the surface of agricultural plants and pasture.

The release of radionuclides into the environment after a nuclear accident, human health can be affected in several ways: external exposure, internal exposure due to inhalation of radionuclides, and internal exposure due to ingestion of radionuclides from contaminated food.[71] The subsequent fate of the radionuclide is determined by its chemical and physical nature and whether it is absorbed and metabolized by plants or animals. Natural radioactivity may become a concern when ores containing radioactive substances are mined and processed. The products or wastes may concentrate the radionuclides. Examples of this are uranium tailings, phosphate rock waste, or slags from phosphorus production. Radium may enter the food chain, when it dissolves in ground water and is taken up through plant roots.

Nuclear reactors normally release radioactive noble gases that do not contaminate foods. Reactors do contain large inventories of fission products, transuranics, and other activation products. Accidental releases can contaminate vegetation by deposition of particles on leaves and soil, or through water. Gaseous releases would most likely involve the volatile elements such as iodine and tritium, or those with volatile precursors, such as strontium-90 and cesium-137.[37,38]

The mechanisms affecting the behaviors of radioactive materials are very complicated. They are related to a wide range of research fields, including physics, chemistry, and biology. In addition, the land usage modes, which include forests, urban areas, farm fields, and rice paddy fields, have a great influence on the behavior of radionuclides. The lifestyle of the public, such as their intake of foods, also affects the level of doses. The most important transfer pathway to animals is the ingestion of contaminated feed, soil, and drinking water. Radionuclide intake via soil can be significant, but the availability for absorption of soil-associated radionuclides is often low. Hence, it is the ingestion of contaminated feed and processes influencing absorption and retention that usually determines the radionuclide content of animals.[44]

Figure 9.1 shows major transfer and exposure pathways of the radionuclides released from the nuclear facilities into the atmosphere. In the early stage after an accident, external exposure from the radioactive plume and internal exposure by inhalation of the plume are dominant exposure pathways. When the release of radionuclides from the nuclear plants becomes negligible, external exposure from radionuclides deposited in the soil or on buildings is dominant. Internal exposure by the ingestion of contaminated food may be dominant in areas where the air dose rate is relatively low.[40,102]

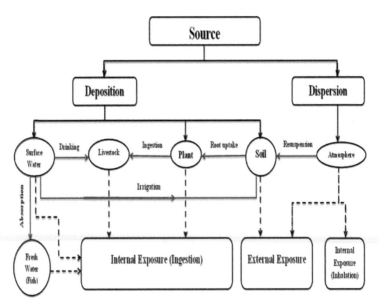

FIGURE 9.1 The major transfer and exposure pathways of radionuclides released from nuclear facilities.
Source: Adapted from Takashi (2014, Outline of the Environmental Monitoring of Tepco's Fukushima Daiichi Nuclear Power Plant Accident. In *Radiation Monitoring and Dose Estimation of the Fukushima Nuclear Accident*; Takashi, S. Ed.; Springer: Tokyo, Heidelberg, New York, Dordrecht, London. http://link.springer.com/book/10.1007%2F978-4-431-54583-5).

The contamination of fruit following a release to atmosphere can be the result of various processes: (1) direct deposition to exposed fruit surfaces, absorption by the fruit skin, and transport to the interior; (2) deposition to exposed plant surfaces (directly from the atmosphere or as a result of resuspension), absorption to interior and translocation to fruit; (3) for perennial plants: remobilization of radionuclides from the leaves to the over-wintering organs prior to leaf drop, followed by retranslocation from storage organs to other plant components at the resumption of growth; (4) deposition to soil, vertical migration in the soil profile, root uptake and transfer to the fruit. The main processes involved in the transfer of radionuclides to fruits are shown in Figure 9.2.[73]

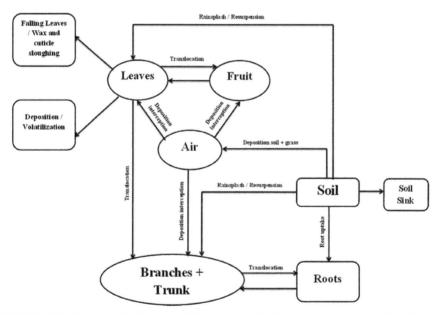

FIGURE 9.2   The transfer of radionuclides in plants (fruits, leaves, branches, and trunk).
Source: Adapted from Carini (2009, Transfer to Fruits. In *Quantification of Radionuclide Transfer in Terrestrial and Freshwater Environments for Radiological Assessments*; IAEA TECDOC-1616, Ed.; International Atomic Energy Agency (IAEA): Vienna, Austria; pp 311–330. www-pub.iaea.org/MTCD/publications/PDF/te_1616_web.pdf).

Relative significances of each pathway after release of radionuclides depend upon the radionuclide, the kind of crop, the stage of plant development, and the season at time of deposition. Furthermore, in agricultural ecosystems, the transfer of radionuclides to fruit is generally affected by human intervention, which includes horticultural practices, intended to modify physiology and translocation of nutrients, to achieve early cropping, high, early, and sustained yield, and high fruit quality.[10]

9.4 CONTAMINATED FOODS

The security of food safety for consumers and producers of various agricultural products is one of the most important issues from an agricultural perspective. Governments have tried to amend standards and enforce much lower radioactivity levels in food, including water and milk. However, some products had higher radioactivity levels than those allowed by the new standards, although they were not regarded as contaminated according to the provisional regulation levels.[78] As mentioned by UNSCEAR,[104] series of natural radionuclide sources (uranium, radium, and thorium occur in three natural decay series, headed by uranium-238, thorium-232, and uranium-235, respectively. In nature, the radionuclides in these three series are approximately in a state of secular equilibrium, in which the activities of all radionuclides within each series are nearly equal.) in contaminated foods are tabulated in Table 9.2.

9.4.1 MILK

Milk is one of the few foods produced over large areas and collected on a daily basis. Its composition is almost identical all over the world, and it is easy to collect a representative sample that can be analyzed in liquid or dried form. Milk is likely to be contaminated by radioactive iodine and cesium within the first days after a release of volatile radionuclides. Contamination of milk will be greatest when cows are grazing during the fallout period, but even when cows are kept indoors, contamination of milk may occur by inhalation of radionuclides or ingestion of radionuclides in drinking water and contaminated feed. Milk from goats and sheep, because of their grazing habits, should be checked periodically over a longer period.[49]

Radioactive iodine (I-131) is rapidly transferred to milk from contaminated feed where it can accumulate in the thyroid gland. Iodine which is inhaled or swallowed will also concentrate in the thyroid gland. Both exposure routes increase the risk of thyroid cancer. Eighty percent of ingested plant associated radiocesium is absorbed from the ruminant gut and is subsequently transported to all soft tissues, milk, urine, and feces.[70] ^{134}Cs and ^{137}Cs

TABLE 9.2 Concentration of Uranium and Thorium Series Radionuclides in Food and Drinking Water (mBq/kg).

Country	Concentrations (mBq/kg)									References
	238U	230Th	226Ra	210Pb	210Po	232Th	228Ra	228Th	235U	
Milk products										
USA	0.7	0.4	5.7	11		0.27			0.05	Fisenne et al.,[32] Morse and Welford[76]
China	13		6	16	13	1.2	21		0.6	Zhu et al.[115]
India	17				15					Dang et al.,[20] Khandekar[62]
Japan	0.55		12			0.29				Servant and Delapart[95]
Italy			3–19							Mastinu and Santaroni[69]
Germany			2–130	5–80						Jaworowski,[60] Muth et al.[77]
Poland	2.6	1.2	10	16	1.2					Pietrzak-Flis et al.[86,87]
Romania			0.9–44	13–140						Botezatu,[6] RSRP[89]
UK	0.1–4.9		<0.4–200	20–220				56		Bradley[7]
Meat products										
USA	0.8–2.3	0.5–3	20	18		0.3–2			0.02	Fisenne et al.,[32] Morse and Welford[76]
China	10		41	140	120	4.3	120		0.5	Zhu et al.[115]
India					440					Khandekar[62]
Japan	13		36			2.3				Servant and Delapart[95]
Germany	1–20		30–220	100–1000	37–4000					Glöbel and Muth,[35] Muth et al.[77]
Poland	1.6–5.6	0.7–3	11–19	98–105	99–102	0.5–3.6				Pietrzak-Flis et al.[86,87]
Romania			2–30	15–19	38–110					Botezatu,[6] RSRP[89]
UK	4.9		2.6–74	40–3700	62,067,000			22–93		Bradley[7]

TABLE 9.2 (Continued)

Country	Concentrations (mBq/kg)									References
	238U	230Th	226Ra	210Pb	210Po	232Th	228Ra	238Th	235U	
Grain products										
USA	3–23	0.9–10	7–100	33–81		0.1–2.8			0.1–1.3	Fisenne et al.,[32] Morse and Welford[76]
China	9.8		17	34	42	13	38		0.5	Zhu et al.[115]
India	7.4				15–120					Dang et al.,[20] Khandekar[62]
Japan	1.2		14			1.2				Servant and Delapart[95]
Germany	20–400		20–2900	40–4000	37–1900					Glöbel and Muth[35]
Poland	4.7–11	1.4–17	80–110	110–160	90–140	2.0–21				Pietrzak-Flis et al.[86,87]
Romania	6.1–85		30–90	49–59	20–360	1.6–33				Botezatu,[6] RSRP[89]
UK			0.7–5200	56–120	27–260	12		180–2300		Bradley[7]
Leafy vegetables										
USA	24	20	56	41		18			1.2	Fisenne et al.,[32] Morse and Welford[76]
China	16		75	360	430	23	220		0.7	Zhu et al.[115]
India	61–72				320					Dang et al.,[20] Khandekar[62]
Germany	6–2200		6–1150	4–4100	4–7400					Glöbel and Muth,[35] Muth et al.[77]
Italy			27–44							De Bortoli and Gaglione[22]
Poland	14–15	6–9	37–43	43–51	40	4.7				Pietrzak-Flis et al.[86,87]
UK	80–380		2.2–170	16–3300	37–3300					Bradley[7]
Root vegetables and fruits										
USA	0.9–7.7	0.2–1.1	7–47	8–150		0.08–1.4			0.1	Fisenne et al.,[32] Morse and Welford[76]

TABLE 9.2 (Continued)

Country	Concentrations (mBq/kg)									References
	238U	230Th	226Ra	210Pb	210Po	232Th	228Ra	238Th	235U	
China	13		63	27	29	4.7	110		0.6	Zhu et al.[115]
India	0.4–77				16–140					Dang et al.,[20] Khandekar[62]
Japan	26		11			2.3				Servant and Delapart[95]
Germany	10–2900		5–9400	20–5200						Glöbel and Muth,[35] Muth et al.[77]
Italy			14–25							De Bortoli and Gaglione[22]
Poland	0.9–10	0.7–7.5	11–215	28–210	0.7–7.1					Pietrzak-Flis et al.[86,87]
Romania	6–120		9–190	12–140	0.4–2.1		22			Botezatu,[6] RSRP[89]
UK	6		9–41							Bradley[7]
Fish products										
USA	13–1900	1.2–29	14–1800	150–55,000	1.2–30				0.4–90	Fisenne et al.,[32] Morse and Welford[76]
China	12		39	3500	4900	1.3	320		0.5	Zhu et al.[115]
France			37							Pellerin et al.[84]
Germany			100–7400	20–4400	50–5200					Glöbel and Muth,[35] Muth et al.[77]
Poland			28–43	81–93	3100–3800					Pietrzak-Flis et al.[87]
Portugal					80–120,000					Carvalho[11]
UK	2.5		8.5–2100	180–4800	60–53,000			56–700		Bradley[7]
Drinking water										
USA	0.3–77	0.1	0.4–1.8	0.1–1.5		0.05	0–0.5		0.04	Cothern and Lappenbusch,[16] Fisenne et al.,[32] McCurdy and Mellor[66]

TABLE 9.2   *(Continued)*

Country	Concentrations (mBq/kg)									References
	^{238}U	^{230}Th	^{226}Ra	^{210}Pb	^{210}Po	^{232}Th	^{228}Ra	^{238}Th	^{235}U	
China	0.1–700		0.2–120			0.04–12				NEPA[83]
India	0.09–1.5									Dang et al.[20]
Finland	0.5–150,000		10–49,000	0.2–21,000	0.2–7600		18–570			Asikainen,[2] Salonen[93]
France	4.4–930		7–700			0–4.2				Pellerin et al.[84]
Germany	0.4–600		1–1800	0.2–200	0.1–200					Glöbel and Muth[35]
Italy	0.5–130		0.2–1200							Sgorbati and Forte[97]
Poland	7.3	1.4	1.7–4.5	1.6	0.5	0.06				Pietrzak-Flis et al.[86,87]
Romania	0.4–37		0.7–21	7–44	7–44	0.04–9.3				Botezatu,[6] RSRP[89]
Swizerland	0–1000		0–1500				0–200		0–50	SFOPH[96]
Spain	3.7–4.4		<20–4000							Soto et al.[100]
UK			0–180	40–200	40–200					Bradley[7]

Source: Adaptetd from UNSCEAR (2000, Sources and Effects of Ionizing Radiation. *Report to the General Assembly of the United Nations with Scientific Annexes B. United Nations Sales Publication E.00.IX.3. New York.* http://www.unscear.org/unscear/en/publications/2000_1.html).

are the largest contributed radionuclides in milk and milk products. Radiation dose from the consumption of milk at average rates was assessed for various age groups. In 2009, the maximum dose was to 1-year-old infants. For the range of radionuclides analyzed, the dose was less than 0.005 mSv.[88]

In a study of Manabe et al.,[67] the changes in the radioactivity of [134]Cs and ^{137}Cs in milk produced by cows given pasture that was contaminated with these radioactive nuclides caused by the Fukushima Daiichi NPP accident on March 11, 2011, were examined between May 16 and June 26, 2011. The study revealed that when cows received radiocesium contamination of less than 300 Bq/kg, they produced milk contaminated with less than 50 Bq/kg radiocesium.

9.4.2 MEAT

After nuclear accidents, meat becomes one of the main sources of radiocesium. Contamination of meat is mainly the result of animal grazing, but contaminated drinking water might also be an important pathway. Inhalation of radiocesium is not likely to be a significant pathway to meat.[49] Meat sampling should normally be done in such a way that the composite sample is representative of a large number of animals, although after heavy fallout screening measurements of individual animals may be necessary.[45]

9.4.3 FISH

Radiocesium is considered to behave like potassium in living organisms after being absorbed into the body because of its similar biochemical properties. Similar to terrestrial animals, aquatic animals are at a high risk of radioactive contamination because radioactive pollutants enter rivers and the ocean. In particular, marine teleosts are vulnerable to the absorption of higher levels of minerals dissolved in seawater because they drink a large amount of ambient seawater to compensate for their osmotic water loss.[28,68]

Following an accident, contamination of fish in nutrient deficient lakes may constitute a particularly significant pathway for the uptake of radiocesium by humans. Obtaining a representative sample from an area containing many lakes may require some compromise since collection of samples from a large number of the lakes may be impracticable. Ocean fish will not take up as much radiocesium as fresh-water fish because of the dilution through the depth of the ocean and the effective dilution associated with the high potassium content in the water, but particulate-associated radionuclides can be enriched to high levels. Mussels like *Mytilus edulis*, some species of macro algae, and other filter feeders quickly take up the contaminants from the sea water and can also be used as biological indicators.[49]

Radioactivity was detected in fish caught in the coastal areas near Fukushima Daiichi.[9] Higher concentrations of radiocesium (^{134}Cs and ^{137}Cs combined) were detected in plaice (*Paralichthys olivaceus*) caught in August (58–590 Bq/kg) compared with in April (approximately 82 Bq/kg) in Fukushima Prefecture.[33,111] Thus, the government declared that fisheries products intended for sale should not contain a level exceeding 500 Bq/kg at the market in Japan, which had not been the case since 1940 after World War II. Certain fish were found to contain radiocesium at levels of >500 Bq/kg, such as plaice and sea bass (*Lateolabrax japonicus*). Recently, the government decided to reduce the limit for radiocesium in fish sold in the market to <100 Bq/kg.[111]

In further investigations on removing radiocesium from contaminated fish,[110] three washes with 0.1% NaCl solution effectively removed the radiocesium from contaminated fish meat. However, extensive washing of small pieces of meat may also remove the flavor attributable to amino acids, nucleotides, and other small compounds present in surimi, thereby yielding a tasteless product. Thus, appropriate conditions should be determined for a balance between risk reduction and food quality. This can be applied in the production of surimi-based products and other processed seafood such as boiled, dried, or seasoned products.

9.4.4 RICE

Although most of the radiocesium fallout that deposited in paddy fields after the Fukushima nuclear disaster in March 2011 was expected to be bound to clay in the soil resulting in a very low soil-to-plant transfer function, a radiocesium contamination level of >500 Bq/kg was detected in brown rice grown in several hilly areas of Fukushima Prefecture in the autumn of the same year.[82]

Radiocesium deposited in soil sticks to soil particles; however, a small portion is redistributed into a soil solution that is readily absorbed by plants. Uptake and transport of cesium (Cs) by plants is an important determinant of the degree of radiocesium contamination of foods. Meanwhile, due to high temperature, organic matter decomposition is enhanced greatly in paddy fields when the air temperature exceeds 30°C, which could have released radiocesium from radiocesium-contaminated organic matter.[34] Over 100 rice varieties were grown using highly contaminated soil and their radiocesium uptake capacities were measured during the vegetative stage. The radiocesium uptake during the vegetative phase was generally higher in *japonica* varieties compared with *indica* varieties.

Cesium (Cs) is an alkaline metal that is taken up by plants in the form of Cs^+. Potassium (K) is an essential plant nutrient that is a major component of fertilizers. Cesium transport has been discussed in conjunction with K transporters. It is well established that the mechanism of Cs absorption by a root is similar to the mechanism of K absorption, following Michaelis–Menten kinetics.[92,112] Inhibition of Cs absorption by the addition of K has also been shown in hydroponics and soil cultures,[99] suggesting that Cs is taken into plants through K transporters.

Cesium is soluble in water as a univalent cation. In general, the cation is fixed on any solid surface with a negative electric charge, such as soil particles or organic matter in soil. The fixed cation is exchanged by any other cation; in a manner that the concentrations of the fixed cations are in equilibrium with the concentrations of other cations in the soil water. Plant roots can easily absorb this form of Cs. This adhesion of Cs on soil solids is referred to as "weak fixation" in the present study. On the other hand, Cs strongly fixes to other types of clay crystals such as silicate sheets. The strongly fixed Cs ions on these clay particles are rarely replaced by other cations as soluble ions, and thus are seldom absorbed by plant roots.[63,98]

9.5   DETERMINING LEVELS OF RADIONUCLIDES IN FOODS

The recommendations advise that health risk to the public be averted by limiting the radiation dose received as a result of consumption of accidentally contaminated food. This will be accomplished by (1) setting limits, called Derived Intervention Levels (DILs) on the radionuclide activity concentration permitted in human food, and (2) taking protective actions (PAs) to reduce the amount of contamination. DILs are limits on the concentrations permitted in human food distributed in commerce. They are established to prevent consumption of undesirable amounts of radionuclides and have units of radionuclide activity per kilogram of food, that is, becquerels per kilogram, Bq/kg (previously used units—pCi/kg).[72] DILs apply during the first year after an accident. If there is concern that food will continue to be significantly contaminated beyond the first year, the long-term circumstances need to be evaluated to determine whether the DILs should be continued or if other guidance may be more applicable. PAs would be initiated subject to evaluation of the situation and would continue until, in the absence of the actions, the concentrations remain below the DILs. PAs can be taken to avoid the amount of contamination that could become incorporated in human food and animal feeds, or to delay consumption of human food and animal feeds suspected of being contaminated until the concentration of contamination has been determined, or to reduce the amount of contamination in human food and animal feeds.[13]

The 1982 FDA recommendations established two levels of Protective Action Guides (PAGs). PAGs were defined as "projected dose commitment values to individuals in the general population that warrant PA following a release of radioactive material." *The lower level*, called the preventive PAG, was a projected dose commitment of 5 mSv (0.5 rem) to the whole body, active bone marrow, or any other organ except the thyroid, or a projected dose commitment of 15 mSv (1.5 rem) to the thyroid. The preventive PAG was associated with low-impact PAs (e.g., placing dairy cows on stored feed). *The upper level*, called the emergency PAG, was a projected dose commitment of 50 mSv (5 rem) to the whole body, active bone marrow, or any other organ except the thyroid, or a projected dose commitment of 150 mSv (15 rem) to the thyroid. The emergency PAG was associated with higher impact PAs (e.g., diversion of fresh milk to cheese or milk powder).[13,25,52,54,57]

Regarding to all previous data, DILs could be calculated from the following equation:

$$\text{DILs (Bq/kg)} = \frac{\text{PAG (mSv)}}{f \times \text{Food intake (kg)} \times \text{DC}}$$

where PAG is the Protective Action Guide; DC is the dose coefficient, the radiation dose received per unit of activity ingested (mSv/Bq); f is the fraction of the food intake assumed to be contaminated.

Food Intake = Quantity of food consumed in an appropriate period of time (kg).

Note: The FDA DILs provide a large margin of safety for the public because each DIL is set according to a conservatively safe scenario for the most vulnerable group of individuals. In addition, PA would be taken if radionuclide concentrations were to reach or exceed a DIL at any point in time, even though such concentrations would need to be sustained throughout the relevant extended period of time for the radiation dose to actually reach the PAG. In practice, when FDA DILs are used, radiation doses to the vast majority of the affected public would be very small fractions of the PAG. As a result, future adjustments in the absolute values of the PAGs would not necessarily require proportionate modifications in the DILs. Any modification of the DILs would depend on a review of all aspects of the conservatively safe scenario and how the DILs are applied. So, food with concentrations below the DILs is permitted to move in commerce without restriction. Food with concentrations at or above the DILs is not normally permitted into commerce.[106] However, state and local officials have flexibility in whether or not to apply restrictions in special circumstances, such as permitting use of food by a population group with a unique dependency on certain food types.

Following the Chernobyl accident in 1986, FDA and the Food Safety and Inspection Service (FSIS) of the US Department of Agriculture established DILs for application to imported foods under their respective regulatory control. The FDA DILs were called "Levels of Concern" (LOCs) and the FSIS DILs were called "Screening Values." Food containing concentrations below the LOCs and Screening Values was allowed to be imported into the United States.[30,31]

A comparison of the resulting recommended "Action Levels" is given below in Table 9.3, which shows that although the approaches taken by different organizations may vary, there is a general consistency in the recommendations.[14,15,39] However, Action Level in Canadian guidelines is the same as DILs in United States Food and Drug Administration.

Numbers within solid outlined box indicates that contributions from radionuclides are summed. Dotted lines indicate summed values with different action levels. [a]Codex Alimentarius Commission, [b]International Atomic Energy Agency, [c]Council of European Communities, [d]World Health Organization. Contribution from all detected radionuclides is summed over all food and radionuclide groups, [e]US Food and Drug Administration, and [f]International Commission of Radiological Protection: applies to a single food stuff, with range of optimized values given as 1000–10,000 Bq/kg (beta/gamma emitters); 10–100 Bq/kg (alpha emitters).

DILs recommended by the US FDA have been derived for those radionuclides, under various accident scenarios, that are expected to deliver the major portion of the radiation dose from ingestion during the first year following an emergency. For each class of radionuclide, DILs have been calculated for the six ICRP age groups (Tables 9.A1–9.A6) based on the total annual dietary intake for each age group, and assuming that contamination would occur in 30% of the dietary intake. An exception was made for ^{131}I in the diets of the 3-month and 1-year age groups, where the entire intake over a 60-day period was assumed to be contaminated. Dose coefficients were taken from ICRP Publication 56.[56] A single DIL was selected for each radionuclide group based on the most limiting age group for the radionuclide group.

TABLE 9.3   The Action Levels in Different Organizations.

Radionuclides	Internal recommendations for equivalent food groups (Bq/kg)							
	Health Canada	CODEX[a]/IAEA[b]	CEC[c]		WHO[d]		US FDA[e]	ICRP[f]
	Fresh liquid milk	Milk and infant foods	Dairy	Infant foods	Milk	Infant foods	Total diet	Single food stuff
134Cs	300					—	1200	
137Cs	300		1000	400		1800		
103Ru	1000	1000					6800	10³–10⁴
106Ru	100				4500	—	450	
89Sr	300		125	75		—	—	
90Sr	30	100				160	160	
131I	100		500	150		1600	170	
241Am	1	1	20	1	45	—	2	10–100
239/239Pu	1					7		
Other commercial foods and beverages								
134Cs	1000			n/a				
137Cs	1000		1250					
103Ru	1000	1000			3500–			
106Ru	300				35,000	n/a		
89Sr	1000		750					
90Sr	100	100						
131I	1000		2000					
241Am	10	10	80		35–			
239/239Pu	10				350			
Public drinking water								
134Cs	100							
137Cs	100		1000			700	700	
103Ru	1000	1000			700			
106Ru	100							
89Sr	300		125					
90Sr	30	100				160		
131I	100		500					
241Am	1	1	20		7	7		
239/239Pu	1							

Source: Adapetd from Health Canada (2000, *Canadian Guidelines for the Restriction of Radioactively Contaminated Food and Water Following a Nuclear Emergency: Guidelines and Rationale.* Minister of Public Works and Government Services Canada, 35 pp. http://www. hc-sc.gc.ca/ewh-semt/pubs/contaminants/emergency-urgence/index-eng.php).

After the Fukushima accident in March 2011, transitional measures provided in the Japanese legislation adopted by the Japanese authorities on February 24, 2012 as regards the maximum levels for the sum of cesium-134 and cesium-137 as mentioned by EU;[27] (1) milk and dairy products, mineral water, and similar drinks that are manufactured and/or processed before March 31, 2012 shall not contain more than 200 Bq/kg of radioactive cesium. Other foods that are manufactured and/or processed before March 31, 2012 shall not contain more than 500 Bq/kg of radioactive cesium, except products made from rice and soybean. (2) Products made from rice that are manufactured, and/or processed before September 30, 2012 shall not contain more than 500 Bq/kg of radioactive cesium. (3) Soybean harvested and placed on the market before December 31, 2012 shall not contain more than 500 Bq/kg of radioactive cesium. (d) Products made from soybean that are manufactured and/or processed before December 31, 2012 shall not contain more than 500 Bq/kg of radioactive cesium.

9.6   NUCLEAR ACCIDENTS AND RADIONUCLIDE CONTAMINATION OF FOODS

Many hypotheses exist about the release of radionuclides in the environment after nuclear accidents. Venting of scrammed and uncontrolled releases from damaged fuel located within central pool are two possible sources of releasing radionuclides from the Fukushima Daiichi nuclear facility. Figure 9.3 provides a graphical overview of the assumed mechanism for radionuclides release from the vented Fukushima Daiichi.

The mechanism involves five steps: (1) volatilization of radionuclides within the fuel matrix, (2) migration of those volatile species through the interstices of the fuel to the cladding wall, (3) escape of the volatile radionuclides through damaged portions of the cladding, (4) bubbling of those volatile contaminants through the confined coolant during venting, and (5) migration of these radionuclides to the environment, where all volatile radionuclides immediately condense upon cooling and mixing with the atmosphere and precipitate without significant fractionation had occurred between differing elements from the point of release to the point of measurement is likely only valid at short distances from the source. On the other hand, both of the two dominant possible sources of radionuclide releases in the environment are identified using activity ratios of $^{136}Cs/^{137}Cs$ and $^{134}Cs/^{137}Cs$. Seven isotopes are the most active predicted to be residing within the environment after 300 d including, ^{85}Kr, ^{103}Ru, ^{91}Y, ^{127m}Te, ^{125}Sb, ^{151}Sm, and ^{129}Te.[94]

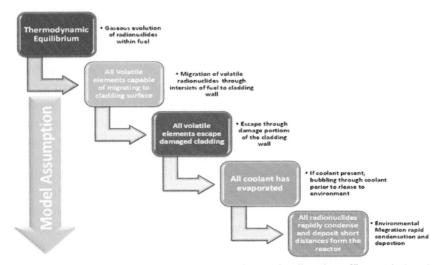

FIGURE 9.3 Major mechanisms of gaseous release of radioactive effluents during the venting contaminant vessels.

Source: Adapted from Schwantes et al. (2012. Analysis of a Nuclear Accident. In: *Fission and Activation Product Releases from the Fukushima Daiichi Nuclear Facility as Remote Indicators of Source Identification, Extent of Release, and State of Damaged Spent Nuclear Fuel*; Pacific Northwest National Laboratory, Ed.; US Department of Energy under Contract, USA. www.pnnl.gov/main/publications/external/technical_reports/PNNL-20912.pdf).

The major radioactive contaminants of nuclear accidents were isotopes of volatile elements; ^{131}I (8-day half-life), ^{132}Te (3-day half-life), ^{134}Cs (2-year half-life), and ^{137}Cs (30-year half-life).[41,101] Codex Alimentarius Commission at its 18th session (Geneva 1989) adopted Guideline Levels (GL) for radio-nuclides in foods following accidental nuclear contamination for use in inter-national trade (CAC/GL 5-1989) applicable for six radionuclides (^{90}Sr, ^{131}I, ^{137}Cs, ^{134}Cs, ^{239}Pu, and ^{241}Am) during one year after the nuclear accident.[15]

The situation with Fukushima differs from that with Chernobyl, in terms of the nature of the soil, breeding species, types of grass or trees, and other aspects. When the nuclear accident at Fukushima is compared with that at Chernobyl, the contaminated area in Fukushima, both low and high in radio-activity, was approximately 6% of that in Chernobyl. The amount of fallout in Fukushima was approximately one-sixth and the distance to which the fallout spread was approximately one-tenth that in Chernobyl.[78]

Nobel gases, ^{131}I, ^{134}Cs, and ^{137}Cs were released in large amount to the envi-ronment, and consequently agriculture land and forests in eastern Japan, due to the accident of Fukushima Nuclear Power Plant (NPP). ^{134}Cs + ^{137}Cs are impor-tant radionuclides that can be used for the assessment of radiation exposure to the human, because it has long half-life, wide distribution in the environment and mainly the adsorption of radiocesium into the soil and binds strongly to clay. The concentration of radiocesium in brown rice produced in several areas of Fukushima Prefecture has exceeded a provisional regulation value.[78,91]

Radionuclides such as Cesium-134 (^{134}Cs), Cesium-137 (^{137}Cs), Strontium-90 (^{90}Sr), and Plutonium-239, and Plutonium-240 ($^{239+240}$Pu), which were released to the environment in the nuclear accident at the Fukushima Daiichi NPP, have been found in bottom ash and fly ash of incinerated wastes produced in Fukushima Prefecture, Japan.[74] ^{134}Cs and ^{137}Cs have been found in sewage sludge produced in a wide area of northern and eastern Japan.[75] Meanwhile, radionuclides Cesium (Cs), Cobalt (Co), Manganese (Mn), Strontium (Sr), and Nikel (Ni) were found in the wastes such as sewage sludge ash and municipal waste ash due to the nuclear accident at the Fukushima Daiichi (NPP).[61] The previous radionuclides could be easily transferred from wastes to soil and plants.

Americium-241 (^{241}Am) is the only radionuclide whose levels are presently increasing with time owing to its ingrowth from the decay of Plutonium-241 (^{241}Pu). The total activity of ^{241}Am in the environment will reach a maximum in the year 2058, after which levels will slowly decline. This peak value is small compared to initial levels of ^{241}Pu. Eventually ^{241}Am will be the most significant remaining radionuclide, albeit at trace levels (Fig. 9.4).[105]

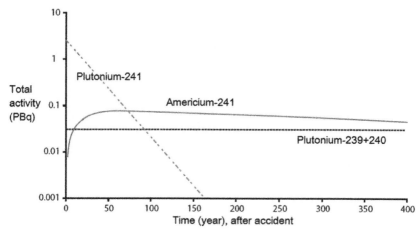

FIGURE 9.4   Total amounts in the environment of various long-lived radionuclides after the accident.

Source: Adapted from UNSCEAR (2011. *Source and Effects of Ionizing Radiation. Volume I: Sources, Volume II: Effects.* United Nations Scientific Committee on the Effects of Atomic Radiation, 2008 Report to the General Assembly with Scientific Annexes C, D and E. United Nations Sales Publication E.11IX.3: New York. www.unscear.org/docs/reports/2008/11-80076_Report_2008_Annex_D.pdf).

9.7 HEALTH RISKS OF RADIONUCLIDES

Exposure to radionuclides can result in their concentration in some tissues of the body which is dependent on the element. These elements will go through gamma, beta, and alpha decays, thereby causing internal damage.[58]

The different metabolisms of each radionuclide after intake can be described by means of the biological half-life, which is always shorter than the physical half-life as it includes both physical decay and biological excretion. Metabolism studies of radio-cesium showed that 10% of the cesium is excreted with a half-life of 2 days, and 90% with a half-life of 110 days.[53,90] Radionuclides of natural origin are ubiquitous and as a consequence are present in all foodstuffs to varying degrees. Radiation doses from the consumption of foodstuffs typically range from a few tens to a few hundreds of microsieverts in a year.[15]

All types of radiation produce the same primary basic physicochemical effects of excitation and ionization within the biological material and they differ only in the spatial distribution and intensity of these effects.[47] Therefore, if the concentration of a radionuclide by a body organ of known mass can be determined from experimental measurement, then the absorbed dose in rads delivered to the organ by the radionuclide can be calculated.

The body organ in which a radionuclide is most concentrated and the associated relative biological efficiency (RBE) dose is highest is, in general, called "the critical organ." Factors other than the RBE dose were considered by the ICRP[50] in choosing the critical organ for a particular radionuclide, such as the essentiality of the organ and its radiosensitivity, but usually the RBE dose was the overriding consideration. Radionuclides are accumulated in different organs as shown in Table 9.4. Maximum permissible concentrations (MPCs) recommended by the ICRP will eventually give rise to a dose rate of 0.1, 0.3, or 0.6 rem/week in the critical organ.

Hence, the basic toxicity grading, based on the arrangement of the radionuclides in order of their most restrictive value of maximum MPC in air (MPC)a for continuous inhalation, takes into account not only the RBE dose rate but also the radiosensitivity of the critical organ. The ICRP give values of the (MPC)a for both soluble and insoluble forms of radionuclides and the more restrictive of these two values is used. The grading obtained in this way is given in Table 9.5, where the radionuclides considered are listed in order of their most restrictive value of (MPC)a.

There are three important ways in which radionuclides normally enter the body—by absorption through the skin, by ingestion, and by inhalation. The radionuclides listed in Table 9.5 have been divided into three main toxicity groups, with a division of the large–medium group into two subgroups to obtain a toxicity classification.[47]

TABLE 9.4 Distribution of the Selected Radionuclides in the Target Organs.

Radionuclide	Target organ	Reference
Americium	Bone and liver	Coughtrey et al.[18]
Cesium	All soft tissue and milk	Coughtrey[19]
Iodine	Thyroid gland and milk	Lengemann et al.[65]
Plutonium	Liver and bone	Coughtrey et al.[17]
Ruthenium	Kidney	Howard et al.[46]
Strontium	Bone and milk	Coughtrey[19]

Source: Adapted from Howard et al. (2009. Transfer to Animals. *IAEA-TECDOC-1616*. www-pub.iaea.org/MTCD/publications/PDF/te_1616_web.pdf).

TABLE 9.5 The Most Restrictive Maximum Permissible Concentration (MPC) in $\mu Ci/cm^3$ Value and the Maximum Permissible Intake (MPI) in μCi.

High toxicity	(MPC)a ($\mu Ci/cm^3$) 10^{-10} and less	(MPI) (μCi) 1 and less
Pa^{231}, Cf^{249}, Pu^{239}, Pu^{240}, Pu^{242}, Th^{232}, Pu^{238}, Ac^{227}, Th^{230}, Np^{237}, Am^{241}, Am^{243}, Cm^{243}, Cf^{250}, U^{232}, Ra^{226}, U^{238}, Pu^{241}, Pb^{210}, U^{230}, U^{233}, U^{234}, U^{235}, U^{236}, Th^{227}, Po^{210}, Ra^{223}, and Sr^{90}		
Medium toxicity	2×10^{-10}–10^{-6}	1–10^4
Upper subgroup A	2×10^{-10}–10^{-8}	1–10^2
Ra^{224}, Pa^{230}, I^{129}, Bi^{210}, Na^{22}, Co^{60}, Ag^{110m}, I^{126}, I^{131}, Cs^{134}, Eu^{152}, Cs^{137}, Bi^{207}, Pd^{212}, Ir^{192}, Ca^{45}, Mn^{54}, Zr^{95}, Cd^{115}, Te^{127m}, Te^{129m}, I^{133}, and Ba^{140}		
Lower subgroup B	2×10^{-8}–10^{-6}	10^2–10^4
P^{32}, V^{48}, Fe^{59}, Co^{58}, Ni^{63}, Zn^{65}, Rb^{86}, Tc^{99}, Cd^{109}, Sn^{113}, Pm^{147}, Hg^{203}, As^{76}, Y^{90}, Zr^{97}, Ag^{105}, Cs^{135}, Eu^{155}, K^{42}, As^{74}, Se^{75}, Sr^{85}, Zr^{93}, I^{135}, Na^{24}, Mn^{52}, Ce^{141}, Pr^{142}, Ir^{194}, Ca^{47}, Co^{57}, Ga^{72}, Br^{82}, Cd^{115}, Cs^{136}, Ag^{111}, I^{132}, Au^{198}, Ir^{190}, Mn^{56}, Ni^{59}, Kr^{87}, I^{134}, Si^{31}, Fe^{55}, Pd^{103}, Au^{199}, Hg^{197m}, Cu^{64}, Hg^{197}, C^{14}, X^{135}, and Cs^{131}		
Low toxicity	2×10^{-6}	10^4
H^3, Zn^{69}, Ge^{71}, Cs^{134m}, Tc^{99m}, Co^{58m}, Kr^{85}, Xe^{133}, Os^{191m}, Xe^{131m}, Tc^{96m}, and Rh^{103m}		

MPI, The maximum permissible intake in microcuries in 1 year.

Source: Adapted from IAEA (1963. *Basic Safety Standards for Radiation Protection*. www.iaea.org/inis/collection/NCLCollectionStore/_Public/24/072/24072024.pdf).

The most radioactive isotopes of elements (radionuclides) which naturally present in the environment including those found in our bodies, food, and water are as follow:

Iodine: Contamination of milk by short-lived radioiodine isotopes is a major source of potential ingested exposure after many types of nuclear accident. It is therefore important that emergency preparedness plans for this eventuality are well informed and appropriately focused.[8]

Isotopes of iodine (^{131}I, ^{132}I, including those arising from the decay of Tellurium-132 (^{132}Te) and ^{133}I are likely to be important components of the release from a severe accident. Radioactive iodines can give rise to both external exposure and internal exposure (from inhalation and ingestion).[113] Iodine-131 is an important isotope of iodine and is a major product from uranium, plutonium, and indirectly from thorium fission, comprising nearly 3% of the total products of fission (by weight). Iodine-131 has a radioactive decay half-life of 8 days. Due to its mode of beta decay (90% beta and 10% gamma), iodine-131 results in mutation and cell death through its ability to penetrate between 0.6 and 2 mm.[58] Absorption of radioiodine in the gut is complete, regardless of source of dietary iodine intake rates, and there is a subsequent rapid transfer to the thyroid and milk.[5,42,109] Small incidental doses of iodine-131 are considered to be the major cause of increased thyroid cancers after accidental nuclear contamination. These cancers result from residual tissue radiation damage and usually appear years after exposure, long after the I^{131} has decayed.

Potassium-40: Potassium-40 is a naturally occurring radioactive isotope of potassium. Two stable (nonradioactive) isotopes of potassium exist, potassium-39 and potassium-41. Potassium-40 (^{40}K) is one of the most common radionuclides in food. In general, ^{40}K is the most commonly occurring natural radioisotope. In milk, levels of ^{40}K measure around 50 Bq/L, and for meat, bananas, and other potassium rich products, levels may measure at several hundreds Bq/kg. Potassium-40 behaves in the environment the same as other potassium isotopes, being assimilated into the tissues of all plants and animals through normal biological processes. It is the predominant radioactive component in human tissues and in most food.[107,114]

Potassium-40 can be taken into the body by drinking water, eating food, or breathing air. Once taken in, potassium-40 behaves in the body in the same manner as other potassium isotopes. Humans require potassium to sustain biological processes, with most (including potassium-40) being almost completely absorbed upon ingestion, moving quickly from the gastrointestinal tract to the bloodstream. The potassium-40 that enters the bloodstream after ingestion or inhalation is quickly distributed to all organs and tissues.[85]

Cesium: Cesium-137 (Cs-137) is a radioisotope and unstable chemical. Its radioactivity is measured by the number of atoms disintegrating per unit time. A Cs-137 atom emits radiation in the form of medium energy gamma rays, and to a lesser extent, high-energy beta particles, which disrupt molecules in cells and deposits energy in tissues, causing damage. Cs-137 is used in medical therapy for oncology, industrial radiography, radiation gauges, food irradiators, and soil testing. It is also a potential plume component following a nuclear power plant incident. Because it readily bonds with chlorides, cesium is typically found as cesium chloride, which is also a powder and is quite soluble in water.[12]

The biological behavior of cesium is similar to potassium. After entering the body, cesium is distributed somewhat uniformly throughout the body with higher concentrations in muscle tissues and lower concentrations in bones. The biological half-life of cesium is rather short at about 110 days. Soil, water, vegetation, aquatic life, and air contaminated with cesium-137 can affect animals and/or humans through their consumption/inhalation.[58]

External exposure to large amounts of Cs-137 can cause burns, acute radiation sickness, and even death. Symptoms typical of cutaneous radiation syndrome (initial dermal erythema and subsequent ulceration) occurred among Russian military recruits who were accidentally exposed to a sealed source of [137]Cs.[12,36] Some of the exposed men also described symptoms of nausea, vomiting, and headache, which occurred at the onset of the dermal effects. Exposure to Cs-137 can increase the risk for cancer because of exposure to high-energy gamma radiation. Internal exposure to Cs-137, through ingestion or inhalation, allows the radioactive material to be distributed in the soft tissues, especially muscle tissue, exposing these tissues to the beta particles and gamma radiation and increasing cancer risk.[79] About 80% of ingested plant-associated cesium is absorbed from the ruminant gut and is subsequently transported to all soft tissues, milk, urine, and feces.[3,70] The muscle and milk represent the most important sources of radiocesium from animal products entering the human food chain.

Strontium-90: Sixteen major radioactive isotopes of strontium exist, but only strontium-90 has a half-life sufficiently long (29 years) to warrant concern for nuclear facilities such as the US Department of Energy Hanford site. The half-lives of all other strontium radionuclides are less than 65 days. Strontium-90 decays to yttrium-90 by emitting a beta particle, and yttrium-90 decays by emitting a more energetic beta particle with a half-life of 64 h to zirconium-90. The main health concerns for strontium-90 are related to the energetic beta particle from yttrium-90.[81] Radiostrontium absorption from the gut depends upon the calcium requirement and intake of an animal, with mean reported values for the degree of true absorption varying between 12% and 72%.[5] Absorbed radiostrontium is subsequently largely deposited in bone and transferred to milk; the latter forms the main exposure route to humans.

Strontium can be taken into the body by eating food, drinking water, or breathing air. Gastrointestinal absorption from food or water is the principal source of internally deposited strontium in the general population. On average, 30–40% of ingested strontium is absorbed into the bloodstream. The amount absorbed tends to decrease with age and is higher (about 60%) in children in their first year of life. Adults on fasting and low-calcium diets can also increase intestinal absorption to these levels, as the body views strontium as a replacement for calcium. Strontium behaves similarly to calcium (although it is not homeostatically controlled, that is, the body does not actively regulate levels within the cells), but living organisms generally use and retain it less effectively.[4,43] For adults, about 31% of the activity entering the blood (plasma) from the gastrointestinal tract is retained by bone surfaces; the remainder goes to soft tissues or is excreted in urine and feces. Much of the activity initially deposited on bone surfaces is returned to plasma within a few days based on an updated biokinetic model that accounts for redistribution in the body. About 8% of the ingested activity remains in the body after 30 days and this decreases to about 4% after 1 year. This activity is mainly in the skeleton. Strontium is a health hazard only if it is taken into the body. External gamma exposure is not a major concern because strontium-90 emits no gamma radiation, and its decay product yttrium-90 emits only a small amount. Strontium-90 concentrates in bone surfaces and bone marrow, and its relatively long radioactive half-life (29 years) make it one of the most hazardous products of radioactive fallout. Bone tumors and tumors of the blood-cell forming organs are the main health concern. These tumors are associated with the beta particles emitted during the radioactive decay of strontium-90 and yttrium-90. In addition to potential radiogenic effects, strontium has been shown to inhibit calcification and cause bone deformities in animals, notably at high doses.[81]

Plutonium: Plutonium was first made in large quantities by American scientists in the 1940s as part of the Manhattan Project to create the atomic bomb, and this production continued through the Cold War. Plutonium is formed when the nucleus of a uranium atom captures one or more neutrons, changing the atomic structure and creating a new element. This process occurs in nuclear reactors and mainly involves transforming uranium-238 into plutonium. The main plutonium isotopes at Department of Energy environmental management sites are plutonium-238, plutonium-239, plutonium-240, and plutonium-241. Except for plutonium-241, these isotopes decay by emitting an alpha particle. Plutonium-241 decays by emitting a low-energy beta particle to americium-241, an alpha emitting radionuclide with a half-life of 430 years that is much more radiotoxic than its parent.[51]

People may inhale plutonium as a contaminant in dust. It can also be ingested with food or water. Most people have extremely low ingestion and inhalation of plutonium. However, people who live near government weapons production or testing facilities may have increased exposure.[45] Internal exposure to plutonium is an extremely serious health hazard. It generally stays in the body for decades, exposing organs and tissues to radiation and increasing the risk of cancer. Plutonium is also a toxic metal and may cause damage to the kidneys. When plutonium is inhaled, a significant fraction can move from the lungs through the blood to other organs, depending on the solubility of the compound. Little plutonium (about 0.05%) is absorbed from the gastrointestinal tract after ingestion, and little is absorbed through the skin following dermal contact. After leaving the intestine or lung, about 10% clears the body. The rest of what enters the bloodstream deposits about equally in the liver and skeleton, where it remains for long periods of time, with biological retention half-lives of about 20 and 50 years, respectively, per simplified models that do not reflect intermediate redistribution.[55] Radionuclides may be inhaled in different forms, including gaseous compounds, aerosols, and particles. The ability of radionuclides to pass through the pulmonary membranes varies considerably; despite low transfer rates for actinides, such as plutonium, they are often more readily absorbed via the lungs than via the gastrointestinal tract.[44]

Americium-241: Americium is a byproduct of plutonium production activities and results from the successive capture of neutrons by plutonium. The most common isotope is americium-241, a decay product of plutonium-241. Atmospheric testing of nuclear weapons, which ceased worldwide by 1980, generated most environmental americium. Accidents and other releases from weapons production facilities have caused localized contamination. Americium oxide is the most common form in the environment. Average americium-241 levels in surface soil are about 0.01 pCi/g. Americium is typically quite insoluble, although a small fraction can become soluble through chemical and biological processes.[80,85]

The main means of exposure are ingestion of food and water containing americium isotopes and inhalation of americium-contaminated dust. Ingestion is generally the exposure of concern unless there is a nearby source of contaminated airborne dust. Because americium is taken up in the body much more readily if inhaled rather than ingested, both exposure routes can be important. The major health concern is tumors resulting from the ionizing radiation emitted by americium isotopes deposited on bone surfaces and in the liver ().[80,85,108]

Various decontamination methods have been developed and applied in nuclear accidents, but optimum methods for remediation of materials under specific contamination conditions still need to be developed. Numerous candidate radiation countermeasures have been identified in Table 9.6.

TABLE 9.6   Internal Radionuclides Contamination.

Radionuclide	Radionuclides eliminators	Therapy
Cesium[134 or 137]	Ferric hexacyanoferrate (Prussian blue)	Blocking
Iodine[125 or 131]	Saturated solution of potassium iodide	Blocking
	Antithyroid drugs	Mobilizing
Plutonium	Zink or Calcium-DTPA; investigational	Chelating
Strontium[89 or 90]	Aluminum phosphate gel antacids	Decrease absorption
Ingestion	Strontium lactate	
	Oral phosphate	Blocking
	Ammonium chloride or parathyroid extract	Displacement
		Mobilization
Tritium	Force fluids	Dilution
Unknown ingestion		Reduce absorption; consider emetics, lavage, charcoal, laxatives
Others	Bicarbonate, barium sulfate, calcium gluconate, penicillamine, the aluminum antacids, and sodium alginate	

DTPA, Diethylenetriaminepentaacetate.

APPENDIX

1. UNITS

Curie (Ci)—The quantity of radioactive material in which 37 billion radioactive atoms transform per second, which is approximately the activity of 1 g of radium.

Becquerel (Bq)—The unit of radionuclide activity or expectation value of the number of spontaneous nuclear transitions per unit of time. Bq = 1 transition per second. Unit: 1/s The unit of radionuclide activity used in the previous FDA guidance was the Ci. 1 Bq = 27×10^{-12} Ci = 27 pCi.

Gray (Gy)—Unit of absorbed dose. 1 Gy = 1 J/kg; 1 mGy = 10^{-3} Gy. The unit of absorbed dose in previous FDA publications was the rad. 1 Gy = 100 rad; 1 mGy = 0.1 rad.

Sievert (Sv)—Unit of dose equivalent. 1 Sv = 1 J/kg; 1 mSv = 10^{-3} Sv. The unit of dose equivalent used in previous FDA guidance was the rem. 1 Sv = 100 rem; 1 mSv = 0.1 rem.

Rad—Unit of absorbed dose (Gy = 100 rad).

Rem (Rontgen equivalent man): Unit of equivalent dose (Sv = 100 rem).

2. GLOSSARY

Absorbed dose—The ionizing radiation energy absorbed in a material per unit mass. The unit for absorbed dose is the gray (Gy) which is equivalent to J/kg.

Contamination—Radionuclides on or in food or animal feed as a result of an accidental release.

Concentration—Radionuclide activity concentration. Unit: Bq/kg; 1 Bq/kg = 27 pCi/kg.

Derived Intervention Level (DIL)—Concentration derived from the intervention level of dose at which introduction of protective measures should be considered. Unit: Bq/kg.[48]

Dose coefficient (DC)—The conversion coefficient for committed dose equivalent or committed effective dose equivalent per unit intake of radionuclide activity. Unit: Sv/Bq.[56]

Guideline Level (GL)—The maximum levels (Bg/kg) of radionuclides in food or feed commodity which is recommended by Codex Alimentarius Commission (CAC) to be acceptable for commodities moving in the international trade, which have been contaminated following a nuclear or radiological emergency. These guideline levels apply to food after reconstitution or as prepared for consumption, that is, not to dried or concentrated foods and are based on an intervention exemption level of 1 mSv in a year.

Half-life—Time required for a radioactive substance to lose 50% of its activity by radioactive decay.

Level of Concern (LOC)—Concentration in an imported food, set by FDA after the Chernobyl accident, below which unrestricted distribution in US commerce is permitted.

Protective action—Action taken to limit the radiation dose from ingestion by avoiding or reducing the contamination in or on human food and animal feeds.

Protective Action Guide (PAG)—Committed effective dose equivalent or committed dose equivalent to an individual organ or tissue that warrants protective action following a release of radionuclides.

Radioactivity—The emission of alpha particles, beta particles, neutrons and gamma or X-radiation from the transformation of an atomic nucleus.

3. DERIVED INTERVENTION LEVELS (DILS) TABLES

TABLE 9.A1   Dose Coefficient[a] (DC) in mSv/Bq.

Radionuclide	Age group					
	3 month	1 year	5 years	10 years	15 years	Adult
Sr^{90} bone surface	1.0E − 03	7.4E − 04	3.9E − 04	5.5E − 04	1.2E − 03	3.8E − 04
Sr^{90}	1.3E − 04	9.1E − 05	4.1E − 05	4.3E − 05	6.7E − 05	3.5E − 05
I^{131} thyroid	3.7E − 03	3.6E − 03	2.1E − 03	1.1E − 03	6.9E − 04	4.4E − 04
I^{131}	1.1E − 04	1.1E − 04	6.3E − 05	3.2E − 05	2.1E − 05	1.3E − 05
Cs^{134}	2.5E − 05	1.5E − 05	1.3E − 05	1.4E − 05	2.0E − 05	1.9E − 05
Cs^{137}	2.0E − 05	1.1E − 05	9.0E − 06	9.8E − 06	1.4E − 05	1.3E − 05
Ru^{103}	7.7E − 06	5.1E − 06	2.7E − 06	1.7E − 06	1.0E − 06	8.1E − 07
Ru^{106}	8.9E − 05	5.3E − 05	2.7E − 05	1.6E − 05	9.2E − 06	7.5E − 06
Pu^{238} bone surface	1.6E − 01	1.6E − 02	1.5E − 02	1.5E − 02	1.6E − 02	1.7E − 02
Pu^{238}	1.3E − 02	1.2E − 03	1.0E − 03	8.8E − 04	8.7E − 04	8.8E − 04
Pu^{239} bone surface	1.8E − 01	1.8E − 02	1.8E − 02	1.7E − 02	1.9E − 02	1.8E − 02
Pu^{239}	1.4E − 02	1.4E − 03	1.1E − 03	1.0E − 03	9.8E − 04	9.8E − 04
Am^{241} bone surface	2.0E − 01	1.9E − 02	1.9E − 02	1.9E − 02	2.1E − 02	2.0E − 02
Am^{241}	1.2E − 02	1.2E − 03	1.0E − 03	9.0E − 04	9.1E − 04	8.9E − 04

[a]Dose coefficients are from ICRP Publication.[56] The committed effective dose equivalents are computed to age 70 years.

Source: Adapted from USDHHS (1998, *Toxicological profile for ionizing radiation*. Public Health Service, Agency for Toxic Substances and Disease Registry: Georgia, USA. www.fda. gov/downloads/MedicalDevices/.../UCM094513.pdf).

TABLE 9.A2 Annual Dietary of Food Intake (kg/year).[a]

Food class	Age group									
	<1	**1–4**	**5–9**	**10–14**	**15–19**	**20–24**	**25–29**	**30–39**	**40–59**	**60 & up**
Dairy (fresh milk)[b]	208 (99.3)	153 (123)	180 (163)	186 (167)	167 (148)	112 (96.5)	98.2 (79.4)	86.4 (66.8)	80.8 (61.7)	90.6 (70.2)
Egg	1.8	7.2	6.2	7.0	9.1	10.3	10.2	11.0	11.4	10.5
Meat	16.5	33.7	46.9	58.4	69.2	71.2	72.6	73.4	70.7	56.3
Fish	0.3	2.5	4.0	4.6	6.1	6.8	7.6	7.1	8.0	6.3
Produce	56.6	59.9	82.3	96.0	97.1	91.4	99.1	102	115	121
Grain	20.4	57.6	79.0	90.6	89.4	77.3	78.4	73.7	70.2	67.1
Beverage (tap water)	112 (62.3)	271 (159)	314 (190)	374 (226)	453 (243)	542 (240)	559 (226)	599 (232)	632 (268)	565
Misc.	2.0	9.3	13.3	14.8	13.9	10.9	11.9	12.5	13.3	13.0
Total Annual Intake (kg/year)	418	594	726	832	905	922	937	965	1001	930

[a]Computed from daily intake values in grams per day provided in EPA.[24] The total annual intakes are rounded to nearest 1 kg/year, [b]Fresh milk is included in the dairy entry and tap water used for drinking is included in the beverage entry. The total annual intakes (kg/year) for fresh milk and tap water are also each given separately in parentheses.

Source: Adapted from USDHHS (1998, *Toxicological Profile for Ionizing Radiation*. Public Health Service, Agency for Toxic Substances and Disease Registry: Georgia, USA. www.fda. gov/downloads/MedicalDevices/.../UCM094513.pdf).

TABLE 9.A3 Dietary Intakes for ICRP Age Groups.

ICRP age group	Annual intakea (kg)	280-Day intake Ru-103 (kg)	60-day intake I-131 (kg)
3 months	418	320	69
1 year	506	387	83
5 years	660	506	109
10 years	779	597	128
15 years	869	666	143
Adult	943	723	155

[a]The annual dietary intakes for the ICRP age groups were obtained by assigning or averaging the appropriate annual dietary intakes given in Table 9.A2 for the EPA age groups, as follows: 3 months: <1, 1 year: average <1 and 1–4, 5 years: average 1–4 and 5–9, 10 years: average 5–9 and 10–14, 15 years: average 10–14 and 15–19, Adult: average 15–19, 20–24, 25–29, 30–39, 40–59, 60, and up.

Source: Adapted from USDHHS (1998, *Toxicological Profile for Ionizing Radiation*. Public Health Service, Agency for Toxic Substances and Disease Registry: Georgia, USA. www.fda. gov/downloads/MedicalDevices/.../UCM094513.pdf).

TABLE 9.A4 Protective Action Guidelines[a] (PAGs) and Derived Intervention Levels (DILs) (individual radionuclides, by age group).

Radionuclide	PAG (mSv)	Derived intervention levels (Bq/kg)					
		3 months	1 year	5 years	10 years	15 years	Adult
Sr^{90} bone surface[b]	50	400	445	648	389	160	465
Sr^{90}	5	308	362	616	497	286	505
I^{131} thyroid	50	196	167	722	1200	1690	2420
I^{131}	5	659	548	2410	4110	5540	8180
Cs^{134}	5	1600	2190	1940	1530	958	930
Cs^{137}	5	2000	2990	2810	2180	1370	1360
Ru^{103}	5	6770	8410	12,200	16,400	25,000	28,400
Ru^{106}	5	449	621	935	1340	2080	2360
Pu^{238} bone surface	50	2.5	21	17	14	12	10
Pu^{238}	5	3.1	27	25	24	22	20
Pu^{239} bone surface	50	2.2	18	14	13	10	9.8
Pu^{239}	5	2.9	24	23	21	20	18
Am^{241} bone surface	50	2.0	17	13	11	9.1	8.8
Am^{241}	5	3.3	27	25	24	21	20

[a]Derived intervention levels were computed using dose coefficients from Table 9.A1, dietary intakes from Table 9.A3), and [$f = 0.3$ (except for I-131 in infant diets, that is, the 3-month and 1-year age groups), $f = 1.0$ (I-131 in infant diets)]. [b]The observed trend in derived intervention levels for Sr-90 as a function of age, that is, minimum values at 15 years, results primarily from the mass of exchangeable strontium in bone as a function of age.[64]

Source: Adapted from USDHHS (1998, *Toxicological Profile for Ionizing Radiation*. Public Health Service, Agency for Toxic Substances and Disease Registry: Georgia, USA. www.fda. gov/downloads/MedicalDevices/.../UCM094513.pdf).

TABLE 9.A5 Derived Intervention Levels (Bq/kg) (Individual Radionuclides, by Age Group, Most Limiting of Either PAG).

Radionuclide	3 months	1 year	5 years	10 years	15 years	Adult
Sr^{90}	308	362	616	389	160	465
I^{131}	196	167	722	1200	1690	2420
Cs^{134}	1600	2190	1940	1530	958	930
Cs^{137}	2000	2990	2810	2180	1370	1360
Cs-group[a]	1800	2590	2380	1880	1160	1150
Ru^{103}	6770	8410	12,200	16,400	25,000	28,400
Ru^{106}	449	621	935	1340	2080	2360
Pu^{238}	2.5	21	17	14	12	10

TABLE 9.A5 *(Continued)*

Radionuclide	3 months	1 year	5 years	10 years	15 years	Adult
Pu^{239}	2.2	18	14	13	10	9.8
Am^{241}	2.0	17	13	11	9.1	8.8
Pu + Am group[b]	2.2	19	15	13	9.6	9.3

[a]Computed as (DIL for Cs-134 + DIL for Cs-137)/2; [b]computed as (DIL for Pu-238 + DIL for Pu-239 + DIL for Am-241)/3.

Source: Adapted from USDHHS (1998, *Toxicological Profile for Ionizing Radiation*. Public Health Service, Agency for Toxic Substances and Disease Registry: Georgia, USA. www.fda. gov/downloads/MedicalDevices/.../UCM094513.pdf).

TABLE 9.A6 Derived Intervention Levels (Bq/kg) (Radionuclides, Most Limiting of All Diets).

Radionuclide group	Derived intervention levels	Age
Sr^{90}	160	15 years
I^{131}	170	1 year
Cs group	1200	Adult
Ru^{103a}	6800	3 months
Ru^{106a}	450	3 months
Pu + Am group	2	3 months

[a]Due to the large differences in DILs for Ru-103 and Ru-106, the individual concentrations of Ru-103 and Ru-106 are divided by their respective DILs and then summed. The sum must be less than one.

Source: Adapted from USDHHS (1998, *Toxicological Profile for Ionizing Radiation*. Public Health Service, Agency for Toxic Substances and Disease Registry: Georgia, USA. www.fda. gov/downloads/MedicalDevices/.../UCM094513.pdf).

KEYWORDS

- radionuclides
- radioactivity
- radiation energy
- alpha particles
- beta particles
- gamma radiation
- derived intervention levels

REFERENCES

1. Aliyu, A. S.; Evangeliou, N.; Mousseau, T. A.; Wu, J.; Ramli, A. T. An Overview of Current Knowledge Concerning the Health and Environmental Consequences of the Fukushima Daiichi Nuclear Power Plant (FDNPP) Accident. *Environ. Int.* **2015,** *85,* 213–228.

2. Asikainen, M. *Natural Radioactivity of Ground Water and Drinking Water in Finland.* STL-A39, 1982.

3. Beresford, N. A.; Barnett, C. L.; Mayes, R. W.; Pollaris, K.; Vandecasteele, C. M.; Howard, B. J. The Use of An In-Vitro Technique to Predict the Absorption of Dietary Radiocesium by Sheep. *Radiat. Environ. Biophys.* **1995,** *34,* 191–194.

4. Beresford, N. A.; Mayes, R. W.; Hansen, H. S.; Crout, N. M. J.; Hove, K.; Howard, B. J. Generic Relationship between Calcium Intake and Radiostrontium Transfer to Milk of Dairy Ruminants. *Radiat. Environ. Biophys.* **1998,** *37,* 129–131.

5. Beresford, N. A.; Mayes, R. W.; Cooke, A. I.; Barnett, C. L.; Howard, B. J.; Lamb, C. S.; Naylor, G. P. L. The Importance of Source Dependent Bioavailability in Determining the Transfer of Ingested Radionuclides to Ruminant Derived Food Products. *Environ. Sci. Technol.* **2000,** *34,* 4455–4462.

6. Botezatu, E. Contribution of the Dietary Ingestion to the Natural Radiation Exposure of Romanian Population. *J. Hyg. Public Health* **1994,** *44* (1–2), 19–21.

7. Bradley, E. J. *Contract Report.* Natural Radionuclides in Environmental Media. NRPB-M439, 1993.

8. Brenda, J. H.; Nicholas, A. B.; Gabriele. V. Countermeasures for Animal Products: A Review of Effectiveness and Potential Usefulness after an Accident. *J. Environ. Radioactiv.* **2001,** *56,* 115–137.

9. Buesseler, K. O.; Jayne, S. R.; Fisher, N. S.; Rypina, I. I.; Baumann, H.; Baumann, Z.; Breier, C. F.; Douglass, E. M.; George, J.; Macdonald, A. M.; Miyamoto, H.; Nishikawa, J.; Pike, S. M.; Yoshida, S. Fukushima-Derived Radionuclides in the Ocean and Biota Off Japan. Proc. Natl. Acad. Sci. U.S.A. **2012,** *109* (16), 5984–5988.

10. Carini, F. Transfer to Fruits. In *Quantification of Radionuclide Transfer in Terrestrial and Freshwater Environments for Radiological Assessments*; IAEA TECDOC-1616, Ed.; International Atomic Energy Agency (IAEA), Vienna, Austria, 2009; pp 311–330.

11. Carvalho, F. P. ^{210}Po in Marine Organisms: A Wide Range of Natural Radiation Dose Domains. *Radiat. Protect. Dosimetry* **1980,** *24* (1/4), 109–111.

12. CDC (Centers for Disease Control and Prevention). *Radiation Emergencies: Radioisotope Brief: I-131, Cs-137,* Department of Health and Human Services/Centers for Disease Control and Prevention, 2005.

13. CDRH (Center for Devices and Radiological Health, Food and Drug Administration). *Accidental Radioactive Contamination of Human Food and Animal Feeds: Recommendations for State and Local Agencies.* CDRH, Rockville, MD 20850, 1998.

14. Codex. *Proposed Draft Revised Guideline Levels for Radionuclides in Foods for Use in International Trade.* Codex Alimentarius Commission, Joint FAO/WHO Food Standard Program, Codex Committee on Food Additives and Contaminants: Nederlands, 24–28 April 2006.

15. Codex. *Codex General Standard for Contaminants and Toxins in Food and Feed.* Codex Alimentarius Commission (CODEX STN 193-1995), 2009.

16. Cothern, C. R.; Lappenbusch, W. L. Occurrence of Uranium in Drinking Water in the United States. *Health Phys.* **1983,** *45* (1), 89–99.

17. Coughtrey, P. J.; Jackson, D.; Jones, J. H.; Kane, P.; Thorne, M. C. *Radionuclide Distribution and Transport in Terrestrial and Aquatic Ecosystems*. A. A. Balkema: Rotterdam, 1984, vol. 4.

18. Coughtrey, P. J.; Jackson, D.; Jones, J. H.; Thorne, M. C. Radionuclide Distribution and Transport in Terrestrial and Aquatic Ecosystems. A. A. Balkema, Rotterdam, 1984, vol. 5.

19. Coughtrey, P. J. *Radiation Protection: Radioactivity Transfer to Animal Products*; Commission of the European Communities: Luxembourg, **1990**; 145 pp.

20. Dang, H. S.; Pullat, V. R.; Jaiswal, D. D. Daily intake of Uranium by the Urban Indian Population. *J. Radioanal. Nucl. Chem.* **1990**, *138* (1), 67–72.

21. Davis, S. Health Risks Associated with Environmental Radiation Exposures. *J. Radiat. Protect.* **2012**, *32* (1), N21–N25.

22. De Bortoli, M.; Gaglione, P. ^{226}Ra in Environmental Materials and Foods. *Health Phys.* **1972**, *22*, 43–48.

23. Delacroix, D.; Guerre, J. P.; Leblanc, P.; Hickman, C. Radionuclide and radiation protection data. *Radiat. Protect. Dosimetry* **2002**, *98* (1), 9–18.

24. EPA (Environmental Protection Agency). *An Estimation of the Daily Average Food Intake by Age and Sex for Use in Assessing the Radionuclide Intake of Individuals in the General Population*. Office of Radiation Programs: Washington, DC, 1984. EPA 520/1-84-021.

25. EPA (Environmental Protection Agency). Radiation Protection Guidance to Federal Agencies for Occupational Exposure. *Fed. Regis.* **1987**, *52*, 2822–2834.

26. EPA (Ohio Environmental Protection Agency). *Radionuclides in Public Drinking Water, Division of Drinking and Ground Water*. EPA: Columbus, OH, 2005.

27. EU (European Union). *Imposing Special Conditions Governing the Import of Feed and Food Originating in or Consigned from Japan Following the Accident at the Fukushima Nuclear Power Station*. Commission Implementing Regulations (EU) No 322/2014, 2014.

28. Evans, D. H.; Piermarini, P. M.; Choe, K. P. The Multifunctional Fish Gill: Dominant Site of Gas Exchange, Osmoregulation, Acid Base Regulation, and Excretion of Nitrogenous Waste. *Physiol. Rev.* **2005**, *85*, 97–177.

29. FAO (Food and Agriculture Organization). *Organization of Surveys for Radionuclides in Food and Agriculture, Atomic Energy Series No. 4*, FAO: Rome, 1962.

30. FDA (Food and Drug Administration). Radionuclides in Imported Foods: Levels of Concern. Availability of Compliance Policy Guide. *Fed. Regis.* **1986**, *51*, 23155.

31. FDA (Food and Drug Administration). *Radionuclides in Imported Foods—Levels of Concern*. FDA: Washington, DC; Compliance Policy Guide No. 7119.14, 1986.

32. Fisenne, I. M.; Perry, P. M.; Decker, K. M.; Keller, H. W. The Daily Intake of ^{234,235,238}U, ^{228,230,232}Th, and ^{226,228}Ra by New York City Residents. *Health Phys.* **1987**, *53*, 357–364.

33. FRA (Fishery Research Agency). *Report on Research Program for the Effect of Radioactive Materials*, 2012. http://www.jfa.maff.go.jp/j/sigen/housyaseibussitutyousakekka/pdf/h23report_fi nal_1.pdf (in Japanese).

34. Fujiwara. Cesium Uptake in Rice: Possible Transporter, Distribution and Variation. In *Agricultural Implications of the Fukushima Nuclear Accident*; Nakanishi, T. M., Tanoi, K., Eds.; Springer: Tokyo, Heidelberg, New York, Dordrecht, London, 2013; pp 29–35.

35. Glöbel, B.; Muth, H. Natural Radioactivity in Drinking Water, Foodstuffs and Man in Germany. In *Seminar on the Radiological Burden of Man from Natural Radioactivity in the Countries of the European Communities*; CEC Doc. No. V/2408/80, 1980; pp 385–418.

36. Gottlöber, P.; Bezold, G.; Schjar, A.; Weber, L. The Radiation Accident in Georgia: Clinical Appearance and Diagnosis of Cutaneous Radiation Syndrome. *J. Am. Acad. Dermatol.* **2000**, *42*, 453–458.

37. Gupta, D. K.; Chatterjee, S.; Datta, S.; Voronina, A. V.; Walther, C. Radionuclides: Accumulation and Transport in Plants. In *Reviews of Environmental and Contamination Toxicology*; de-Voogt, P., Ed.; Springer: Berlin, 2016; pp 1–22.

38. Harley, N. H. Appendix I. Analysis of Foods for Radioactivity. In *Environmental Contaminants of Food*; Holmes, J. C., Boss, K. S., Heming, J., Eds.; Library of Congress: Washington, DC, USA, 1979; pp 215–227.

39. Health Canada. *Canadian Guidelines for the Restriction of Radioactively Contaminated Food and Water Following a Nuclear Emergency: Guidelines and Rationale*; Minister of Public Works and Government Services Canada, 2000; 35 pp.

40. Hirose, K. Fukushima Daiichi Nuclear Plant accident: Atmospheric and oceanic impacts over the five years. *J. Environ. Radioactiv.* **2016**, *157*, 113–130.

41. Hoeve, J. E. T.; Jacobson, M. Z. Worldwide Health Effects of the Fukushima Daiichi Nuclear Accident. Energy Environ. Sci. **2012**, *5*, 8743–8757.

42. Howard, B. J.; Voigt, G.; Segal, M.; Ward, G. A Review of Countermeasures to Reduce Radioiodine in Milk of Dairy Animals. *Health Phys.* **1996**, *71*, 661–673.

43. Howard, B. J.; Beresford, N. A.; Mayes, R. W.; Hansen, H. S.; Crout, N. M. J.; and Hove, K. The Use of Dietary Calcium Intake of Dairy Ruminants to Predict the Transfer Coefficient of Radiostrontium to Milk. *Radiat. Environ. Biophys.* **1997**, *36*, 39–43.

44. Howard, B. J.; Beresford, N. A.; Barnett, C. L.; Fesenko, S. Radionuclide Transfer to Animal Products: Revised Recommended Transfer Coefficient Values. *J. Environ. Radioactiv.* **2009**, *100* (3), 263–273.

45. Howard, B. J.; Beresford, N. A.; Barnett, C. L.; Fesenko, S. Transfer to Animals. In *Quantification of Radionuclide Transfer in Terrestrial and Freshwater Environments for Radiological Assessments*; IAEA TECDOC-1616, Ed.; International Atomic Energy Agency (IAEA): Vienna, Austria, 2009; pp 267–307.

46. Howard, B. J.; Beresford, N. A.; Barnett, C. L.; Fesenko, S. F. Gastrointestinal Fractional Absorption of Radionuclides in Adult Domestic Ruminants. *J. Environ. Radioactiv.* **2009**, *100* (12), 1069–1078.

47. IAEA (International Atomic Energy Agency). *Basic Safety Standards for Radiation Protection*; IAEA: Vienna, 1963; Safety Series 15. STI/PUB/26, 39 pp.

48. IAEA (International Atomic Energy Agency). *Principles for Establishing Intervention Levels for the Protection of the Public in the Event of a Nuclear Accident or Radiological Emergency*. IAEA: Vienna; Safety Series No. 72, 1985.

49. IAEA (International Atomic Energy Agency). *Measurement of Radionuclides in Food and the Environment (A Guidebook)*. IAEA:Vienna; Technical Reports Series No. 295, 1989.

50. ICRP (International Commission on Radiological Protection). *Report of Committee II on Permissible Dose for Internal Radiation*. Pergamon Press: Oxford, 1959; 233 pp.

51. ICRP (International Commission on Radiological Protection). *The Metabolism of Compounds of Plutonium and Other Actinides*. ICRP Publication 19, May 1972.

52. ICRP (International Commission on Radiological Protection). *Recommendations of the International Commission on Radiological Protection*. Pergamon Press: Oxford; ICRP Publication 26, *Annu. ICRP* **1977**, *1* (3), 1–159.

53. ICRP (International Commission on Radiological Protection). *Limits for Intakes of Radionuclides by Workers*. International Commission on Radiological Protection. *Annu. ICRP* **1979**, *2*, 1–116.

54. ICRP (International Commission on Radiological Protection). *A Compilation of the Major Concepts and Quantities in Use by ICRP.* Pergamon Press: Oxford; ICRP Publication 42, *Annu. ICRP* **1984,** *14* (4), 1–52.

55. ICRP (International Commission on Radiological Protection). *The Metabolism of Plutonium and Related Elements,* ICRP Publication 48, *Ann. ICRP* **1986,** *16* (2/3).

56. ICRP (International Commission on Radiological Protection). *Age-dependent Doses to Members of the Public from Intake of Radionuclides.* Pergamon Press: Oxford; ICRP Publication 56, Part 1; *Annu. ICRP* **1989,** *20* (2).

57. ICRP (International Commission on Radiological Protection). *Recommendations of the International Commission on Radiological Protection.* Pergamon Press: Oxford; ICRP Publication 60; *Ann. ICRP* **1991,** *21* (1–3), 1–236.

58. IUFoST (International Union of Food science and Technology). *Radioactive Fallout from the 2011 Japan Nuclear Plant Accident and Some Recommended Precautions and Countermeasures.* Scientific Information Bulletin: Ontario, Canada, 2011.

59. Jacobi, W. Radiation Exposure and Radiation Risk of the Population from the Chernobyl Accident. *Phys. Blätt.* **1988,** *44,* 240–246.

60. Jaworowski, Z. Radioactive Lead in the Environment and in the Human Body. *At. Energy Rev.* **1969,** *1,* 3–45.

61. Kamei-Ishikawa, N.; Ito, A.; Umita, T. Outline Evaluating Removal of Radionuclides from Landfill Leachate Using Generally Practiced Wastewater Treatment Processes. In *Radiation Monitoring and Dose Estimation of the Fukushima Nuclear Accident;* Takashi, S. Ed.; Springer: Tokyo, Heidelberg, New York, Dordrecht, London, 2014; pp 127–134.

62. Khandekar, R. N. ^{210}Po in Bombay Diet. *Health Phys.* **1977,** *33,* 148–150.

63. Konoplev, A.; Golosov, V.; Laptev, G.; Nanba, K.; Onda, Y.; Takase, T.; Wakiyama, Y.; Yoshimura, K. Behavior of Accidentally Released Radiocesium in Soil–Water Environment: Looking at Fukushima from a Chernobyl Perspective. *J. Environ. Radioactiv.* **2016,** *151* (3), 568–578.

64. Leggett, R. W.; Eckerman, K. F.; Williams, L. R. Strontium-90 in Bone: A Case Study in Age-Dependent Dosimetric Modeling. *Health Phys.* **1982,** *43* (3), 307–322.

65. Lengemann, F. W.; Wentworth, R. A.; Comar, C. L. Physiological and Biochemical Aspects of the Accumulation of Contaminant Radionuclides in Milk. In *Lactation: A Comprehensive Treatise. Nutrition and Biochemistry of Milk/Maintenance;* Larson, B. L., Smith, V. R., Eds.; Academic Press, London, 1974; vol. 3, pp 159–215.

66. McCurdy, D. E.; Mellor, R. A. The Concentration of ^{226}Ra and ^{228}Ra in Domestic and Imported Bottled Waters. *Health Phys.* **1981,** *40,* 250–253.

67. Manabe, N.; Takahashi, T.; Li, J.-Y.; Tanoi, K.; Nakanishi, T. M. Changes in the Transfer of Fallout Radiocesium from Pasture Harvested in Ibaraki Prefecture, Japan, to Cow Milk Two Months after the Fukushima Daiichi Nuclear Power Plant Accident. In *Agricultural Implications of the Fukushima Nuclear Accident;* Nakanishi, T. M., Tanoi, K., Eds.; Springer: Tokyo, Heidelberg, New York, Dordrecht, London, 2013; pp 87–95.

68. Marshall, W. S.; Grosell, M. Ion Transport, Osmoregulation, and Acid–Base Balance. In *The Physiology of Fishes;* Evans, D. H.; Claiborne, J. B., Eds.; 3rd ed. CRC: Boca Raton, FL, 2006; pp 177–230.

69. Mastinu, G. G.; Santaroni, G. P. The Exposure of the Italian Population to Natural Radioactivity in Drinking Water and Food. In *Seminar on the Radiological Burden of Man from Natural Radioactivity in the Countries of the European Communities.* CEC Doc. No. V/2408/80, 1980; pp 349–368.

70. Mayes, R. W.; Beresford, N. A.; Howard, B. J.; Vandecasteele, C. M.; Stakelum, G. The Use of the True Absorption Coefficient as a Measure of the Bioavailability of Radiocesium in Ruminants. *Radiat. Environ. Biophys.* **1996**, *35*, 101–109.

71. Merz, S.; Steinhauser, G.; Hamada, N. Anthropogenic Radionuclides in Japanese Food: Environmental and Legal Implications. *Environ. Sci. Technol.* **2013**, *47*, 1248–1256.

72. Merz, S.; Shozugawa, K.; Steinhauser, G. Analysis of Japanese Radionuclide Monitoring Data of Food before and after the Fukushima Nuclear Accident. *Environ. Sci. Technol.* **2015**, *49* (5), 2875–2885.

73. Mitchell, N. G. Models for Radionuclide Transfer to Fruits and Data Requirements. *J. Environ. Radioactiv.* **2001**, *52* (2–3), 291–307.

74. Ministry of the Environment. *Additional Information of 13th Commission of Safety Assessment for Disaster Waste Disposal*, 2012 (accessed 9 July 2013) (in Japanese).

75. Ministry of Land, Infrastructure, Transport and Tourism. Measurements of Radioactive Concentrations in Sewage Sludge etc., 2013 (accessed 9 July 2013) (in Japanese).

76. Morse, R. S.; Welford, G. A. Dietary Intake of ^{210}Pb. *Health Phys.* **1971**, 21, 53–55.

77. Muth, H.; Rajewsky, B.; Hantke, H. J.; Aurand, K. The Normal Radium Content and the ^{226}Ra/Ca Ratio of Various Foods, Drinking Water and Different Organs and Tissues of the Human Body. *Health Phys.* **1960**, *2*, 239–245.

78. Nakanishi, T. M.; Tanoi, K. *Agricultural Implications of the Fukushima Nuclear Accident*; Springer: Tokyo, Heidelberg, New York, Dordrecht, London, 2013; 204 p.

79. NCRP. National Council on Radiation Protection and Measurements, Cesium-137 from the Environment to Man: Metabolism and Dose. *NCRP Report 52*. NCRP: Washington, DC, January 1977.

80. NCRP (National Council on Radiation Protection and Measurements). *Report 85, Mammography—A User's Guide*. NCRP: Bethesda, MD, 1986.

81. NCRP (National Council on Radiation Protection and Measurements). Some Aspects of Strontium Radiobiology. *NCRP Report No. 110*, Bethesda, MD, August 1991.

82. Nemoto, K.; Abe, J. Radiocesium Absorption by Rice in Paddy Field Ecosystems. In *Agricultural Implications of the Fukushima Nuclear Accident*; Nakanishi, T. M., Tanoi, K., Eds.; Springer: Tokyo, Heidelberg, New York, Dordrecht, London, 2013; pp 19–27.

83. NEPA (National Environmental Protection Agency). *Nationwide Survey of Environmental Radioactivity Level in China (1983–1990)*. 90-S315-206. The People's Republic of China, 1990.

84. Pellerin, P.; Gahinet, M. E.; Moroni, J. P.; Remy, M. L. Some Observations on Natural Radioactivity in Food in France. In: *Seminar on the Radiological Burden of Man from Natural Radioactivity in the Countries of the European Communities*. CEC Doc. No. V/2408/80, 1980; pp 331–348.

85. Peterson, J.; MacDonell, M.; Haroun, L.; Monette, F. *Radiological and Chemical Fact Sheets to Support Health Risk Analyses for Contaminated Areas*. Argonne National Laboratory, Environmental Science Division: Chicago, USA, 2007.

86. Pietrzak-Flis, Z.; Suplinska, M. M.; Rosiak, L. The Dietary Intake of ^{238}U, ^{234}U, ^{230}Th, ^{232}Th, ^{228}Th and ^{226}Ra from Food and Drinking Water by Inhabitants of the Walbrzych Region. *J. Radioanal. Nucl. Chem.* **1997**, *222* (1–2), 189–193.

87. Pietrzak-Flis, Z.; Chrzanowski, E.; Dembinska, S. Intake of ^{226}Ra, ^{210}Pb and ^{210}Po with Food in Poland. *Sci. Total Environ.* **1997**, 203, 157–165.

88. RIFE (Radioactivity in Food and the Environment). Environment Agency, Food Standards Agency, Northern Ireland Environment Agency and Scottish Environment Protection Agency, UK, 2010; pp 1–254.

89. RSRP (Romanian Society for Radiological Protection). In *Natural Radioactivity in Romania*, RSRP: Bucharest. REG Project No. 852/1993, 1999; pp 109–111.

90. Rühm, W.; König, K.; Bayer, A. Long-Term Follow-Up of the [137]Cs Body Burden of Individuals after the Chernobyl Accident—A Means for the Determination of Biological Half-Lives. *Health Phys.* **1999**, *77*, 373–382.

91. Saito, T.; Suzuki, Y.; Fujimura, S.; H. Tsukada, S. Studies on Radiocesium Transfer in Agricultural Plants in Fukushima Prefecture. In *Radiation Monitoring and Dose Estimation of the Fukushima Nuclear Accident*; Takashi, S., Ed.; Springer: Tokyo, Heidelberg, New York, Dordrecht, London; 2014; pp 135–140.

92. Saito, T.; Takahashi, K.; Murakami, T.; Shinano, T. Analysis of Factors Causing High Radiocesium Concentrations in Brown Rice Grown in Minamisoma City. In: *Radiological Issues for Fukushima's Revitalized Future*; Takahashi, T., Ed.; Springer: Tokyo, Heidelberg, New York, Dordrecht, London, 2016; pp 189–198.

93. Salonen, L. [238]U Series Radionuclides as a Source of Increased Radioactivity in Groundwater Originating from Finnish Bedrock. In *Future Groundwater Resources at Risk*; Soukko, J., Ed.; IAHS Publication No. 222. IAHS Press, Oxfordshire, 1994; pp. 71–84.

94. Schwantes, J. M.; Orton, C. R.; Clark, R. A. In *Analysis of a Nuclear Accident: Fission and Activation Product Releases from the Fukushima Daiichi Nuclear Facility as Remote Indicators of Source Identification, Extent of Release, and State of Damaged Spent Nuclear Fuel*. Pacific Northwest National, Laboratory Ed.; US Department of Energy under Contract, USA, 2012.

95. Servant, J.; Delapart, M. Blood Lead and [210]Pb Origins in Residents of Toulouse. *Health Phys.* **1981**, *41*, 483–487.

96. SFOPH (Swiss Federal Office of Public Health). *Environmental Radioactivity and Radiation Exposure in Switzerland*. SFOPH: Bern, 1997.

97. Sgorbati, G.; Forte, M. Determination of [238]Uand [226]Ra Concentrations in Drinking Waters in Lombardia Region, Italy. Communication to UNSCEAR Secretariat, 1997.

98. Shiozawa, S. Vertical Migration of Radiocesium Fallout in Soil in Fukushima. In *Agricultural Implications of the Fukushima Nuclear Accident*; Nakanishi, T. M., Tanoi, K. Eds.; Springer: Tokyo, Heidelberg, New York, Dordrecht, London, 2013; pp 49–60.

99. Smolders, E.; Tsukada, H. The Transfer of Radiocesium from Soil to Plants: Mechanisms, Data, and Perspectives for Potential Countermeasures in Japan. *Integr. Environ. Assess. Manage.* **2011**, *7*, 379–381.

100. Soto, J.; Quindos, L. S.; Diaz-Canej, N. [226]Ra and [222]Rn in Natural Waters in Two Typical Locations in Spain. *Radiat. Prot. Dosim.* **1988**, *24* (1/4), 93–95.

101. Steinhauser, G.; Brandl, A.; Johnson, T. E. Comparison of the Chernobyl and Fukushima Nuclear Accidents: A Review of the Environmental Impacts. *Sci. Total Environ.* **2014**, *470–471*, 800–817.

102. Takashi, T. Outline of the Environmental Monitoring of Tepco's Fukushima Daiichi Nuclear Power Plant Accident. In *Radiation Monitoring and Dose Estimation of the Fukushima Nuclear Accident*; Takashi, S., Ed.; Springer: Tokyo, Heidelberg, New York, Dordrecht, London, 2014.

103. UNSCEAR (United Nations Scientific Committee on the Effects of Atomic Radiation). *Ionizing Radiation: Sources and Effects*, UNSCEAR, United Nations, New York, 1982.

104. UNSCEAR (United Nations Scientific Committee on the Effects of Atomic Radiation). Sources and Effects of Ionizing Radiation. *Report to the General Assembly of the United Nations with Scientific Annexes B*. United Nations Sales Publication E.00.IX.3: New York, 2000.

105. UNSCEAR (United Nations Scientific Committee on the Effects of Atomic Radiation). Source and Effects of Ionizing Radiation. Volume I: Sources, Volume II: Effects. *United Nations Scientific Committee on the Effects of Atomic Radiation, 2008 Report to the General Assembly with Scientific Annexes C, D and E*. United Nations Sales Publication E.11IX.3: New York, 2011.

106. USDHHS (US Department of Health and Human Services). *Accidental Radioactive Contamination of Human Food and Animal Feeds: Recommendations for State and Local Agencies*. Center for Devices and Radiological Health: Rockville, MD, 1998.

107. USDHHS (US Department of Health and Human Services). *Toxicological Profile for Ionizing Radiation*. Public Health Service, Agency for Toxic Substances and Disease Registry: Georgia, USA, 1999.

108. USDHHS (US Department of Health and Human Services). *Toxicological Profile for Americium*. Public Health Service, Agency for Toxic Substances and Disease Registry: Georgia, USA, 2004.

109. Vandecasteele, C. M.; Van Hees, M.; Hardeman, F.; Voigt, G. M.; Howard, B. J. The True Absorption of Iodine and Effect of Increasing Stable Iodine in the Diet. *J. Environ. Radioactiv.* **2000,** *47*, 301–317.

110. Watabe, S.; Matasuoka, Y.; Nakaya, M.; Ushio, H.; Nemoto, Y.; Sato, M.; Tanoi, K.; Nakanishi, T. Removal of Radiocesium Accumulated in Fish Muscle by Washing Aimed for Production of Surimi-Based Products. *Radioisotopes* **2013,** *62*, 31–38.

111. Watabe, S.; Ushio, H.; Ikeda, D. Radiocesium Contamination of Marine Fish Muscle and its Effective Elimination. In *Agricultural Implications of the Fukushima Nuclear Accident*; Nakanishi, T. M., Tanoi, K. Eds.; Springer: Tokyo, Heidelberg, New York, Dordrecht, London, 2013; pp 97–103.

112. White, P. J.; Broadley, M. R. Mechanisms of Cesium Uptake by Plants. *New Phytol.* **2000,** *147*, 241–256.

113. WHO (World Health Organization). *Guidelines for Iodine Prophylaxis following Nuclear Accidents*; WHO/SDE/PHE 99.6: Geneva, 1999; pp 1–30.

114. WHO/FAO (World Health Organization/Food and Agriculture Organization. *Nuclear Accidents and Radioactive Contamination of Foods*, March 2011; pp 1–5. http://www.who.int/foodsafety/fs_management/radionuclides_and_food_300311.pdf.

115. Zhu, H.; Wang, S.; Wei, M. Determinations ^{90}Sr, ^{137}Cs, ^{226}Ra, ^{228}Ra, 210Pb, ^{210}Po Contents in Chinese Diet and Estimations of Internal Doses Due to These Radionuclides. *Radiat. Protect.* **1993,** *13*, 85–92.

CHAPTER 10

METAL TOXICITY IN FOODS

M. CARMEN RUBIO ARMENDÁRIZ[1*],
ARTURO HARDISSON DE LA TORRE[1],
ÁNGEL J. GUTIÉRREZ FERNÁNDEZ[1],
DAILOS GONZÁLEZ WELLER[1,2], SORAYA PAZ MONTELONGO[1],
CONSUELO REVERT GIRONÉS[1], and JOSÉ M. CABALLERO MESA[1]

[1]*Área de Toxicología, Universidad de La Laguna, 38071 La Laguna, Tenerife, Islas Canarias, España*

[2]*Servicio de Inspección Sanitaria y Laboratorio, Área de Salud de Tenerife, Servicio Canario de Salud, Rambla de Santa Cruz, 38006 Santa Cruz de Tenerife, España*

Corresponding author. E-mail: crubiotox@gmail.com

CONTENTS

ABSTRACT

This chapter offers a description of the toxicological importance of metals as well as the risk analysis of metals. The parameters and recommendations used to establish safe intake values for the population are also reviewed. Given that the European Food Safety Agency (EFSA) and other food safety agencies describe cadmium, lead, mercury, arsenic, and aluminum as contaminants in food; the chapter also describes the main characteristics of these metals. For every metal, aspects like distribution in the environment and sources for humans, toxicokinetic and toxicodynamic properties, dietary intakes, and toxic effects are presented.

10.1 INTRODUCTION

Although metals are perhaps the best known toxic agents, interest in them has neither diminished nor has the quest for knowledge of their potential toxic effects and mechanisms of action.[72] Heavy metal exposure continues and is increasing in many parts of the world and is even increasing in some parts of the world, in particular, in less developed countries, though emissions have declined in most developed countries over the last 100 years.[92,94]

Compared with other toxic substances, metals are among the most harmful toxics to living systems. Common mechanisms by which toxic metals may act include inhibition of enzymes, disruption of the structure and/or function of subcellular organelles, interaction with DNA leading to mutagenesis or carcinogenesis, covalent modification of proteins, displacement of other critical metals in various metal dependent proteins, and inhibitory or stimulatory effects on the regulation of expression of various proteins. Competition between metals for complexation sites in proteins involved in electron transfer reactions may lead to toxicity.[83] Binding with protein sites by displacing original metals from their natural binding sites causing malfunctioning of cells and ultimately toxicity. Oxidative deterioration of biological macromolecules is primarily due to binding of heavy metals to the DNA and nuclear proteins.[92] Some metals act to generate free radicals, which lead to damage and subsequent degradation of critical cellular proteins, membranes, and organelles.[83] While for mercury and cadmium, the primary route for their toxicity is depletion of glutathione and bonding to sulfhydryl groups of proteins, for arsenic is thought to bind directly to critical thiols; however, other mechanisms, involving formation of hydrogen peroxide under physiological conditions, have been proposed.[214] Redox-inactive metals, such as lead, cadmium, mercury, and others deplete cells' major antioxidants, particularly thiol-containing antioxidants and enzymes, may cause an increase in production of reactive oxygen species (ROS) which can overwhelm cells' intrinsic antioxidant defenses and result in a condition known as "oxidative stress" and display various dysfunctions due to lesions caused by ROS to lipids, proteins, and DNA. Metal-induced oxidative stress in cells can be partially responsible for the toxic effects of heavy metals.[45] Generally speaking, the ionic forms are usually more harmful to humans than the metal itself or element itself.[194] These diverse mechanisms of actions allow metals to damage various organs including the kidney, nervous system, respiratory system, endocrine, and reproductive systems.[83]

It is well established that for the general population dietary intake (both food and water) is the major source of exposure to metals. Moreover, food processing and packaging may also contribute to the presence of metals in foodstuffs.[52,64,153,174]

10.2 INTAKE ASSESSMENT AND RISK ESTIMATION

It is essential to estimate the dietary intake of metals derived from food consumption to evaluate the risks and the food safety.[146,239] The risk analysis has been defined as a process consisting of three components: risk assessment, risk management, and risk communication. The risk assessment, a scientifically based process, consists of the following steps: (1) hazard identification, (2) hazard characterization, (3) exposure assessment, and (4) risk characterization. The risk (a function of the probability of an adverse health effect and the severity of that effect, consequential to a hazard in food) assessment is based on establishing the likelihood of adverse effects on human health from exposure to hazards (biological, chemical, or physical agents in, or condition of, food with the potential to cause an adverse health effect) through food. The exposure assessment of food metals is the qualitative and/or quantitative evaluation of the likely intake of metals via food as well as exposure from other sources if relevant. The exposure assessment of the dietary intake of a metal and its toxicological evaluation based on tolerable intakes is needed to assess these risks in a population or community and should take into consideration the occurrence and concentrations of the metal in the diet of the studied community, the consumption patterns of the foods containing the metal and the likelihood of high consumers and of the metal being present in these foods at high levels.[208,224,239]

The World Health Organization (WHO) supports TDSs (total diet studies) as one of the most cost-effective means for assuring that people are not exposed to unsafe levels of toxic chemicals through foods TDS provides "closer-to-real" estimates of exposure to hazardous materials through analysis on table-ready (cooked) samples of foods would be the solution to more comprehensive food safety management.[103] TDS are understood to be made to know what the dietary exposure to various substances in a given population is, both nutrients and contaminants, in food and to evaluate their safety and quality.[115,117,142,208,226,235] The TDS is an ongoing FDA program that monitors levels of about 800 contaminants and nutrients in the average US diet.[71,128]

Availability of data on food composition, contamination, and consumption and advances in both analytical (better detection and quantification limits and better accuracy values and reproducibility techniques, which results in less uncertainty of methods), and scientific knowledge about the influence of metals have on health, have influenced the concept and the setting of reference values, such as tolerable intakes. Based on these factors, it has been possible to set recommendations and evaluations for metals, by means of:

- Tolerable daily intake (TDI): Estimate of the total amount of a chemical contained in food and/or drinking water, expressed relative to body mass (mg/kg), which can be ingested daily over a lifetime set by the Scientific Committee on Food of the European Community (European Union), expressed in mg/person, assuming a body weight of 60 kg.
- Provisional tolerable weekly intake (PTWI) or provisional maximum tolerable daily intake (PMTDI): These are estimates based on health effects of intakes associated with negligible risk. Tolerable intakes refer to weekly intervals, by how much the products considered as contaminants can accumulate in the body for a certain time. The PMTDI is reserved for food contaminants that according to available data do not accumulate in the body.
- Benchmark dose lower limit: The lower one-sided confidence limit of the benchmark dose (BMD) for a predetermined level of response, called the benchmark response (BMR), such as a 5% or 10% incidence of an effect. It is determined by dose–response modeling of toxicological data.

According to the European Food Safety Agency (EFSA) and other food safety agencies, metals such as arsenic, cadmium, lead, and mercury are naturally occurring chemical compounds that can occur as residues in food. People can be exposed to these metals by ingesting contaminated food or water and their accumulation in the body can lead to harmful effects over time. Therefore, and taking into account all the above, this chapter will address various aspects of the following toxic metals in food: cadmium, lead, mercury, arsenic, and aluminum.

10.3   CADMIUM

Cadmium is a metal in group 12 of the periodic table; specifically, it is element 48 of that table with an atomic weight of 112.4 AMU (atomic mass units). It has an oxidation state of +2 and forms numerous inorganic compounds such as chlorides or sulfates, among others.[163] Along with lead and mercury, it is characterized as being one of the most toxic and harmful elements to living beings due to its exposure cause a broad range of adverse health effects in humans and animals.[236] That is why it is subjected to continuous scientific reviews by the EFSA, United Nations for Food and Agriculture/World Health Organization (FAO/WHO), Spanish Consumer Agency for Food Safety and Nutrition (AECOSAN) and Agency for Toxic Substances and Disease Registry United States (ATSDR).[101,122,132,175]

Cadmium's toxicological significance lies in its ability to persist in the environment and its long biological half-life (20–30 years in the human body), low rate of excretion from the body, and storage predominantly in soft tissues (primarily, liver and kidneys).[161,210]

10.3.1   Cd DISTRIBUTION IN THE ENVIRONMENT AND SOURCES FOR HUMANS

Cadmium is a toxic heavy metal, found in all ecosystems around the world and introduced into the biosphere by various anthropogenic activities.[178,183] Its content in the earth's crust is estimated at around 0.10–015 mg/kg in inorganic form, mainly in the form of sulfide minerals associated with the minerals such as zinc, lead, and copper.[140]

Its specific toxicity is closely related to its concentration and chemical form.[28,37,101,175] It also has a biological half-life of 10–20 years and can accumulate in significant amounts in animals and humans.[183]

Cadmium arrives in the environment from two sources: a natural source, of a geobiological origin, caused by the decomposition of rocks, forest fires and volcanic derivations, and an anthropogenic source, which releases and disperses far superior levels to those of the former. In addition, it can be discharged into the environment by the combustion of fuel and coal, the incineration of rubbish and organic waste, indiscriminate use of phosphate fertilizers, tobacco smoke, and by improper management when treating sewage sludge from sewage treatment plants and smelting industries. Both the natural and anthropogenic processes can contribute to cadmium pollution of the environment and therefore the food chain[5,28,32,37,58,76,88,94,101,122,149,155,175,239]

Human beings are exposed to the metal via inhalation, mainly in the case of smokers, as well as the aspiration of fumes and fine particles dispersed in the air from areas near industrial and mining centers.[76,94,239]

However, foods are the main source of cadmium exposure in the nonsmoking general population; hence, their intake from food is considered a risk that must be taken into account in the guidelines of the risk analysis.[101] Cadmium contamination in food is regarded as the main environmental source of nonoccupational exposure.[75]

Once cadmium enters the tissues of plants and animals, it goes into the food chain, as plants and animals are the primary source of food, the main vehicles by which it reaches humans, most notably cereals and vegetables.[46,58,94,122,155,165,187]

It should be noted that cadmium is present in many foods but at low concentrations. Cereals and cereal products, vegetables, nuts and legumes, potatoes and meat, entrails and organs, meat products, and fish are the food groups that contribute mainly to cadmium exposure in the diet, mainly because of their high consumption.[16,28,32,61,101,112,113,122,125,142,146,155,165,187,215]

10.3.2 Cd TOXICOKINETICS AND TOXICODYNAMICS

The average cadmium content in the human body, for an adult of 70 kg, is around 40 mg. The main routes of absorption are inhalation (about 25–50%), gastrointestinal (3–5%), and to a lesser degree by dermal absorption. Oral absorption is conditioned in turn by various nutritional factors, such as a low body iron deposit, exceptional physiological situations such as pregnancy and previous pathologies, which are situations that can increase absorption. On the other hand, the presence of high calcium, iron, zinc, and fiber can inhibit cadmium content.

When its concentration is low, most of the cadmium is retained in the intestinal mucosa bound to metallothioneins and is eliminated by desquamation during the process of the renewal of the epithelium. Faced with high oral concentrations, the ability of metallothioneins is exceeded to fix cadmium and the free metal is able to pass directly into the blood, where it enters with the metabolism of zinc, iron, copper, and calcium.[5,28,126,124,156,188]

Once absorbed, the average plasma concentration is 0.1–0.7 µg/dL, cadmium circulates bound to metallothioneins or albumin to the liver and kidneys where it is stored. These organs accumulates between 40% and 80% of the organic cadmium with a very long biological half-life ranging from 10 to 30 years because the body does not have homeostatic mechanisms that keep their levels constant. Bioaccumulation is directly proportional to the duration of exposure and administered dose.[23,88,91,101,119,126,132,155,175,188,197,216] Cadmium has the ability to accumulate in organs such as the pancreas, lungs, testes, and the salivary glands.[122,175] Its elimination is very slow and mainly occurs in feces and to a lesser extent through urine.[76,127,197]

10.3.3   Cd INTAKE ASSESSMENT

As for the risk characterization for the danger to public health, after continuing evaluations in 2009, the various competent international organizations on food safety and chemical hazards set the tolerable weekly intake (TWI) at 2.5 µg/kg bw/week. In 2011, the Joint FAO/WHO Expert Committee on Food Additives (JECFA) revised its previous assessment and established a provisional tolerable monthly intake PTMI of 25 µg/kg bw.[225]

Table 10.1 shows the weekly intakes of cadmium in different countries assuming a body weight of 60 kg.

TABLE 10.1   Cadmium Intake in Adults in Different Countries.

Country	Intake (µg/kg bw/week) (µg/week)	References
Canada	1.61	Dabeka et al.[29]
USA	1.33–3.27	Coni et al.[26]
Croatia	2.02	Sapunar-Postruznik et al.[180]
Slovenia	1.53	Erzen et al.[47]
Italy	3.3	Coni et al.[26]
Spain (Catalonia) (male adults)	109.9 µg/week (15.7 µg/day)	Llobet et al.[122]
	60.9 µg/week (8.7 µg/day)	Perello et al.[153]
Spain (Canary Islands)	1.3	Rubio et al.[175]
Ireland (adults)	1.1–1.5	FSAI[55]
	97.5th percentile: 2.3–3.0	

TABLE 10.1   *(Continued)*

Country	Intake (μg/kg bw/week) (μg/week)	References
Ireland (children)	1.7–2.3	FSAI[55]
	97.5th percentile: 3.3–4.1	
UK	84 μg/week	Ysart[235]
Total diet study	12 μg/day	
Belgium	0.98	Vromman et al.[217]
	95th percentile: 2.02	
Belgium	2.33	EFSA[37]
Bulgaria	1.89	
Czech Republic	2.37	
Denmark	2.26	
Finland	1.95	
France	2.27	
Germany	2.96	
Hungary	2.16	
Iceland	2.08	
Ireland	2.54	
Italy	2.05	
Netherlands	2.25	
Norway	2.31	
Slovakia	2.29	
Sweden	2.32	
UK	2.15	
USA	1.5	
Italy	From fish = 0.04–0.32	Barone et al.[11]
	From cephalopods = 0.07–0.27	
	From crustaceans = 0.05–0.11	
Greenland	16.73	Johansen et al.[97]
Philippines (women)	1.63	Zhang et al.[238]
Korea	2.47	Moon et al.[139]
China	Mean: 89.6 μg/week (12.8 μg/day)	He et al.[75]
	Men: 98 μg/week (14.0 μg/day)	
	Women: 83.3 μg/week (11.9 μg/day)	
China	1.82	Tang et al.[201]

TABLE 10.1 *(Continued)*

Country	Intake (µg/kg bw/week) (µg/week)	References
Thailand (Bangkok)	Males who never smoked: 0–392 µg/week (0–56 µg/day)	Satarug et al.[184]
	Females who never smoked: 147–189 µg/week (21–27 µg/day)	
Thailand (Mae Sot)	Males who never smoked: 1316–1568 µg/week (188–224 µg/day)	
	Females who never smoked: 693–791 µg/week (99–113 µg/day)	
Japan (Shiga Prefecture) (women)	1991: 259 µg/week (37 µg/day)	Tsuda et al.[206]
Duplicate portion method	1992: 189 µg/week (27 µg/day)	
Japan (Shiga Prefecture) (women)	1991: 224 µg/week (32 µg/day)	
Market basket method	1992: 245 µg/week (35 µg/day)	
Egypt University students living at hostels of Beni-Suef University (BSU)	4.99	Hassan et al.[74]
Iran (Shiraz)	4.27	Rahmdel et al.[159]
	61% PTWI	

Two percent of the Belgian adult population has a dietary Cd intake above the recent TWI of 2.5 µg/kg bw established by EFSA in 2009. Cereal products and potatoes contribute for more than 60% to Cd intake.[217] Cd was detected in 43% of all samples analysed in Ireland, where cereals (39% and 48%) and vegetables (36% and 30%) were the major contributing sources in adults and children, respectively.[55]

10.3.4 Cd TOXIC EFFECTS

Clinical manifestations produced as a result of the toxicity of this element are associated with an etiology which covers a series of mainly cardiovascular and renal diseases.[94,158,208,235]

Cadmium is a fundamentally toxic element for the kidney. Various transporting systems have been identified to play a role in renal cadmium accumulation, renal cadmium transport and cadmium-induced nephrotoxicity under different conditions.[229] It damages proximal renal tubules causing nephritic dysfunction, which causes an increased excretion of low molecular weight proteins. After prolonged and/or high exposure, damage intensifies diminishing the glomerular filtration rate, thereby leading to chronic renal failure, the first step to a multisystem failure. It can also cause bone demineralization, either directly or indirectly as a result of renal dysfunction damage. It is genotoxic by the induction of oxidative stress and inhibition of DNA repair. Dietary-Cd intake estimated from food frequency questionnaires correlates only minimally with U-Cd biomarker, and its use as a Cd exposure indicator may be of limited utility in epidemiologic studies.[209] If the kidney's cadmium binding sites all become saturated after chronic high-level exposure, renal dysfunction results and urine cadmium levels increase dramatically. In that setting, urine levels reflect recent exposure rather than total body burden. Renal dysfunction is considered unlikely when urinary cadmium levels are less than 10 µg/g creatinine.[164] Elevated blood cadmium levels confirm recent acute exposure but do not correlate with body burden or clinical outcome and should not be used to determine the need for treatment. The 95% confidence limit for blood cadmium levels in the United States for healthy nonexposed, nonsmokers is 0.4 µg/L. The US Occupational Safety and Health Administration (OSHA) considers a whole blood level of 5 µg/L or higher hazardous.

The International Agency for Research on Cancer (IARC) has categorized cadmium as a Class 1 human carcinogen,[84,86] EPA has classified cadmium as a Group B1 or "probable" human carcinogen and the American Conference of Governmental Industrial Hygienists considers cadmium a suspected human carcinogen. Cd is statistically associated with the prevalence and higher incidence of lung, endometrial, bladder, and breast cancer.[4,5,28,31,32,37,76,91,94,101,158,175,239] Dietary cadmium exposure may have a role in prostate cancer development.[99]

Other clinical manifestations produced as a result of its toxicity include cardiovascular disease, hypertension, gastrointestinal damage (nausea, vomiting), lung damage (cough, shortness of breath, respiratory tract irritation, edema, cell proliferation, and fibrosis),[32,64,94,158,208,235] bone disease (pains, fractures, osteomalacia, and osteoporosis),[94] abnormalities in the reproductive system (carcinogenicity, mutagenicity, and teratogenicity).[6,88,101,127,239] It is well known that moderate-to-high dose exposure to cadmium (1 mg/kg for 5 days/week for 6 weeks by i.p.) affects steroid synthesis in reproductive organs in female rats.[38] Given the spectrum of human cancers associated with Cd exposure, the ability of Cd to adversely affect steroid hormones deserves further study in humans.

10.4 LEAD

Lead is an element with atomic number 82 belonging to group 14 of the periodic table and atomic weight of 207.2 AMU. It is not a good conductor either for heat or electricity and its oxidation states are 0, +2, and +4, the +2 state being the most common in inorganic compounds.[163]

Lead forms numerous salts, oxides, and organometallic compounds and has a high ability to persist in the environment and cumulative power in living tissue, and this effect is magnified along the food chain.[2,88,93,173,196]

10.4.1 Pb DISTRIBUTION IN THE ENVIRONMENT AND SOURCES FOR HUMANS

Lead is present in an inorganic form in the earth's crust, with an average concentration of 13 mg/kg in the forms of igneous, metamorphic, and sedimentary rocks (in concentrations ranging from 10 to 20 mg/kg). This metal is not abundant in nature in its pure state, but combined with other metals such as silver and zinc. It is present in alloys such as galena (lead and sulfur), the cerussite (lead carbonate), anglesite (lead sulfate), and pyromorphite (lead phosphate).

Its presence in the earth's crust is due to the mining of lead from the ground and various anthropogenic activities such as smelting of minerals, the emission of highly polluting urban gases, uncontrolled deposition of municipal solid waste, and incomplete combustion of fuels, among others.[5,28,49,87,91, 94,122,136,165,239] Ceramics and glasses made with lead are also additional sources of this metal.[94,167] Lead and its derivatives are ubiquitous, being located in the soil, air, and water, and can be carried by primary source foods (vegetables and meats), which contain it, up to the food chain. Lead enters the body orally (food and drinking water), via respiratory, or cutaneous (skin) channels (through the air, dust, and soil elements, important sources especially in the infant stages).[49,94,136,165] It should be noted that the presence of lead in foods and in the environment has declined significantly since the ban on the use of tetraethyl lead (antiknock) in petrol.

Cereal products contribute most to dietary lead exposure, while dust and soil can be important nondietary sources in children.[39] Foods that contribute most to the lead intake because of their high consumption are as follows: cereals and derivatives (16.1%), followed by milk and dairy products (10.4%), soft drinks (10.2%) and vegetables and vegetable products (8.4%). Bread and buns stand out in each group (8.5%), tea (6.2%), water consumption (6.1%), potatoes and potato products (4.9%), fermented milk (4.2%), and beer and derivatives (4.1%). Entrails and fish also contribute to the intake of this element, although the contribution of each varies depending on the age of the individuals who consume them.[39,58,112,113,122,132,142,146,167,165,239]

In Ireland, lead was detected above the LOD in 29% of all samples analyzed in the last TDS carried out by the food safety authority of Ireland in the period 2012–2014 and alcoholic beverages, cereals, and vegetables were the major contributors (28%, 22%, and 12% of total intake, respectively) in adults. Childrens' cereals, beverages, and vegetables were found to be the major contributors (37%, 19%, and 22% of total intake, respectively).[55]

Lead levels in the Canadian TDS conducted from 2003 to 2007 ranged from less than 0.1 µg/kg in natural spring water to 392 µg/kg in herbs and spices. In Canada, during the years 2003–2007 lead levels were consistently highest in herbs and spices (ranging from 292 to 392 µg/kg), with the next highest food product being salt (ranging from 41.5 to 202 µg/kg). The food groups contributing most to the Canadian dietary intake of lead since 2004 are beverages (e.g., beer, wine, coffee, tea, soft drinks), cereal-based foods, and vegetables.[78,79]

FDA completed in 2010 its check of amounts of lead in some specific brands of commercial juice and food products that contain fruit. None of the products exceeded current tolerable intake levels for lead that FDA had established to protect consumers from the known effects of lead such as impaired cognitive development in children. For juice, FDA stated in "Guidance for Industry: Juice HACCP Hazards and Controls" that lead levels in juice above 50 ppb may constitute a health hazard. Where it has not established regulations, FDA assesses the significance of lead levels found in food on a case-by-case basis. None of the juices or other foods tested by FDA in the July 2010 survey contained levels of lead above FDA's current tolerable intake levels.

10.4.2   Pb TOXICOKINETICS AND TOXICODYNAMICS

The average lead content in a human being of 70 kg oscillates is around 122 mg. Absorption in the gastrointestinal tract in an adult is approximately

20–30%, whereas in children this increases to 50%, ages when it is retained up to 30% more.[49,94] After ingestion, the plasma concentration is 15–40 μg/dL, circulating 95–99% bound to hemoglobin in red blood cells. About 10% is distributed in soft tissues and 90% in the bones and can even reach 94% if the blood concentrations of the metal are high. This can happen to the body during the growth and development of children, constituting a significant source of long-term lead exposure in blood.[129] The half-life of lead is 25 days in blood, 40 days in soft tissues, and 30 years in bone.[5,49,88,94,126,132,136,144,165,197]

The liver is the organ responsible for metabolizing lead for its later removal via the bile duct, unless the metal concentrations are very high, in which case sweat, saliva, and urine act as excretion routes. Approximately, 80% of the unprocessed metal is excreted by urine and secondarily via feces, hair, and nails.

From the toxicodynamics perspective, lead competes with essential metals such as calcium, iron, zinc, and copper, modifying their tissue concentrations. Lead in turn acts by inhibiting the ATPase enzyme of the Na/K pump, thereby increasing cell permeability and interferes with DNA and RNA synthesis. In addition, it inhibits the synthesis of the heme group and of hemoglobin.

10.4.3 Pb INTAKE ASSESSMENT

The value of the PTWI for lead is set at 25 μg/kg bw/week, although since 2011, this value has been considered obsolete until a new PTWI value considered protective for the health of people is set.[66] Both the EFSA and the JECFA have expressed concern about the possible negative effect of lead in neural development of infants, children, and fetuses at current levels of exposure through diet, which is why they recommend taking steps to identify the main sources of lead in the diet and identify methods to reduce this dietary exposure. The EFSA CONTAM Panel concluded that the PTWI of 25 μg/kg bw is no longer appropriate as there is no evidence for a threshold for critical lead-induced effects. In adults, children and infants the margins of exposures were such that the possibility of an effect from lead in some consumers, particularly in children from 1 to 7 years of age, cannot be excluded. Protection of children against the potential risk of neurodevelopmental effects would be protective for all other adverse effects of lead, in all populations.[39]

Table 10.2 shows the weekly lead intakes in different countries assuming a body weight of 60 kg.

TABLE 10.2 Lead Intake in Adults in Different Countries.

Country	Intake (µg/kg bw/week) (µg/week)	Reference
Europe (adults)	2.52–8.68	EFSA[39]
	High consumers: 17.01	
Europe (infants)	1.47–6.58	
Europe (children)	5.6–21.7	
	High consumers: 38.57	
Holland	3.73	Van Dokkum et al.[213]
Sweden	1.98	Becker and Kumpulainen[13]
China	12.11	Yang et al.[227]
Canada	2.8	Dabeka and McKenzie[30]
Canada	0.7	Health Canada[79]
Croatia	11.68	Sapunar-Postruznik et al.[180]
Finland	1.42	Tahvonen and Kumpulainen[200]
Taiwan (Tainan)	2.57	Ikeda et al.[89]
USA	1.98	Thomas et al.[203]
Denmark (1983–1987)	4.9	Larsen et al.[112]
Denmark (1993–1997)	2.1	
Slovenia	7.04	Erzen et al.[47]
Spain (Catalonia)	22.89 µg/week (3.27 µg/day)	Llobet et al.[122]
Adult male	58.8 µg/week (8.4 µg/day)	Perello et al.[153]
Spain (Canary Islands)	8.49	Rubio et al.[174]
France	0.65	Leblanc et al.[113]
(children 3–14 years of age)[b]		
France	2.1	
UK	0.7	FSA[54]
Ireland (adults)	0.28–0.84	FSAI[55]
	97.5th percentile: 0.77–1.54	
Ireland (children)	0.28–1.19	
	97.5th percentile: 0.63–1.89	
Portugal (Aveiro)	1.54–24.5	Coelho et al.[25]
China	3.85	Tang et al.[201]
Egypt University students living at hostels of Beni-Suef University (BSU)	5.38	Hassan et al.[74]
Iran (Shiraz)	78.29	Rahmdel et al.[159]
	313% PTWI	

[b]Assuming a body weight of 20 kg.

10.4.4   Pb TOXIC EFFECTS

Developmental neurotoxicity in young children and cardiovascular effects and nephrotoxicity in adults has been identified by EFSA[39] as the critical effects for the risk assessment. The respective BMDLs derived from blood lead levels in µg/L (corresponding dietary intake values in µg/kg bw/day) were: developmental neurotoxicity $BMDL_{01}$, 12 (0.50); effects on systolic blood pressure $BMDL_{01}$, 36 (1.50); effects on prevalence of chronic kidney disease $BMDL_{10}$, 15 (0.63).

Lead neurological effects in children have been documented at exposure levels once thought to cause no harmful effects (<10 µg/dL).[22] Blood lead concentrations under 10 µg/dL, down to 1–2 µg/dL have been associated with neurodevelopmental, neurodegenerative, cardiovascular, renal, and reproductive effects. Of these, the strongest association at the lower levels is for neurodevelopmental effects, specifically the reduction of IQ score and attention-related behaviors.[79]

The IARC classifies inorganic lead compounds as probably carcinogenic to humans (Group 2A).[104] Prenatal exposure to low lead levels (e.g., maternal BLLs of 14 µg/dL) may increase the risk of reduced birth weight and premature birth.

In the bloodstream, it acts by inhibiting the synthesis of hemoglobin, thereby reducing the half-life of erythrocytes. It is also able to cross the placenta and the blood–brain barrier, where it concentrates in the gray matter.[2,5,28,49,87,94,132,165,175,196]

Chronic lead poisoning presents diverse symptoms reflecting its action in different organs. Thus, it can cause subacute and chronic encephalopathy with cognitive and mood impairment in the central nervous system and causes headaches, fatigue, insomnia, irritability, loss of libido, ataxia, coma, and convulsions. Lead encephalopathy is much rarer in adults than in children. Adults have increased resistance to the development of lead encephalopathy due to the capacity of the mature adult brain to sequester lead away from its mitochondrial site of action within cerebral and cerebellar neurons.[162] Lead encephalopathy may occur at extremely high BLLs, for example, 460 µg/dL.[9] In the peripheral nervous system, it produces peripheral polyneuropathy, mainly in the limbs, which begins with muscle aches. In fact, persons with extremely long-term lead intoxication manifested a mild sensory and autonomic polyneuropathy which probably reflects a direct neurotoxic effect of lead.[172]

As for the digestive system lead induces anorexia, constipation, and, in its most critical phase, "lead colic" (intense abdominal cramps with nausea and vomiting). On the renal level, lead is able to accumulate in proximal tubular cells producing kidney failure, hypertension, and gout. Some studies show the incidence of changes in the endocrine and reproductive system (miscarriage, birth defects, premature birth) as well as cardiotoxicity and having mutagenic effects.[4,34,49,64,88,91,94,235,239]

10.5   MERCURY

10.5.1   Hg PROPERTIES

Mercury is number 80 in the periodic table and its atomic weight is 200.59 AMU.[163] It is a liquid metal at room temperature, being the only metal that remains liquid at 0°C.[59] It has the peculiarity of dissolving many metals forming amalgams, except with iron, which is why the latter is used in the manufacture of packaging used in transport and conservation.

Mercury has been found naturally since the beginning of the existence of the planet and is part of the lithosphere, hydrosphere, atmosphere, and biosphere in low amounts. It was one of the first metals used by man, since both its healing and toxic properties were known in ancient times. The environmental presence of this metal has been increasing as a result of human activities, where it is found in the air, food, soil, and water.[147] In fact, it is estimated that approximately 10,000 t of the metal are discharged each year into the environment as a result of anthropogenic activities.[73]

10.5.2   Hg DISTRIBUTION IN THE ENVIRONMENT AND SOURCES FOR HUMANS

Mercury is found in most natural resources as elemental or metallic mercury (Hg^0), inorganic mercury (mercury salts) and organic mercury, with methylmercury being the organic form of mercury in the food chain.[40]

The environmental presence of this metal has grown with by human activities.[147] Mercury is capable of causing adverse effects on health (e.g., neurological, nephrological, immunological, cardiac, motor, reproductive, and even genetic problems) and the environment because of its toxicity, long-range transport in the atmosphere, persistence in the environment and its ability to bioaccumulate in ecosystems. The toxic effects of mercury imposed on humans and other organisms are dependent on many factors including the chemical form, the amount, the exposure pathway, and differences in vulnerability between exposed subjects.[3,105]

When entering aquatic ecosystems, certain bacteria can cause the formation of methylmercury by the methylation of environmental inorganic mercury, with being this species being the most toxic and bioaccumulative organic form of this element.[3,141] Mercury of industrial origin is the most dangerous source of mercury in the environment (resulting from chlor-alkali industry plants, paper mills, and purification of gold).

The sources of exposure for humans are mostly occupational (e.g., mining, mercury refining, battery production, lamps, thermometers, fungicides, explosives, pigments, dental amalgams used in dentistry, photographs, etc.), although there are also domestic sources of exposure such as broken thermometers, mirrors, toys, and batteries; drug sources of exposure such as diuretics, laxatives, disinfectants and spermicides, and food sources, mainly contaminated fish.[3,109,194,220]

Mercury is a toxic element which causes a wide range of poisonings that occur accidentally, caused either by environmental or occupational exposure, or due to the consumption of contaminated food, especially seafood, with the main source of human exposure to mercury methylmercury coming from fish with widely varying levels between different species, which are higher in predatory fish. For this reason, although seafood provides important nutritional benefits, hazards from contaminants such as mercury and methylmercury must be monitorized.[3,42,70,121,141,153,182] It is important to note that the ability of shellfish to accumulate mercury makes them good bioindicators of contamination by this element.[154] In Korea, the fish and shellfish group contributed most to total dietary intake at 76%.[110] In Ireland, for both adults and children, white fish was found to be the main contributor to mercury intake.[55]

Rice is also known to accumulate methylmercury (MeHg); therefore, commercial rice products sold in Europe have been investigated finding that 30% of all commercial market rice products exceeded 10% of the PTWI calculated for toddlers or 13% of products for adults with rice based diet. MeHg ranged from 0.11 to 6.45 µg/kg with an average value of 1.91 ± 1.07 µg/kg. Total Hg ranged from 0.53 to 11.1 µg/kg with an average of 3.04 ± 2.07 µg/kg.[18]

10.5.3 Hg TOXICOKINETICS AND TOXICODYNAMICS

As regards the digestive tract, metallic mercury is hardly absorbed by this route; inorganic derivatives (ionic forms) are 70% absorbed while organic forms (mainly methylmercury) are almost entirely absorbed and faster than inorganic derivatives.[40,141]

From a human health perspective, it is the amount of methylmercury, rather than total mercury, that is, of most interest, since methylmercury is much more readily absorbed into the human bloodstream. As a result, in the absence of detailed information on mercury speciation, it is simply assumed, for the purposes of health risk assessments, that 100% of total mercury is in the methylated form as methylmercury.[77] After absorption, mercury is actively transported to different tissues after the formation of a methylmercury–cysteine complex.[141]

In blood, inorganic mercury is distributed between plasma and erythrocytes and is more present in the plasma, while methylmercury accumulates mostly (>90%) in erythrocytes. Unlike inorganic derivatives, methylmercury is able to enter the hair follicle, cross the placental barrier and the blood–brain barrier, allowing its accumulation in the hair, the fetus, and the brain.[40]

The plasma half-life of metallic mercury is estimated to be approximately 23–40 days, and that of mercury's organic compounds, such as methylmercury, to be 40–50 days.[220]

The excretion of inorganic mercury occurs primarily through urine, while the main route of excretion of methylmercury is through feces, with a lower percentage being eliminated through the hair and urine. The levels of total mercury in urine could be a good biomarker for exposure to inorganic and elemental mercury but not for methylmercury.[40,160]

10.5.4 Hg INTAKE ASSESSMENT

Fish meat is the food with the highest contribution of methylmercury, followed by fish products. Tuna, swordfish, cod, whiting, and pike are the fish which contribute most to dietary exposure to methylmercury in adult age groups, while in the case of children, hake can also be included in addition to those mentioned above.[40]

Several studies assume that 100% of the mercury in fish and shellfish products is present as methylmercury.[150] Levels can be variable, even among fish of the same species. For example, in sablefish, the percentage of total mercury that was in the organic form ranged from 81% to 95%.[21] In samples of various species of tuna, the portion of total mercury present as methylmercury ranged from 70% to 77%[230] and 61% to 94%. In swordfish, the percentage of total mercury that was in the organic form varied from 43% to 76% and 72%.[230] A wide range of percentages of total mercury that was in the organic form (30% to 79%) were also found in canned tuna. In Canada, swordfish is the only fish for which regular weekly consumption would result in a PDI that exceeds the pTDI (% pTDI of 143%).[77] According to Ilmiawati et al.,[90] the exposure to MeHg in Japanese children is above the US EPA RfD, which, in part, is explained by their consumption of large predatory fish. For methylmercury, some studies indicate that beneficial effects related to long-chain omega 3 fatty acids present in fish may have previously led to an underestimation of the potential adverse effects of methylmercury in fish.[40]

In 2012, the EFSA evaluated the PTWI of 1.6 µg/kg of body weight for methylmercury and 4 µg/kg of body weight for inorganic mercury established by the Committee JECFA[95] and set a TWI of methylmercury at 1.3 µg/kg of body weight, primarily based on prenatal neurodevelopmental toxicity, although it has kept the TWI of inorganic mercury at 4 µg/kg of body weight.

Table 10.3 shows the weekly intakes of mercury in different countries assuming a body weight of 60 kg.

TABLE 10.3 Mercury Intake in Adults in Different Countries.

Country	Intake (µg/kg bw/week) (µg/week)	Reference
China	5.92	Yang et al.[227]
Canada	1.87	Chan et al.[23]
USA	0.28–0.56	Dougherty et al.[35]
Croatia	2.33	Blanusa and Juresa[16]
Australia	0.07–0.63	FSANZ[56]
France	7.35	Noël et al.[146]
Chile	0.49	Muñoz et al.[142]
Korea	0.21	Lee et al.[114]
Korea	16.8 µg/week	Kwon et al.[110]
	From fish and shellfish: 1 µg/kg bw/week	
Lebanon	0.28	Nasreddine et al.[145]
Spain (Canary Islands)	0.65	Rubio et al.[176]
Spain (Catalonia)	Hg: 70 µg/week	Perello et al.[153]
Male adults	MeHg: 51.1 µg/week	
UK	21 µg/week	Ysart et al.[235]
	0.14–0.55	Rose et al.[171]
Ireland (adults)	0.17	FSAI[55]
Ireland (children)	0.14	
Italy	From fish = 0.07–1.44	Barone et al.[11]
	From cephalopods = 0.05–0.15	
	From crustaceans = 0.04–0.08	

TABLE 10.3   *(Continued)*

Country	Intake (µg/kg bw/week) (µg/week)	Reference
Australia	0.21–0.35	FSANZ[57]
China	0.63	Sun et al.[199]
France	0.16–1.39	Arnich et al.[7]
Norway	0.35	Jenssen et al.[96]
China	0.98	Tang et al.[201]
Japan (women Shiga Prefecture)	1991: 30.1 µg/week	Tsuda et al.[206]
	1992: 24.5 µg/week	
Duplicate portion method		
Japan (Women Shiga Prefecture)	1991: 30.1 µg/week	
	1992: 69.3 µg/week	
Market-basket method		
Egypt University students	2.76	Hassan et al.[74]
Beni-Suef University (BSU)		

There are currently global efforts to establish food consumption guidelines to reduce the mercury exposure and institute mercury management practices to reduce mercury emissions.[150]

10.5.5 Hg TOXIC EFFECTS

The toxicity of this element varies depending on the compound (chemical form), route of exposure and sensitivity of the person.[160,231] The toxicity increases with lipid solubility and the alkyl forms are more toxic than the inorganic and elemental forms.[194] IARC has concluded that methylmercury compounds are possibly carcinogenic to humans (Group 2B).[85]

Mercury is a neurotoxin that affects the nervous system whose action is more damaging in the early stages of human development as it affects brain development and also affects the gastrointestinal tract and renal function.[3,123,143,190]

The characteristic clinical manifestations of the ingestion of inorganic mercury compounds are initially acute gastroenteritis syndrome followed by renal failure and uremia and anuria ulcerative hemorrhagic colitis.

Methylmercury may induce the production of neurological effects on the fetus, since it is able to cross the placental barrier, thereby causing alterations in the normal brain development of infants and children and, at higher levels, may induce neurological changes in adults and suppression of the immune system.[3,143,160,189]

Death resulting from organic mercury ingestion has been amply documented following outbreaks of poisoning (Minamata disease) after consumption of methylmercury-contaminated fish in Minamata and after consumption of grains contaminated with methyl- and ethylmercury in Iraq.

10.6   ARSENIC

Arsenic is in group 15 of the periodic table. It is a metalloid or semimetal with atomic number 33 and an atomic weight of 74.9 AMU. It has oxidation states of −3, 0, +3, and +5,[163] depending on the toxicity of these.[67] It is characterized as a major toxic due to its ability to accumulate in the food chain, which implies a potential danger to human and animal health. Acute exposure to high doses has lethal effects on organisms and chronic exposure to low doses is associated with mutagenic and carcinogenic problems, poor body condition, etc.[124]

10.6.1   DISTRIBUTION IN THE ENVIRONMENT AND SOURCES FOR HUMANS

Arsenic is a widely found toxic element in nature, both in natural sources and anthropogenic activities, and it is found in both the environment and in living organisms. It can be found in various chemical and oxidation states forms, which determines its toxicity, with the inorganic forms being more toxic than the organic forms.[14,38,63,80,81,179,181] Its wide distribution implies serious harmful effects on various aquatic and terrestrial organisms and ultimately on human health.[219]

Basically, the three most important routes of arsenic exposure for humans are work, food, and medication.[179] Arsenic can be introduced into foods by environmental pollution, or at any stage of the food chain, and from production to consumption. Fish, seafood, and meats are the main sources of this element in the diet.[138,222]

The USA's TDS (1991–1997) data showed that the highest As concentration was found in seafood, followed by rice/rice cereal, mushrooms, and poultry. The proportion of organic As in fish, which is less toxic, is usually higher than inorganic As.[38] The inorganic forms account for 2.9–26.0% of total arsenic in fish, whereas the equivalent values for organic arsenic are 74–90%. More detailed studies show that the percentage of arsenic in fresh seafood (specifically arsenobetaine) is high (81% of the total) but decreases in frozen and canned foods.[80] There does not appear to be any particular type of fish that contains higher levels of arsenic and biomagnification in aquatic food chains has not been observed. A UK multielement survey of fish (1998), levels of total arsenic in the most commonly consumed fish (cod, haddock, salmon, tuna) in the United Kingdom were in the range of 1.9–8.4 mg/kg fresh weight, with a mean of 4.6 mg/kg.[27] In Hong Kong, rice is the major contributor of the dietary exposure to inorganic arsenic.[50]

Many arsenic compounds are soluble in water. Arsenic pollution in water is mainly due to industrial waste, and it could result from both, oxidative and reductive dissolution.[181] Basically, what appears is inorganic arsenic and this is a threat to more than 150 million people living in countries like Bangladesh, Cambodia, China, India, Laos, Myanmar, Nepal, Pakistan, Taiwan, and Vietnam. In addition, endemic areas with dangerously high levels of arsenic in the water have also been reported in Canada, Germany, and Argentina.[179,191]

A provisional guideline value for arsenic in drinking water was set at 10 mg/L.[221] Arsenicism could be associated with water supplies containing an upper arsenic concentration of 1 mg/L or greater and that a concentration of 0.1 mg/L may give rise to presumptive signs of toxicity. Assuming a daily water consumption of 1.5 L, JECFA concluded that intakes of 1.5 mg/day of inorganic arsenic were likely to result in chronic arsenic toxicity and daily intakes of 0.15 mg may also be toxic in the long term to some individuals.[27]

An index dose (applied to a single source of contaminant and is defined as a level at which the risk is considered minimal, but there is a requirement that exposure from each individual source should be as low as reasonably practicable) of 0.3 mg/kg bw/day was recommended for inorganic arsenic, based on the EU/WHO drinking water guideline of 10 mg/L, assuming consumption of 2 L water/day.[27]

10.6.2 TOXICOKINETICS AND TOXICODYNAMICS

Between 21% and 40% of total dietary arsenic occurs in inorganic forms.[233] Soluble species of inorganic arsenic are rapidly absorbed after its ingestion in proportions of 95%. Pentavalent arsenic is better absorbed through the intestines and trivalent forms are more soluble in lipid membranes. The absorption of organic compounds exceeds 70%.[38,44]

Arsenic is widely distributed throughout the body and is initially accumulated in organs such as the liver, kidney, and lung and finally in the hair, nails, teeth, skin, upper gastrointestinal tract, epididymis, thyroid glands, lens, and skeleton and is able to cross the placental barrier.[38,194]

The body has the ability to transform arsenic into methylated species, which are organic arsenical compounds, and which are therefore less toxic than inorganic forms and are easily eliminated by the body.[211] Although there are experimental studies suggesting that this biomethylation, namely the production of methylated metabolites containing trivalent arsenic, is a process that activates this element as a toxin and a carcinogen.[198]

The most important arsenic elimination route is the renal route where 35% of arsenic is eliminated in 24 h, but it can also be eliminated from the gastrointestinal tract, skin, hair, nails, and sweat.[44,194]

Both pentavalent and trivalent forms of inorganic arsenic will cause an uncoupling of oxidative phosphorylation. The pentavalent forms will compete with the phosphate groups as a substrate in various enzymatic reactions; in one of them, there is a binding to glyceraldehyde-3P-dehydrogenase enzyme, forming a labile arcilarseniate which prevents ATP synthesis. This process is known as arsenolysis. Trivalent forms, however, have a high affinity for lipoic acid, which is a coenzyme of dihydrolipoamide-transacetilase, an enzyme involved in mitochondrial respiration, which will produce alterations in carbohydrate metabolism by blocking different oxidative enzymes that need this coenzyme such as pyruvate dehydrogenase and alpha-ketoglutarate dehydrogenase. They will also uncouple the respiratory chain by decreasing intracellular levels of NADH, thereby producing a deficit of ATP by inhibition of its synthesis.

As a toxicological guidance value, a $BMDL_{0.5}$ has been set at 3.0 μg/kg bw/day (2–7 μg/kg bw/day based on the range of estimated total dietary exposure).

10.6.3 AS INTAKE ASSESSMENT

To correctly estimate the risk of arsenic from food sources, it is important to determine the bioavailability of inorganic arsenic forms in them since considering the total arsenic regardless of inorganic species would lead to a substantial overestimation of the health-related risk from dietary exposure of this element. It is also noteworthy that the intake of inorganic arsenic from drinking water and food varies according to the geographic levels of arsenic.[17,38]

The PTWI of total arsenic has been set by the JECFA at 15 µg of As/kg bw/week, a value that is above intakes established in Europe, although the existence of new studies reporting that inorganic arsenic causes lung cancer, urinary tract cancer, and skin cancer, along with a wide range of adverse effects at exposures lower than those revised by the JECFA suggest that there is a need to consider reviewing this PTWI value depending on the form of arsenic (Table 10.4).

TABLE 10.4 Arsenic Intake in Adults in Different Countries.

Country	Dietary intake/exposure (µg/kg bw/week) (µg/week)	Reference
Europe (19 countries)	**Inorganic arsenic 0.91–3.92 µg/kg bw/week**	EFSA[38]
Ireland (adults)	**Total As**	FSAI[55]
	Average consumers: 4.9–6.3	
	Above average consumers: 27.3–29.4	
	Inorganic As	
	Average: 0.07–0.14	
	Above average consumers: 0.42–0.56	
Ireland (children)	**Total As**	FSAI[55]
	Average consumers: 4.2–6.3	
	Above average consumers: 20.3–23.1	
	Inorganic As	
	Average consumers: 0.21–0.35	
	Above average consumers: 0.91–0.98	
UK	455 µg/week (0.065 mg/day)	Ysart et al.[235]
UK	**Total As**	COT[27]
	Upper bound population: 0.83 mg/kg bw/day	
	High levels: 4.4 mg/kg bw/day	
	In As	
	Upper bound mean: 0.07–0.2 mg/kg bw/day	
	Upper bound high level: 0.13–0.34 mg/kg bw/day	
	Upper bound population mean exposure to inorganic arsenic was 0.09 mg/kg bw/day	
Spain (Catalonia)	As: 216 µg/day	Perello et al.[153]
Male adults	InAs: 2.6 µg/day	

TABLE 10.4   *(Continued)*

Country	Dietary intake/exposure (µg/kg bw/week) (µg/week)	Reference
USA US Department of Agriculture's 1987–1988 Nationwide Food Consumption Survey	Total arsenic µg/day: 1 for infants 23 for toddlers 20 for 6-year-old children 13 for 10-year-old children 15 for 14–16-year-old boys 21 for 14–16-year-old girls 57 for 25–30-year-old men 28 for 25–30-year-old women 47 for 40–45-year-old men 37 for 40–45-year-old women 92 for 60–65-year-old men 72 for 60–65-year-old women 69 for 70-year-old men 42 for 70-year-old women	Tao and Bolger[202]
USA	InAs: 8.3–14 µg/day	Yost et al.[233]
Canada	InAs: 4.8–12.7 µg/day	Yost et al.[233]
Hong Kong	InAs: Average consumers: 0.22 µg/kg bw/day High consumers: 0.38 µg/kg bw/day Male: 0.23 µg/kg bw/day Female: 0.21 µg/kg bw/day	FEHD[50]
Japan (Shiga Prefecture) Women	**Duplicate portion method** 1991: 260 µg/day 1992: 210 µg/day **Market basket method** 1991: 160 µg/day 1992: 280 µg/day	Tsuda et al.[206]
Egypt University students living at hostels of Beni-Suef University (BSU)	4.77 µg/kg bw/week	Hassan et al.[74]

The Food Standards Agency (FSA) has recently completed a Total Diet Study (TDS) of total and inorganic arsenic levels in food, which was carried out between 1999 and 2002 in the UK.[27] In the United Kingdom, fish was the major contributor to dietary exposure to total arsenic providing 4.6 mg/kg bw/day for the high-level adult consumers of fish. The miscellaneous cereals food group was the major contributor to inorganic arsenic, providing up to 0.064 mg/kg bw/day for the high-level adult consumer.[27] Later, in 2006, this Committee On Toxicity stated that the dietary exposure to organic arsenic is unlikely to constitute a risk to health in the United Kingdom, but the exposure to inorganic arsenic should be as low as reasonably practicable (ALARP).

10.6.4 TOXIC EFFECTS

Exposure to this element, both from direct consumption of water or through contaminated food is associated with adverse effects on human health, in fact, arsenic has been associated with persuading a variety of complications in body organ systems: integumentary, nervous, respiratory, cardiovascular, hematopoietic, immune, endocrine, hepatic, renal, reproductive system, and development.[1,191] Its toxicity and bioavailability depend on its speciation, which in turn depends on microbial transformation reactions, which include reduction reactions, oxidation, and methylation.[118] Some of the more note-worthy toxicological effects of As are gastrointestinal symptoms (hemor-rhagic gastroenteritis, nausea, vomiting, garlic odor in the breath, abdominal pain, and watery diarrhea), renal toxicity (serious damage can be caused in renal capillaries, tubules, and glomeruli, varying degrees of necrosis and tubular degeneration), neurological effects (delirium, disorientation, agitation, encephalopathy, seizures, painful paresthesia, muscle weakness, paralysis, neuromuscular respiratory failure, and coma), blood affectation (affects bone marrow and alters the cellular composition of the blood, in addition, there is a possibility of anemia, mild-to-moderate leucopenia and eosinophilia), and liver toxicity (adipose infiltration, central necrosis, and cirrhosis).[106,181,194]

As already mentioned above, arsenic is a carcinogen, inorganic arsenic is classified as a group I carcinogen.[17] Strikingly, arsenic has been able to induce epigenetic changes (in utero) and genetic mutations (a leading cause of cancer) in the body.[1] Epidemiological studies support a causal relation-ship between As exposure and skin cancer. There may be two cell types of skin cancer induced by arsenic: basal cell carcinomas and squamous cell carcinomas arising in keratotic areas. Basal cell cancers usually only produce local invasion, but squamous cell carcinomas can emit distant metastases.

Occupational exposure to airborne arsenic can also be linked with lung cancer, which is normally in a form little differentiated from epidermoid carcinoma. Other cancers that have been linked to exposure to As are liver angiosarcoma, lymphoma, leukemia, and nasopharyngeal, kidney, and bladder cancers. Studies on the mutagenic effects have generally been negative.

Excessive occupational exposure to As appears to have effects on reproduction and teratogenicity.[36,107,194]

In acute arsenic poisoning, death is usually due to cardiovascular collapse and hypovolemic shock. The fatal human dose for ingested arsenic trioxide is 70–180 mg or about 600 μg/kg. The minimal lethal dose is in the range of 1–3 mg/kg.[8]

10.7 ALUMINUM

Aluminum is an element with atomic weight of 26.98 and number 13 in the periodic table. It is extremely resistant to corrosion and its main inorganic compounds are oxide, hydroxide, sulfate, fluoride, and chloride.[163]

It is abundantly distributed in our environment and is the most abundant element in the earth's crust[33,82,102,157] and the third most abundant in nature after oxygen and silicon.[177,193] Aluminum is present in soil, minerals, and rocks and even in water and food and does not seem to have any role in human and animal biology.[193]

10.7.1 AL DISTRIBUTION IN THE ENVIRONMENT AND SOURCES FOR HUMANS

The main sources of aluminum for humans are drinking water and food consumption.[53,193,232] Aluminum is naturally present in food or is a result of the addition of additives or through the contact of food with packaging, containers, foil, or cookware containing this metal.[68,131,151,193] However, the amount of aluminum that comes from the diet is small compared to that present in certain pharmaceutical preparations such as antacids.[193]

The content of this metal in most food does not exceed 10 mg/kg, and the most usual concentrations are between 0.1 and 1 mg/kg.[195] Some vegetables (spices, herbs, tea leaves) contain more aluminum than animal foods. The content varies depending on plant species and soil pH.[15]

This element is mainly consumed through cereals, cheese, and salt, although, as noted above, certain types of tea and spices that naturally contain high levels of aluminum. Therefore, it is necessary to take into account that certain foods are a significant source of aluminum for children.[148,195] The main dietary aluminum exposure sources are fried twisted cruller, leaf vegetables, and bean products.[228]

In China, high aluminum levels were found in jellyfish (mean 4862 mg/kg), laver (mean 455.2 mg/kg), and fried twisted cruller (mean 392.4 mg/kg), therefore, jellyfish is the main Al contributor, providing 37.6% of the daily intake.[237] The Al content in processed foods and unprocessed foods in Japan ranged from 0.40 to 21.7 mg/kg and from 0.32 to 0.54 mg/kg, respectively. The Japanese daily dietary Al intake from processed foods was much larger than that from unprocessed foods and the highest consumer Al exposure value (>P95) of the young children group exceeded the PTWI.[185] In Australia, the major food group with the highest contribution to estimated aluminum dietary exposures was cakes and biscuits for all age groups assessed (32–48%) probably because of the presence of aluminum-containing additives such as sodium aluminum phosphate, which are commonly used as leavening agents.[51]

10.7.2 AL TOXICOKINETICS AND TOXICODYNAMICS

The human being has effective barriers (skin, lungs, gastrointestinal tract) to reduce the absorption of aluminum from water, food, medicine, and air.[193]

Although the inhalation route does not affect the general population, mining and smelting workers, among others, could be exposed to toxic levels of this metal through dust and aerosols.[33] It is estimated that approximately 3% of the aluminum enters the bloodstream from the lungs.[98]

Aluminum ingested through the diet can be absorbed through the gastrointestinal tract,[134] albeit at low rates (0.1–1%). While part of the absorption occurs in the stomach, most occurs in the intestine. Ionic aluminum is actively absorbed by the presence of an active transporter (via iron absorption) in the intestine, whereas elemental aluminum does this by diffusion.[33] It is important to note speciation, as this affects its absorption, as well as a wide range of substances including calcium and citrate iron.[69,212]

Once absorbed, 80–90% is bound to an iron protein carrier (transferrin) and possibly to albumin. The remaining 10–20% forms small molecular complexes, particularly with citrate and phosphate groups.[33] Aluminum can be accumulated in various organs such as the brain, spleen, lungs, liver, and bone in both adults and children.[108,134,208]

Most of the aluminum which humans intake in the diet is eliminated in the urine and to a lesser extent in feces.[131,193]

The lowest LOELs for Al of different studies in mice, rats, and dogs were in the range of 50–75 mg/kg bw/day.

10.7.3 AL INTAKE ASSESSMENT

The aluminum level that has been established as PTWI has been set at 1 mg/kg bw/week.[223]

Table 10.5 shows the weekly intakes of aluminum in different countries.

TABLE 10.5 Aluminum Intake in Adults in Different Countries.

Country	Intake (mg/kg bw/week)	Reference
Holland	0.36	Van Dokkum et al.[213]
Holland	0.36	Ellen et al.[43]
Taiwan	0.47–1.05	Liu et al.[120]
China	1.05–1.40	Wang et al.[218]
China (Shenzhen)	1.263	Yang et al.[228]
	0–2 age: 3.356	
	3–13 age: 3.248	
China, Zhejiang Province	1.15	Zhang et al.[237]
USA	0.76–0.81	Pennington and Schoen[152]
Australia	Lower bound exposures: 0.35–0.84	Food Standards Australia New Zealand[51]
	Upper bound exposures: 0.35–0.91	
Germany	0.58–1.17	Schaller et al.[186]
Italy	0.29–0.74	Gramiccioni et al.[67]
India	0.75	Tripathi et al.[204]
France	0.49	Biego et al.[15]
Belgium	0.21	Fekete et al.[48]
Canada	0.72	Soliman and Zikovsky[192]
UK	1.28	Ysart et al.[234]
	0.40	Ysart et al.[235]
India	0.75	Tripathi et al.[205]

TABLE 10.5 *(Continued)*

Country	Intake (mg/kg bw/week)	Reference
France	0.24	Noël et al.[146]
	0.19	Leblanc et al.[113]
Japan (Tokyo)	1.41	Aung et al.[10]
Spain (Canary Islands)	1.19	González-Weller et al.[64]
Ireland	0.35	FSAI[55]

In Australia, the estimated dietary exposures were under the PTWI for all population groups assessed except for 2–5-year-old 90th percentile consumers who had an estimated exposure of 110% of the PTWI. This small exceedance is unlikely to represent a major public health and safety issue.[51] In Belgium, at the 98.2th percentile the dietary Al exposure reached 0.144 mg/kg bw/day (0.113 and 0.031 mg/kg bw/day, respectively) what exceeds the PTWI indicating that a well-defined subgroup of the population might be at risk.[49] Children in Shenzhen (China) also have high dietary aluminum exposure risk.[228]

10.7.4 AL TOXIC EFFECTS

In humans, aluminum is potentially neurotoxic.[65] It is widely accepted that this element is a recognized neurotoxin, and that it could cause cognitive deficiency and dementia when it enters the brain and may have various adverse effects on central nervous system.[102] Such health effects can be divided into three categories: neurological disorders, cognitive decline, and dementia or Alzheimer's disease.[170] Alzheimer's disease is characterized by two major anatomo-pathological features, intraneuronal protein clusters composed of paired helical filaments consisting of hyperphosphorylated forms of tau protein [neurofibrillary tangles (NFTs)], and extracellular protein aggregates (senile plaques). The hypothesis that aluminum could be involved in the pathogenesis of Alzheimer's disease arose when this element was detected in senile plaques and NFTs in brain tissue from patients with this disease.[53]

Although the results of molecular epidemiological investigations have suggested an association between chronic exposure to aluminum and risk of Alzheimer's disease (this association is stronger for exposure to aluminum from drinking water compared to food), some people who are exposed chronically to this element, either from through water and/or food, have not shown any pathology of this disease, apparently because their gastrointestinal barrier is more effective.[53,100]

Several studies have shown that large doses decrease the absorption of phosphorus in humans and laboratory animals.[193]

It should be noted that there are risk groups particularly sensitive to toxic effects of aluminum: people with chronic renal failure, children with immature or damaged kidney function and premature babies.[82,134]

It is noteworthy that there is some controversy with regard to the possible link between water intake and the onset of Alzheimer's disease, since the role of aluminum in the pathogenesis of this and other mental degenerative diseases is not entirely clear.[137] It seems that there is not enough epidemiological evidence to prove this relationship.[170] It was demonstrated that any risk of Alzheimer's by ingestion of this metal at concentrations less than 2 mg/L from drinking water is minimal,[133] although it has found that cognitive decline with time was greater in subjects with a higher daily intake of aluminum from drinking water (≥ 0.1 mg/day) or higher geographic exposure to aluminum, and in another study, it was reported that high concentrations in the drinking water itself may be a risk factor for developing of this disease.[169]

In the case of the relationship between the intake of aluminum and the development of Alzheimer's, Rogers and Simon[166] suggested that this could have an influence during the development of this disease, although more research of this kind are needed to corroborate or not this study.

As demonstrated in different studies on occupational exposure to aluminum, this exposure has an adverse effect on the status of certain essential elements in humans, with the consequent impact on cellular enzymes and metabolic processes.[135,168] A study by Lerner[116] in 2007 describes aluminum as a potential environmental factor in the induction and development of Crohn's disease.

Al is also implicated as a toxic agent in the etiology of various diseases such as encephalopathy caused by dialysis, anemia, amyotrophic lateral sclerosis, osteomalacia, and parkinsonism-dementia.[12,82,111] Aluminum can also cause toxicity in the bones since it can be incorporated into the bones and causes physiochemical mineral dissolution as well as cell-mediated bone resorption.[108] Apart from that, it can produce inhibition of hydroxyapatite formation (due to this element can bind to the phosphate groups of hydroxyapatite crystals of the calcified bone matrix), suppression of proliferation, and inhibition of the activity of bone cells, leading to decreased bone formation and mineralization.[24,130]

Therefore and in spite of the fact that Al is an element that is associated with a variety of neurological disorders with an increase in the formation of ROS,[20] the exact mechanism or mechanisms by which aluminum produces toxicity are not entirely clear, with the most accepted mechanisms being the enhancement of oxidative and inflammatory phenomena resulting in damage in different tissues.[12] Strong evidence is there that Al complexes with superoxide anion forming aluminum superoxide anion, which is a more potent oxidant than superoxide anion on its own and promotes the formation of hydrogen peroxide and hydroxyl radicals that contribute to an oxidizing environment.[108]

As a summary, metals as contaminants in foods still pose significant health risks upon acute and chronic exposure to high levels. The CNS is particularly vulnerable to metals like Pb, Hg, and Cd which retain themselves in the brain for longer period of time and are one of the causes of neurotoxicity. The brain readily accumulates metals, which under physiologic conditions are incorporated into essential metalloproteins required for neuronal health and energy homeostasis. The neurologic pathology and physiology following Al, As, Pb, and Hg exposures, among others, have been highlighted as classic examples of metal-induced neurotoxicity.[19,60] The knowledge of dose–effect/response relations has been useful to control nephrotoxic effects of these metals through a "biological monitoring of exposure approach." Combining food survey and food contamination detection provide a reliable estimation for environmental contamination exposure.[75]

KEYWORDS

- cadmium
- lead
- mercury
- arsenic
- aluminum
- toxic effects
- intake
- risk assessment

REFERENCES

1. Abdul, K. S.; Jayasinghe, S. S.; Chandana, E. P.; Jayasumana, C.; De Silva, P. M. Arsenic and Human Health Effects: A Review. *Environ. Toxicol. Pharmacol.* **2015**, *40* (3), 828–846.

2. Abou Donia, M. A. Lead Concentrations in Different Animals Muscles and Consumable Organs at Specific Localities in Cairo. *Glob. Vet.* **2008**, *2*, 280–284.

3. Agamuthu, P. Mercury—The Real Story. *Waste Manage. Res.* **2013**, *31*, 233–234.

4. Alturiqi, A. S.; Albedair, L. A. Evaluation of Some Heavy Metals in Certain Fish, Meat and Meat Products in Saudi Arabian Markets. *Egyptian J. Aquat. Res.* **2012**, *38*, 45–49.

5. Andrée, S.; Jira, W.; Schwind, K. H.; Wagner, H.; Schwägele, F. Chemical Safety of Meat and Meat Products. *Meat Sci.* **2010**, *86*, 38–48.

6. Alexander, J.; Benford, D.; Cockburn, A.; Cravedi, J. P.; Dogliotti, E.; Di Domenico, A.; et al. Scientific Opinion Cadmium in Food. *EFSA J.* **2009**, *980*, 1–139.

7. Arnich, N.; Sirot, V.; Riviere, G.; Jean, J.; Noël, L.; Guérin, T.; Leblanc, J. C. Dietary Exposure to Trace Elements and Health Risk Assessment in the 2nd French Total Diet Study. *Food Chem. Toxicol.* **2012**, *50*, 2432–2449.

8. ATSDR. *Toxicological Profile for Lead*. Agency for Toxic Substances and Disease Registry (US), 2007.

9. Agency for Toxic Substances and Disease Registry (ATSDR). *Toxicological Profiles*. Agency for Toxic Substances and Disease Registry, 2005.

10. Aung, N. N.; Yoshinaga, J.; Takahashi, J. I. Dietary Intake of Toxic and Essential Trace Elements by the Children and Parents Living in Tokyo Metropolitan Area, Japan. *Food Addit. Contam.* **2006**, *23*, 883–894.

11. Barone, G.; Storelli, A.; Garofalo, R.; Busco, V. P.; Quaglia, N. C.; Centrone, G.; Storelli, M. M. Assessment of Mercury and Cadmium via Seafood Consumption in Italy: Estimated Dietary Intake (EWI) and Target Hazard Quotient (THQ). *Food Addit. Contam., A: Chem. Anal. Control Expo. Risk Assess.* 2015, *32* (8), 1277–1286. DOI:10 .1080/19440049.2015.1055594.

12. Becaria, A.; Campbell, A.; Bondy, S. C. Aluminum as a Toxicant. *Toxicol. Ind. Health* **2002**, *18*, 309–320.

13. Becker, W.; Kumpulainen, J. Contents of Essential and Toxic Mineral Elements in Swedish Market-Basket Diets in 1987. *Br. J. Nutr.* **1991**, *66*, 151–160.

14. Bhattacharya, P.; Welch, A. H.; Stollenwerk, K. G.; McLaughlin, M. J.; Bundschuh, J.; Panaullah, G. Arsenic in the Environment: Biology and Chemistry. *Sci. Total Environ.* **2007**, *379*, 109–120.

15. Biego, G. H.; Joyeux, M.; Hartemann, P.; Debry, G. Daily Intake of Essential Minerals and Metallic Micropollutants from Foods in France. *Sci. Total. Environ.* **1998**, *217*, 27–36.

16. Blanusa, M.; Juresa, D. Lead, Cadmium, and Mercury Dietary Intake in Croatia. *Arh. Hig. Rad. Toksikol.* **2001**, *52*, 229–237.

17. Brandon, E. F.; Janssen, P. J.; de Wit-Bos, L. Arsenic: Bioaccessibility from Seaweed and Rice, Dietary Exposure Calculations and Risk Assessment. *Food Addit. Contam. A: Chem. Anal. Control Expo. Risk Assess.* **2014**, *31*, 1993–2003.

18. Brombach, C. C.; Manorut, P.; Kolambage-Dona, P. P.; Ezzeldin, M. F.; Chen, B.; Corns, W. T.; Feldmann, J.; Krupp, E. M. Methylmercury Varies more than One Order of Magnitude in Commercial European Rice. *Food Chem.* **2017**, *214*, 360–365. DOI:10.1016/j.foodchem.2016.07.064.

19. Caito, S.; Aschner, M. Neurotoxicity of Metals. *Handb. Clin. Neurol.* **2015**, *131*, 169–189. DOI:10.1016/B978-0-444-62627-1.00011-1.

20. Campbell, A. The Potential Role of Aluminium in Alzheimer's Disease. *Nephrol. Dial Transpl.* **2002**, *17* (Suppl. 2), 17–20.

21. Canadian Food Inspection Agency (CFIA). Contaminants—Western Area—Mercury Sampling. *HC/CFA Fish Chemistry Working Group Meeting*, February 26, 2003.

22. Canfield, R. L.; Kreher, D. A.; Cornwell, C.; Henderson, Jr.; C. R. Low-Level Lead Exposure, Executive Functioning, and Learning in Early Childhood. *Child Neuropsychol.* **2003**, *9* (1), 35–53.

23. Chan, H. M.; Kim, C.; Khoday, K.; Receveur, O.; Kuhnlen, H. V. Assessment of Dietary Exposure to Trace Metals in Baffin Inuit Food. *Environ. Health Perspect.* **1995**, *103*, 740–746.

24. Chappard, D.; Bizot, P.; Mabilleau, G.; Hubert, L. Aluminum and bone: Review of New Clinical Circumstances Associated with Al^{3+} Deposition in the Calcified Matrix of Bone. *Morphologie* **2016**, *100*, 95–105.

25. Coelho, S. D.; Pastorinho, M. R.; Itai, T.; Isobe, T.; Kunisue, T.; Nogueira, A. J.; Tanabe, S.; Sousa, A. C. Lead in Duplicate Diet Samples from an Academic Community. *Sci. Total Environ.* **2016**, *573*, 603–607. DOI:10.1016/j.scitotenv.2016.08.133.

26. Coni, E.; Baldini, M.; Stacchini, P.; Zanasi, F. Cadmium Intake with Diet in Italy: A Pilot Study. *J. Trace Elem. Electrolytes Health Dis.* **1992**, *6*, 175–181.

27. COT (Committee on Toxicity of Chemicals in Food, Consumer Products and the Environment, UK). *Statement on Arsenic in Food: Results of the 1999 Total Diet Study*, 2003.

28. Cressey, P.; Vannoort, R.; Fowles, J. *1997/98 New Zealand Total Diet Survey. Parts 2: Elements. Selected Contaminants and Nutrients.* Ministry of Health: Wellington, New Zealand, 2000.

29. Dabeka, R. W.; McKenzie, A. D.; Lacroix, G. M. Dietary Intakes of Lead, Cadmium, Arsenic and Fluoride by Canadian Adults: a 24-hour Duplicate Diet Study. *Food Addit. Contam.* **1987**, *4*, 89–101.

30. Dabeka, R. W.; McKenzie, A. D. Survey of Lead, Cadmium, Fluoride, Nickel, and Cobalt in Food Composites and Estimation of Dietary Intakes of these Elements by Canadians in 1986–1988. *J. AOAC Int.* **1995**, *78*, 897–909.

31. Damin, I. C. F.; Silva, M. M.; Vale, M. G. R.; Welz, B. Feasibility of Using Direct Determination of Cadmium and Lead in Fresh Meat by Electrothermal Atomic Absorption Spectrometry for Screening Purposes. *Spec. Acta—B* **2007**, *62*, 1037–1045.

32. De Meeûs, C.; Eduljee, G. H.; Hutton, M. Assessment and Management of Risks Arising from Exposure to Cadmium in Fertilisers. I. *Sci. Total Environ.* **2002**, *291*, 167–187.

33. DeVoto, E.; Yokel, R. A. The Biological Speciation and Toxicokinetics of Aluminum. *Environ. Health Perspect.* **1994**, *102*, 940–951.

34. Dilek, D.; Kadiriye, U. Comparative Study of Trace Elements in Certain Fish, Meat and Meat Products. *Meat Sci.* **2006**, *74*, 255–260.

35. Dougherty, C. P.; Henricks Holtz, S.; Reinert, J. C.; Panyacosit, L.; Axelrad, D. A.; Woodruff, T. J. Dietary Exposures to Food Contaminants across the United States. *Environ. Res.* **2000**, *84*, 170–185.

36. Dueñas, A.; Martín, J. C.; González, M. A. *Arsénico*. In *Intoxicaciones agudas en medicina de urgencia y cuidados críticos*; Aparicio, M., Armentia, A., Belcher, L., et al., Ed.; Masson: Barcelona, 2001; pp 171–174.

37. EFSA. Scientific Opinion of the Panel on Contaminants in the Food Chain. Cadmium in food. *EFSA J.* **2009**, *980*, 1–139.

38. EFSA Panel on Contaminants in the Food Chain. Scientific Opinion on Arsenic in Food. *EFSA J.* **2009,** *7*, 1351 [199 pp].

39. EFSA Panel on Contaminants in the Food Chain (CONTAM). Scientific Opinion on Lead in Food. *EFSA J.* **2010,** *8* (4), 1570 [151 pp].

40. EFSA (European Food Safety Authority). Cadmium Dietary Exposure in the European Population. *EFSA J.* **2012,** *10* (1), 2551 [37 pp].

41. EFSA Panel on Contaminants in the Food Chain (CONTAM). Scientific Opinion on the Risk for Public Health Related to the Presence of Mercury and Methylmercury in Food. *EFSA J.* **2012,** *10* (12), 2985 [241 pp].

42. EFSA Scientific Committee. Statement on the Benefits of Fish/Seafood Consumption Compared to the Risks of Methylmercury in Fish/Seafood. *EFSA J.* **2015,** *13*, 3982 [36 pp].

43. Ellen, G.; Egmond, E.; Van Loon, J. W.; Sahertian, E. T.; Tolsma, K. Dietary Intakes of Some Essential and Non-essential Trace Elements, Nitrate, Nitrite and *N*-Nitrosamines, by Dutch Adults: Estimated via a 24-hour Duplicate Portion Study. *Food Addit. Contam.* **1990**, *7*, 207–221.

44. Ellenhorn, J.; Schonwald, S.; Ordog, G.; Wasserberger, J. *Ellenhorn's Medical Toxicology.* William & Wilkins: Philadelphia, PA, 1997.

45. Ercal, N.; Gurer-Orhan, H.; Aykin-Burns, N. Toxic Metals and Oxidative Stress, Part I: Mechanisms Involved in Metal-Induced Oxidative Damage. *Curr. Top. Med. Chem.* **2001**, *1* (6), 529–539.

46. Eriksen, K. T.; Halkjær, J.; Sørensen, M.; Meliker, J. R.; McElroy, J. A.; Tjønneland, A.; Raaschou-Nielsen, O. Dietary Cadmium Intake and Risk of Breast, Endometrial and Ovarian Cancer in Danish Postmenopausal Women: A Prospective Cohort Study. *PLoS ONE* **2014**, *9*, e100815.

47. Erzen, I.; Ursic, S.; Bosnjak, K. Assessment of Dietary Intake of Cadmium, Lead and Mercury via Foods of the Plant and Animal Origin in Slovenia. *Med. Arch.* **2002**, *56*, 105–109.

48. Fekete, V.; Vandevijvere, S.; Bolle, F.; Van Loco, J. Estimation of Dietary Aluminum Exposure of the Belgian Adult Population: Evaluation of Contribution of Food and Kitchenware. *Food Chem. Toxicol.* **2013**, *55*, 602–608.

49. Ferrer, A. Metal Poisoning. Unidad de Toxicología Clínica. Hospital Clínico Universitario de Zaragoza. *ANALES Sis San Navarra* **2003**, *26*, 141–153.

50. FEHD (Centre for Food Safety of the Food and Environmental Hygiene Department (FEHD) of the Government of the Hong Kong Special Administrative Region). *The First Hong Kong Total Diet Study: Inorganic Arsenic*, 2012.

51. Food Standards Australia New Zealand. *24th Australian Total Diet Study*. 2014. file:///C:/Users/Usuario/Documents/ASUS%20ANTIGUO/Mis%20documentos/bibliografia/metales/plomo/Australian%20TDS%202014.pdf.

52. Frías, I.; Rubio, C.; González-Iglesias, T.; Gutiérrez, A. J.; González-Weller, D.; Hardisson, A. Metals in Fresh Honeys from Tenerife Island, Spain. *Bull. Environ. Contam. Toxicol.* **2008**, *80*, 30–33.

53. Frisardi, V.; Solfrizzi, V.; Capurso, C. et al. Aluminum in the Diet and Alzheimer's Disease: From Current Epidemiology to Possible Disease-Modifying Treatment. *J. Alzheimers Dis.* **2010**, *20*, 17–30.

54. FSA (Food Standards Agency). Survey on Measurement of the Concentrations of Metals and Other Elements from the 2006 UK Total Diet Study, 2009. http://www.food. gov.uk/multimedia/pdfs/fsis0909metals.pdf.

55. FSAI (Food Safety Authority of Ireland). *Report on a Total Diet Study Carried Out by the Food Safety Authority of Ireland in the Period 2012–2014*, 2016.

56. FSANZ (Food Standards Australia New Zealand). *20th Australian Total Diet Study*, 2003. http://www.foodstandards.gov.au/_srcfiles/Final_20th_Total_Diet_Survey.pdf.

57. FSANZ (Food Standards Australia New Zealand). *23rd Australian Total Diet Study*, 2011. http://www.foodstandards.gov.au/_srcfiles/FSANZ%2023rd%20ATDS_v5.pdf.

58. Galal-Gorchev, H. Dietary Intake, Levels in Food and Estimated of Lead, Cadmium, and Mercury. *Food Addit. Contam.* **1997**, *10*, 115–128.

59. García Ariño, C. *Síntesis diagnóstica de las enfermedades profesionales producidas por metales, productos químicos y agentes vivos.* MAPFRE: Madrid, 1996.

60. Gilani, S. R.; Zaidi, S. R.; Batool, M.; Bhatti, A. A.; Durrani, A. I.; Mahmood, Z. Report: Central Nervous System (CNS) Toxicity Caused by Metal Poisoning: Brain as a Target Organ. *Pak. J. Pharm. Sci.* **2015**, *28* (4), 1417–1423.

61. Gimou, M. M.; Pouillot, R.; Charrondiere, U. R.; Noël, L.; Guérin, T.; Leblanc, J. C. Dietary Exposure and Health Risk Assessment for 14 Toxic and Essential Trace Elements in Yaoundé: the Cameroonian Total Diet Study. *Food Addit. Contam., A Chem. Anal. Control Expo. Risk Assess.* **2014**, *31*, 1064–1080.

62. González Weller, D.; Rubio Armendáriz, C.; Revert Jirones, C.; Hardisson de la Torre, A. El arsénico y la salud pública. *Atención Farmacéut.* **2003**, *julio–agosto*, 227–232.

63. González Weller, D.; Rubio, C.; Revert, C.; Hardisson, A. El arsénico en los alimentos. *Alimentaria* **2003**, *347*, 21–27.

64. González-Weller, D.; Karlsson, L.; Caballero, A.; Hernández, F.; Gutiérrez, A. J.; et al. Lead and Cadmium in Meat and Meat Products Consumed by a Spanish Population (Tenerife Island, Spain). *Food Addit. Contam.* **2006**, *23*, 757–763.

65. González-Weller, D.; Gutiérrez, A. J.; Rubio, C.; Revert, C.; Hardisson, A. Dietary Intake of Aluminum in a Spanish Population (Canary Islands). *J. Agric. Food Chem.* **2010**, *58*, 10452–10457.

66. Gonzalez-Weller, D.; Rubio, C.; Gutiérrez, A. J.; Pérez, B.; Hernández-Sánchez, C.; Caballero, J. M.; Revert, C.; Hardisson, A. Dietary Content and Evaluation of Metals in Four Types of Tea (White, Black, Red and Green) Consumed by the Population of the Canary Islands. *Pharm. Anal. Acta* **2015**, *6*, 1–10.

67. Gramiccioni, L.; Ingrao, G.; Milana, M. R.; Santaroni, P.; Tomassi, G. Aluminium Levels in Italian Diets and in Selected Foods from Aluminium Utensils. *Food Addit. Contam.* **1996**, *13*, 767–774.

68. Greger, J. L. Dietary and Other Sources of Aluminium Intake. *Ciba Found Symp.* **1992**, *169*, 26–35.

69. Greger, J. L.; Sutherland, J. E. Aluminum Exposure and Metabolism. *Crit. Rev. Clin. Lab. Sci.* **1997**, *34*, 439–474.

70. Gribble, M. O.; Karimi, R.; Feingold, B. J.; et al. Mercury, Selenium and Fish Oils in Marine Food Webs and Implications for Human Health. *J. Mar. Biol. Assoc. UK* **2016**, *96*, 43–59.

71. Gunderson, E. L. FDA Total Diet Study, July 1986–April 1991, Dietary Intakes of Pesticides, Selected Elements, and Other Chemicals. *J. AOAC Int.* **1995**, *78* (6), 1353–1363.

72. Gutiérrez, A. J.; González-Weller, D.; González, T.; Burgos, A.; Lozano, G.; et al. Content of Toxic Heavy Metals (Hg, Pb, Cd) in Canned Variegated Scallops (*Chlamys varia*). *J. Food Prot.* **2007**, *70*, 2911–2915.

73. Hardisson, A.; Rubio, C.; Gutiérrez, A.; Jalili, A.; Hernández-Sánchez, C.; Lozano, G.; Revert, C.; Hernández-Armas, J. Total Mercury in Aquaculture Fish. *Pol. J. Environ. Stud.* **2012**, *21*, 79–85.

74. Hassan, A. R.; Zeinhom, M. M.; Abdel-Wahab, M. A.; Tolba, M. H. Heavy Metal Dietary Intake and Potential Health Risks for University Hostel Students. *Biol. Trace Elem. Res.* **2016**, *170* (1), 65–74.

75. He, P.; Lu, Y.; Liang, Y.; et al. Exposure Assessment of Dietary Cadmium: Findings from Shanghainese over 40 years, China. *BMC Public Health* **2013**, *13*, 590.

76. Hellström, L.; Persson, B.; Brudin, L.; Grawé, K. P.; Öborn, I.; Järup, L. Cadmium Exposure Pathways in a Population Living near a Battery Plant. *Sci. Total Environ.* **2007**, *373*, 447–455.

77. Health Canada. *Human Health Risk Assessment of Mercury in Fish and Health Benefits of Fish Consumption.* Bureau of Chemical Safety Food Directorate Health Products and Food Branch, 2007. http://hc-sc.gc.ca/fn-an/pubs/mercur/merc_fish_poisson_e.html.

78. Health Canada. *Food and Nutrition—Lead*, 2011. http://www.hc-sc.gc.ca/fnan/securit/chem-chim/environ/lead_plomb-eng.php.

79. Health Canada. *Final Human Health State of the Science Report on Lead*, 2013. http://www.hc-sc.gc.ca/ewh-semt/pubs/contaminants/dhhssrl-rpecscepsh/index-eng.php.

80. Herce-Pagliai, C.; Cameán, A.; Repetto, M. Interés toxicológico de la especiación de arsénico. *Rev. Toxicol.* **1998**, *15*, 3–11.

81. Hettick, B. E.; Cañas-Carrell, J. E.; French, A. D.; Klein, D. M. Arsenic: A Review of the Element's Toxicity, Plant Interactions, and Potential Methods of Remediation. *J. Agric. Food Chem.* **2015**, *63*, 7097–7107.

82. Hewitt, C. D.; Savory, J.; Wills, M. R. Aspects of Aluminum Toxicity. *Clin. Lab. Med.* **1990**, *10*, 403–422.

83. Hollenberg, P. F. Introduction: Mechanisms of Metal Toxicity Special Issue. *Chem. Res. Toxicol.* **2010**, *23*, 292–293.

84. IARC (International Agency for Research on Cancer). *Summaries & Evaluations Cadmium and Cadmium compounds*; 1993; vol. 58, p 119. Available at http://www.inchem.org/documents/iarc/vol58/mono58-3.html.

85. IARC (International Agency for Research on Cancer). *Summaries & Evaluations Mercury and Mercury Compounds*; 1993, vol. 58, p 239. Available at http://www.inchem.org/documents/iarc/vol58/mono58-3.html.

86. IARC. *IARC Monographs on the Evaluation of Carcinogenic Risks to Humans Volume 100C Arsenic, Metals, Fibres and Dusts*, 2012. Available at http://monographs.iarc.fr/ENG/Monographs/vol100C/.

87. Ibrahim, D.; Froberg, B.; Wolf, A.; Rusyniak, D. E. Heavy Metal Poisoning: Clinical presentations and pathophysiology. *Clin. Lab. Med.* **2006**, *26*, 67–97.

88. Ihedioha, J. N.; Okoye, C. O. B. Cadmium and Lead Levels in Muscle and Edible Offal of Cow Reared in Nigeria. *Bull. Environ. Contam. Toxicol.* **2012**, *88*, 422–427.

89. Ikeda, M.; Zhang, Z. W.; Moon, C. S.; Imai, Y.; Watanabe, T.; Shimbo, S.; Guo, Y. L. Background exposure of general population to cadmium and lead in Tainan city, Taiwan. *Arch. Environ. Contam. Toxicol.* **1996**, *30*, 121–126.

90. Ilmiawati, C.; Yoshida, T.; Itoh, T.; et al. Biomonitoring of Mercury, Cadmium, and Lead Exposure in Japanese Children: A Cross-Sectional Study. *Environ. Health Prevent. Med.* **2015**, *20* (1), 18–27.

91. Irfana, M.; Iqbal, S.; Nagra, S. A. Minerals in Meat. *Int. J. Agric. Biol.* **2004**, *6*, 816–841.

92. Jaishankar, M.; Tseten, T.; Anbalagan, N.; Mathew, B. B.; Beeregowda, K. N. Toxicity, Mechanism and Health Effects of Some Heavy Metals. *Interdiscipl. Toxicol.* **2014**, *7* (2), 60–72. DOI:10.2478/intox-2014–0009.

93. Jarzyńska, G.; Falandysz, J. Selenium and 17 Other Largely Essential and Toxic Metals in Muscle and Organ Meats of Red Deer (*Cervus elaphus*)—Consequences to Human Health. *Environ. Int.* **2011**, *37*, 882–888.

94. Järup, L. Hazards of Heavy Metals Contamination. *Br. Med. Bull.* **2003**, *68*, 167–182.

95. JECFA (Joint FAO/WHO Expert Committee on Food Additives). *Summary and Conclusions of the Sixty-First Meeting of the Joint FAO/WHO Expert Committee on Food Additives (JECFA)*, 2003; pp 18–22.

96. Jenssen, M. T.; Brantsaeter, A. L.; Haugen, M.; Meltzer, H. M.; Larssen, T.; Kvalem, H. E.; Birgisdottir, B. E.; Thomassen, Y.; Ellingsen, D.; Alexander, J.; Knutsen, H. K. Dietary Mercury Exposure in a Population with a Wide Range of Fish Consumption—Self-capture of Fish and Regional Differences are Important Determinants of Mercury in Blood. *Sci. Total Environ.* **2012**, *439*, 220–229.

97. Johansen, P.; Pars, T.; Bjerregaard, P. Lead, Cadmium, Mercury and Selenium Intake by Greenlanders from Local Marine Food. *Sci. Total Environ.* **2000**, *245*, 187–194.

98. Jones, K. C.; Bennett, B. G. Exposure of Man to Environmental Aluminium—An Exposure Commitment Assessment. *Sci. Total Environ.* **1986**, *52*, 65–82.

99. Julin, B.; Wolk, A.; Johansson, J.-E.; Andersson, S.-O.; Andrén, O.; Åkesson, A. Dietary Cadmium Exposure and Prostate Cancer Incidence: A Population-Based Prospective Cohort Study. *Br. J. Cancer* **2012**, *107* (5), 895–900.

100. Kandimalla, R.; Vallamkondu, J.; Corgiat, E. B.; Gill, K. D. Understanding Aspects of Aluminum Exposure in Alzheimer's Disease Development. *Brain Pathol.* **2016**, *26*, 139–154.

101. Karavoltsos, S.; Sakellari, A.; Dimopoulos, M.; Dasenakis, M.; Scoullos, M. Cadmium Content in Foodstuffs from the Greek Market. *Food Addit. Contam.* **2002**, *19*, 954–962.

102. Kawahara, M.; Kato-Negishi, M. Link between Aluminum and the Pathogenesis of Alzheimer's Disease: The Integration of the Aluminum and Amyloid Cascade Hypotheses. *Int. J. Alzheimers Dis.* **2011**, *2011*, 276393.

103. Kim, H. C.; Jang, T. W.; Chae, H. J.; et al. Evaluation and Management of Lead Exposure. *Ann. Occup. Environ. Med.* **2015**, *27*, 30.

104. Kim, C.; Lee, J.; Kwon, S.; Yoon, H.-J. Total Diet Study: For a Closer-to-Real Estimate of Dietary Exposure to Chemical Substances. *Toxicol. Res.* **2015**, *31* (3), 227–240.

105. Kim, K. H.; Kabir, E.; Jahan, S. A. A Review on the Distribution of Hg in the Environment and its Human Health Impacts. *J. Hazard Mater.* **2016**, *306*, 376–385.

106. Klaassen, C. D. In *Metales pesados y sus antagonistas*. In *Las bases farmacológicas de la terapéutica*; Hardman, J., Limbird, L., Molinoff, P., Ruddon, R., Goodman, A., Ed.; Goodman & Gilman, McGraw-Hill Interamericana: México, 1996; pp 1755–1779.

107. Klaassen, C. D.; Watkins, J. B. *Manual de toxicología*. McGraw-Hill Interamericana: México, 2001.

108. Kumar, V.; Gill, K. D. Aluminium Neurotoxicity: Neurobehavioural and Oxidative Aspects. *Arch. Toxicol.* **2009**, *83*, 965–978.

109. Kurasaki, M.; Hartoto, D. I.; Saito, T.; Suzuki-Kurasaki, M.; Iwakuma, T. Metals in Water in the Central Kalimantan, Indonesia. *Bull. Environ. Contam. Toxicol.* **2000,** *65,* 591–597.

110. Kwon, Y. M.; Lee, H. S.; Yoo, D. C.; Kim, C. H.; Kim, G. S.; Kim, J. A.; Lee, Y. N.; Kim, Y. S.; Kang, K. M.; No, K. M.; Paek, O. J.; Seo, J. H.; Choi, H.; Park, S. K.; Choi, D. M.; Kim, D. S, Choi, D. W. Dietary Exposure and Risk Assessment of Mercury from the Korean Total Diet Study. *J. Toxicol. Environ. Health. A* **2009,** *72* (21–22), 1484–1492.

111. Langauer-Lewowicka, H. Aluminum Neurotoxicity. *Neurol. Neurochir. Pol.* **1994,** *28,* 221–226.

112. Larsen, E. H.; Andersen, N. L.; Moller, A.; Petersen, A.; Mortensen, G. K.; Petersen, J. Monitoring the Content and Intake of Trace Elements from Food in Denmark. *Food Addit. Contam.* **2002,** *19,* 33–46.

113. Leblanc, J.; Guérin, T.; Noël, L.; Calamassi-Tran, G.; Volatier, J. L.; Verger, P. Dietary Exposure Estimates of 18 Elements from the 1st French Total Diet Study. *Food Addit. Contam.* **2005,** *22,* 624–641.

114. Lee, H. S.; Cho, Y. H.; Park, S. O.; Kye, S. H.; Kim, B. H.; Hahm, T. S.; Kim, M.; Lee, J. O.; Kim, C. Dietary Exposure of the Korean Population to Arsenic, Cadmium, Lead and Mercury. *J. Food Compost. Anal.* **2006,** *19,* S31–S37.

115. Lee, J.-G.; Kim, S.-H.; Kim, H.-J.; Yoon, H.-J. Total Diet Studies as a Tool for Ensuring Food Safety. *Toxicol. Res.* **2015,** *31* (3), 221–226.

116. Lerner, A. Aluminum is a Potential Environmental Factor for Crohn's Disease Induction: Extended Hypothesis. *Ann. N.Y. Acad. Sci.* **2007,** *1107,* 329–345.

117. Li, X.; Liu, Q.; Liu, L.; Wu, Y. Application of the Data from China Total Diet Study to Assess the Distribution of Lead Exposure in Different Age-Gender Population Groups. *J. Hyg. Res.* **2012,** *41,* 379–384.

118. Lièvremont, D.; Bertin, P. N.; Lett, M. C. Arsenic in Contaminated Waters: Biogeochemical Cycle, Microbial Metabolism and Biotreatment Processes. *Biochimie* **2009,** *91,* 1229–1237.

119. Linder, M. C. *Nutritional Biochemistry and Metabolism: With Clinical Applications.* Elsevier: New York, 2008; pp 603.

120. Liu, S. M.; Chung, C.; Chan, C. C. Daily Dietary Intake of Pratas Islands in the South China Sea. *J. Radioanal. Nucl. Chem.* **1992,** *162,* 363.

121. Llobet, J. M.; Granero, S.; Schuhmacher, M.; Corbella, J.; Domingo, J. L. Biological Monitoring of Environmental Pollution and Human Exposure to Metals in Tarragona, Spain. IV. Estimation of the Dietary Intake. *Trace Elem. Electr.* **1998,** *15,* 136–141.

122. Llobet, J. M.; Falcón, G.; Casas, C.; Teixidó, A.; Domingo, J. L. Concentrations of Arsenic, Cadmium, Mercury and Lead in Common Foods and Estimated Daily Intake by Children, Adolescents, Adults and Senior of Catalonia, of Spain. *J. Agric. Food Chem.* **2003,** *51,* 838–842.

123. Llop, S.; Guxens, M.; Murcia, M.; Lertxundi, A.; Ramon, R.; Riaño, I.; Rebagliato, M.; Ibarluzea, J.; Tardon, A.; Sunyer, J.; Ballester, F.; INMA Project. Prenatal Exposure to Mercury and Infant Neurodevelopment in a Multicenter Cohort in Spain: Study of Potential Modifiers. *Am. J. Epidemiol.* **2012,** *175,* 451–465.

124. López Alonso, M.; Miranda, M.; Castillo, C.; Hernández, J.; Benedito, J. L. Interacción entre metales tóxicos y esenciales en ganado vacuno de Galicia. *Rev. Toxicol.* **2002,** *19,* 69–72.

125. López Alonso, M.; Benedito, J. L.; Miranda, M.; Castillo, C.; Hernández, J.; Shore, R. F. Arsenic, Cadmium, Lead, Copper and Zinc in Cattle from Galicia, NW Spain. *Sci. Tot. Environ.* **2000**, *246*, 237–248.

126. López Alonso, M.; Prieto, M. F.; Miranda, M.; Castillo, C.; Hernández, J.; Benedito, J. L. Interactions between Toxic (As, Cd, Hg and Pb) and Nutritional Essential (Ca, Co, Cr, Cu, Fe, Mn, Mo, Ni, Se, Zn) Elements in the Tissues of Cattle from NW Spain. *Biometals* **2004**, *17*, 389–397.

127. López-Artíguez, M.; Repetto, M. *Estado actual de la toxicología del cadmio.* In *Toxicología Avanzada*; Repetto, M., Ed.; Díaz de Santos: Madrid, 1995; pp 393–423.

128. MacIntosh, D. L.; Spengler, J. D.; Ozkaynak, H.; Tsai, L.; Ryan, P. B. Dietary Exposures to Selected Metals and Pesticides. *Environ. Health Perspect.* **1996**, *104* (2), 202–209.

129. Mahaffey, K.; McKinney, J.; Reigart, J. R. Lead and Compounds. In *Environmental Toxicants, Human Exposure and their Health Effects*; Lippmann, M., Ed.; John Wiley and Sons: New York, 2000; pp 481–482.

130. Malluche, H. H. Aluminium and Bone Disease in Chronic Renal Failure. *Nephrol. Dial Transpl.* **2002**, *17* (Suppl. 2), 21–24.

131. Marcus, D. L.; Wong, S.; Freedman, M. L. Dietary Aluminium and Alzheimer's Disease. *J. Nutr. Elderly* **1992**, *12*, 55–61.

132. Mariusz, R. The Analysis of Correlations between the Age and the Level of Bioaccumulation of Heavy Metals in Tissues and the Chemical Composition of Sheep Meat from the Region in SE Poland. *Food Chem. Toxicol.* **2009**, *47*, 1117–1122.

133. Martyn, C. N.; Coggon, D. N.; Inskip, H.; Lacey, R. F.; Young, W. F. Aluminum Concentrations in Drinking Water and Risk of Alzheimer's Disease. *Epidemiology* **1997**, *8*, 281–286.

134. Meiri, H.; Banin, E.; Roll, M. Aluminium Ingestion—Is It Related to Dementia? *Rev. Environ. Health* **1991**, *9*, 191–205.

135. Metwally, F. M.; Mazhar, M. S. Effect of Aluminium on the Levels of Some Essential Elements in Occupationally Exposed Workers. *Arh. Hig. Rada. Toksikol.* **2007**, *58*, 305–311.

136. Miranda, M.; López Alonso, M.; Castillo, C.; Hernández, J.; Prieto, F.; Benedito, J. L. Some Toxic Elements in Liver, Kidney and Meat from Calves Slaughtered in Asturias (Northern Spain). *Eur. Food Res. Technol.* **2003**, *216*, 284–289.

137. Molloy, D. W.; Standish, T. I.; Nieboer, E.; Turnbull, J. D.; Smith, S. D.; Dubois, S. Effects of Acute Exposure to Aluminum on Cognition in Humans. *J. Toxicol. Environ. Health A* **2007**, *70*, 2011–2019.

138. Molin, M.; Ulven, S. M.; Meltzer, H. M.; Alexander, J. Arsenic in the Human Food Chain, Biotransformation and Toxicology—Review Focusing on Seafood Arsenic. *J. Trace Elem. Med. Biol.* **2015**, *31*, 249–259.

139. Moon, C. S.; Zhang, Z. W.; Shimbo, S.; Watanabe, T.; Moon, D. H.; Lee, C. U.; Lee, B. K.; Ahn, K. D.; Lee, S. H.; Ikeda, M. Dietary Intake of Cadmium and Lead among the General Population in Korea. *Environ. Res.* **1995**, *71*, 46–54.

140. Morrow, H. Cadmium and Cadmium Alloys. In *Encyclopedia of Chemical Technology*; Kirk-Othmer, Ed.; John Wiley & Sons, Inc.: New York, 2001; p 471–507.

141. Mozaffarian, D. Fish, Mercury, Selenium and Cardiovascular Risk: Current Evidence and Unanswered Questions. *Int. J. Environ. Res. Public Health* **2009**, *6*, 1894–1916.

142. Muñoz, O.; Bastias, J. M.; Araya, M.; Morales, A.; Orellana, C.; Rebolledo, R.; et al. Estimation of the Dietary Intake of Cadmium, Lead, Mercury and Arsenic by the

Population of Santiago (Chile) Using a Total Diet Study. *Food Chem. Toxicol.* **2005,** *43*, 1647–1655.

143. Myers, G. J.; Thurston, S. W.; Pearson, A. T.; Davidson, P. W.; Cox, C.; Shamlaye, C. F.; Cernichiari, E.; Clarkson, T. W. Postnatal Exposure to Methyl Mercury from Fish Consumption: A Review and New Data from the Seychelles Child Development Study. *Neurotoxicology* **2009,** *30*, 338–349.

144. Najarnezhad, V.; Jalizadeh-Amin, G.; Anassori, E.; Zeinali, V. Lead and Cadmium in Raw Buffalo, Cow and Ewe Milk from West Azerbaijan, Iran. *Food Addit. Contam., B: Surveill.* **2015,** *25*, 1–5.

145. Nasreddine, L.; Hwalla, N.; El Samad, O.; Le Blanc, J. C.; Hamze, M.; Sibiril, Y.; Parent-Massin, D. Dietary Exposure to Lead, Cadmium, Mercury and Radionuclides of an Adult Urban Population in Lebanon: A Total Diet Study Approach. *Food Addit. Contam.* **2006,** *23*, 579–590.

146. Noël, L.; Leblanc, J. C.; Guerín, T. Determination of Several Elements in Duplicate Meals from Catering Establishments Using Closed Vessel Microwave Digestion with Inductively Couple Plasma Mass Spectrometry Detection: Estimation of Daily Dietary Intake. *Food Addit. Contam.* **2003,** *20*, 44–56.

147. Nriagu, J. O. A History of Global Metal Pollution. *Science* **1996,** *272*, 223–224.

148. Ogimoto, M.; Suzuki, K.; Haneishi, N.; Kikuchi, Y.; Takanashi, M.; Tomioka, N.; Uematsu, Y.; Monma, K. Aluminium Content of Foods Originating from Aluminium-Containing Food Additives. *Food Addit. Contam., B: Surveill.* **2016,** *9* (3), 185–190.

149. Onianwa, P. C.; Lawal, J. A.; Ogunkeye, A. A.; Orejimi, B. M. Cadmium and Nickel Composition of Nigerian Foods. *J. Food Compos. Anal.* **2000,** *13*, 961–969.

150. Park, J. H.; Hwang, M. S.; Ko, A.; Jeong, D. H.; Kang, H. S.; Yoon, H. J.; Hong, J. H. Total Mercury Concentrations in the General Korean Population, 2008–2011. *Regul. Toxicol. Pharmacol.* **2014,** *70* (3), 681–686.

151. Pennington, J. A. T. Aluminium Content of Foods and Diets. *Food Addit. Contam.* **1987,** *5*, 161–232.

152. Pennington, J. A.; Schoen, S. A. Estimates of Dietary Exposure to Aluminium. *Food Addit. Contam.* **1995,** *12*, 119–128.

153. Perello, G.; Llobet, J. M.; Gómez-Catalán, J.; et al. Human Health Risks Derived from Dietary Exposure to Toxic Metals in Catalonia, Spain: Temporal Trend. *Biol. Trace Elem. Res.* **2014,** *162* (1–3), 26–37.

154. Pérez, D. Mercury Levels in Mole Crabs *Hippa cubensis, Emerita brasiliensis, E. portoricensis* and *Lepidopa richmondi* (Crustacea: Decapoda: Hippidae) from a Sandy Beach at Venezuela. *Bull. Environ. Contam. Toxicol.* **1999,** *63*, 320–326.

155. Petersson, K.; Thierfelder, T.; Jorhem, L.; Oskarsson, A. Cadmium Levels in Kidneys from Swedish Pigs in Relation to Environmental Factors—Temporal and Spatial Trends. *Sci. Total Environ.* **1997,** *208*, 111–122.

156. Pilarczyk, R.; Wójcik, J.; Czerniak, P.; Sablik, P.; Pilarczyk, B.; Tomza-Marciniak, A. Concentrations of Toxic Heavy Metals and Trace Elements in Raw Milk of Simmental and Holstein—Friesian Cows from Organic Farm. *Environ. Monit. Assess.* **2013,** *185*, 8383–8392.

157. Piña, R. G.; Cervantes, C. Microbial Interactions with Aluminium. *Biometals* **1996,** *9*, 311–316.

158. Prieto Méndez, J.; González Ramírez, C. A.; Román Gutiérrez, A. D.; Prieto García, F. Plant Contamination and Phytotoxicity Due to Heavy Metals from Soil and Water. *Trop. Subtrop. Agroecosyst.* **2009,** *10*, 29–44.

159. Rahmdel, S.; Abdollahzadeh, S. M.; Mazloomi, S. M.; Babajafari, S. Daily Dietary Intakes of Zinc, Copper, Lead, and Cadmium as Determined by Duplicate Portion Sampling Combined with either Instrumental Analysis or the Use of Food Composition Tables, Shiraz, Iran. *Environ. Monit. Assess.* **2015**, *187* (5), 349.

160. Raimann, X.; Rodríguez, L.; Chávez, P.; Torrejón, C. Mercurio en pescados y su importancia en la salud. *Rev. Med. Chil.* **2014**, *142*, 1174–1180.

161. Rani, A.; Kumar, A.; Lal, A.; Pant, M. Cellular Mechanisms of Cadmium-Induced Toxicity: A Review. *Int. J. Environ. Health Res.* **2014**, *24*, 378–399.

162. Rao, J. V. B.; Vengamma, B.; Naveen, T.; Naveen, V. Lead Encephalopathy in Adults. *J. Neurosci. Rural Pract.* **2014**, *5*, 161–163.

163. Reilly, C. *Metal Contamination of Food*, 3rd ed.; Blackwell Science Ltd: London, UK, 2002.

164. Roels, H. A.; Hoet, P.; Lison, D. Usefulness of Biomarkers of Exposure to Inorganic Mercury, Lead, or Cadmium in Controlling Occupational and Environmental Risks of Nephrotoxicity. *Ren. Fail.* **1999**, *21* (3–4), 251–262.

165. Rodríguez, M. A.; Navarro, M.; Cabrera, C.; López, M. C. Elementos tóxicos en alimentos, bebidas y envases. *Alimentaria* **2001**, *322*, 23–32.

166. Rogers, M. A.; Simon, D. G. A Preliminary Study of Dietary Aluminium Intake and Risk of Alzheimer's Disease. *Age Ageing* **1999**, *28*, 205–209.

167. Rojas, E.; Herrera, L. A.; Poirier, L. A.; Ostrosky-Wegman, P. Are Metals Dietary Carcinogens? *Mut. Res.* **1999**, *443*, 157–181.

168. Röllin, H. B.; Theodorou, P.; Kilroe-Smith, T. A. The Effect of Exposure to Aluminium on Concentrations of Essential Metals in Serum of Foundry Workers. *Br. J. Ind. Med.* **1991**, *48*, 243–246.

169. Rondeau, V.; Commenges, D.; Jacqmin-Gadda, H.; Dartigues, J. F. Relation between Aluminum Concentrations in Drinking Water and Alzheimer's Disease: An 8-year Follow-Up Study. *Am. J. Epidemiol.* **2000**, *152*, 59–66.

170. Rondeau, V. A Review of Epidemiologic Studies on Aluminium and Silica in Relation to Alzheimer's Disease and Associated Disorders. *Rev. Environ. Health* **2002**, *17*, 107–121.

171. Rose, M.; Baxter, M.; Brereton, N.; Baskaran, C. Dietary Exposure to Metals and Other Elements in the 2006 UK Total Diet Study and Some Trends over the Last 30 Years. *Food Addit. Contam., A: Chem. Anal. Ctrl. Expo. Risk Assess.* **2010**, *27*, 1380–1404.

172. Rubens, O.; Logina, I.; Kravale, I.; Eglîte, M.; Donaghy, M. Peripheral Neuropathy in Chronic Occupational Inorganic Lead Exposure: A Clinical and Electrophysiological Study. *J. Neurol. Neurosurg. Psychiatry* **2001**, *71*, 200–204.

173. Rubio, C.; Frías, I.; Hardisson, A. Toxicología del plomo y su presencia en los alimentos. *Alimentaria* **1999**, *305*, 77–86.

174. Rubio, C.; González-Iglesias, T.; Revert, C.; Reguera, J. I.; Gutiérrez, A. J.; Hardisson, A. Lead Dietary Intake in a Spanish Population (Canary Islands). *J. Agric. Food. Chem.* **2005**, *53*, 6543–6549.

175. Rubio, C.; Hardisson, A.; Reguera, J. I.; Revert, C.; Lafuente, M. A.; González-Iglesias, T. Cadmium Dietary Intake in the Canary Islands, Spain. *Environ. Res.* **2006**, *100*, 123–129.

176. Rubio, C.; Gutiérrez, A.; Burgos, A.; Hardisson, A. Total Dietary Intake of Mercury in the Canary Islands, Spain. *Food Addit. Contam., A: Chem. Anal. Control Expo. Risk Assess.* **2008**, *25*, 946–952.

177. Saiyed, S. M.; Yokel, R. A. Aluminium Content of Some Foods and Food Products in the USA, with Aluminium Food Additives. *Food Addit. Contam.* **2005,** *22,* 234–244.

178. Sandbichler, A. M.; Höckner, M. Cadmium Protection Strategies—A Hidden Trade-Off? *Int. J. Mol. Sci.* **2016,** *17,* 139.

179. Sanz-Gallén, P.; Nogué, S.; Corbella, J. Metales. In *Toxicología clínica*; Marruecos, L., Nogué, S., Nolla, J.; Ed.; Springer-Verlag Ibérica: Barcelona, 1993; pp 275–291.

180. Sapunar-Postruznik, J.; Bazulic, D.; Kubala, H.; Balin, L. Estimation of Dietary Intake of Lead and Cadmium in the General Population of the Republic of Croatia. *Sci. Total Environ.* **1996,** *177,* 31–35.

181. Sarkar, A; Paul, B. The Global Menace of Arsenic and Its Conventional Remediation—A Critical Review. *Chemosphere* **2016,** *158,* 37–49.

182. Savoir, J.; Wills, M. R. Trace Metals: Essential Nutrients or Toxins. *Clin. Chem.* **1992,** *38,* 1565–1573.

183. Satarug, S.; Garrett, S. H.; Sens, M. A.; Sens, D. A. Cadmium, Environmental Exposure, and Health Outcomes. *Environ. Health. Perspect.* **2010,** *118,* 182–190.

184. Satarug S, Swaddiwudhipong W, Ruangyuttikarn W, Nishijo M, Ruiz, P. Modeling Cadmium Exposures in Low- and High-Exposure Areas in Thailand. *Environ. Health Perspect.* **2013,** *121* (5), 531–536.

185. Sato, K.; Suzuki, I.; Kubota, H.; et al. Estimation of Daily Aluminum Intake in Japan Based on Food Consumption Inspection Results: Impact of Food Additives. *Food Sci. Nutr.* **2014,** *2* (4), 389–397.

186. Schaller, K. H.; Letzel, S.; Angerer, J. *Aluminium: Handbook of Metals in Clinical and Analytical Chemistry*. Marcel Dekker: New York, 1995.

187. Schwarz, M. A.; Lindtner, O.; Blume, K.; Heinemeyer, G.; Schneider, K. Cadmium Exposure from Food: The German LExUKon Project. *Food Addit. Contam., A: Chem. Anal. Control Expo. Risk Assess.* **2014,** *31,* 1038–1051.

188. Sedki, A.; Lekouch, N.; Gamon, S.; Pineau, A. Toxic and Essential Trace Metals in Muscle, Liver and Kidney of Bovines from Polluted Area of Morocco. *Sci. Total Environ.* **2003,** *317,* 201–205.

189. Selin, N. E.; Sunderland, E. M.; Knightes, C. D.; Mason, R. P. Sources of Mercury Exposure for US Seafood Consumers: Implications for Policy. *Environ. Health Perspect.* **2010,** *118,* 137–143.

190. Sierra, A.; Hardisson, A. La contaminación química de los alimentos. Aditivos alimentarios. In *Medicina Preventiva y Salud pública*, 9th ed.; Piédrola, G., Domínguez, M., Cortina, P., et al., Eds.; Salvat: Barcelona, 1991; pp 293–303.

191. Singh, R.; Singh, S.; Parihar, P.; Singh, V. P.; Prasad, S. M. Arsenic contamination, Consequences and Remediation Techniques: A Review. *Ecotoxicol. Environ. Saf.* **2015,** *112,* 247–270.

192. Soliman, K.; Zikovsky, L. Concentrations of Al in Montreal and its Daily Dietary Intake. *J. Radioanal. Nucl. Chem.* **1999,** *212,* 807–809.

193. Soni, M. G.; White, S. M.; Flamm, W. G.; Burdock, G. A. Safety Evaluation of Dietary Aluminium. *Regul. Toxicol. Pharmacol.* **2001,** *33,* 66–79.

194. Soria, M. L.; Repetto, G.; Repetto, M. Revisión general de la toxicología de los metales. In *Toxicología Avanzada*; Camean, A., López-Artiguez, M., Martínez, D., et al., Eds.; Díaz de Santos, S. A.: Madrid, 1995; pp 293–358.

195. Starska, K. Aluminum in Food. *Rocz. Panstw. Zakl. Hig.* **1993,** *44,* 55–63.

196. Stavreva-Veselinonska, S.; Živanović, J. Lead Concentrations in Different Animal, Tissues, Muscles and Organs at Specific Localities in Probiŝtip and Its Surroundings. *Nat. Monteneg.* **2010**, *10*, 161–168.

197. Stevens, J. B. Disposition of Toxic Metals in the Agricultural Food Chain. 2. Steady-State Bovine Tissue Biotransfer Factors. *Environ. Sci. Technol.* **1992**, *26*, 1915–1921.

198. Styblo, M.; Drobna, Z.; Jaspers, I.; Lins, S.; Thomas, D. J. The Role of Biomethylation in Toxicity and Carcinogenecity of Arsenic: A Research Update. *Environ. Health. Perspect.* **2002**, *110*, 767–771.

199. Sun, J.; Wang, C.; Song, X.; Wu, Y.; Yuan, B.; Liu, P. Dietary Intake of Mercury by Children and Adults in Jinhu Area of China. *Int. J. Hyg. Environ. Health* **2011**, *214*, 246–250.

200. Tahvonen, T.; Kumpulainen, J. Contents of Lead and Cadmium in Selected Fish Species Consumed in Finland in 1993–1994. *Food Addit. Contam.* **1996**, *13*, 647–654.

201. Tang, J.; Huang, Z.; Pan, X. D. Exposure Assessment of Heavy Metals (Cd, Hg, and Pb) by the Intake of Local Foods from Zhejiang, China. *Environ. Geochem. Health* **2014**, *36*, 765–771.

202. Tao, S.; Bolger, P. M. Dietary Arsenic Intakes in the United States: FDA Total Diet Study, September 1991–December 1996. *Food Addit. Contam.* **1999**, 16 (11), 465–472.

203. Thomas, K. W.; Pellizzari, E. D.; Berry, M. R. Population-Based Dietary Intakes and Tap Water Concentrations for Selected Elements in the EPA Region V National Human Exposure Assessment Survey (NHEXAS). *J. Expo. Sci. Environ. Epidemiol.* **1999**, *9*, 402–413.

204. Tripathi, R. M.; Raghunath, R.; Krishnamoorthy, T. M. Dietary Intake of Heavy Metals in Bombay City, India. *Sci. Total Environ.* **1997**, *208*, 149–159.

205. Tripathi, R. M.; Mahapatra, S.; Raghunath, R.; Vinod Kumar, A.; Sadasivan, S. Daily Intake of Aluminium by Adult Population of Mumbai, India. *Sci. Total. Environ.* **2002**, *299*, 73–77.

206. Tsuda, T.; Inoue, T.; Kojima, M.; Aoki, S. Market Basket and Duplicate Portion Estimation of Dietary Intakes of Cadmium, Mercury, Arsenic, Copper, Manganese, and Zinc by Japanese Adults. *J. AOAC Int.* **1995**, *78* (6), 1363–1368.

207. Turconi, G.; Minoia, C.; Ronchi, A.; Roggi, C. Dietary Exposure Estimates of Twenty-One Trace Elements from a Total Diet Study Carried Out in Pavia, Northern Italy. *Br. J. Nutr.* **2009**, *101*, 1200–1208.

208. Uchida, H.; Nagai, M. Intakes and Health Effects of Aluminum. Is Aluminum a Risk Factor for Alzheimer's Disease? *Nihon Koshu Eisei Zasshi* **1997**, *44*, 671–681.

209. Vacchi-Suzzi, C.; Eriksen, K. T.; Levine, K.; et al. Dietary Intake Estimates and Urinary Cadmium Levels in Danish Postmenopausal Women; Lof, M., Ed.; *PLoS ONE* **2015**, 10 (9), e0138784.

210. Vahter, M.; Berglung, M.; Nermell, B.; Akesson, A. Bioavailability of Cadmium from Shellfish and Mixed Diet in Women. *Toxicol. Appl. Pharmacol.* **1996**, *136*, 332–341.

211. Vahter, M. Mechanisms of Arsenic Biotransformation. *Toxicology* **2002**, *181–182*, 211–217.

212. Van der Voet, G. B. Intestinal Absorption of Aluminium. *Ciba Found. Symp.* **1992**, *169*, 109–117.

213. Van Dokkum, W.; de Vos, R. H.; Muys, T.; Wesstra, J. A. Minerals and Trace Elements in Total Diets in The Netherlands. *Br. J. Nutr.* **1989**, *61*, 7–15.

214. Valko, M.; Morris, H.; Cronin, M. T. Metals, Toxicity and Oxidative Stress. *Curr. Med. Chem.* **2005**, *12* (10), 1161–208.

215. Vince, C.; Ramos, G.; Ablan-Lagman, M. C. Heavy Metal Levels in Mud Crabs (*Scylla* spp.) from East Bataan Coast. *Environ. Sci. Pollut. Res. Int.* **2015,** *22,* 6359–6363.

216. Vos, G.; Lammers, H.; Kan, C. A. Cadmium and Lead in Muscle Tissue and Organs of Broilers, Turkeys and Spent Hens in Mechanically Deboned Poultry Meat. *Food Addit. Contam.* **1990,** *7,* 83–91.

217. Vromman, V.; Waegeneers, N.; Cornelis, C.; De Boosere, I.; Van Holderbeke, M.; Vinkx, C.; Smolders, E.; Huyghebaert, A.; Pussemier, L. Dietary Cadmium Intake by the Belgian Adult Population. *Food Addit. Contam. A: Chem. Anal. Control Expo. Risk Assess.* **2010,** *27* (12), 1665–1673.

218. Wang, L.; Su, D. Z.; Wang, Y. F. Studies on the Aluminium Content in Chinese Foods and the Maximum Permitted Levels of Aluminum in Wheat Flour Products. *Biomed. Environ. Sci.* **1994,** *7,* 91–99.

219. Wang, S.; Mulligan, C. N. Occurrence of Arsenic Contamination in Canada: Sources, Behavior and Distribution. *Sci. Total Environ.* **2006,** *366,* 701–721.

220. Weiner, J. A.; Nylander, M. The Relationship between Mercury Concentration in Human Organs and Different Predictor Variables. *Sci. Total Environ.* **1993,** *138,* 101–115.

221. WHO (World Health Organization). *Guidelines for Drinking-Water Quality, Vol. 1. Recommendations*; WHO: Geneva, 1993.

222. WHO (World Health Organization). *Arsenic. Guidelines for Drinking-Water Quality. Health Criteria and Other Supporting Information*, 2nd ed. WHO: Geneva, 1996; vol 2, pp 156–167.

223. WHO (World Health Organization). List of Substances Scheduled for Evaluation and Request for Data. In: *Sixty-Seventh Meeting of the Joint FAO/WHO Expert Committee on Food Additives*. WHO: Rome, 2006.

224. WHO (World Health Organization). Chapter 2: Risk Assessment and Its Role in Risk Analysis. *Principles and Methods for the Risk Assessment of Chemicals in Food*. Environmental Health Criteria 240, 2009.

225. WHO (Joint FAO/WHO Expert Committee on Food Additives). Evaluation on Certain Food Additives and Contaminants. In: *73rd Report of the Joint FAO/WHO Expert Committee on Food Additive. WHO Technical Report Series 960.* FAO/WHO: Rome, Italy, 2011.

226. WHO (World Health Organization). *Total Diet Studies: A Recipe for Safer Food*, 2015. Available from http://www.who.int/entity/foodsafety/chem/TDS_recipe_2005_en.pdf.

227. Yang, H. F.; Luo, X. Y.; Shen, W.; Zhou, Z. F.; Jin, C. Y.; Yu Liang, C. S. National Food Contamination Monitoring Programmes—Levels of Mercury, Lead and Cadmium in Chinese Foods. *Biomed. Environ. Sci.* **1994,** *7,* 362–368.

228. Yang, M.; Jiang, L.; Huang, H.; et al. Dietary Exposure to Aluminium and Health Risk Assessment in the Residents of Shenzhen, China, Sun, Q., Ed. *PLoS ONE* **2014,** 9 (3), e89715.

229. Yang, H.; Shu, Y. Cadmium Transporters in the Kidney and Cadmium-Induced Nephrotoxicity. *Int. J. Mol. Sci.* **2015,** *16,* 1484–1494.

230. Yamashita, Y.; Omura, Y.; and Okazaki, E. Total Mercury and Methylmercury Levels in Commercially Important Fishes in Japan. *Fish. Sci.* **2005,** *71,* 1029–1035.

231. Ye, B. J.; Kim, B. G.; Jeon, M. J.; Kim, S. Y.; Kim, H. C.; Jang, T. W.; Chae, H. J.; Choi, W. J.; Ha, M. N.; Hong, Y. S. Evaluation of Mercury Exposure Level, Clinical Diagnosis and Treatment for Mercury Intoxication. *Ann. Occup. Environ. Med.* **2016,** *28,* 5.

232. Yokel, R. A.; McNamara, P. J. Aluminium Toxicokinetics: An Update Mini-Review. *Pharmacol. Toxicol.* **2001,** *88,* 159–167.

233. Yost, L. J.; Schoof, R. A.; Aucoin, R. Intake of Inorganic Arsenic in the North American Diet. *Hum. Ecol. Risk Assess.: Int. J.* **1998,** 4 (1), 137–152.

234. Ysart, G.; Miller, P.; Crews, H.; Robb, P.; Baxter, M.; De L'Argy, C.; Lofthouse, S.; Sargent, C.; Harrison, N. Dietary Exposure Estimates of 30 elements from the UK Total Diet Study. *Food Addit. Contam.* **1999,** *16,* 391–403.

235. Ysart, G.; Miller, P.; Croasdale, M.; Crews, H.; Robb, P.; Baxter, M.; et al. 1997 UK Total Diet Study—Dietary Exposures to Aluminium, Arsenic, Cadmium, Chromium, Copper, Lead, Mercury, Nickel, Selenium, Tin and Zinc. *Food Addit. Contam.* **2000,** *17,* 775–786.

236. Zhai, Q.; Narbad, A.; Chen, W. Dietary Strategies for the Treatment of Cadmium and Lead Toxicity. *Nutrients* **2015,** *7,* 552–571.

237. Zhang, H.; Tang, J.; Huang, L.; Shen, X.; Zhang, R.; Chen, J. Aluminium in Food and Daily Dietary Intake Assessment from 15 Food Groups in Zhejiang Province, China. *Food Addit. Contam., B: Surveill.* **2016,** *9* (2), 73–78.

238. Zhang, Z. W.; Subida, R. D.; Agetano, M. G.; Nakatsuka, P.; Inoguchi, N.; Watanabe, T.; Shimbo, S.; Higashikawa, K.; Ikeda, M. Non-occupational Exposure of Adult Women in Manila, the Philippines, to Lead and Cadmium. *Sci. Total. Environ.* **1998,** *215,* 157–165.

239. Zukowska, J.; Biziuk, M. Methodological Evaluation of Method for Dietary Heavy Metal Intake. *J. Food Sci.* **2008,** *73,* 21–29.

CHAPTER 11

TOXICOLOGICAL RISKS OF WASTE BURNING RESIDUES IN FOODS: A VIEW ON LOW-INCOME COUNTRIES

ILARIA PROIETTI[1,2*] and ALBERTO MANTOVANI[2,3]

¹European Commission, Joint Research Centre (JRC), Economics of Agriculture—Sustainable Resources, Edificio Expo. C/Inca Garcilaso 3, 41092 Seville, Spain

²Noodles Onlus, Nutrition & food safety and wholesomeness, Rome, Italy

³Istituto Superiore di Sanità, Viale Regina Elena 299, 00161 Rome, Italy

*Corresponding author. E-mail: ilaria.proietti@ec.europa.eu

CONTENTS

ABSTRACT

Garbage management is an issue of growing concern nowadays due to the rapid growth of urban populations and changing consumption patterns. The environmental implications and impact on health and consequences on food safety associated with waste management need an urgent attention, particularly in low-income countries where waste removal and management facilities are often inadequate or missing.

In response to the lack of infrastructures, people often use a hazardous practice of burning to eliminate solid waste. Nevertheless, burning garbage is a source of several and diverse pollutants, including dioxins and related compounds (polychlorinated dibenzo-*p*-dioxins and furans, polychlorinated biphenyls, polycyclic aromatic hydrocarbons, polybrominated diphenyl ethers), particulate chloride, carbon monoxide, and toxic elements, these latter also as nanoparticles. Hazardous residues of burning can affect human health directly, through inhalation, or indirectly, through ingestion of contaminated food. Food can be contaminated directly through the fallout of smoke and ash on food displayed on street vending stalls; in the meanwhile, long-range environmental pollution of food chains may occur through the uptake of hazardous substances by crops from polluted soil and groundwater or their carryover to animals and animal products via contaminated feed.

The aim of this chapter is to review the toxicological risks of residues in foods from domestic and outdoor solid waste burning in the low-income countries scenarios. In particular, we will explore the potential for food contamination and the associated health hazards from the burning by-products of different solid waste not only such as rubber and leather, plastics, metal, electronic waste but also organic residues, including agricultural residues and food waste. A special attention is given to vulnerable groups of population, that is, young children, pregnant women, the elderly, or immunocompromised people that are exposed to these pollutants.

11.1 INTRODUCTION

Despite a slowdown in the rate of population growth, world population is estimated to reach 9.7 billion of people by 2050, according to the latest UN projections.[72] This constant rise of population would inevitably entrain several global consequences, including the increase in world production of garbage. Production and distribution of solid waste will not be homogeneous among and within countries: because of the sheer weight of people agglomeration, garbage production is considerably higher in urban areas than in the rural ones. Moreover, the production of urban waste has continued increasing over time due to higher incomes and changing attitudes toward the use of foods and consumer products. The World Bank estimates that from 2002 to 2012, the production of solid waste nearly doubled in urban areas: from 680 million tons per year to more than 1.3 billion tons per year, and this quantity is expected to increase to 2.2 billion tons by 2025.[85] As a consequence of such big volumes, a sound management of solid wastes represents one of the biggest challenges of the urban growth phenomenon.

The garbage management is challenging due to the large volumes produced, as well as because of the costs it implies. In 2010, the expenditures for managing municipal waste worldwide were about $205.4 billion, and in 2025, they are expected to increase to $375 billion. These figures are especially worrying for low-income countries as their total costs for solid waste management in 2025 are estimated to increase by more than fivefold.[85]

For example, municipal solid waste management represents a crucial environmental problem facing African cities, where it represents the largest budget item in many cases.[55,70,85] Despite the fact that garbage management represents the biggest demand on municipal budget, only a small volume of the waste is collected, often irregularly.[1,70] To overcome the severe lack of adequate infrastructures to treat solid waste, urban people in low-income countries often burn their domestic wastes without awareness of the potential harmful consequences of this practice for their own health and for the safety of their living environment. Furthermore, even when waste is collected and brought to dump sites, it is not uncommon for it to be burned by open uncontrolled fires.[58,82]

11.2   THE OPEN BURNING

According to Lemieux et al.,[44] open burning is defined as the unenclosed combustion of materials in an environment. The main concern of this practice is the emissions of a range of toxic substances directly into the atmosphere both because of the kind of toxicants produced and because it usually occurs in areas where a big number of people live. Open burning is relevant to a wide variety of activities and wastes, including household waste, agricultural and industrial wastes, as well as waste produced from other commercial and institutional sources.[51] Within the various components of solid garbage, those most widely released in the open environment generally include electrical and electronic equipment, health-care waste, organic waste, such as food and garden waste, wood, textiles, disposable diapers, plastics, and metals.[29,35]

The content and the volume of solid waste often differ between high-income and low-income countries and even between regions and munici-palities. Generally, the waste produced in low-income countries is primarily composed by organic waste and the volume depends on a number of factors including the consumption patterns, standard of living, and the commercial and institutional activities.[4,14] The chemical composition of the garbage clearly determines the typology of the compounds released during the combustion. However, some groups of major toxicants produced as burning by-products are consistently identified: dioxins and related compounds [polychlorinated dibenzo-p-dioxins (PCDD) and furans (PCDF), dioxin-like polychlorinated biphenyls (DL-PCB), particulate chloride, polycyclic aromatic hydrocar-bons (PAHs), carbon monoxide, and toxic elements, these latter also as nanoparticles. Whereas chlorine and carbon monoxide are important mainly for people directly exposed to the fumes, the other pollutants lead to much broader environmental pollution problems. Toxic emissions are then intensi-fied when waste is piled, resulting in a long-lasting smoldering phase that may be responsible of the production and release of high concentrations of products of incomplete combustion including persistent organic pollutants.[47] It has been estimated that nonindustrial emission sources of pollutants such as landfill fires and backyard barrel burning may become the major cause of the overall annual emission of PCDD/Fs in Europe.[75]

Also in many developing countries, the open burning represents one of the largest sources of toxic substances, such as PCDD/Fs;[76] according to a compilation of inventories pursuant to the Stockholm Convention on Persis-tent Organic Pollutants (POPs), open burning of waste or biomass represents the main source of unintentionally generated POPs (with PCDD/Fs featuring prominently) in developing countries.[23,71] A more recent compilation of 61 national inventories found that the open burning represents at least around 30% of the PCDD/PCDF emissions for approximately 75% of the countries and over 80% for 25% of the countries.[24]

The Stockholm Convention on POPs was adopted in Stockholm on 2001, and it entered into force on 2004 with the objective "to protect human health and the environment from persistent organic pollutants."[63] The convention currently lists 26 chemicals whose production and use need to be restricted, eliminated, or the unintentional releases be reduced. In the list PCDDs, PCDFs, and polychlorinated biphenyls (PCBs) are included (Table 11.1).

TABLE 11.1 POPs in the Stockholm Convention.

POP	Pesticide	Industrial chemical	Unintentional production
To be eliminated			
Aldrin	X		
Chlordane	X		
Dieldrin	X		
Endrin	X		
Heptachlor	X		
HCB	X	X	X
Mirex	X		
Toxaphene	X		
PCB		X	X
Chlordecone	X		
Hexabromobiphenyl		X	
HBCD		X	
Hexabromodiphenyl ether and heptabromodiphenyl ether		X	
Hexachlorobutadiene		X	
Alpha hexachlorocyclohexane	X		
Beta hexachlorocyclohexane	X		
Lindane	X		
Pentachlorobenzene	X	X	X
Pentachlorophenol and its salts and esters	X		
Polychlorinated naphthalenes		X	X
Technical endosulfan and its related isomers	X		
Tetrabromodiphenyl ether and pentabromodiphenyl ether		X	
To be restricted			
DDT	X		
Perfluorooctane sulfonic acid, its salts and perfluorooctane sulfonyl fluoride		X	
To be reduced			
PCDD			X
PCDF			X

HCB, Hexachlorobenzene, *PCBs*, polychlorinated biphenyls, *HBCD*, hexabromocyclodo-decane, *PCDD*, polychlorinated dibenzo-*p*-dioxins, *PCDF*, polychlorinated dibenzofurans. Adapted from Stockholm Convention.[63]

In the convention, the health concerns resulting from local exposure to POPs are mentioned as primary reasons behind the adoption of the convention itself, especially in developing countries. The developing countries are indeed most at risk not only because of the lack of adequate infrastructures for the collection, removal, and disposal of waste, but also because of people's common practice of waste burning and the kind of garbage burnt.[71,76]

Exposure to toxic by-products of garbage burning occurs through three major routes: people might be exposed to contaminants by inhaling the toxic products of combustion, via dermal absorption, or via ingestion of contaminated food and water.[79] Breathing of contaminated air is the most common way that chemicals, either gaseous or produced as dusts of respirable size (particles below 10 μm), enter the body. However, atmospheric deposition on the soil and surface water also represents a pivotal source of contamination, especially for compounds able to persist and/or bioaccumulate; soil and surface water are the living substrata for the plants growing on the polluted sites, thus, the entrance door of pollutants into the human food production chains.[19,21,79]

11.3 BURNING OF SOLID WASTE: THE EMERGING ISSUE OF E-WASTE

Depending on the substrate, different chemicals are released during the combustion. The most common examples of frequently burnt solid waste in low-income countries are mainly represented by agricultural and household waste. In addition, low-income countries have become recently a common destination for the disposal of obsolete electronic devices like computers, televisions, audio and stereo equipment, telephones and mobile phones, printers, leading to a new emerging environmental health issue: the e-waste. In fact, e-waste may easily release into the environment many toxic components, both as substances present in the materials, such as brominated flame retardants [in particular, polybrominated diphenyl ethers (PBDEs)] and toxic metals (Cu, Cd, Hg but also Al, Ni, and Mn) and as combustion by-products, such as PAHs, PCDD/F, and DL-PCB.[27,69,84] As a consequence, people may be exposed to toxic releases directly, via local or occupational exposure, and indirectly through the food chain. Many of these multiple contaminants are able to build-up a body burden and to expose the next generation via transplacental transfer as well as passage to the breast milk.[27]

A high number of studies showed the contribution of the e-waste combustion to the environmental exposure to certain chemicals of high concern. Due to the high temperature during the combustion process, dioxins and related compounds, heavy metals, and nanosized airborne pollution are created and released in the environment, representing an important source of contamination for humans and animals via air, soil, water, and food. The unfiltered burning and recycling of e-waste is considered as likely responsible for the production and release of metal nanoparticles along with volatile organic compounds and hydrocarbons.[40] Fu et al.[28] investigated the concentration of toxic metals in soils and food (vegetables, rice, and wide plants) in Chinese sites, where e-wastes were regularly burned. The results showed that the soil had high concentrations of Cd, Cu, Pb, and Zn; in particular, Cd and Cu in soil samples were 4.0 and 2.0-folds, respectively, the maximum allowable concentrations in China. Moreover, the concentrations of Cd and Pb in most analyzed rice samples exceeded the food safety limit of China.

In a research conducted in Bangalore and Chennai (India), trace elements in soil samples from e-waste recycling sites have been analyzed. The researchers found higher concentrations of Cu, Zn, Ag, Cd, In, Sn, Sb, Hg, Pb, and Bi in soil from e-waste recycling sites compared to reference sites. In particular, the levels of Cu, Sb, Hg, and Pb in some soils exceeded the values proposed by US Environmental Protection Agency.[31] Considering that Hg and Pb, in particular, are recognized as major human toxicants and the soil is the root of agricultural productions, these data point out a potential significant risk for food safety.

The recycling and disposal of e-waste causes an increasing concern not only in Asia but also in Africa, where the import of e-waste has been increasing.[61] A study in Accra (Ghana) examined the concentrations of a number of dioxin-related compounds (DRCs) including, brominated and mixed halogenated ones (PBDD/Fs, PXDD/Fs) as well as PCDD/Fs and DL-PCBs, in soils from the Agbogbloshie informal e-waste recycling site. Results showed that PCDF and PBDF concentrations in open burning areas (520 and 3800 ng/g dry, respectively) were among the highest reported in soils from informal e-waste sites and suggested that PBDE-containing plastics and poly(vinyl chloride) (PVC) were the principal substrates burnt.[69]

Also in Nigeria, the practice of open burning of e-waste is quite common with severe contamination as a consequence. Sindiku et al.[62] sampled and analyzed the PBDD/F content in e-waste plastics from cathode ray tubes. They detected very high concentration of PBDD/Fs (mainly of PBDFs) in brominated flame retardant-containing plastic; this result is alarming, when considering that 140 t of PBDEs have been estimated to be subjected to open burning in Nigeria.[7]

Compared to the emerging e-waste, the combustion of agricultural waste represents an established major source of contamination of the environment and hence of animal and vegetal products. Agricultural waste, which includes both organic and chemical products, encompasses all type of wastes produced during farming activities. Open burning of agricultural residues, including straw, husks, and grass silage, is a common practice in many rural areas with the main aim to prepare land for the following crop harvest.[39] Nevertheless, as large uncertainties exist about the typology of the substrate and fraction of residues subjected to burning,[36] it is difficult to estimate the most relevant combustion by-products. Further to the environmental concerns, due to the emissions of methane and nitrogen oxides, another major alarm related to the agriculture waste is the health threat due to the formation of low molecular weight PAHs during the incomplete combustion.[13,37] In their study, Agarwal et al.[2] analyzed surface soils (0–5 cm) from urban and rural agricultural sites in Delhi. PAHs concentration were greatly different among the sites, ranging from 830 to 3880 µg/kg (dry wt.), with predominance of low molecular weight PAHs. Moreover, PAHs values in the urban agricultural sites were found to be 2–5 times higher as compared to the rural sites, suggesting biomass and fossil fuel combustion as the main sources of these chemicals. Wevers et al.[77] evaluated the PCDD/F air emission factors of yard wastes (trimmings and fallen leaves) burned in different environment, like open pile and in galvanized drums. The analysis on the smoke showed that garden wastes combusted in an open pile released 4.4 ng TEQ (toxic equivalency factor [TEF])/kg, while garden wastes burnt in galvanized drums had higher emission rates ranging 4.7–20 ng TEQ/kg.

Moreover, the practice of burning treated plants, as a field remediation measure, black plastic film, and plastic bags delivering residues of agrochemicals constitutes another common environmental source of toxic substances like DRCs.[9,44] There is no doubt that agricultural waste contains pesticide residues; it is not known whether such residues may result, upon combustion, in specific toxic by-products. A recent draft guidance by the European Food Safety Authority points to the need of assessing the potential health risks of environmental by-products of pesticides with chemical and toxicological properties differing from the parent compounds.[18]

Open burning of household waste represents a common practice people use to manage garbage in the domestic environment. Emissions from open uncontrolled burning of household waste represent another significant source of PAHs, PCBs, PCDDs/Fs, PBDDs/Fs, and PBDEs, especially when products containing brominated flame retardants are burnt.[30,33,52] The combustion of domestic waste has been reproduced in a laboratory by Nakao et al.[50] using a small home waste incinerator and the flue gas analyzed. In this study eight different waste samples, including paper, building materials, nonchlorinated and chlorinated plastics, and electric wire were burned. The results showed that copper wire, chlorinated plastics as well as its precursors such as polychlorinated benzenes, polychlorinated phenols, polychlorinated diphenyl ethers, and PCBs increased the release of PCDD/Fs in both the flue gas and ash. In a laboratory study, simulating conditions of open fire combustion, Valavanidis et al.[73] detected high molecular weight PAHs and some lithophilic elements, such as Na, Ca, Mg, Si, and Al, at high concentrations in the residue solid ashes of plastics. In particular, PVC resulted to be the main source of toxicants as compared to other types of plastic.

Gullett et al.[30] investigated the PBDE, PCDD/F, and PBDD/F emissions from open burning of domestic wastes in Mexico. The measurement revealed that emissions of PBDE congeners averaged 724 µg/kg C_{burned} and were thought to likely originate from brominated flame retardants contained in the burnt materials. Emission factors of PBDD/Fs showed an average of 470 ng TEQ/kg C_{burned} and may originate as reaction products of PBDE combustion. Finally, PCDD/F emission factors averaged 823 ng TEQ: this value is around 2000 times higher than that from modern municipal waste combustors.

11.4 ROUTES OF HUMAN EXPOSURE

During the process of combustion, volatile and semivolatile compounds are transported in the atmosphere and deposited on the top-soil level. Releases in the atmosphere, deposition on the soil as well as water runoff from polluted soil are the sources of air, soil, and water contamination.

The occupational exposure to toxicants is normally limited to specific subgroups of the population, while the environmental exposure affects the general population. People can be affected directly and indirectly through inhalation of contaminated air and ingestion of polluted food, respectively.

While inhalation of air contributes to a small amount of the total intake, it is estimated that food represents the major route of human exposure to persistent pollutants produced by combustion.[12,81]

EFSA[15] computed the relative contribution to PCBs exposure from air, soil, skin, and food. Exposure to PCBs via inhalation was estimated to be 0.05–0.5 ng/kg bw/day for children and 0.03–0.3 ng/kg bw/day for adults. Soil or dust contribution to the total PCBs exposure would be around 0.06–0.6 ng/kg bw/day for children, for whom nondietary soil ingestion does occur. Dermal absorption was estimated to be approximately 5 pg/kg bw/day for children and 0.76 pg/kg bw/day for adults. Compared with the exposure from these sources, EFSA concluded that exposure from food was around 3–4 orders of magnitude higher. Also for PCDD/F food accounts for the 90% of the daily human exposure and the 90% of this percentage normally comes from foods of animal origin.[34,81]

The contamination of food can be either direct, for example, through deposition on exposed edible parts of vegetables, or result from the accumulation of the toxicants along the food chain. Therefore, pollution from open burning may concern the food chain from the preharvesting to the consumption phase, for instance, when food is displayed and sold in the open-air like in the street-food vending.

One of the key aspects of the POPs is their hydrophobic nature and low solubility that give them the ability to resist to biodegradation and bioaccumulate in the environment, representing a long-term threat to human health.

PCBs include 209 congeners, of which only a few are PCB-DL while the vast majority, including the most persistent congeners (e.g., PCB 153, 180) have different toxicological characteristics. These PCBs were identified as one of the priority POPs in the Stockholm Convention of 2000[63] because of their environmental persistence and toxicity for animals and humans they were already banned in industrialized countries since the 1970s and 1980s.

Even though the commercial production has been prohibited since long, due to the past widespread use in electric transformers and its current use in some industries,[46] significant amount of PCBs has been found in wastes (mainly e-waste)[27] and subsequently in the environment as well as in animals and humans. PCBs have been detected also in animal products and in human tissues, milk, and blood.[53,66] The highest concentrations of PCBs are generally found in fish and seafood, followed by milk, eggs, and meat and meat products. Foods of plant origin instead (cereals, fruits, and vegetables) contain only low amounts of PCBs.[15,81] The Stockholm Convention[63] on POPs has identified seven PCBs that can be used to characterize the contamination by these toxicants. Six of these seven are NDL-PCB congeners, and they were chosen because they are found at high concentrations in the environment, in food, or in human fluids/tissues.[71] Although NDL-PCBs account for the majority of the total PCB contamination in food, the evaluation of their effects results to be hard since no distinction can be made between the effects caused by NDL-PCBs and those caused by DL-PCBs or dioxins.[81]

The atmosphere represents the first route of exposure. When wastes are burned, toxic compounds are discharged into the air, which then get deposited on the soil. The plants and trees growing on the neighboring sites could be the first recipient of these substances.

Edible plants can represent then a pivotal route of human exposure and the first route of contamination for grazing animals. Through the vegetable feed, animals can take up also contaminants that may accumulate in their liver and fat tissue, or be excreted in edible products, such as milk and eggs. In its evaluation on the levels of dioxins and dioxin-like PCBs in liver from sheep and deer, EFSA estimated that a regular consumption of these food items would result in about 20% increased exposure to dioxins and dioxin-like PCBs for adults. In particular, the frequent consumption of sheep liver by children and women of child-bearing age may represent a potential health concern.[16] Also, milk and eggs are especially important for POPs, because they are produced continuously by live animals, are rich in lipid, and can be considered as an excretion pathways for POP-exposed animals. Products from animals directly exposed to environmental emissions are more liable to contamination. The contribution of eggs to the total dioxin intake, for instance, has been estimated to ranges from 2% to 8% in European countries.[6,41,45] However, not all eggs are the same. In their study on dioxin content in eggs from free-range hens, Air et al.[3] found a mean dioxin level of 9 pgI-TE/g, about sevenfold higher than the average level of dioxin in eggs from caged animals and threefold higher than the EU maximum level. Although a direct correlation is not straightforward, the presence of dioxins in eggs from free-range or organic hens seems to be mainly related to feed and soil contamination. In its study on eggs contamination in an area of intensive industrial and domestic activities, Pussemier et al.[57] reported a remarkable similarity in PCDD/F pattern between soil samples and home-produced eggs, suggesting soil to be a major source of dioxin contamination for laying hens.

In addition to POPs, other compounds are characterized by their persistence in the environment, their ability to accumulate in organisms and enter the food chain. They are called persistent toxic substances (PTSs) and, like POPs, may pose a serious threat to humans and the environment. This category of chemicals are not yet specifically defined but it could include, for instance, mercury, cadmium, lead, PAHs, and PBDEs.[80] PBDEs are compounds recently emerged as a major environmental pollutant of health concern, being endocrine disruptors and neurotoxins. They have been used as a flame-retardant and are found in consumer goods such as paints, plastics, textiles, televisions, building materials, airplanes, and automobiles. Due to their toxicity and persistence in the environment, the industrial production of some PBDEs is restricted under the Stockholm Convention[63]. However,

PBDE bioaccumulate and their contamination of certain foods, such as large fish and dairy products, must not be overlooked.[17] A research conducted in a major e-waste recycling area in China showed that locally produced food (including meat, fish, and eggs) was highly contaminated by PBDE, and this contamination was extended to adult and young children. The highest level of contamination was reported in children showing an estimated median PBDE exposure of 614.1 ng/kg bw/day (normalized to body weight).[43]

According to many studies carried out to evaluate the human dietary intake, milk and dairy products result as another major contributor of PCDD/Fs uptake. Many researches were conducted in particular on PCDD/Fs and PCBs content in cow milk worldwide.[20,25,42,59,60] Kunisue et al.[42] found that concentrations of PCDDs, PCDFs, DL-, and non-DL PCB in bovine milk in India were approximately fourfold higher in an area close to one dumping site compared with those from reference sites. In the same way, women living in the same dumping area showed levels of the investigated POPs in their breast milk two to fourfold higher than those women living near reference sites and in other Asian developing countries (Cambodia and Vietnam). The fact that lactating women bear the burden of high levels of toxicants in their milk represents a concern because of the transfer of these lipophylic substances to their infant(s) through breastfeeding.[27,42] Indeed, breast feeding allows fat-soluble pollutants, accumulated in the maternal adipose tissue to be excreted; thus, the protection of breastfeeding is a component of the sustainable food safety framework, where the wholesome diet in the present generation reduces the health risks for the generation(s) to come.[26]

The aquatic environment represents an additional route of exposure. After their deposition on the soil, volatile and semivolatile compounds may also flow into groundwater, surface water, or other water bodies like rivers and lakes and pollute the environment and the aquatic species living there. Many of the toxicants cited in this chapter are persistent and bioaccumulate in the soil as well as in the water, especially in the groundwater, where they may persist for decades and contaminate fishes, mollusks, and other aquatic animals, according to the environmental food webs: in general, in an exposed ecosystem, large predatory species are more liable to contamination.[48] In a number of researches on bioaccumulation factors in aquatic organisms, seafood products showed to be prone to accumulate a number of POPs, including PCBs, PCDDs/Fs, and BFRs.[12,19,48] Therefore, the water itself and all the aquatic species living in polluted water bodies might show high concentration of toxicants and the population living close the water basin may be affected by eating the contaminated aquatic species or by using the polluted water for irrigation or directly for drinking or cooking.

As already mentioned, food can be contaminated along the entire food chain, not only at its first phases (through contaminated raw food items) but also during the production process and even at the consumption phase. Volatile and semivolatile compounds and polluted water can enter and contaminate the food chain at every stage, from the production to the consumption process. Concerning the latter, food can be exposed to airborne chemicals released during the open burning. This is particularly recurrent whether open food-vending activities, like a market or street food stands or bar/restaurants in the open air, are situated in the area surrounding the burning area. Street vending food, in particular, is a common activity worldwide, especially in low-income countries where it represents an important source of nourishment, generates income and employment, and contributes to the empowerment of women involved in the vending.[11,22,54] Food sold in the street is particularly susceptible to contamination. Displayed food is not often stored or covered and so protected against the volatile substances that may deposit on the surface of foods. Moreover, inadequate waste facilities may lead vendors to burning the food residues and/or the plastic materials of packaging just by the vending sites generating severe health hazards.[56]

11.4.1 ADVERSE HEALTH EFFECTS

The main concern related to POPs and PTSs is their ability to resist to biodegradation and bioaccumulate due to their hydrophobicity and low solubility. Their ability to accumulate in the adipose tissue of living organisms for long period leads to their biomagnification in the food chain; in other words, they can pass from one species to the next through the food chain.

Although many of the POPs included in the Stockholm convention[63] are no longer produced, they can still be found in the environment due to their characteristic persistency. This is particularly common in low-income countries where POPs production, use and release have only recently been restricted.[19] In addition to the intentionally produced chemicals already banned but still used in some countries, unintentionally produced chemicals, such as dioxins resulting from industrial processes and combustion, can also seriously contaminate the environment.

POPs and PTSs are highly toxic to the environment, animals, and humans and are able to produce adverse health effects at low doses. As for many other environmental pollutants, it is difficult to associate a specific persistent organic pollutant with a disease. This is furthermore more challenging if considering that they are rarely formed as single compound and that field studies struggle to prove the clear evidence of cause and effect.[79]

Even though it is difficult to point out specific diseases, there is sufficient evidence that exposure to POPs and PTSs can lead to an increased risk of a serious health effects, including decreased reproduction and growth rate, endocrine, immune, and reproductive systems imbalance, cancers, teratogenicity and neurobehavioral effects.[63,79,80] The complex nature of chemical mixtures complicates the risk evaluation for humans. So, to facilitate risk assessment and control of exposure to these toxicants, the concept of TEFs has been developed. The use of the TEF approach is based on the assumptions that PCDDs, PCDFs, and DL-PCBs have a common mechanism of action (interaction with the aryl hydrocarbon receptor) and their biochemical and toxicological effects are directly related to their concentrations in tissues and not to the daily dose. According to this approach, the toxicity of the chemicals in the series is related to that of the 2,3,7,8-tetrachlorodibenzodioxin (TCDD), one of the most potent and well-known dioxins. This approach enables also to express the daily intake in units of TCDD equivalents (TEQs) for comparison to the tolerable daily intake (TDI) of TCDD, ranging 1–4 pg/kg of body weight as established by WHO.[78]

Brominated dioxin-like chemicals should also be included in the AhR-based TEF approach.[74] Although the non-DL PCB occurs together, there is no standardized test to assess a possible cumulative risk. In vitro studies indicate that also the nondioxin-like PCB may be grouped according to their molecular mechanisms in clusters, such as the "estrogenic" and the "highly persistent-cytochrome P-450 inducers."[32,64] PBDEs also occur together and share common mechanisms.[17]

Vulnerable groups of population, and in particular young children and pregnant women, deem special attention due to their health condition, stage of life, as well as the chemicals' long-term or transgenerational effects. Exposure of pregnant women to toxicants results in a double impact: an impairment of the maternal health condition as well as of that of future generations.[63]

Exposure appears to be highest in young children because compared to their body mass, they have relatively higher intake of water and food as well as a relatively greater breathing volume; moreover, nondietary ingestion of soil and dust may be considered as a children-specific exposure route. Due to their stage of development, fetuses and, to a lesser extent, infants may experience important disorders in the functioning of organs and systems still not completely developed. Toxic substances, like PCBs, can show a significant transplacental transfer and enter into the fetal blood stream.[38,65] Prenatal exposure to PCBs can also affect neurobehavioral development in fetus[83] and according to Boucher et al.[8] is determinant of a relatively specific cognitive profile of impairments. Under this respect, it is noteworthy that PCBs may impair thyroid function, a mechanism that can, at least, contribute to neurobehavioral effects. Moreover, fetus exposure to POPs appears to cause also reduced birth weight.[5,49] The maternal household environment, in particular, seems to have a strong influence on newborn birth weight. In their study, Amegah et al.[5] found a reduction of 178 g reduction in birth weight for pregnant women exposed to garbage burning by-products in Ghana. Reduced birth weight is an adverse birth outcome per se as well as a risk factor for long-term health problems in adulthood such as metabolic syndrome.

In regard to POP exposure, breastfed infants are a special case. In contaminated areas, they are usually more exposed than adults because of the additional concentration of toxic substances in the breast milk. Toms et al.[67,68] estimated a possible PBDE concentration in breastfed babies up to 440 ng/kg/day compared to 1 ng/kg/day for adults.

Analysis conducted on breast milk of women at an e-waste recycling site in China revealed an estimated daily intake of PCDD/Fs in breastfed infants significantly higher, exceeding the WHO total daily dietary intake for adults by 25 times.[10] Breastfeeding is an important factor in PBDE exposure, too.[17] Since breastfeeding is a highly beneficial practice, the safety of maternal food prior and during breastfeeding is paramount to protect the full benefits for the mother–infant dyad.

11.5   CONCLUSIONS AND RECOMMENDATIONS

Due to the inadequacy of the infrastructures for the collection, removal and disposal of waste, the lack of education and information about the risks incurred and the common practice of different populations, the burning of waste represents a big health concern worldwide and, in particular, in low-income countries. The open burning of garbage represents indeed a source of a number of hazardous toxic chemicals, like dioxins, PCBs, PAHs, PCDDs/Fs, PBDEs, and heavy metals, and in developing countries, in particular, the main source of unintentionally generated POPs.

The health concerns resulting from local exposure to the substances emitted during the combustion of waste are enormous and with long-term effects and depend on the chemical composition of the garbage, the magnitude of exposure (especially in some cultures where it is a commonly accepted practice) and the vulnerability of the population exposed.

Depending on the substrate, different by-products are released during the combustion. The most common and at the same time hazardous examples of frequently burnt solid waste are represented by e-waste, agricultural, and household waste. Low-income countries have become a common destination for the disposal of obsolete electronic and electric waste mainly computers, televisions, and mobile phones and people use to burn them as disposal practice. This practice is causing serious health and environmental concerns because of the toxic components released during combustion, such as dioxins, PCB, brominated flame retardants, PAHs, and heavy metals. Burning agriculture waste, including yard wastes, treated plants, black plastic film, and plastic bags delivering residues of agrochemicals, represents an additional common health threat due to the formation of low molecular weight PAHs dioxin and dioxin-like compounds during the incomplete combustion. Moreover, people especially in many low-income countries use to burn household waste to manage garbage in the domestic environment. Emissions from open uncontrolled burning of household waste represent another alarming health issue due to the release of toxic chemicals, including PAHs, PCBs, PCDDs/Fs, PBDDs/Fs, and PBDEs.

Hazardous residues of burning can affect human health via different routes of exposure like direct contact or through ingestion of contaminated food. Food represents the major route of human exposure and its contamination can occur during the entire food chain from the plants growing, due to soil contamination, to the consumption phase, for the fall out of volatile compounds onto food surface. Due to the lipophilicity of these compounds, foods from animal origin, in particular high-fat foods, represent a major source of exposure. Fish, mammals (and their products), and humans are high up the food chain and so accumulate those toxicants at high concentrations.

Attention need to be posed to the effects the common practice of burning the garbage causes. Vulnerable groups of population and, in particular, young children and pregnant women deem special attention due to their health condition, stage of life, as well as toxicants long-term or transgenerational effects. In the case of pregnant women, the impact of toxicants on the health results magnified: an impairment of the maternal health condition and through them, of future generations, while young children may experience important disorders in the functioning of organs and systems still not completely developed.

The open burning of waste represents indeed a big health concern for the entire population and, in particular, for the most vulnerable groups. Long-term effect measures are pivotal to reduce the consequences of the exposure to the by-products produced during the combustion for the present generation and for the one to come. First of all, it is necessary to reduce the production of waste by encouraging the recycling. It would also be important improve the quality of waste by minimizing the content of substances releasing toxic residues during the combustion, advocating the elimination of POPs and the support of the Stockholm Convention. Educating people on the health risks they incur by burning waste would represent an essential preventive measure to strongly reduce the burning of garbage.

It would also be appropriate to put in place methods able to reduce the concentration of these toxic substances in animals and their products. Observance of good agricultural practices and good animal feeding practices, for instance, will contribute to the reduction of toxicant concentrations in food for human consumption. Concerning grazing animals, it would also be important to avoid the pasturing of animals in contaminated lands, if detected, or near open burning sites. At the same time, people could reduce the exposure to these toxicants along the food chain by limiting the consumption of foods that contributes most greatly to their exposure, fatty foods above all.

KEYWORDS

- **garbage management**
- **hazardous substances**
- **food contamination**
- **environmental health**
- **developing countries**

REFERENCES

1. Achankeng, E. Globalization, Urbanization and Municipal Solid Waste Management in Africa. *African Studies Association of Australasia and the Pacific 2003 Conference Proceedings—African on a Global Stage*, 2003.
2. Agarwal, T.; Khillare, P. S.; Shridhar, V.; Ray, S. Pattern, Sources and Toxic Potential of PAHs in the Agricultural Soils of Delhi, India. *J. Hazard. Mater.* **2009**, *163* (2–3), 1033–1039.
3. Air, V.; Pless-Mulloli, T.; Schilling, B.; Paepke, O. Environmental Non-Feed Contributors to PCDD/PCDF in Free-Range Allotment Poultry Eggs: Many Questions and Some Answers. *Organohal. Compd.* **2003**, *63*, 126–129.

4. Aleluia, J. O.; Ferrão, P. Characterization of Urban Waste Management Practices in Developing Asian Countries: A New Analytical Framework Based on Waste Characteristics and Urban Dimension. *Waste Manage.* **2016,** *58,* 415–429.

5. Amegah, A. K.; Jaakkola, J. J.; Quansah, R.; Norgbe, G. K.; Dzodzomenyo, M. Cooking Fuel Choices and Garbage Burning Practices as Determinants of Birth Weight: A Cross-Sectional Study in Accra, Ghana. *Environ. Health* **2012,** *17* (11), 78.

6. Baars, A. J.; Bakker, M. I.; Baumann, R. A.; Boon, P. E.; Freijer, J. I.; Hoogenboom, L. A.; Hoogerbrugge, R.; van Klaveren, J. D.; Liem, A. K.; Traag, W. A.; de Vries, J. Dioxins, Dioxin-Like PCBs and Non-dioxin-Like PCBs in Foodstuffs: Occurrence and Dietary Intake in The Netherlands. *Toxicol Lett.* **2004,** *151* (1), 51–61.

7. Babayemi, J.; Sindiku, O.; Osibanjo, O.; Lundstedt, S.; Weber, R. Material Flow and Substance Flow Analysis of POP-PBDEs in Nigeria and the Risk of Dioxin Formation and Release. *Organohal. Compd.* **2014,** *76,* 1453–1456.

8. Boucher, O.; Muckle, G.; Bastien, C. H. Prenatal Exposure to Polychlorinated Biphenyls: A Neuropsychologic Analysis. *Environ. Health Perspect.* **2009,** *117,* 7–16.

9. Brambilla, G.; Cherubini, G.; De Filippis, S.; Magliuolo, M.; di Domenico, A. Review of Aspects Pertaining to Food Contamination by Polychlorinated Dibenzodioxins, Dibenzofurans, and Biphenyls at the Farm Level. *Anal. Chim. Acta* **2004,** *514,* 1–7.

10. Chan, J. K.; Xing, G. H.; Xu, Y.; Liang, Y.; Chen, L. X.; Wu, S. C.; Wong, C. K.; Leung, C. K.; Wong, M. H. Body Loadings and Health Risk Assessment of Polychlorinated Dibenzo-*p*-dioxins and Dibenzofurans at an Intensive Electronic Waste Recycling Site in China. *Environ. Sci. Technol.* **2007,** *41,* 7668–7674.

11. Chukuezi, C. O. Food Safety and Hygienic Practices of Street Food Vendors in Owerri, Nigeria. *Stud. Sociol. Sci.* **2010,** *1,* 50–57.

12. Domingo, J. L.; Bocio, A. Levels of PCDD/PCDFs and PCBs in Edible Marine Species and Human Intake: A Literature Review. *Environ. Int.* **2007,** *33* (3), 397–405.

13. Drescher, S.; Zurbrügg, C. Decentralised Composting: Lessons Learned and Future Potentials for Meeting the Millennium Development Goals. In *CWG—WASH Workshop 2006,* Kolkata, India, 1–5 February 2006.

14. EAWAG. Global Waste Challenge, Situation in Developing Countries. In *Global Waste Challenge: Situation in Developing Countries*; EAWAG, Ed.; Pamphlet January 2008, EAWAG: Dubendorf, 2008; 12 pp.

15. EFSA. Opinion of the Scientific Panel on Contaminants in the Food Chain on a Request from the Commission Related to the Presence of Non-dioxin-Like Polychlorinated Biphenyls (PCB) in Feed and Food. *EFSA J.* **2005,** *284,* 1–137.

16. EFSA. Scientific Opinion on the Risk to Public Health Related to the Presence of High Levels of Dioxins and Dioxin-Like PCBs in Liver from Sheep and Deer. *EFSA J.* **2011,** *9* (7), 2297 [71 pp].

17. EFSA. Scientific Opinion on Polybrominated Diphenyl Ethers (PBDEs) in Food. *EFSA J.* **2011,** *9* (5), 2156 [274 pp].

18. EFSA. *Guidance on the Establishment of the Residue Definition for Dietary Risk Assessment,* 2016. Available at https://www.efsa.europa.eu/sites/default/files/consultation/160307_4.pdf (accessed: 2 August 2016).

19. EPA. Persistent Organic Pollutants: A Global Issue. *A Global Response,* 2009. Available at https://www.epa.gov/international-cooperation/persistent-organic-pollutants-global-issue-global-response (accessed: 28 July 2016).

20. Esposito, M.; Cavallo, S.; Serpe, F. P.; D'Ambrosio, R.; Gallo, P.; Colarusso, G.; Pellicanò, R.; Baldi, L.; Guarino, A.; Serpe, L. Levels and Congener Profiles of

Polychlorinated Dibenzo-*p*-dioxins, Polychlorinated Dibenzofurans and Dioxin-Like Polychlorinated Biphenyls in Cow's Milk Collected in Campania, Italy. *Chemosphere* **2009**, *77* (9), 1212–1216.

21. Estrellan, C. R.; Iino, F. Toxic Emissions from Open Burning. *Chemosphere* **2010**, *80* (3), 193–207.

22. FAO. *Improving the Nutritional Quality of Street Foods to Better Meet the Micronutrient Needs of School Children in Urban Areas*, FAO, Nutrition and Consumer Protection Division: Rome, 2006.

23. Fiedler, H. National PCDD/PCDF Release Inventories under the Stockholm. Convention on Persistent Organic Pollutants. *Chemosphere* **2007**, *67*, S96–S108.

24. Fiedler, H.; Solorzano Ochoa, G.; Yu, G.; Zhang, T.; Marklund, S.; Lundin, L. *Hazardous Chemicals from Open Burning of Waste in Developing Countries—Final Report*. United Nations Environment Programme, Division of Technology, Industry and Economics, Chemicals Branch, 2010.

25. Focant, J. F.; Pirard, C.; Massard, A. C.; de Pauw, E. Survey of Commercial Pasteurised Cows' Milk in Wallonia (Belgium) for the Occurrence of Polychlorinated Dibenzo-*p*-dioxins, Dibenzofurans and Coplanar Polychlorinated Biphenyls. *Chemosphere* **2003**, *52*, 725–733.

26. Frazzoli, C.; Petrini, C.; Mantovani, A. Sustainable Development and Next Generation's Health: A Long-Term Perspective about the Consequences of Today's Activities for Food Safety. *Ann. Ist. Super. Sanit.* **2009**, *45* (1), 65–75.

27. Frazzoli, C.; Orisakwe, O. E.; Dragone, R.; Mantovani, A. Diagnostic Health Risk Assessment of Electronic Waste on the General Population in Developing Countries Scenarios. *Environ. Impact Assess. Rev.* **2010**, *30*, 388–399.

28. Fu, J. Zhou, Q.; Liu, J.; Liu, W.; Wang, T.; Zhang, Q.; Jiang, G. High Levels of Heavy Metals in Rice (*Oryza sativa* L.) from a Typical E-Waste Recycling Area in Southeast China and its Potential Risk to Human Health. *Chemosphere* **2008**, *71*, 1269–1275.

29. Guerrero, L.; Maas, G.; Hogland, W. Solid Waste Management Challenges for Cities in Developing Countries. *Waste Manage.* **2013**, *33* (1), 220–232.

30. Gullett, B. K.; Wyrzykowska, B.; Grandesso, E.; Touati, A.; Tabor, D. G.; Ochoa, G. S. PCDD/F, PBDD/F, and PBDE Emissions from Open Burning of a Residential Waste Dump. *Environ. Sci. Technol.* **2010**, *44*, 394–399.

31. Ha, N. N.; Agusa, T.; Ramu, K.; Tu, N. P. C.; Murata, S.; Bulbule, K. A.; Parthasaraty, P.; Takahashi, S.; Subramanian, A.; Tanabe, S. Contamination by Trace Elements at E-Waste Recycling Sites in Bangalore, India. *Chemosphere* **2009**, *76* (1), 9–15.

32. Hamers, T.; Kamstra, J. H.; Cenijn, P. H.; Pencikova, K.; Palkova, L.; Simeckova, P.; Vondracek, J.; Andersson, P. L.; Stenberg, M.; Machala, M. In Vitro Toxicity Profiling of Ultrapure Non-dioxin-Like Polychlorinated Biphenyl Congeners and their Relative Toxic Contribution to PCB Mixtures in Humans. *Toxicol Sci.* **2011**, *121*, 88–100.

33. Hedman, B.; Näslund, M.; Nilsson, C.; Marklund, S. Emissions of Polychlorinated Dibenzodioxins and Dibenzofurans and Polychlorinated Biphenyls from Uncontrolled Burning of Garden and Domestic Waste (Backyard Burning). *Environ. Sci. Technol.* **2005**, *39* (22), 8790–8796.

34. Huisman, M.; Eerenstein, S. E. J.; Koopman-Esseboom, C.; Brouwer, M.; Fidler, V.; Muskiet, F. A. J.; Sauer, P. J. J.; Boersma, E. R. Perinatal Exposure to Polychlorinated Biphenyls and Dioxins through Dietary Intake. *Chemosphere* **1995**, *31* (10), 4273–4287.

35. IPCC. Chapter 10—Waste Management. In *Climate Change 2007, Mitigation. Contribution of Working Group III to the Fourth Assessment Report of the Intergovernmental*

Panel on Climate Change; Bogner, J., Metz, B., Davidson, O. R., Bosch, P. R., Dave, R., Meyer, L. A., Eds.; Cambridge University Press: Cambridge, 2007.

36. Jain, N.; Bhatia, A.; Pathak, H. Emission of Air Pollutants from Crop Residue Burning in India. *Aerosol Air Qual. Res.* **2014**, *14*, 422–430.

37. Kakareka, S. V.; Kukharchyk, T. I. PAH Emission from the Open Burning of Agricultural Debris. *Sci. Total Environ.* **2003**, *308*, 257–261.

38. Kawashiro, Y.; Fukata, H.; Omori-Inoue, M.; Kubonoya, K.; Jotaki, T.; Takigami, H.; Sakai, S.; Mori, C. Perinatal Exposure to Brominated Flame Retardants and Polychlorinated Biphenyls in Japan. *Endocr. J.* **2008**, *55*, 1071–1084.

39. Keshtkar, H.; Ashbaugh, L. L. Size Distribution of Polycyclic Aromatic Hydrocarbon Particulate Emission Factors from Agricultural Burning. *Atmos. Environ.* **2007**, *41* (13), 2729–2739.

40. Kharlamova, G.; Kirillova, N. Nanomaterials in Environmental Contamination: Their Nanotoxicological Peculiarities. In *Technological Innovations in Sensing and Detection of Chemical, Biological, Radiological, Nuclear Threats and Ecological Terrorism, NATO Science for Peace and Security Series: Chemistry and Biology*; Vaseashta, A., et al., Eds.; Springer Science+Business Media B.V.: Dordrecht, 2011; pp 131–140.

41. Kiviranta, H.; Hallikainen, A.; Ovaskainen, M. L.; Kumpulainen, J.; Vartiainen, T. *Dietary Intakes of Polychlorinated Dibenzo-p-dioxins, Dibenzofurans and Polychlorinated Biphenyls in Finland. Food Additives and Contaminants*; Kiviranta, H., et al., Eds.; Dietary Intakes of Polychlorinated Dibenzo-*p*-dioxins, Dibenzofurans and Polychlorinated Biphenyls in Finland. *Food Addit.* **2001**, *18* (11), 945–953.

42. Kunisue, T.; Watanabe, M.; Iwata, H.; Subramanian, A.; Monirith, I.; Minh, T. B.; Baburajendran, R.; Tana, T. S.; Viet, P. H.; Prudente, M.; Tanabe, S. Dioxins and Related Compounds in Human Breast Milk Collected Around Open Dumping Sites in Asian Developing Countries: Bovine Milk as a Potential Source. *Arch. Environ. Contam. Toxicol.* **2004**, *47*, 414.

43. Labunska, I.; Harrad, S.; Wang, M.; Santillo, D.; Johnston, P. Human Dietary Exposure to PBDEs around E-Waste Recycling Sites in Eastern China. *Environ. Sci. Technol.* **2014**, *48* (10), 5555–5564.

44. Lemieux, P. M.; Lutes, C. C.; Santoianni, D. A. Emissions of Organic Air Toxics from Open Burning: A Comprehensive Review. *Progr. Energy Combust.* Sci. **2004**, *30* (1), 1–32.

45. Llobet, J. M.; Domingo, J. L.; Bocio, A.; Casas, C.; Teixidó, A.; Müller, L. Human Exposure to Dioxins through the Diet in Catalonia, Spain: Carcinogenic and Non-carcinogenic Risk. *Chemosphere* **2003**, *50*, 1193–1200.

46. Ludewig, G.; Lehmann, L.; Esch, H.; Robertson, L. W. Metabolic Activation of PCBs to Carcinogens In Vivo—A Review. *Environ. Toxicol. Pharmacol.* **2008**, *25*, 241–246.

47. Lundin, L.; Gullett, B.; Carroll, Jr., W. F.; Touati, A.; Marklund, S.; Fiedler, H. The Effect of Developing Nations' Municipal Waste Composition on PCDD/PCDF Emissions from Open Burning. *Atmos. Environ.* **2013**, *79*, 433–441.

48. Manahan, S. E. *Environmental Chemistry*, eighth ed. CRC Press LLC: Boca Raton, FL, 2005.

49. Murphy, L. E.; Gollenberg, A. L.; Buck Louis, G. M.; Kostyniak, P. J.; Sundaram, R. Maternal Serum Preconception Polychlorinated Biphenyl Concentrations and Infant Birth Weight. *Environ. Health Perspect.* **2010**, *118*, 297–302.

50. Nakao, T.; Aozasa, O.; Ohta, S.; Miyata, H. Formation of Toxic Chemicals Including Dioxin-Related Compounds by Combustion from a Small Home Waste Incinerator. *Chemosphere* **2006**, *62*, 459–468.

51. NDDH-DAQ (North Dakota Department of Health, Division of Air Quality). *North Dakota Air Pollution Control Rules*, 2007 (Chapter 33-15-04).
52. Ni, H.-G.; Lu, S.-Y.; et al. Brominated Flame Retardant Emissions from the Open Burning of Five Plastic Wastes and Implications for Environmental Exposure in China. *Environ. Pollut.* **2016**, *214*, 70–76.
53. Noakes, P. S.; Taylor, P.; Wilkinson, S.; Prescott, S. L. The Relationship between Persistent Organic Pollutants in Maternal and Neonatal Tissues and Immune Responses to Allergens: A Novel Exploratory Study. *Chemosphere* **2006**, *63*, 1304–1311.
54. Ohiokpehai, O. Nutritional Aspects of Street Foods in Botswana. *Pak. J. Nutr.* **2003**, *2* (2), 76–81.
55. Parrot, L.; Sotamenou, J.; et al. Municipal Solid Waste Management in Africa: Strategies and Livelihoods in Yaounde, Cameroon. *Waste Manage.* **2009**, *29* (2), 986–995.
56. Proietti, I.; Frazzoli, C.; Mantovani, A. Identification and Management of Toxicological Hazards of Street Foods in Developing Countries. *Food Chem. Toxicol.* **2014**, *63*, 143–152.
57. Pussemier, L.; L. Mohimont, et al. Enhanced Levels of Dioxins in Eggs from Free Range Hens: A Fast Evaluation Approach. *Talanta* **2004**, *63* (5), 1273–1276.
58. Ramaswami, A.; Baidwan, N. K.; et al. Exploring Social and Infrastructural Factors Affecting Open Burning of Municipal Solid Waste (MSW) in Indian Cities: A Comparative Case Study of Three Neighborhoods of Delhi. *Waste Manage Res.* **2016**, *34* (11), 1164–1172.
59. Rappolder, M.; Bruders, N.; Schroter-Kermani, C. Comparison of Congener Patterns and TEQs in Environmental and Human Samples. *Organohal. Compd.* **2005**, *67*, 2086–2089.
60. Schmid, P.; Gujer, E.; Zennegg, M.; Studer, C. Temporal and Local Trends of PCDD/F Levels in Cow's Milk in Switzerland. *Chemosphere* **2003**, *53*, 129–136.
61. Schmidt, C. W. Unfair Trade: E-Waste in Africa. Environ. *Health Perspect.* **2006**, *114*, A232–A235.
62. Sindiku, O.; Babayemi, J. O.; Tysklind, M.; Osibanjo, O.; Weber, R.; Watson, A.; Schlummer, M.; Lundstedt, S. Polybrominated Dibenzo-*p*-dioxins and Dibenzofurans (PBDD/Fs) in E-Waste Plastic in Nigeria. Environ. Sci. Pollut. Res. **2015**, *22* (19), 14515–14529.
63. Stockholm Convention. *Stockholm Convention on Persistent Organic Pollutants*, 2001. Full Text for Download in English. http://www.pops.int/documents/convtext/convtext_en.pdf.
64. Tait, S.; La Rocca, C.; Mantovani, A. Exposure of Human Fetal Penile Cells to Different PCB Mixtures: Transcriptome Analysis Points to Diverse Modes of Interference on External Genitalia Programming. Reprod. Toxicol. **2011**, *32*, 1–14.
65. Tan, J.; Loganath, A.; Chong, Y. S.; Obbard, J. P. Exposure to Persistent Organic Pollutants in Utero and Related Maternal Characteristics on Birth Outcomes: A Multivariate Data Analysis Approach. *Chemosphere* **2009**, *74*, 428–433.
66. Todaka, T.; Hirakawa, H.; Kajiwara, J.; Hori, T.; Tobiishi, K.; Yasutake, D.; Onozuka, D.; Sasaki, S.; Miyashita, C.; Yoshioka, E.; Yuasa, M.; Kishi, R.; Iida, T.; Furue, M. Relationship between the Concentrations of Polychlorinated Dibenzo-*p*-dioxins, Polychlorinated Dibenzofurans, and Polychlorinated Biphenyls in Maternal Blood and those in Breast Milk. *Chemosphere* **2010**, *78*, 185–192.
67. Toms, L. M.; Harden, F.; Paepke, O.; Hobson, P.; Ryan, J. J.; Mueller, J. F. Higher Accumulation of Polybrominated Diphenyl Ethers in Infants than in Adults. *Environ Sci. Technol.* **2008**, *42*, 7510–7515.

68. Toms, L. M.; Hearn, L.; Kennedy, K.; Harden, F.; Bartkow, M.; Temme, C.; et al. Concentrations of Polybrominated Diphenyl Ethers (PBDEs) in Matched Samples of Human Milk, Dust and Indoor Air. *Environ. Int.* **2009**, *35*, 864–869.

69. Tue, N. M.; Goto, A.; Takahashi, S.; Itai, T.; Asante, K. A.; Kunisue, T.; Tanabe, S. Release of Chlorinated, Brominated and Mixed Halogenated Dioxin-Related Compounds to Soils from Open Burning of E-Waste in Agbogbloshie (Accra, Ghana). *J. Hazard. Mater.* **2016**, *302*, 151–157.

70. UN-HABITAT. *Collection of Municipal Solid Waste in Developing Countries*. United Nations Human Settlements Programme (UN-HABITAT): Nairobi, 2010.

71. UNEP. Guidance on the Global Monitoring Plan for Persistent Organic Pollutants. Stockholm Convention on Persistent Organic Pollutants (UNEP/POPS/COP.6/INF/31), 2013.

72. United Nations, Department of Economic and Social Affairs, Population Division. *World Population Prospects: The 2015 Revision*. United Nations: New York, 2015.

73. Valavanidis, A.; Iliopoulos, N.; Gotsis, G.; Fiotakis, K. Persistent Free Radicals, Heavy Metals and PAHs Generated in Particulate Soot Emissions and Residue Ash from Controlled Combustion of Common Types of Plastic. *J. Hazard. Mater.* **2008**, *156* (1), 277–284.

74. Van den Berg, M.; Denison, M. S.; Birnbaum, L. S.; Devito, M. J.; Fiedler, H.; Faландysz, J.; Rose, M.; Schrenk, D.; Safe, S.; Tohyama, C.; Tritscher, A.; Tysklind, M.; Peterson, R. E. Polybrominated Dibenzo-*p*-dioxins, Dibenzofurans, and Biphenyls: Inclusion in the Toxicity Equivalency Factor Concept for Dioxin-Like Compounds. *Toxicol Sci.* **2013**, *133*, 197–208.

75. Vassiliadou, I.; Papadopoulos, A.; Costopoulou, D.; Vasiliadou, S.; Christoforou, S.; Leondiadis, L. Dioxin Contamination after an Accidental Fire in the Municipal Landfill of Tagarades, Thessaloniki, Greece. *Chemosphere* **2009**, *74* (7), 879–884.

76. Waleij, A.; Edlund, C.; Holmberg, M.; Lesko, B.; Liljedahl, B.; Lindblad, A.; Melin, L.; Normark, M.; Sandström, B.; Sedig, M.; Sundström, S.; Westerdahl, K. S. *SUDAN Environmental and Health Risks to Personnel to be Deployed to Sudan—Pre-deployment Assessment*. FOI—Swedish Defence Research Agency, NBC Defence: Umeå, Stockholm, 2004.

77. Wevers, M.; De Fre, R.; Desmedt, M. Effect of Backyard Burning on Dioxin Deposition and Air Concentrations. *Chemosphere* **2004**, *54*, 1351–1356.

78. WHO. *Assessment of the Health Risk of Dioxins: Re-evaluation of the Tolerable Daily Intake (TDI)*. WHO Consultation: Geneva, Switzerland, 1998.

79. WHO. *Children's Health and the Environment. WHO Training Package for the Health Sector*. World Health Organization, 2008.

80. WHO. *Persistent Organic Pollutants: Impact on Child Health*. World Health Organization, 2010.

81. WHO. Safety Evaluation of Certain Food Additives and Contaminants. Supplement 1: Non-dioxin-Like Polychlorinated Biphenyls. *WHO Food Additives Series: 71-S1*, 2016.

82. Wiedinmyer, C.; Yokelson, R. J.; Gullett, B. K. Global Emissions of Trace Gases, Particulate Matter, and Hazardous Air Pollutants from Open Burning of Domestic Waste. *Environ. Sci. Technol.* **2014**, *48* (16), 9523–9530.

83. Wilhelm, M.; Wittsiepe, J.; Lemm, F.; Ranft, U.; Kramer, U.; Furst, P.; Roseler, S. C.; Greshake, M.; Imohl, M.; Eberwein, G.; Rauchfuss, K.; Kraft, M.; Winneke, G. The Duisburg Birth Cohort Study: Influence of the Prenatal Exposure to PCDD/Fs and

Dioxin-Like PCBs on Thyroid Hormone Status in Newborns and Neurodevelopment of Infants until the Age of 24 Months. *Mutat. Res.* **2008,** *659*, 83–92.

84. Wong, M.; Wu, S.; Deng, W.; Yu, X.; Luo, Q.; Leung, A.; Wong, C.; Luksemburg, W.; Wong, A. Export of Toxic Chemicals: A Review of the Case of Uncontrolled Electronic-Waste Recycling. *Environ. Pollut.* **2007,** *149*, 131–140.

85. World Bank. What a Waste—A Global Review of Solid Waste Management. *Urban Development Series Knowledge Papers*. World Bank, 2012. Available at http://go.worldbank.org/BCQEP0TMO0.

CHAPTER 12

INTRODUCTION TO EPITHELIAL-TO-MESENCHYMAL TRANSITION AND ITS ROLE AS POTENTIAL MARKER IN FOOD TOXICOLOGY

LUDOVIC PEYRE[1*] and MAEVA GIRAUDO[2]

¹UMR 1331 TOXALIM (Research Center in Food Toxicology), Institut National de la Recherche Agronomique (INRA), Laboratory of Xenobiotic's Cellular and Molecular Toxicology, 400 route des Chappes, BP 167, 06903 Sophia-Antipolis Cedex, France

²Environment and Climate Change Canada, Aquatic Contaminants Research Division, 105 McGill Street, Montreal, QC, Canada H2Y 2E7

*Corresponding author. E-mail: ludomailinra@gmail.com

CONTENTS

ABSTRACT

Epithelial-to-mesenchymal transition (EMT) is a process that occurs during the early (fibrosis) and late (metastasis) stages of carcinogenesis. On the other hand, EMT is a well-known physiological multistep phenomenon that leads to embryonic development and tissue repair. It is defined as the loss of epithelial characteristics such as cell adherent junctions, cell polarity and the gain of mesenchymal properties, thereby conferring to cells the ability to modify extracellular matrix, to become resistant to apoptosis, to migrate, and finally to invade the adjacent tissues. During the two last decades, many environmental xenobiotics were found to mimic endogenous signals (hormones, cytokines, etc.) leading to the activation of EMT. Thanks to the use of EMT biomarkers as predictive tool, it is now possible to identify in the same time the teratogenic and pro-tumoral effects of food contaminants.

12.1 EPITHELIAL-TO-MESENCHYMAL TRANSITION HISTORY

Epithelial-to-mesenchymal transition (EMT) is a complex biological process that has long been associated with different cellular phenomenon such as transformation, diffusion, and transdifferentiation, that is, the transition of already differentiated cells into other differentiated cells. Examples of such confusing association are numerous in the literature and represent an era in biology when the lack of appropriate tools made it difficult to detect and isolate epithelial, mesenchymal, and stem cell markers. Yet, despite the shortcomings, developmental biologists have been interested in this process for almost a century. In 1908, the Canadian zoologist Franck Rattray Lillie published the first description of a related-EMT phenomenon in "The Development of the Chick".[1] At the same time, the German biologist Theodor Boveri predicted the functional role of stroma in tumor development as one of the crucial events involved in cancer progression.[2] However, it was not until the end of 1960 that Elizabeth Hay from Harvard Medical School first described EMT during the early stages of embryonic development.[3] Her observations were confirmed 20 more years later by her student Gary Greenburg in his now famous experiment where he observed that chicken embryo epithelial cells deposited on a collagen matrix had adopted a mesenchymal phenotype.[4] In the meantime, Jean-Paul Thiery and his team from the French National Center for Scientific Research found that cultures of rat bladder carcinoma cells were able to transform into invasive mesenchymal tumor cells and then return to their original state.[5] As it was a fairly new phenomenon not yet fully understood, the term "plasticity" was commonly used for describing this cellular process. Jean-Paul Thiery, now recognized as an EMT pioneer, considers that the first true characterization of EMT was made by Michael Stoker and Michael Perryman in 1985, who identified scattered factors from fibroblasts responsible for the increasing migration of epithelial cells.[6,7] Finally in the 1990s, EMT became clearly associated with developmental mechanisms, tumor progression, and ultimately with metastasis.[8] Although this process was well described in a large number of in vitro models, especially thanks to Hay's studies,[9,10] it was not until 2008 that the first evidence of an EMT was observed in vivo in breast cancer.[11] Subsequently, further study of this unique phenomenon led to substantial progress in cancer treatment, through the understanding of EMT-induced resistance of tumor cells to therapeutic agents involved in cancer recurrence. Today, more than 14,000 articles have been published about EMT in different research areas including developmental biology, clinical pathology, environmental sciences, and toxicology. Interestingly, a paradox exists in food toxicology, where chemical pollutants,[12] toxins,[13] and metabolites[14] that are able to induce EMT in the body have antagonistic effects compared to natural EMT-inhibitor compounds such as curcumin,[15] vitamin D,[16] or polyphenols.[17] In this chapter, we will summarize the current knowledge about EMT processes and their involvement in diseases associated with food toxicology.

12.2 GENERAL MECHANISMS

The EMT is a complex process involving a cascade of cellular and molecular events. It is characterized by the loss of epithelial functions such as dissociation/degradation of tight, adherent junctions (AJs), and desmosomes. Altogether, these events lead to the ultimate loss of basolateral polarity while cells acquire new functions with the gain of mesenchymal characters.[18,19] Various cellular events have been observed during EMT such as changes in cytoskeleton organization with stress fibers formation, rearrangement of focal adhesions and modification of extracellular matrix (ECM).[20] In addition, cells embrace migratory and invasive abilities, increase aerobic glycolysis, and acquire resistance to apoptosis (Fig. 12.1). The EMT can be induced by different endogenous stimuli such as hormones in homeostatic conditions but also by infectious agents, toxins, chemicals, oxidative stress, inflammation (e.g., cytokines production), hypoxia, and after changes in ECM components. This cellular transition is facilitated by local hormonal

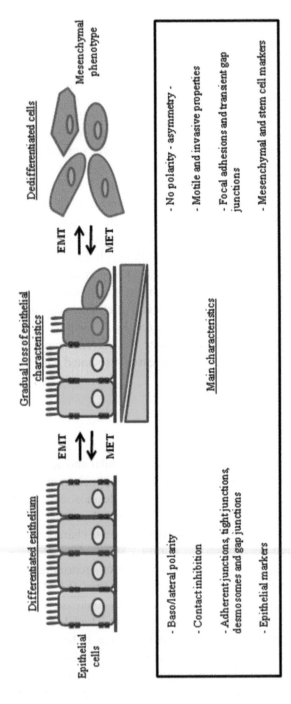

FIGURE 12.1   Evolution of cell characteristics during epithelial-to-mesenchymal transition (EMT).

secretion such as transforming growth factor beta (TGF-β), epidermal growth factor (EGF), insulin-like growth factor (IGF), or fibroblast growth factor (FGF) but also requires the continuous action of inducers [i.e., epidermal growth factor receptor, Hedgehog, Wingless integration site (Wnt)/β-catenin, TGF-β/TGF-βR, and ECM/β1-integrin signaling] to prevent the return of cells to their original phenotype.[21] The EMT is a reversible process that occurs during embryogenesis, wound healing, fibrosis, and metastasis, with tissue-dependent differences in the signaling pathways involved and the markers expressed.[22] Interestingly, cells that undergo EMT have pluripotent stem cells characteristics allowing them to leave the tumor, invade other tissue, and settle via a mesenchymal-to-epithelial transition (MET) before resuming proliferation. These cells are known to be resistant to therapeutic treatments and to be responsible for relapses.[23] The role of the areca nut in the pathogenesis of oral submucous fibrosis is a good example to illustrate the general mechanisms of EMT.[24] This inflammatory precancerous fibrotic condition has recently been described as a consequence of a TGF-β-dependent signaling pathway activation in response to areca nut, and leading to the activation of fibroblasts and to the expression of fibrotic markers such as a-SMA and collagen. On the opposite, astragalin, a kaempferol-3-*O*-glucoside found in persimmon leaves and green tea seeds, showed antifibrotic properties in the respiratory tract and the lungs by inhibiting EMT induced by ROS.[25] Altogether, these data reveal complex interactions between natural compounds found in food, but they also point out the fragility of the balance between positive and negatives effects when the number and the concentration of EMT inducers increase, especially through food chemical contamination.

12.3 DETAILED EMT EVENTS

12.3.1 *LOSS OF EPITHELIAL MARKERS AND FUNCTIONS*

12.3.1.1 *CELL–CELL INTERACTIONS*

During the first step of EMT, epithelial cells progressively lose the expression of key proteins participating in homeostasis, especially those involved in cell–cell interactions and in ECM-cell exchanges. In the literature, many examples of dietary phytochemicals or pesticides have been reported to modulate the expression of cell junction markers, resulting in the perturbation of cell–cell interactions. For example, while the diterpene carnosol[26] or the mixture of luteolin and quercetin[27] have protective effects against EMT-induced metastasis, many organochlorine pesticides were described to increase loss of cell–cell junctions.[28,29] Moreover, the loss of cell adhesion has been shown to correlate with migration, invasion, and finally dissemination inside the body.[30] In such conditions, a decrease of the four main adhesive and communicative junctions occurs: AJs, tight junctions (TJs), gap junctions (GJs), and desmosomes. The AJs are responsible for the stable contact between epithelial cells through the action of cadherins, a group of transmembrane glycoproteins that mediate calcium-dependent homophilic cell-to-cell adhesion.[31,32] The E-cadherin appears to be the major component of AJs and is linked to the cytoskeleton via intracellular molecules involved in AJ stabilization and called catenins (i.e., α/β-catenins and P-120-catenin). The TJs are mainly composed of transmembrane proteins called claudins and occludins, localized in the apical-lateral part of epithelial cells where they play an important role in cell polarity.[33] They separate the apical and basolateral cell surface domains and inhibit solute and water flow through the paracellular space.[34] The GJs are transmembrane channels made of two pores called connexons. Each connexon is composed of a ring of six connexin-type proteins, which allow the selective control of small molecules trafficking (<1 kDa) between adjacent cells.[35] Finally, desmosomes form attachment points between the side edges of adjacent epithelial cells and cadherin-like molecules such as desmoglein and desmocollin. They bind to cytokeratin fibers through transitional proteins such as desmoplakin and γ-catenin to form desmosomal plate.[36]

12.3.1.2   DUAL ROLE OF SENESCENCE

Lessons learned from cancer research allowed biologists to acknowledge the remarkable cell strategies applied in the case of unsuitable events such as the loss of cell contact, telomere dysfunction, and oncogenic, oxidative, or genotoxic stresses, which all lead to the inhibition of abnormal proliferation. In early stages of tumor progression, the cancer safeguard program can engage an apoptotic process, induce a transient growth arrest, or enter senescence.[37] This latter response induces a cessation of cell growth followed by numerous metabolic changes and an inability of cells to proliferate despite the presence of growth factors and nutrients. It results in an increase in cell volume and other morphological changes of the cytoplasm and the nucleus. Although senescence is usually considered as a tumor-suppressor mechanism,[38,39] it may also be associated with events facilitating tumor progression and leading to the metastatic state. Indeed, senescence is prevented during EMT. This phenomenon seems to be partly related to the overexpression of twist family BHLH transcription factors (TWIST1/2) that are found in many cancers and known to interfere with zinc finger E-box-binding homeobox (ZEB1),[40] P53, and retinoblastoma (Rb) proteins.[41] Paradoxically, senescence plays also an important role in modulating the interaction between stromal cells of the microenvironment. For example, the increase of senescence in fibroblasts localized in tumors can result in the gain of pro-inflammatory abilities. These cells called "senescence-associated secretory phenotype" release interleukins targeting neighboring epithelial cells and leading to tumor progression via the induction of the EMT processes.[42]

12.3.1.3   RESISTANCE TO ANOÏKIS

An increase of cell survival has been described during EMT. In the epithelial tissue, numerous interactions are necessary for cell homeostasis such as cell/cell interactions controlled by cadherins and cell/ECM interactions where integrins play an important role.[43] Together, these interactions induce the activation of different signaling cascades that regulate the expression of anti-apoptotic genes. During the progressive loss of adherence in homeostatic conditions, cells can enter into an apoptosis-like process called anoïkis. During EMT, however, the loss of epithelial cell adherence happens without triggering apoptosis through a phenomenon called resistance to anoïkis.[44] This well-known physiological process allows cell cycle control and the intestinal epithelium renewal.[45] Chemical substances such as paraquat,[46] nitric oxide exposure,[47] or high-glucose uptake[48] are known to promote such a resistance to anoïkis. As in apoptosis, anoïkis induces the activation of caspases and endonucleases, DNA fragmentation, modulation of B-cell lymphoma-2 (Bcl2)/B-cell lymphoma extralarge (Bcl-xl) or Bcl2-associated X protein (Bax)/Bcl-xl balance, and the activation of intrinsic (e.g., mitochondrial permeabilization with cytochrome C release), and extrinsic (e.g., Fas ligand and tumor necrosis factor-receptor (TNF-R) activation) signals. Other molecular actors are also involved in anoïkis such as the tumor suppressor p16[INK4a], a cyclin-dependent kinase inhibitor that controls cell cycle progression. It is worth nothing that cells undergoing EMT are very resistant to cancer treatments such as chemotherapy.[49] This resistance is in part regulated by Ras-extracellular signal-regulated kinase (Ras/ERK), phosphatidylinositol-3-kinase (PI3K)/Akt, and nuclear factor-kappa B (NF-κB) pathways, responsible for anti-apoptotic gene expression and cell survival.[50] As apoptosis-resistant tumor cells are often linked to a high metastasis potential, EMT is now described as an important event during cancer progression.

12.3.2   GAIN OF MESENCHYMAL MARKERS

12.3.2.1   CELL JUNCTIONS

Mesenchymal cells are nonpolarized cells, which are able to migrate individually and to invade due to the lack of intercellular connections and to the specific enzymatic secretion resulting in ECM degradation.[51] They show close similarities to the organization and localization of fibroblasts. For example, mesenchymal cells have low levels of cytokeratins in contrast to epithelial cells.[52] While E-cadherin is considered by specialists as the most expressed and well-characterized EMT epithelial marker, N-cadherin was described as one of the major mesenchymal marker. This protein is usually found in premigrating cells such as neuronal and endothelial cells, in fibroblasts, osteoblasts (OBs), myocytes, oocytes, Sertoli cells, spermatids, and in the cartilage. N-cadherin physically interacts with cells from the stromal and endothelial compartments and with the FGFR1, leading to a continuous activation of the mitogen-activated protein kinases (MAPKs)/ERK pathway. It was shown that the increase in N-cadherin expression is concomitant with the inhibition of E-cadherin gene expression, which confers migration abilities to the cells[53] as it was recently described in the case of Benzophenone-1 and 4-*tert*-octylphenol, two suspected endocrine disruptors (EDs), in human ovarian cancer cells.[54] Consequently, the E-cadherin/N-cadherin switch is now recognized as the most efficient tool for studying EMT. OB-cadherin (i.e., cadherin-11) has also been described as an EMT predictor but data are conflicting and depend on the cell model studied. OB-cadherin is mainly expressed in the mesenchymal tissue and seems to interact with the TGF-β-signaling pathway. The role of this marker was first described during wound healing, then in fibrosis promotion by facilitating the differentiation of resident tissue fibroblasts into myofibroblasts.[55]

12.3.2.2   TRANSCRIPTION FACTORS PRODUCTION

A wide number of factors were first described for their role in embryogenesis and have since then been associated with EMT. Snail1 and 2 (also known as Slug), Twist1 and 2, and ZEB1 and 2 are some of these deciding factors whose role depends on the cell type.

Snail genes are strong repressors able to negatively regulate the expression of the EMT markers claudin, occludin, and E-cadherin.[56] They are regulated by FGF, hepatocyte-growth factor (HGF), Wnt, TGF-β, bone-morphogenetic protein (BMP), EGF, Notch, stem cell factor, integrins, cytokines, and by estrogens.[57] An increase of *snail* gene expression in different types of cancer was correlated with the downregulation of other epithelial markers such as cytokeratins or desmoplakin, but also with the decrease of proliferative and survival gene expression (e.g., *Cyclin D, Cyclin-dependent kinase 4, P53, BH3 interacting-domain, Caspases*). The expression of *snail* is also positively correlated with the increase of mesenchymal markers such as fibronectin and vitronectin, and associated with changes in cell shape and movement through the positive regulation of metalloproteases (MMPs) and RhoB.[58] Finally, *snail* gene expression is correlated with invasiveness during cancer progression leading to metastasis.[59]

Twist1 and 2 are two basic helix–loop–helix transcription factors found overexpressed in many cancers and correlated with metastasis.[60] The expression of these two proteins is controlled by signal factors (e.g., FGF, IGF1, EGF, BMP, Wnt) and cytokines (e.g., TGF-β, TNF-α, interferon alpha, interleukin 17 (IL-17), by transcription factors [e.g., msh homeobox 2 (MSX2), v-myc avian myelocytomatosis viral oncogene neuroblastoma-derived homolog (N-MYC), NF-κB, proto-oncogene tyrosine-protein kinase 1 (SRC-1), signal transducers and activators of transcription 3 (STAT-3)], and by stress conditions (e.g., hypoxia, mechanical compression). Twist 1 and 2 regulate E-cadherin expression, inhibit the activation of the onco-suppressive pathway Myc/ADP-ribosylation factor (ARF)/p53, and modulate the activities of myogenic differentiation (MyoD), myocyte enhancer factor 2 (MEF-2), runt-related transcription factor 1 and 2 (RUNX1/2), peroxisome proliferator-activated receptor gamma coactivator 1 alpha (PGC1-α) and NF-κB transcription factors.[41] Twist1 and 2 have embryonic, tissue specification and oncogenic (i.e., chemoresistance, angiogenesis) properties, which make them essential during EMT.

ZEB1 and 2 are mostly expressed in mesenchymal and neuronal progenitor cells and found to be mutated in developmental diseases.[61] ZEB1 and 2 factors are regulated by TGF-β/Smads, Ras/MAPK, hypoxia-inducible factor (Hif)-1α, Rb, Nf-κB, and Wnt/β-catenin-signaling pathways. They repress E-cadherin gene expression but also the expression of genes involved in epithelial homeostasis such as TJ protein coding gene *ZO-3* and *plakophilin-2*. They activate MMP family members, increase stem cells markers such as Kruppel-Like Factor 4 (KLF4), and are also able to repress the miR-200 family.[57]

Other important transcription factors involved in EMT have been characterized.[62–68] Among them, osteopontin-A (OPN), Hif-1α, SRY-related HMG-box (SOX), Gli, c-MYC, octamer-binding transcription factor 4 (OCT-4), NANOG, T-cell factor 3 and 4 (TCF-3/4), MSX2, DNA-binding protein inhibitor 1 (Id-1), Smad2/Smad4/lymphoid enhancer factor 1 (Lef-1), NF-κB, forkhead box protein 2 (FOX-C2), estrogen receptor alpha (ER-α), CArG box-binding factor A (CBF-A)/KRAB-associated protein 1 (KAP-1)/ fibroblast transcription site-1 (FTS-1), and Ets-1 have been shown to modulate the expression of specific markers and to be involved in different cellular processes such as the loss of AJs, actin disassembly, ECM reorganization, migration, and invasion.

12.3.2.3 INTERACTION WITH THE EXTRACELLULAR MATRIX

During EMT, a switch in ECM proteins synthesis leads to the increase of interstitial components such as type 1 collagen, fibronectin, secreted protein acidic and rich in cysteine (SPARC) and tenascin-C.[69] By influencing cell behavior and function, they become essential for driving migration, adhesion, invasion, and proliferation and can be considered as both EMT and tumor progression markers.[70]

In the meantime, there is a change in the expression of integrins with the increase of α5β1, αvβ6, and αvβ3 forms.[71] They are transmembrane proteins, which belong to the heterodimeric receptors family and form a large complex at the cytoplasmic membrane with different proteins [e.g., integrin-linked kinase (ILK), Src, paxilline, focal adhesion kinase (FAK)]. Integrins connect ECM proteins to cytoskeleton components. When they are over expressed with their specific ligands such as fibronectin and vitronectin, they play a central role in migration, invasion, and proliferation.[72] Integrin α5β1 is the most described of this class of proteins and is necessary for TGF-β-dependent EMT.[73]

The ILK, a multifunctional receptor and intracellular protein linked to integrins, is also used as an EMT marker. This protein regulates a variety of cellular processes including cell growth, proliferation, apoptosis, survival, differentiation, migration, and invasion.[74] It is found overexpressed in a wide number of cancers and is also a marker of poor prognosis. This kinase is located in focal adhesions and is able to activate the PI3K/AKT pathway by mediating integrin-signaling activation.[75]

The proteolytic enzymes MMPs can be secreted or associated with the cell membrane. Their main functions are the degradation and remodeling of ECM components. Consequently, they are essential for cell proliferation, differentiation, and morphogenesis during embryogenesis, but also in the maintenance of tissue homeostasis and in the control of dynamic morphogenic processes in adults such as wound healing. The MMPs are under the control of TWIST-1 factor and tissue inhibitor of metalloproteinase protein inhibitors.[57] Because mesenchymal cells secrete more MMPs than epithelial cells, they are used as EMT indicators.[76]

12.3.2.4 CHANGES IN ACTIN CYTOSKELETON: ROLE IN CELL MIGRATION AND INVASION

Vimentin, an important stress fibers component, is probably one of the most used EMT markers. This protein belongs to the intermediate filaments family and is mostly induced in mesodermal cells during development, in precursor cells and in fibroblasts. Vimentin expression is associated in cancer with an invasive phenotype and a poor prognosis,[77] making a good EMT marker. However, because vimentin can be expressed in stressed cells and during transitory migration of epithelial cells,[76] it must be used with caution like the other traditional EMT indicators.

Cell migration is a three-step process, depending first on stimulus perception from cellular receptors [i.e., G-protein-coupled receptor and receptor tyrosine kinase (RTK)], then on signal transduction by intracellular pathways (i.e., AKT/PI3K, Src/FAK, janus kinase (JAK)/STAT and ERK/MAPK), which ultimately lead to the last step of cell orientation and movement.[78] This multistep process occurs thanks to proteins that are controlling actin/myosin polymerization and filament assembly. Among them, Rho GTPases activation (i.e., Rho, Rac, and cell-division control protein 42 Cdc42) appears to be an essential phenomenon in cytoskeleton dynamic. These proteins play a central role in contractile forces organization, stress fibril formation and are also responsible for substrate adhesion during cell movement.[79] The complexity of Rho GTPase activities during cellular motility depends on three groups of regulatory proteins: guanine nucleotide exchange factors, GTPase-activating proteins, and guanine nucleotide dissociation inhibitors. It also relies on the interaction with serine/threonine p21-activated kinases, Wiskott–Aldrich syndrome protein, rho-associated protein kinase (ROCK), P120-catenin, actin-related protein 2 and 3 (Arp2/3) and mDia2 protein activation.[80] During EMT, the increase of Rho GTPase activity is linked to the gain of migratory properties thanks to the upregulation of actin turnover, focal adhesion formation, and cytoskeletal rearrangement.[81]

Finally, MMPs activities and integrin receptors play also a central role during the invasion process. Invasive cells must leave the tumor and cross the endothelial barrier to reach the blood and lymphatic circulation (i.e., extravasation), which requires a close collaboration with endothelial cells, macrophage, and other ECM cells. Invasion also allows mesenchymal cells to infiltrate and colonize the neighboring tissues (i.e., intravasation). This was confirmed a few years ago via circulating tumor cells (CTC). Indeed, CTC were found in blood of patients with highly expressed EMT markers,[82] which underlines the importance of EMT during cancer progression. Finally, the newly installed cancer cells can then become proliferative after the initiation of a reverse EMT-process, also known as MET, and the re-expression of epithelial markers, which represents a major step during metastasis.[83]

12.3.2.5   CELL SIGNALING DURING EMT

Two types of cell signaling exist during EMT: one microenvironment-dependent extrinsic signal (e.g., low injury pressure, injury) and one intrinsic signal depending on internal signals (e.g., insulin, FGF, HGF) and on transduction molecule mutations. Several feedback loops and many crosstalks exist between both signaling pathways and are detailed below.[84]

The ERK/MAPK pathway is a major activation cascade, which contains numerous kinases including RAS, RAF, MEK, ERK1, ERK2, and P90RSK. It controls a wide number of downstream transcription factors that regulate the expression of genes involved in differentiation, proliferation, survival, motility, adhesion, invasion, and survival.[85] In many cancers, this pathway is constitutively active, thus making the ERK/MAPK pathway a target of interest in the search for treatment preventing the disease development.[86] The ERK/MAPK is important for EMT because it contributes to the maintenance of an undifferentiated/mesenchymal state in tumor cells. It also cooperates with other pathways, mainly the TGFβ/SMAD pathway, to upregulate the expression of EMT-related genes such as *Snail1/2, Zeb1/2*, and *Twist1/2*. Finally, it regulates the expression of ECM proteases (e.g., MMPs) and modulates the activation of the cytoskeleton Rho/Rac pathway.[87]

The PI3K/AKT pathway is involved in cell morphology, survival, migration, and invasion. The serine/threonine kinase AKT is a downstream effector of PI3K. This pathway needs the activation of RTKs or growth factor receptors [e.g., IGF, FGF, HGF, EGF, Crypto (CFC)] in order to activate the specific intermediates phosphatidylinositol (3,4,5)-trisphosphate (PIP3), 3-phosphoinositide dependent protein kinase-1 (PDK1), glycogen synthase kinase 3 beta (GSK3-β) and mammalian target of rapamycin (mTOR) through phosphorylation. Similarly to ERK/MAPK, PI3K/AKT has an important role during mesoderm formation and was also found activated during epithelial cancers.[88] The AKT can be activated by ILK, depending on integrin interaction with the ECM proteins collagen or fibronectin.[89] The PI3K/AKT pathway is a major signaling cascade during EMT and a potential target in cancer therapy through its interaction with NF-κB, NOTCH, TGF-β, and WNT/β-catenin pathways and via the modulation of Snail expression leading to the down regulation of E-cadherin.[90]

The WNT/β-catenin signaling pathway targets cell growth and survival, proliferation, and the expression of invasion-related genes. This is an essential pathway during early stage of embryogenic development (i.e., formation of the primary axis), as well as in stem cells regulation necessary for tissue regeneration.[91] Crosstalk between the TGF-β and Wnt-signaling pathways have been identified. The WNT/β-catenin signaling pathway is found constitutively active in numerous cancers, and its activation is linked to the loss of cell junctions during EMT.[92] Indeed, the destabilization of cadherin/catenin complex leads to an increased level of a β-catenin pool in the cytoplasm and in the nucleus where it is ultimately associated with a transcription factor complex (TCF/LEF) that regulates the expression of genes involved in EMT [e.g., *S100a4*, *MMP7*, *c-Myc*, *CD44*, *cyclin D1* (*CCND1*), *Wnt-inducible-signaling pathway protein 1* (*WISP-1*), *vascular endothelial growth factor*, *fibronectin-1*, *TWIST* and *SNAIL*].[93,94]

The TGF-β/SMAD is another important pathway involved in the control of homeostasis, differentiation, and apoptosis in normal cells. It contributes to EMT during development, fibrosis and cancer where it is associated with cell migration. The TGF-β signaling pathway regulates cytoskeleton reorganization and protein expression through the activation of Smads, RhoA/ROCK, PI3K/AKT, ERK1/2, and p38MAPK, thus impacting cell motility, contractility, and shape.[95] Moreover, it regulates the production of ECM proteins such as collagens, fibronectin, and MMPs,[96] and the expression of the master transcription factors Snail, Zeb, FOXC2, and Twist.[97]

Finally, many other pathways are also involved in the EMT signaling such as Jagged/Notch,[98,99] Hedgehog,[100] Src/Fak,[101] protein kinase C (PKC),[102] IL6-/STAT3,[103] and HIF-1α[104] pathways (Table 12.1).

TABLE 12.1   Nonexhaustive List of EMT Markers.

Epit helial markers	Mesenchymal markers
Transmembrane proteins	**Transmembrane proteins**
E-cadherin	N-cadherin
VE-cadherin	Cadherin 11 (OB-cadherin)
Claudins	Integrins α5β1, αvβ6, αvβ3
Occludins	
Desmoplakin	
Plakophilin	
ZO-1	
Crumbs3	
Cytoskeleton	**Cytoskeleton**
Cytokeratins	Vimentin, α-SMA
	Rho, ILK, and FAK activities
	Myosins, tenascin C, fascin
	SPARC, S100A4 (FSP1)
Extracellular matrix	**Extracellular matrix**
Laminin 1	Laminin 5
Mucins	Fibronectin
WISP-2	Vitronectin
	Collagens I, III
	MMPs
Transcription factors	**Transcription factors**
	Snail/Slug, ZEB1/2, Twist1/2
	Id-1, Smad2/3/4, HIF-1α
	OPN, Ets-1, FOXC2, ERα
	β-Catenin/WNT, NF-kB
	CBF-A/KAP-1, TCF3/4
miRNA	**miRNA**
miR-200, -203, -205	miR-21
miR-1, -20, -29b, -30a	miR-130, -206
miR-192/215, miR-34	miR-221/222
Cell cycle	**Cell cycle**
	Cyclins A, D, E

Proteins, factors, and miRNAs involved in the maintaining of epithelial characteristics (left) and in the change into a mesenchymal phenotype (right) are indicated.

12.4 EMT MARKERS: WHY ARE THEY HELPFUL TOOLS FOR TOXICOLOGY?

With the advent of the chemical industry in the 1950s and the growing needs of new technologies, the production and release of environmental contaminants [e.g., pesticides, drugs, plastics, polychlorinated biphenyls (PCBs), dioxins] has increased dramatically worldwide. More than 100,000 molecules have been authorized and are currently used in a wide variety of domestic and industrial applications. However, current evaluation tools and regulations have not progressed as quickly as the chemical innovations themselves (e.g., nanotechnologies, *Bacillus thuringiensis* (*Bt*) pesticides, additives), whereas compelling bodies of evidence have continued to accumulate linking the effect of contaminants to the development of pathologies. It is now well documented that the increase rate of some diseases (e.g., allergies, obesity, cancer, neurodegenerative symptoms) cannot be the sole consequence of genetic variation but is in fact associated with the ubiquitous presence of contaminants in all environmental compartments (i.e., water, air, and sediments). Through their position at the top of the food chain, humans are particularly exposed to these persistent molecules such as organochlorine pesticides and PCBs and are vulnerable to their cumulative effects in mixture. In order to enhance traditional toxicological testing methods, EMT makes an ideal, consensual, reliable, and inexpensive tool to assess the effects of environmental contaminants. It has been proven to be particularly robust, predictive with good reproducibility between in vitro and in vivo models in carcinogenesis studies and can be used to evaluate contaminant effects on various processes such as development (embryogenesis), fibrosis (predisposing ground for more serious diseases), and the progression of cancer leading to metastasis (Fig. 12.2A and B).

As EMT was originally described in distinct and independent fields of biology, it has been classified into three different types in order to better illustrate the whole mechanism depending on the biological context.[22] Type-1 EMT is called mesenchymal and happens during embryogenesis and physiological development. Type-1 EMT induces a mesenchymal step from the blastoderm through EMT, before transforming into secondary epithelia thanks to a MET process. The EMT/MET cycles are then repeated to allow full organ development.[105] During type-1 EMT, *sox* gene family involved in migration allows the generation of mature cells and tissues similar to what happens during the neural crest formation,[106] the gastrulation,[107] and during the establishment of heart.[108] Hence, the use of EMT biomarkers as predictive tool to investigate the developmental effects of chemicals could be of

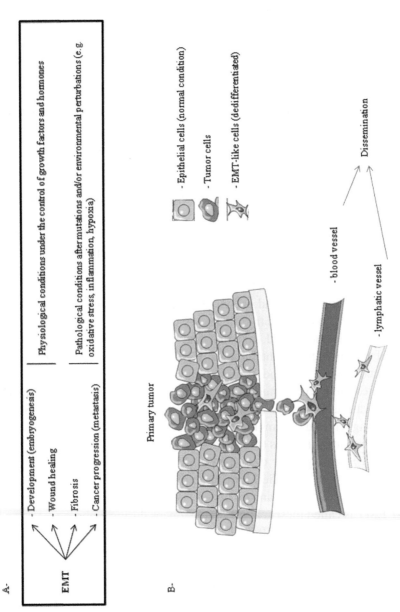

FIGURE 12.2 (A) Physiological and pathological involvement of epithelial-to-mesenchymal transition (EMT). (B) Proposed mechanism of the EMT process during tumor progression and leading to metastasis.

interest. To illustrate this idea, Doi and colleagues demonstrated that pregnant female rats exposed to 100 mg of intragastrically administered Nitrofen, a widely used herbicide in the late 20th century, showed an induction of the expression of *iroquois* genes during fetal early lung development (day 15 of gestation), resulting in pulmonary hypoplasia via lung dismorphogenesis.[109] Another example is the impact of bisphenol A (BPA), a leachable monomer of polymerized polycarbonate plastics used for a wide range of applications including food packaging, baby bottles, epoxy resins, and dental sealants.[110] Chronic (2 months) exposure of males rats to 1 and 0.1 mg/kg of oral BPA per day caused moderate corporal veno-occlusive dysfunction through the induction of the EMT process.[111] Other studies using mice epicardial cell cultures exposed for 24-h to 1.34 μM or 0.13 μM of the arsenic metabolites arsenite and monomethylarsonous acid, respectively, showed a potential disruption of cardiac EMT, predisposing to cardiovascular disorders.[112] Arsenic can also block developmental EMT gene programming in murine coronary progenitor cells exposed for 18 h to 1.3 μM of arsenite.[113] Finally, a very interesting study pointed out the impact of environmentally persistent free radicals (EPFRs) on neonatal mouse lungs.[114] Authors showed that 24-h exposure to low doses (20 μg/cm²) of EPFRs consisting of 1,2-dichlorobenzene chemisorbed onto a silica/Cu(II)O substrate-induced EMT in murine airway epithelial cells, thus providing an explanation for epidemiological evidence supporting particulate matter exposure and increased risk of asthma. The EMT seems therefore to be a key marker for early effect studies of drugs and environmental contaminants on development.

The type-2 EMT is called fibroblast and happens on secondary epithelial or endothelial cells during tissue regeneration (wound healing), inflammation, and fibrosis.[22] During fibrosis, myofibroblasts and fibroblasts modulate ECM composition by secreting enzymes and factors (e.g., MMPs, S100a4, platelet-derived growth factor, TGF-β) and by inducing the expression of high levels of ECM proteins (e.g., collagens, fibronectin, and tenascin), which accumulate and modify cell/ECM interactions. These events are mainly under the control of the TGF-β pathway, which was described as a key regulator during type-2 EMT. Fibrosis is linked to chronic inflammation in response to virus infection, toxic agents, or to autoimmune responses,[115] and leads to the increase of excessive protein deposition. Correlations between environmental exposition to xenobiotics and fibrosis have been reported in the literature, such as for paraquat, an organic heterocyclic herbicide widely used in agriculture. This compound is known to be associated with acute lung injury and was shown to induce EMT leading to pulmonary fibrosis,[46,116] and in lung tissues of rats up to 21 days after a single intraperitoneal injection of 15 mg/kg.[117] Another study using Wistar rat liver revealed the profibrotic effects of a 56-day dietary exposure to sublethal doses (15–30 mg/kg/day) of ametryn, a triazine herbicide.[118] Organochlorine pesticides can also induce EMT in human primary hepatocytes thus predisposing to fibrosis in vitro (see Fig. 12.3).[29] Finally, TCDD (2,3,7,8-tetrachlorodibenzo-p-dioxin) showed profibrotic properties in liver of mice after 6 weeks of treatment with weekly intraperitoneal injections of 25 µg/kg by activating EMT through the interaction with the aryl hydrocarbon receptor.[119]

Type-3 EMT is linked to cancer progression and is a crucial step in the gain of migratory and invasive characteristics.[22] As in type-1 EMT, an EMT/MET switch allows the mesenchymal cancer cells to colonize new organs by forming secondary tumor nodules (i.e., distal metastatic formation). Constitutive genetic mutations and epigenetic changes (e.g., DNA methylation, modulation of miRNA) are necessary for sustaining EMT activation. Type-3 EMT is responsible for the metastatic behavior of epithelial carcinoma by inducing the expression of specific markers such as Nanog, Oct-4, CD44, SOXs, and epithelial cell adhesion molecule (EpCam) in tumor cells. Hypoxia is one of the strongest signals for tumor progression and is considered as an aggravating factor allowing proliferation, survival, angiogenesis, and invasion through the activation of the HIF factors. The HIF-1α stability triggers EMT processes via the activation of the TGF-β, NF-κB, and NOTCH-signaling pathways.[120] Type-3 EMT induction by environmental toxicants has also been reported in the literature. Wang and colleagues showed that the nicotine-derived nitrosamine ketone (NNK) found in tobacco leaves and cigarette smoke was able to induce the migration of gastric cancer cells in a dose-dependent concentration manner via an EMT initiation process after 24-h of treatment with 1–10 µM of NNK.[121] In another study, treatment with 10 nM to 1 µM of p,p'-DDT (dichlorodiphenyltrichloroethane), a persistent organic pollutant (POP), was found to decrease HepG2 cell adhesion over a period of 6 days of exposure, along with the concomitant increase of cell/matrix adhesions and ROS-mediated JAK/STAT3 pathway.[122] The organochlorine pesticide and POP endosulfan was also able to induce an EMT-like phenomenon and to increase the anoïkis resistance in the same hepatoblastoma cell line after 48-h of treatment with doses ranging from 10 to 20 µM, thus conferring procarcinogenic potential to the liver.[28]

EDs such as natural hormones and synthetic molecules that mimic endogenous hormone action are able to induce EMT, making it a powerful tool to study their effects. Among the numerous chemicals identified as potential ED, BPA has drawn toxicologists' attention. Human exposure to BPA through food and water has become a great concern worldwide.[123]

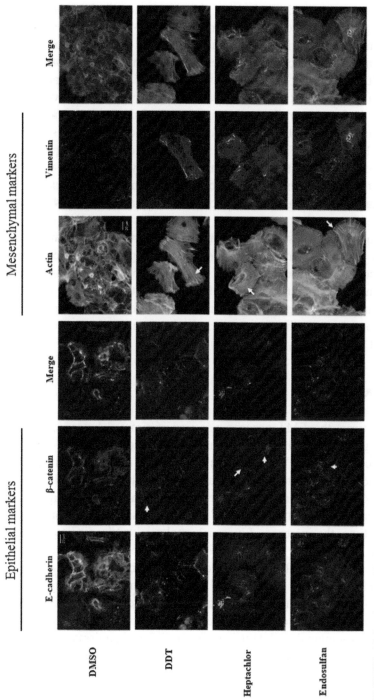

FIGURE 12.3 Organochlorines disrupt adherent junctions, disorganize the actin cytoskeleton, and increase the expression of vimentin in human hepatocytes.[29] (Reprinted from Zucchini-Pascal, N.; Peyre, L.; de Sousa, G.; Rahmani, R. Organochlorine Pesticides Induce Epithelial to Mesenchymal Transition of Human Primary Cultured Hepatocytes. *Food Chem. Toxicol.* **2012,** *50* (11), 3963–3970. Copyright 2012 with permission from Elsevier.)

Recently, it was shown that BPA was able to promote EMT processes in estrogen negative breast cancer cells treated with 1–10 nM of BPA for 24 h,[124] in human endometrial carcinoma cells after 14-day exposure to 10 nM[125] and in ovarian cancer cells treated for 24–48 h with 1 μM of BPA.[126] These results suggest that BPA is a possible type-3 EMT inducer, and additional data from different studies indicate that it could also be involved in type-1 EMT in female gerbils exposed in utero for 7 days to 40 μg/kg/day[127] and in mammary glands from mouse fetuses exposed to low doses of BPA (250 ng/kg BW/day).[128] Finally, it was shown that a diet-borne exposure of male Sprague–Dawley rats to 5–100 mg BPA/kg/day for 30 days was able to induce a type-2 EMT through cardiac fibrosis,[129] and through endometrial periglandular fibrosis in adult mice exposed to dietary BPA (0.004–40 mg/kg/day).[130]

Altogether, these results underline the importance of using EMT markers to provide information on both proteratogenic and procarcinogenic effects of new molecules and substances of concern for human and environmental health. The use of EMT as a toxicological tool can be particularly interesting regarding the effects of food-borne exposure to chemical contaminants that are found in high amount in human diet such as BPA and pesticides.

12.5 CONCLUSION

For a long time, the EMT process was underestimated by toxicologists, despite the growing number of data published on the loss of GJs[131–134] and changes in cell plasticity[135–137] in vitro and in vivo in response to chemicals. It took decades for EMT to be investigated in a developmental and carcinogenic context. EMT can easily be studied by following the disappearance of epithelial markers under the control of specific transcription factors,[138] and in the meantime, the increase of mesenchymal markers leading to the gain of migratory and invasive properties. The importance and relevance of such markers are now undeniable in the field of ecotoxicology and human toxicology. EMT represents a powerful tool that would only require a handful of analytical methods such as Western blot, quantitative real-time PCR (qRT-PCR) and immunofluorescence microscopy to become a useful indicator of food and environmental pollutant impacts on living organisms, from early development stages to the occurrence of pathology such as fibrosis and cancer. It can also be used in association with modern technologies such as high-content screening and can be completed by additional common markers such as apoptosis, cell cycle, or signaling pathways.[12] Future research efforts are now focused on the validation of EMT markers as toxicological assays. This could help pharmaceutical companies during early drug discovery and development and could benefit academic laboratories by providing reliable and inexpensive methods to assess the effect of pollutants.

KEYWORDS

- **epithelial-to-mesenchymal transition**
- **predictive tool**
- **cancer**
- **development**
- **dietary phytochemicals**
- **pesticides**

REFERENCES

1. Lillie, F. R. *The Development of the Chick: An Introduction to Embryology. Henry Holt & Co., 1908. Anat. Rec.* **1909,** *3* (3), 141–144.
2. Wunderlich, V. JMM—Past and Present. Chromosomes and Cancer: Theodor Boveri's Predictions 100 Years Later. *J. Mol. Med. (Berl.)* **2002,** *80* (9), 545–548.
3. Hay, E. D. Organization and Fine Structure of Epithelium and Mesenchyme in the Developing Chick Embryo. In *Epithelial-Mesenchymal Interactions*; Fleischmajer, R., Billingham, R. E., Eds.; Williams and Wilkins: Baltimore, MD, 1968; pp 31–55.
4. Greenburg, G.; Hay, E. D. Epithelia Suspended in Collagen Gels Can Lose Polarity and Express Characteristics of Migrating Mesenchymal Cells. *J. Cell Biol.* **1982,** *95* (1), 333–339.
5. Boyer, B.; Tucker, G. C.; Valles, A. M.; Gavrilovic, J.; Thiery, J. P. Reversible Transition towards a Fibroblastic Phenotype in a Rat Carcinoma Cell Line. *Int. J. Cancer Suppl.* **1989,** *4*, 69–75.
6. Stoker, M.; Perryman, M. An Epithelial Scatter Factor Released by Embryo Fibroblasts. *J. Cell Sci.* **1985,** *77*, 209–223.
7. Thiery, J. P. Epithelial–Mesenchymal Transitions in Development and Pathologies. *Curr. Opin. Cell Biol.* **2003,** *15* (6), 740–746.
8. Savagner, P.; Boyer, B.; Valles, A. M.; Jouanneau, J.; Thiery, J. P. Modulations of the Epithelial Phenotype during Embryogenesis and Cancer Progression. *Cancer Treat. Res.* **1994,** *71*, 229–249.
9. Hay, E. D. Role of Cell–Matrix Contacts in Cell Migration and Epithelial–Mesenchymal Transformation. *Cell Differ. Dev.* **1990,** *32* (3), 367–375.

10. Hay, E. D. An Overview of Epithelio-Mesenchymal Transformation. *Acta Anat. (Basel)* **1995,** *154* (1), 8–20.

11. Trimboli, A. J.; Fukino, K.; de Bruin, A.; Wei, G.; Shen, L.; Tanner, S. M.; Creasap, N.; Rosol, T. J.; Robinson, M. L.; Eng, C.; Ostrowski, M. C.; Leone, G. Direct Evidence for Epithelial–Mesenchymal Transitions in Breast Cancer. *Cancer Res.* **2008,** *68* (3), 937–945.

12. Peyre, L.; Zucchini-Pascal, N.; de Sousa, G.; Luzy, A. P.; Rahmani, R. Potential Involvement of Chemicals in Liver Cancer Progression: An Alternative Toxicological Approach Combining Biomarkers and Innovative Technologies. *Toxicol. In Vitro* **2014,** *28* (8), 1507–1520.

13. Takano, H.; Takumi, S.; Ikema, S.; Mizoue, N.; Hotta, Y.; Shiozaki, K.; Sugiyama, Y.; Furukawa, T.; Komatsu, M. Microcystin-LR Induces Anoikis Resistance to the Hepatocyte Uptake Transporter OATP1B3-Expressing Cell Lines. *Toxicology* **2014,** *326*, 53–61.

14. Elamin, E.; Masclee, A.; Troost, F.; Dekker, J.; Jonkers, D. Activation of the Epithelial-to-Mesenchymal Transition Factor Snail Mediates Acetaldehyde-Induced Intestinal Epithelial Barrier Disruption. *Alcohol Clin. Exp. Res.* **2014,** *38* (2), 344–353.

15. Liang, Z.; Xie, W.; Wu, R.; Geng, H.; Zhao, L.; Xie, C.; Li, X.; Zhu, M.; Zhu, W.; Zhu, J.; Huang, C.; Ma, X.; Wu, J.; Geng, S.; Zhong, C.; Han, H. Inhibition of Tobacco Smoke-Induced Bladder MAPK Activation and Epithelial–Mesenchymal Transition in Mice by Curcumin. *Int. J. Clin. Exp. Pathol.* **2015,** *8* (5), 4503–4513.

16. Fischer, K. D.; Hall, S. C.; Agrawal, D. K. Vitamin D Supplementation Reduces Induction of Epithelial–Mesenchymal Transition in Allergen Sensitized and Challenged Mice. *PLoS ONE* **2016,** *11* (2), e0149180.

17. Huang, S. F.; Horng, C. T.; Hsieh, Y. S.; Hsieh, Y. H.; Chu, S. C.; Chen, P. N. Epicatechin-3-Gallate Reverses TGF-Beta1-Induced Epithelial-to-Mesenchymal Transition and Inhibits Cell Invasion and Protease Activities in Human Lung Cancer Cells. *Food Chem. Toxicol.* **2016,** *94*, 1–10.

18. Savagner, P. Leaving the Neighborhood: Molecular Mechanisms Involved during Epithelial–Mesenchymal Transition. *Bioessays* **2001,** *23* (10), 912–923.

19. Topcul, M.; Cetin, I. Clinical Significance of Epithelial–Mesenchymal Transition and Cancer Stem Cells. *J BUON* **2016,** *21* (2), 312–319.

20. Savagner, P. The Epithelial–Mesenchymal Transition (EMT) Phenomenon. *Ann. Oncol.* **2010,** *21* (Suppl. 7), vii89–92.

21. Mimeault, M.; Batra, S. K. Interplay of Distinct Growth Factors during Epithelial Mesenchymal Transition of Cancer Progenitor Cells and Molecular Targeting as Novel Cancer Therapies. *Ann. Oncol.* **2007,** *18* (10), 1605–1619.

22. Kalluri, R. EMT: When Epithelial Cells Decide to Become Mesenchymal-Like Cells. *J. Clin. Invest.* **2009,** *119* (6), 1417–1419.

23. Micalizzi, D. S.; Farabaugh, S. M.; Ford, H. L. Epithelial–Mesenchymal Transition in Cancer: Parallels between Normal Development and Tumor Progression. *J. Mammary Gland Biol. Neoplasia* **2010,** *15* (2), 117–134.

24. Pant, I.; Kumar, N.; Khan, I.; Rao, S. G.; Kondaiah, P. Role of Areca Nut Induced TGF-beta and Epithelial–Mesenchymal Interaction in the Pathogenesis of Oral Submucous Fibrosis. *PLoS ONE* **2015,** *10* (6), e0129252.

25. Cho, I. H.; Choi, Y. J.; Gong, J. H.; Shin, D.; Kang, M. K.; Kang, Y. H. Astragalin Inhibits Autophagy-Associated Airway Epithelial Fibrosis. *Respir. Res.* **2015,** *16*, 51.

26. Vergara, D.; Simeone, P.; Bettini, S.; Tinelli, A.; Valli, L.; Storelli, C.; Leo, S.; Santino, A.; Maffia, M. Antitumor Activity of the Dietary Diterpene Carnosol against a Panel of Human Cancer Cell Lines. *Food Funct.* **2014,** *5* (6), 1261–1269.

27. Lin, Y. C.; Tsai, P. H.; Lin, C. Y.; Cheng, C. H.; Lin, T. H.; Lee, K. P.; Huang, K. Y.; Chen, S. H.; Hwang, J. J.; Kandaswami, C. C.; Lee, M. T. Impact of Flavonoids on Matrix Metalloproteinase Secretion and Invadopodia Formation in Highly Invasive A431-III Cancer Cells. *PLoS ONE* **2013,** *8* (8), e71903.

28. Peyre, L.; Zucchini-Pascal, N.; de Sousa, G.; Rahmani, R. Effects of Endosulfan on Hepatoma Cell Adhesion: Epithelial–Mesenchymal Transition and Anoikis Resistance. *Toxicology* **2012,** *300* (1–2), 19–30.

29. Zucchini-Pascal, N.; Peyre, L.; de Sousa, G.; Rahmani, R. Organochlorine Pesticides Induce Epithelial to Mesenchymal Transition of Human Primary Cultured Hepatocytes. *Food Chem. Toxicol.* **2012,** *50* (11), 3963–3970.

30. Le Bras, G. F.; Taubenslag, K. J.; Andl, C. D. The Regulation of Cell–Cell Adhesion during Epithelial–Mesenchymal Transition, Motility and Tumor Progression. *Cell Adh. Migr.* **2012,** *6* (4), 365–373.

31. Wheelock, M. J.; Shintani, Y.; Maeda, M.; Fukumoto, Y.; Johnson, K. R. Cadherin Switching. *J. Cell Sci.* **2008,** *121* (Pt. 6), 727–735.

32. Miyamoto, Y.; Sakane, F.; Hashimoto, K. N-Cadherin-Based Adherens Junction Regulates the Maintenance, Proliferation, and Differentiation of Neural Progenitor Cells during Development. *Cell Adh. Migr.* **2015,** *9* (3), 183–192.

33. Runkle, E. A.; Mu, D. Tight Junction Proteins: From Barrier to Tumorigenesis. *Cancer Lett.* **2013,** *337* (1), 41–48.

34. Kojima, T.; Takano, K.; Yamamoto, T.; Murata, M.; Son, S.; Imamura, M.; Yamaguchi, H.; Osanai, M.; Chiba, H.; Himi, T.; Sawada, N. Transforming Growth Factor-beta Induces Epithelial to Mesenchymal Transition by Down-regulation of Claudin-1 Expression and the Fence Function in Adult Rat Hepatocytes. *Liver Int.* **2008,** *28* (4), 534–545.

35. Langlois, S.; Cowan, K. N.; Shao, Q.; Cowan, B. J.; Laird, D. W. The Tumor-Suppressive Function of Connexin43 in Keratinocytes Is Mediated in Part via Interaction with Caveolin-1. *Cancer Res.* **2010,** *70* (10), 4222–4232.

36. Cotrutz, C. E.; Abuelba, H.; Olinici, D.; Petreus, T. E-cadherin Expression in Invasive Ductal Carcinoma Associates Ultrastructural Changes in Desmosomes Structure. *Rom. J. Morphol. Embryol.* **2012,** *53* (3 Suppl.), 731–734.

37. Ansieau, S.; Caron de Fromentel, C.; Bastid, J.; Morel, A. P.; Puisieux, A. Role of the Epithelial–Mesenchymal Transition during Tumor Progression. *Bull. Cancer* **2010,** *97* (1), 7–15.

38. Ben-Porath, I.; Weinberg, R. A. When Cells Get Stressed: An Integrative View of Cellular Senescence. *J. Clin. Invest.* **2004,** *113* (1), 8–13.

39. Marthandan, S.; Menzel, U.; Priebe, S.; Groth, M.; Guthke, R.; Platzer, M.; Hemmerich, P.; Kaether, C.; Diekmann, S. Conserved Genes and Pathways in Primary Human Fibroblast Strains Undergoing Replicative and Radiation Induced Senescence. *Biol. Res.* **2016,** *49* (1), 34.

40. Kishi, S.; Bayliss, P. E.; Hanai, J. A prospective Epigenetic Paradigm between Cellular Senescence and Epithelial–Mesenchymal Transition in Organismal Development and Aging. *Transl. Res.* **2015,** *165* (1), 241–249.

41. Ansieau, S.; Morel, A. P.; Hinkal, G.; Bastid, J.; Puisieux, A. TWISTing an Embryonic Transcription Factor into an Oncoprotein. *Oncogene* **2010,** *29* (22), 3173–3184.

42. Laberge, R. M.; Awad, P.; Campisi, J.; Desprez, P. Y. Epithelial–Mesenchymal Transition Induced by Senescent Fibroblasts. *Cancer Microenviron.* **2012,** *5* (1), 39–44.
43. Giehl, K.; Menke, A. Microenvironmental Regulation of E-Cadherin-Mediated Adherens Junctions. *Front. Biosci.* **2008,** *13,* 3975–3985.
44. Chiarugi, P.; Giannoni, E. Anoikis: A Necessary Death Program for Anchorage-Dependent Cells. *Biochem. Pharmacol.* **2008,** *76* (11), 1352–1364.
45. Hofmann, C.; Obermeier, F.; Artinger, M.; Hausmann, M.; Falk, W.; Schoelmerich, J.; Rogler, G.; Grossmann, J. Cell–Cell Contacts Prevent Anoikis in Primary Human Colonic Epithelial Cells. *Gastroenterology* **2007,** *132* (2), 587–600.
46. Yamada, A.; Aki, T.; Unuma, K.; Funakoshi, T.; Uemura, K. Paraquat Induces Epithelial–Mesenchymal Transition-Like Cellular Response Resulting in Fibrogenesis and the Prevention of Apoptosis in Human Pulmonary Epithelial Cells. *PLoS ONE* **2015,** *10* (3), e0120192.
47. Chanvorachote, P.; Pongrakhananon, V.; Chunhacha, P. Prolonged Nitric Oxide Exposure Enhances Anoikis Resistance and Migration through Epithelial–Mesenchymal Transition and Caveolin-1 Upregulation. *Biomed. Res. Int.* **2014,** *2014,* 941359.
48. Kwon, T.; Youn, H.; Son, B.; Kim, D.; Seong, K. M.; Park, S.; Kim, W.; Youn, B. DANGER Is Involved in High Glucose-Induced Radioresistance through Inhibiting DAPK-Mediated Anoikis in Non-Small Cell Lung Cancer. *Oncotarget* **2016,** *7* (6), 7193–7206.
49. Mitra, A.; Mishra, L.; Li, S. EMT, CTCs and CSCs in Tumor Relapse and Drug-Resistance. *Oncotarget* **2015,** *6* (13), 10697–10711.
50. Paoli, P.; Giannoni, E.; Chiarugi, P. Anoikis Molecular Pathways and Its Role in Cancer Progression. *Biochim. Biophys. Acta* **2013,** *1833* (12), 3481–3498.
51. Hay, E. D. The Mesenchymal Cell, its Role in the Embryo, and the Remarkable Signaling Mechanisms that Create It. *Dev. Dyn.* **2005,** *233* (3), 706–720.
52. Strutz, F.; Zeisberg, M.; Ziyadeh, F. N.; Yang, C. Q.; Kalluri, R.; Muller, G. A.; Neilson, E. G. Role of Basic Fibroblast Growth Factor-2 in Epithelial–Mesenchymal Transformation. *Kidney Int.* **2002,** *61* (5), 1714–1728.
53. Gheldof, A.; Berx, G. Cadherins and Epithelial-to-Mesenchymal Transition. *Prog. Mol. Biol. Transl. Sci.* **2013,** *116,* 317–336.
54. Shin, S.; Go, R. E.; Kim, C. W.; Hwang, K. A.; Nam, K. H.; Choi, K. C. Effect of Benzophenone-1 and Octylphenol on the Regulation of Epithelial–Mesenchymal Transition via an Estrogen Receptor-Dependent Pathway in Estrogen Receptor Expressing Ovarian Cancer Cells. *Food Chem. Toxicol.* **2016,** *93,* 58–65.
55. Schneider, D. J.; Wu, M.; Le, T. T.; Cho, S. H.; Brenner, M. B.; Blackburn, M. R.; Agarwal, S. K. Cadherin-11 Contributes to Pulmonary Fibrosis: Potential Role in TGF-Beta Production and Epithelial to Mesenchymal Transition. *FASEB J.* **2012,** *26* (2), 503–512.
56. Ohkubo, T.; Ozawa, M. The Transcription Factor Snail Downregulates the Tight Junction Components Independently of E-Cadherin Downregulation. *J. Cell Sci.* **2004,** *117* (Pt. 9), 1675–1685.
57. Tania, M.; Khan, M. A.; Fu, J. Epithelial to Mesenchymal Transition Inducing Transcription Factors and Metastatic Cancer. *Tumour Biol.* **2014,** *35* (8), 7335–7342.
58. Barrallo-Gimeno, A.; Nieto, M. A. The Snail Genes as Inducers of Cell Movement and Survival: Implications in Development and Cancer. *Development* **2005,** *132* (14), 3151–3161.

59. Fabregat, I.; Malfettone, A.; Soukupova, J. New Insights into the Crossroads between EMT and Stemness in the Context of Cancer. *J. Clin. Med.* **2016,** *5* (3). DOI:10.3390/jcm5030037.

60. Kang, Y.; Massague, J. Epithelial–Mesenchymal Transitions: Twist in Development and Metastasis. *Cell* **2004,** *118* (3), 277–279.

61. Zheng, H.; Kang, Y. Multilayer Control of the EMT Master Regulators. *Oncogene* **2014,** *33* (14), 1755–1763.

62. Hugo, H. J.; Kokkinos, M. I.; Blick, T.; Ackland, M. L.; Thompson, E. W.; Newgreen, D. F. Defining the E-Cadherin Repressor Interactome in Epithelial–Mesenchymal Transition: The PMC42 Model as a Case Study. *Cells Tissues Organs* **2011,** *193* (1–2), 23–40.

63. Katoh, M. Integrative Genomic Analyses of ZEB2: Transcriptional Regulation of ZEB2 Based on SMADs, ETS1, HIF1alpha, POU/OCT, and NF-kappaB. *Int. J. Oncol.* **2009,** *34* (6), 1737–1742.

64. Kothari, A. N.; Arffa, M. L.; Chang, V.; Blackwell, R. H.; Syn, W. K.; Zhang, J.; Mi, Z.; Kuo, P. C. Osteopontin-A Master Regulator of Epithelial-Mesenchymal Transition. *J. Clin. Med.* **2016,** *5* (4). DOI:10.3390/jcm5040039.

65. Liang, H.; Zhang, Q.; Lu, J.; Yang, G.; Tian, N.; Wang, X.; Tan, Y.; Tan, D. MSX2 Induces Trophoblast Invasion in Human Placenta. *PLoS ONE* **2016,** *11* (4), e0153656.

66. Mimeault, M.; Batra, S. K. Altered Gene Products Involved in the Malignant Reprogramming of Cancer Stem/Progenitor Cells and Multitargeted Therapies. *Mol. Aspects Med.* **2014,** *39*, 3–32.

67. Nawshad, A.; Medici, D.; Liu, C. C.; Hay, E. D. TGFbeta3 Inhibits E-Cadherin Gene Expression in Palate Medial-Edge Epithelial Cells through a Smad2–Smad4–LEF1 Transcription Complex. *J. Cell Sci.* **2007,** *120* (Pt. 9), 1646–1653.

68. 68. Venkov, C. D.; Link, A. J.; Jennings, J. L.; Plieth, D.; Inoue, T.; Nagai, K.; Xu, C.; Dimitrova, Y. N.; Rauscher, F. J.; Neilson, E. G. A Proximal Activator of Transcription in Epithelial-Mesenchymal Transition. *J. Clin. Invest.* **2007,** *117* (2), 482–491.

69. Nguyen, Q. D.; De Wever, O.; Bruyneel, E.; Hendrix, A.; Xie, W. Z.; Lombet, A.; Leibl, M.; Mareel, M.; Gieseler, F.; Bracke, M.; Gespach, C. Commutators of PAR-1 Signaling in Cancer Cell Invasion Reveal an Essential Role of the Rho–Rho Kinase Axis and Tumor Microenvironment. *Oncogene* **2005,** *24* (56), 8240–8251.

70. De Wever, O.; Pauwels, P.; De Craene, B.; Sabbah, M.; Emami, S.; Redeuilh, G.; Gespach, C.; Bracke, M.; Berx, G. Molecular and Pathological Signatures of Epithelial–Mesenchymal Transitions at the Cancer Invasion Front. *Histochem. Cell Biol.* **2008,** *130* (3), 481–494.

71. Mamuya, F. A.; Duncan, M. K. aV Integrins and TGF-beta-Induced EMT: A Circle of Regulation. *J. Cell Mol. Med.* **2012,** *16* (3), 445–455.

72. Yoshida, T.; Akatsuka, T.; Imanaka-Yoshida, K. Tenascin-C and Integrins in Cancer. *Cell. Adh. Migr.* **2015,** *9* (1–2), 96–104.

73. Bianchi, A.; Gervasi, M. E.; Bakin, A. Role of Beta5-Integrin in Epithelial–Mesenchymal Transition in Response to TGF-beta. *Cell Cycle* **2010,** *9* (8), 1647–1659.

74. Chen, D.; Zhang, Y.; Zhang, X.; Li, J.; Han, B.; Liu, S.; Wang, L.; Ling, Y.; Mao, S.; Wang, X. Overexpression of Integrin-Linked Kinase Correlates with Malignant Phenotype in Non-Small Cell Lung Cancer and Promotes Lung Cancer Cell Invasion and Migration via Regulating Epithelial-Mesenchymal Transition (EMT)-Related Genes. *Acta Histochem.* **2013,** *115* (2), 128–136.

75. Liu, Y. New Insights into Epithelial–Mesenchymal Transition in Kidney Fibrosis. *J. Am. Soc. Nephrol.* **2010,** *21* (2), 212–222.

76. Chai, J. Y.; Modak, C.; Mouazzen, W.; Narvaez, R.; Pham, J. Epithelial or Mesenchymal: Where to Draw the Line? *Biosci. Trends* **2010**, *4* (3), 130–142.
77. Satelli, A.; Li, S. Vimentin in Cancer and its Potential as a Molecular Target for Cancer Therapy. *Cell Mol. Life Sci.* **2011**, *68* (18), 3033–3046.
78. Entschladen, F.; Zanker, K. S.; Powe, D. G. Heterotrimeric G Protein Signaling in Cancer Cells with Regard to Metastasis Formation. *Cell Cycle* **2011**, *10* (7), 1086–1091.
79. Yilmaz, M.; Christofori, G. EMT, the Cytoskeleton, and Cancer Cell Invasion. *Cancer Metastasis Rev.* **2009**, *28* (1–2), 15–33.
80. Zhou, H.; Kramer, R. H. Integrin Engagement Differentially Modulates Epithelial Cell Motility by RhoA/ROCK and PAK1. *J. Biol. Chem.* **2005**, *280* (11), 10624–10635.
81. Sun, C. K.; Ng, K. T.; Lim, Z. X.; Cheng, Q.; Lo, C. M.; Poon, R. T.; Man, K.; Wong, N.; Fan, S. T. Proline-Rich Tyrosine Kinase 2 (Pyk2) Promotes Cell Motility of Hepatocellular Carcinoma through Induction of Epithelial to Mesenchymal Transition. *PLoS ONE* **2011**, *6* (4), e18878.
82. Kallergi, G.; Papadaki, M. A.; Politaki, E.; Mavroudis, D.; Georgoulias, V.; Agelaki, S. Epithelial to Mesenchymal Transition Markers Expressed in Circulating Tumour Cells of Early and Metastatic Breast Cancer Patients. *Breast Cancer Res.* **2011**, *13* (3), R59.
83. Puisieux, A. Role of Epithelial-Mesenchymal Transition in Tumor Progression. *Bull. Acad. Natl. Med.* **2009**, *193* (9), 2017–2032; discussion 2032–2034.
84. Zhang, J.; Tian, X. J.; Xing, J. Signal Transduction Pathways of EMT Induced by TGF-beta, SHH, and WNT and Their Crosstalks. *J. Clin. Med.* **2016**, *5* (4). DOI:10.3390/jcm5040041.
85. Smith, B. N.; Burton, L. J.; Henderson, V.; Randle, D. D.; Morton, D. J.; Smith, B. A.; Taliaferro-Smith, L.; Nagappan, P.; Yates, C.; Zayzafoon, M.; Chung, L. W.; Odero-Marah, V. A. Snail Promotes Epithelial Mesenchymal Transition in Breast Cancer Cells in Part via Activation of Nuclear ERK2. *PLoS ONE* **2014**, *9* (8), e104987.
86. Neuzillet, C.; Tijeras-Raballand, A.; de Mestier, L.; Cros, J.; Faivre, S.; Raymond, E. MEK in Cancer and Cancer Therapy. *Pharmacol. Ther.* **2014**, *141* (2), 160–171.
87. Maurer, G.; Tarkowski, B.; Baccarini, M. Raf Kinases in Cancer-Roles and Therapeutic Opportunities. *Oncogene* **2011**, *30* (32), 3477–3488.
88. Larue, L.; Bellacosa, A. Epithelial–Mesenchymal Transition in Development and Cancer: Role of Phosphatidylinositol 3'-Kinase/AKT Pathways. *Oncogene* **2005**, *24* (50), 7443–7454.
89. Fuchs, B. C.; Fujii, T.; Dorfman, J. D.; Goodwin, J. M.; Zhu, A. X.; Lanuti, M.; Tanabe, K. K. Epithelial-to-Mesenchymal Transition and Integrin-Linked Kinase Mediate Sensitivity to Epidermal Growth Factor Receptor Inhibition in Human Hepatoma Cells. *Cancer Res.* **2008**, *68* (7), 2391–2399.
90. Ginnebaugh, K. R.; Ahmad, A.; Sarkar, F. H. The Therapeutic Potential of Targeting the Epithelial–Mesenchymal Transition in Cancer. *Expert Opin. Ther. Targets* **2014**, *18* (7), 731–745.
91. Garber, K. Drugging the Wnt Pathway: Problems and Progress. *J. Natl. Cancer Inst.* **2009**, *101* (8), 548–550.
92. Gonzalez, D. M.; Medici, D. Signaling Mechanisms of the Epithelial–Mesenchymal Transition. *Sci. Signal.* **2014**, *7* (344), re8.
93. Ghahhari, N. M.; Babashah, S. Interplay between MicroRNAs and WNT/β-catenin Signalling Pathway Regulates Epithelial–Mesenchymal Transition in Cancer. *Eur J Cancer* **2015**, *51* (12), 1638–1649.

94. Medici, D.; Hay, E. D.; Olsen, B. R. Snail and Slug Promote Epithelial–Mesenchymal Transition through Beta-Catenin-T-Cell Factor-4-Dependent Expression of Transforming Growth Factor-Beta3. *Mol. Biol. Cell* **2008**, *19* (11), 4875–4887.

95. Nalluri, S. M.; O'Connor, J. W.; Gomez, E. W. Cytoskeletal Signaling in TGFbeta-Induced Epithelial–Mesenchymal Transition. *Cytoskeleton (Hoboken)* **2015**, *72* (11), 557–569.

96. Leask, A.; Abraham, D. J. TGF-beta Signaling and the Fibrotic Response. *FASEB J.* **2004**, *18* (7), 816–827.

97. Derynck, R.; Muthusamy, B. P.; Saeteurn, K. Y. Signaling Pathway Cooperation in TGF-Beta-Induced Epithelial–Mesenchymal Transition. *Curr. Opin. Cell Biol.* **2014**, *31*, 56–66.

98. Chen, X.; Xiao, W.; Wang, W.; Luo, L.; Ye, S.; Liu, Y. The Complex Interplay between ERK1/2, TGFbeta/Smad, and Jagged/Notch Signaling Pathways in the Regulation of Epithelial-Mesenchymal Transition in Retinal Pigment Epithelium Cells. *PLoS ONE* **2014**, *9* (5), e96365.

99. Fender, A. W.; Nutter, J. M.; Fitzgerald, T. L.; Bertrand, F. E.; Sigounas, G. Notch-1 Promotes Stemness and Epithelial to Mesenchymal Transition in Colorectal Cancer. *J. Cell Biochem.* **2015**, *116* (11), 2517–2527.

100. Tang, C.; Mei, L.; Pan, L.; Xiong, W.; Zhu, H.; Ruan, H.; Zou, C.; Tang, L.; Iguchi, T.; Wu, X. Hedgehog Signaling through GLI1 and GLI2 Is Required for Epithelial–Mesenchymal Transition in Human Trophoblasts. *Biochim. Biophys. Acta* **2015**, *1850* (7), 1438–1448.

101. Wilson, C.; Nicholes, K.; Bustos, D.; Lin, E.; Song, Q.; Stephan, J. P.; Kirkpatrick, D. S.; Settleman, J. Overcoming EMT-Associated Resistance to Anti-Cancer Drugs via Src/FAK Pathway Inhibition. *Oncotarget* **2014**, *5* (17), 7328–7341.

102. Kinehara, M.; Kawamura, S.; Mimura, S.; Suga, M.; Hamada, A.; Wakabayashi, M.; Nikawa, H.; Furue, M. K. Protein Kinase C-Induced Early Growth Response Protein-1 Binding to SNAIL Promoter in Epithelial–Mesenchymal Transition of Human Embryonic Stem Cells. *Stem Cells Dev.* **2014**, *23* (18), 2180–2189.

103. Chen, W.; Gao, Q.; Han, S.; Pan, F.; Fan, W. The CCL2/CCR2 Axis Enhances IL-6-Induced Epithelial–Mesenchymal Transition by Cooperatively Activating STAT3-Twist Signaling. *Tumour Biol.* **2015**, *36* (2), 973–981.

104. Joseph, J. V.; Conroy, S.; Pavlov, K.; Sontakke, P.; Tomar, T.; Eggens-Meijer, E.; Balasubramaniyan, V.; Wagemakers, M.; den Dunnen, W. F.; Kruyt, F. A. Hypoxia Enhances Migration and Invasion in Glioblastoma by Promoting a Mesenchymal Shift Mediated by the HIF1alpha-ZEB1 Axis. *Cancer Lett.* **2015**, *359* (1), 107–116.

105. Choi, S. S.; Diehl, A. M. Epithelial-to-Mesenchymal Transitions in the Liver. *Hepatology* **2009**, *50* (6), 2007–2013.

106. Sakai, D.; Suzuki, T.; Osumi, N.; Wakamatsu, Y. Cooperative Action of Sox9, Snail2 and PKA Signaling in Early Neural Crest Development. *Development* **2006**, *133* (7), 1323–1333.

107. Thiery, J. P.; Acloque, H.; Huang, R. Y.; Nieto, M. A. Epithelial–Mesenchymal Transitions in Development and Disease. *Cell* **2009**, *139* (5), 871–890.

108. Person, A. D.; Klewer, S. E.; Runyan, R. B. Cell biology of cardiac cushion development. *Int. Rev. Cytol.* **2005**, *243*, 287–335.

109. Doi, T.; Lukosiute, A.; Ruttenstock, E.; Dingemann, J.; Puri, P. Expression of Iroquois Genes is Up-regulated during Early Lung Development in the Nitrofen-Induced Pulmonary Hypoplasia. *J. Pediatr. Surg.* **2011**, *46* (1), 62–66.

110. Helies-Toussaint, C.; Peyre, L.; Costanzo, C.; Chagnon, M. C.; Rahmani, R. Is Bisphenol S a Safe Substitute for Bisphenol A in Terms of Metabolic Function? An In Vitro Study. *Toxicol. Appl. Pharmacol.* **2014**, *280* (2), 224–235.

111. Kovanecz, I.; Gelfand, R.; Masouminia, M.; Gharib, S.; Segura, D.; Vernet, D.; Rajfer, J.; Li, D. K.; Kannan, K.; Gonzalez-Cadavid, N. F. Oral Bisphenol A (BPA) Given to Rats at Moderate Doses Is Associated with Erectile Dysfunction, Cavernosal Lipofibrosis and Alterations of Global Gene Transcription. *Int. J. Impot. Res.* **2014**, *26* (2), 67–75.

112. Huang, T.; Barnett, J. V.; Camenisch, T. D. Cardiac Epithelial–Mesenchymal Transition Is Blocked by Monomethylarsonous Acid (III). *Toxicol. Sci.* **2014**, *142* (1), 225–238.

113. Allison, P.; Huang, T.; Broka, D.; Parker, P.; Barnett, J. V.; Camenisch, T. D. Disruption of Canonical TGFbeta-Signaling in Murine Coronary Progenitor Cells by Low Level Arsenic. *Toxicol. Appl. Pharmacol.* **2013**, *272* (1), 147–153.

114. Thevenot, P. T.; Saravia, J.; Jin, N.; Giaimo, J. D.; Chustz, R. E.; Mahne, S.; Kelley, M. A.; Hebert, V. Y.; Dellinger, B.; Dugas, T. R.; Demayo, F. J.; Cormier, S. A. Radical-Containing Ultrafine Particulate Matter Initiates Epithelial-to-Mesenchymal Transitions in Airway Epithelial Cells. *Am. J. Respir. Cell Mol. Biol.* **2013**, *48* (2), 188–197.

115. Lamouille, S.; Xu, J.; Derynck, R. Molecular Mechanisms of Epithelial–Mesenchymal Transition. *Nat. Rev. Mol. Cell Biol.* **2014**, *15* (3), 178–196.

116. Xie, L.; Zhou, D.; Xiong, J.; You, J.; Zeng, Y.; Peng, L. Paraquat Induce Pulmonary Epithelial–Mesenchymal Transition through Transforming Growth factor-Beta1-Dependent Mechanism. *Exp. Toxicol. Pathol.* **2016**, *68* (1), 69–76.

117. Han, Y. Y.; Shen, P.; Chang, W. X. Involvement of Epithelial-to-Mesenchymal Transition and Associated Transforming Growth Factor-Beta/Smad Signaling in Paraquat-Induced Pulmonary Fibrosis. *Mol. Med. Rep.* **2015**, *12* (6), 7979–7984.

118. Santos, T.; Cancian, G.; Neodini, D. N.; Mano, D. R.; Capucho, C.; Predes, F. S.; Barbieri, R.; Oliveira, C. A.; Pigoso, A. A.; Dolder, H.; Severi-Aguiar, G. D. Toxicological Evaluation of Ametryn Effects in Wistar Rats. *Exp. Toxicol. Pathol.* **2015**, *67* (10), 525–532.

119. Pierre, S.; Chevallier, A.; Teixeira-Clerc, F.; Ambolet-Camoit, A.; Bui, L. C.; Bats, A. S.; Fournet, J. C.; Fernandez-Salguero, P.; Aggerbeck, M.; Lotersztajn, S.; Barouki, R.; Coumoul, X. Aryl Hydrocarbon Receptor-Dependent Induction of Liver Fibrosis by Dioxin. *Toxicol. Sci.* **2014**, *137* (1), 114–124.

120. Mimeault, M.; Batra, S. K. Hypoxia-Inducing Factors as Master Regulators of Stemness Properties and Altered Metabolism of Cancer- and Metastasis-Initiating Cells. *J. Cell Mol Med* **2013**, *17* (1), 30–54.

121. Wang, W.; Chin-Sheng, H.; Kuo, L. J.; Wei, P. L.; Lien, Y. C.; Lin, F. Y.; Liu, H. H.; Ho, Y. S.; Wu, C. H.; Chang, Y. J. NNK Enhances Cell Migration through Alpha7-Nicotinic Acetylcholine Receptor Accompanied by Increased of Fibronectin Expression in Gastric Cancer. *Ann. Surg. Oncol.* **2012**, *19* (Suppl. 3), S580–S588.

122. Jin, X.; Chen, M.; Song, L.; Li, H.; Li, Z. The Evaluation of *p,p'*-DDT Exposure on Cell Adhesion of Hepatocellular Carcinoma. *Toxicology* **2014**, *322*, 99–108.

123. Kubwabo, C.; Kosarac, I.; Stewart, B.; Gauthier, B. R.; Lalonde, K.; Lalonde, P. J. Migration of Bisphenol A from Plastic Baby Bottles, Baby Bottle Liners and Reusable Polycarbonate Drinking Bottles. *Food Addit. Contam., A: Chem. Anal. Control Expo. Risk Assess.* **2009**, *26* (6), 928–937.

124. Zhang, X. L.; Wang, H. S.; Liu, N.; Ge, L. C. Bisphenol A Stimulates the Epithelial Mesenchymal Transition of Estrogen Negative Breast Cancer Cells via FOXA1 Signals. *Arch. Biochem. Biophys.* **2015**, *585*, 10–16.

125. Wang, K. H.; Kao, A. P.; Chang, C. C.; Lin, T. C.; Kuo, T. C. Bisphenol A-Induced Epithelial to Mesenchymal Transition is Mediated by Cyclooxygenase-2 Up-regulation in Human Endometrial Carcinoma Cells. *Reprod. Toxicol.* **2015**, *58*, 229–233.

126. Kim, Y. S.; Hwang, K. A.; Hyun, S. H.; Nam, K. H.; Lee, C. K.; Choi, K. C. Bisphenol A and Nonylphenol Have the Potential to Stimulate the Migration of Ovarian Cancer Cells by Inducing Epithelial–Mesenchymal Transition via an Estrogen Receptor Dependent Pathway. *Chem. Res. Toxicol.* **2015**, *28* (4), 662–671.

127. Rodriguez, D. A.; de Lima, R. F.; Campos, M. S.; Costa, J. R.; Biancardi, M. F.; Marques, M. R.; Taboga, S. R.; Santos, F. C. Intrauterine Exposure to Bisphenol A Promotes Different Effects in both Neonatal and Adult Prostate of Male and Female Gerbils (*Meriones unguiculatus*). *Environ. Toxicol.* **2015**. DOI:10.1002/tox.22176.

128. Wadia, P. R.; Cabaton, N. J.; Borrero, M. D.; Rubin, B. S.; Sonnenschein, C.; Shioda, T.; Soto, A. M. Low-Dose BPA Exposure Alters the Mesenchymal and Epithelial Transcriptomes of the Mouse Fetal Mammary Gland. *PLoS ONE* **2013**, *8* (5), e63902.

129. Hu, Y.; Zhang, L.; Wu, X.; Hou, L.; Li, Z.; Ju, J.; Li, Q.; Qin, W.; Li, J.; Zhang, Q.; Zhou, T.; Xu, C.; Fang, Z.; Zhang, Y. Bisphenol A, an Environmental Estrogen-Like Toxic Chemical, Induces Cardiac Fibrosis by Activating the ERK1/2 Pathway. *Toxicol. Lett.* **2016**, *250–251*, 1–9.

130. Kendziorski, J. A.; Belcher, S. M. Strain-Specific Induction of Endometrial Periglandular Fibrosis in Mice Exposed during Adulthood to the Endocrine Disrupting Chemical Bisphenol A. *Reprod. Toxicol.* **2015**, *58*, 119–130.

131. Alink, G. M.; Sjogren, M.; Bos, R. P.; Doekes, G.; Kromhout, H.; Scheepers, P. T. Effect of Airborne Particles from Selected Indoor and Outdoor Environments on Gap-Junctional Intercellular Communication. *Toxicol. Lett.* **1998**, *96–97*, 209–213.

132. Cowles, C.; Mally, A.; Chipman, J. K. Different Mechanisms of Modulation of Gap Junction Communication by Non-genotoxic Carcinogens in Rat Liver In Vivo. *Toxicology* **2007**, *238* (1), 49–59.

133. Jeong, S. H.; Habeebu, S. S.; Klaassen, C. D. Cadmium Decreases Gap Junctional Intercellular Communication in Mouse Liver. *Toxicol. Sci.* **2000**, *57* (1), 156–166.

134. Nomata, K.; Kang, K. S.; Hayashi, T.; Matesic, D.; Lockwood, L.; Chang, C. C.; Trosko, J. E. Inhibition of Gap Junctional Intercellular Communication in Heptachlor- and Heptachlor Epoxide-Treated Normal Human Breast Epithelial Cells. *Cell Biol. Toxicol.* **1996**, *12* (2), 69–78.

135. Hruba, E.; Vondracek, J.; Libalova, H.; Topinka, J.; Bryja, V.; Soucek, K.; Machala, M. Gene Expression Changes in Human Prostate Carcinoma Cells Exposed to Genotoxic and Nongenotoxic Aryl Hydrocarbon Receptor Ligands. *Toxicol. Lett.* **2011**, *206* (2), 178–188.

136. McCullough, K. D.; Coleman, W. B.; Ricketts, S. L.; Wilson, J. W.; Smith, G. J.; Grisham, J. W. Plasticity of the Neoplastic Phenotype In Vivo is Regulated by Epigenetic Factors. *Proc. Natl. Acad. Sci. U.S.A.* **1998**, *95* (26), 15333–15338.

137. Murphy, K. J.; Fox, G. B.; Kelly, J.; Regan, C. M. Influence of Toxicants on Neural Cell Adhesion Molecule-Mediated Neuroplasticity in the Developing and Adult Animal: Persistent Effects of Chronic Perinatal Low-Level Lead Exposure. *Toxicol. Lett.* **1995**, *82–83*, 271–276.

138. Lee, J. Y.; Kong, G. Roles and Epigenetic Regulation of Epithelial–Mesenchymal Transition and its Transcription Factors in Cancer Initiation and Progression. *Cell Mol Life Sci* **2016**, *73* (24), 4643–4660.

CHAPTER 13

REGULATION AND MONITORING OF PESTICIDE RESIDUES IN WATER AND FOOD IN BRAZIL

LUCIANO ZANETTI PESSÔA CANDIOTTO[1],
LUNEIA CATIANE DE SOUZA[1], VANESSA JACOB VICTORINO[1,2],
and CAROLINA PANIS[1,3*]

[1]*Group of Advanced Studies in Health Sciences, State University of West Paraná, UNIOESTE, Campus Francisco Beltrão, Paraná, Brazil*

[2]*Laboratory of Immunophatology, State University of Londrina, UEL, Londrina, Paraná, Brazil*

[3]*Laboratory of Inflammatory Mediators, State University of West Paraná, UNIOESTE, Campus Francisco Beltrão, Paraná, Brazil*

Corresponding author. E-mail: carolpanis@sercomtel.com.br

CONTENTS

ABSTRACT

Brazil uses large amounts of pesticides on crops, and the social and environmental issues surrounding the use of these products are complex and multidimensional. In addition to regulating the registration, usage, and monitoring of pesticide residues in water and food, this chapter provides an overview of this issue, based on research showing their harmful impact on human health. This review includes a review of the literature based on search results in PubMed and Scielo, analysis of Brazilian legislation concerning the usage release process and the monitoring of pesticide residues, and a discussion of the results of Pesticide Residue Analysis Program in Food. The results indicate that despite a recent regulation process, Brazil has an ineffective monitoring system, which has been characterized by a few government actions directed to a strict control process and reduction of pesticide use.

13.1 INTRODUCTION

The prevailing agricultural model in most of the world is largely linked to scientific and technological development of various products such as machinery, chemical inputs (fertilizers for soil, insecticides, fungicides, and herbicides), and genetically modified organisms, among others. Called the Green Revolution, the modernization process of agriculture began in the 1960s and since then has been responsible for several productive, economic, environmental, and social changes.

In Brazil, the modernization of agriculture began in the 1970s from government incentives, access to new production technologies, and installation of companies linked to the agricultural sector. Thus, the country has become one of the largest producers and exporters of several food products, especially soybeans, corn, beef, pork, and poultry meat.

The chemicalization process of agriculture and livestock[50] has led to increased use of pesticides in Brazil. Therefore, legal regulations became necessary which were directed toward the registration, use, waste analysis, disposal of packaging, etc. Concomitantly, contrary positions, questions, and debate about the possible disadvantageous and advantageous effects of pesticide usage have been expanded in the country.

Recently, Brazil has become one of the largest consumers of pesticides in the world.[27] Thus, approaches to evaluate the impact of pesticides on the environment and human health had to be implemented. In this context, regulation standards were created regarding the use of pesticide in the country, as well as measures to monitor the presence of residues in consumed water and food. In this chapter, we present an overview of the usage and regulation of these substances in Brazil, emphasizing the monitoring of pesticide residues in food and water, as well as evidence of their harmful impact on human health.

13.2 OVERVIEW OF PESTICIDES USE IN BRAZIL

Even though the government incentives for the purchase of pesticides in Brazil begun in the 1970s, since the 1950s, an increased access to pesticides has been seen. The increased assess to pesticides was due to reducing their costs through tax exemptions and credit lines for rural development. Law No. 3,244, from 1957, exempted the taxation of imported products. In 1959, imported products were also exempt from the Tax on Industrialized Products (IPI); and in 1969, the Tax on Circulation of Goods (ICM). In the 1970s, the Brazilian government had not taxed the purchase of imported pesticides and other tax incentives for the installation of industries, including the pesticide companies. In 1975, the National Plan for Agricultural Pesticides was created with the goal for the expansion of domestic supply of pesticides, studies and actions to control the damage caused by agricultural "pests" and to gain knowledge on the adverse effects of pesticides on human health and the environment. However, regarding human health and the environment, very little was achieved during the 1970s.[77]

These incentives have led to a significant increase in the overall average consumption of pesticides in Brazil. While in 1970, an average of 0.8 kg of active ingredient (a.i.) per hectare (10,000 m[2]) was used, in 1998, use reached 7.0 kg/ha.[95] Between 1991 and 1998, there was a 160% increase in the sale of pesticides in Brazil.[99]

In 1998, the crops that used the greatest amount of pesticides were tomatoes (52.5 kg/ha), potatoes (28.8 kg/ha), citrus (12.4 kg/ha), cotton (5.9 kg/ha), coffee (4.2 kg/ha), sugar cane (2 kg/ha), and soybean (3.2 kg/ha). On average, pesticide use was 2.9 kg/ha in agriculture.[95]

From the 2000s, Brazil has experienced the highest growth rate of world imports of pesticides, becoming the second largest national market. The sales reached US$ 11.5 billion, and Brazil becomes the world's largest importer, with an amount of US$ 3 billion spent in importation in 2013. Between 2000 and 2013, the growth rate of pesticide imports in Brazil was 1.000%, while the growth rate globally was 208%. In 2013, Brazil imported 10% of the world production of pesticides.[77]

Based on the Brazilian Agricultural Census of 2006, Bombardi[12] showed that 27% of the property from 0 to 10 ha; 36% of property of 10–100 ha; and 80% of the properties larger than 100 ha consumed pesticides.

Between 1985 and 2015, there was a 700% increase in the consumption of pesticides in Brazil, and the agricultural sector grew 78% in the same period. Currently, more than 300,000 tons of pesticides *per year* are used in Brazil, whereas 130,000 t are composed of active ingredients. The world average pesticide use is 2.5 million tons per year.[95]

Since 2008, Brazil is the largest consumer of pesticides in the world. Soybeans, sugar cane, corn, and cotton accounted for 80% of pesticides consumed in Brazil in 2013.[40] From 2008, the growth rate of active ingredients imported was 400% and for formulated products was 700%. A total of 936 products were marketed in Brazil, with 90% of the formulated products were made of material from other countries, especially China. In the season between mid-2010 and 2011, 936,000 t of pesticides were sold.[73]

Between 2002 and 2011, Brazil increased pesticides consumption in 42% and fertilizers consumption in 37% (Fig. 13.1). In the same period, the average consumption of pesticides in Brazil increased from 10.5 to 12 L/ha and Brazilian market grew 190%, while the global market grew 93%.[27] Table 13.1 presents the amount of pesticides used in Brazilian crops in the year of 2011.

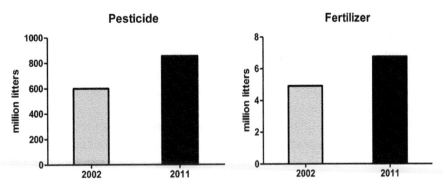

FIGURE 13.1 *Pesticide and fertilizer consumption in Brazil (2002 and 2011). Source:* Adapted from Carneiro, F. F.; Augusto, L. G. S.; Rigotto, R. M.; Friedrich, K.; Búrigo, A. C., Eds; *Dossier ABRASCO: A Warning about the Impacts of Pesticides on Health.* EPSJV: Rio de Janeiro; Expressão Popular: São Paulo, 2015: 52 (accessed September 21, 2015).

In 2012, 45% of all pesticides sold in the country were herbicides, 14% fungicides, and 12% insecticides.[73] About 430 a.i., 750 technical products and 1400 pesticide formulations were authorized by the Ministry of Health (MH), Ministry of Environment (MMA) and recorded in the Ministry of Agriculture, Livestock and Supply (MAPA). Figure 13.2 indicates the growth recorded between 2006 and 2012 in agrochemical sales in Brazil.

TABLE 13.1 Amount of Pesticides per Hectare Used for Farming (2011).

Crop	Herbicides, insecticides, and fungicides (L/ha)
Cotton	28
Citrus	23
Soybean	12
Coffee	10
Wheat	10
Rice	10
Corn	6
Bean	5
Sugar cane	4,8

Source: Adapted from Carneiro, F. F.; Augusto, L. G. S.; Rigotto, R. M.; Friedrich, K.; Búrigo, A. C., Eds; *Dossier ABRASCO: A Warning about the Impacts of Pesticides on Health.* EPSJV: Rio de Janeiro; Expressão Popular: São Paulo, 2015 (accessed September 21, 2015).

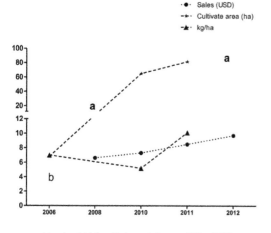

a = data not available for cultivate area in the years 2008 and 2012.
b = data not available for sales in the year 2006.

FIGURE 13.2 Recent developments in pesticide sales (in tons and dollars) and cultivated area in Brazil. *Source:* Vekic, A. The Pesticide Regulation in Brazil: Perceptions and Challenges. *Lecture Slides Given at Seminar "Sector Dialogues on the control and regulation of Pesticides and Biocides,"* Brasilia, 2014.

All this growth recorded in the last 50 years shows that there is a legal and governmental maintenance encouraging the use of pesticides in Brazil. Exemption of IPI remains through Decree No. 2,006/2006. Currently, there is a reduction of 60% on Tax on Movement of Goods (ICM) for pesticides and a proposed constitutional amendment (PEC 491-B 2010) exempting pesticides from all taxes.[77]

13.3 LEGISLATION CONCERNING THE MONITORING OF RESIDUES IN FOOD AND WATER, AND PESTICIDE REGISTRATION

The concept of pesticide standards was first institutionalized in Rio Grande do Sul, a state of the Federative Republic of Brazil, and later incorporated into the country by the Law No. 7,802/1989.[23,25] This was considered a milestone in the regulation of various aspects of pesticide production, use, registration, classification, and monitoring.[82] Law No. 7,802[23,25] provides for research, experimentation, production, packaging and labeling, transport, storage, marketing, commercial advertising, use, import, export, waste disposal and packaging, registration, classification, control, inspection, and surveillance of pesticides, their components and the like, and other measures. It consisted of 23 articles, as it was general law, which was regulated 13 years later, by Decree No. 4,074/2002.[107] Prior to that, changes in six articles of Law No. 7,802/1989 were made by Law No. 9,974/2000.[24]

Law No. 7,802/1989[23,25] set stricter rules for granting registration to pesticides, while the registering company provides information on the agronomic performance and human and environmental toxicity of the product.[79] However, the supervising agencies of government have not been provided with material, human and financial resources necessary for the registration activities and supervision of pesticides.[53]

In Brazilian law, pesticides and related products are defined as

> Products and agents of physical, chemical or biological processes, intended for use in the sectors of production, storage and processing of agricultural products, in the pastures, in the protection of forests, native or planted, and other ecosystems and urban, water environments industrial and intended to change the composition of flora and fauna in order to preserve them from harmful action of living beings considered harmful, as well as substances and products used as defoliants, desiccants, stimulators and growth inhibitors. (Brazil,[107] Art. 1, IV)

Decree No. 4,074/2002[107] was the most comprehensive standard on pesticides existing in Brazil. It had minor changes in 2005, with the promulgation of Decree No. 5,549/2005,[21] and more significant changes were observed in Decree No. 5,981/2006.[19] The Decree No. 4,074[107] indicates a concern of the government and Brazilian society regarding the issue of pesticides.

In addition to the standards previously mentioned, there were Normative Instructions, Resolutions, and Ordinances institutions linked to the federal government that were responsible for the regulation of pesticides. However, it is important to consider difficulties to insuring legal compliance in Brazil. If the pesticide legislation was fulfilled by all links of the chain (manufacturers, traders, inspection agencies, and farmers), the damage caused by them would be considerably minimized.[62] The problems of legal compliance in Brazilian pesticide chain are related to a weak process of surveillance on (1) registration;[77] (2) sales and use of unauthorized pesticides; (3) incorrect packing discard; (4) difficulties to identify human pesticide intoxication; (5) presence of pesticide residues on water[109] and food, etc.

The competent bodies for pesticide regulation in Brazil are as follows: (1) MAPA; (2) MMA by the Brazilian Institute of Environment and Renewable Natural Resources (IBAMA); and (3) the MH, through the National Health Surveillance Agency (ANVISA).

Considering the large number of standards, those most relevant for the purposes of this chapter will be highlighted, that is, the toxicological evaluation of pesticide residues in food; the maximum allowable limits for pesticides in water intended for human consumption; and the registration process for a new pesticide.

13.3.1 TOXICOLOGICAL EVALUATION OF FOOD IN BRAZIL

The body responsible for the process of toxicological evaluation of pesticides and residues of pesticides in food is the ANVISA of the MH. It stands out in this context, the Resolution No. 119/2003 establishing the Program Analysis of Pesticide Residues in Food (PARA); Resolution No. 216/2006, which provides for pesticide residues in plant products; and Resolution No. 48/2008, which provides for administrative procedures for toxicological re-evaluation of technical and formulated products based on active ingredients with health concerns. Due to the increasing use of pesticides for food cultivation in Brazil, in 2001, the PARA was created by ANVISA aiming at permanent check of pesticide residue levels in agricultural products sold in Brazil. The main objective of this program is to prevent the occurrence of potential noncommunicable chronic diseases, such as cancer, infertility, immune disturbs, etc., caused by the gradual intake of pesticides by controlling the marketing of these substances and monitoring of residues in food.[16]

To accomplish this goal, the collections are performed by the State Sanitary Surveillance, which performs the vegetable sampling available for sale on the market and then further sends them for analysis of pesticide levels by government laboratories. From this, the analysis is compared to those established by ANVISA through the Maximum Residue Limits (MRLs) allowed. In this analysis, the types of pesticides present in the samples evaluated are also verified as well as their presence in the list of products allowed by ANVISA. According to ANVISA, the MRL is determined based on the analysis of residues that remain on crops after application of pesticides.[17] The results of these analyzes are compiled in the PARA annual report, which serves as an indicator of the quality of the vegetables consumed by the Brazilian population. Such data have supported the decision taking of ANVISA regarding the types and levels of pesticides to be allowed in agricultural products sold in Brazil and serve as reference to estimate the health risks of consumers. The choice of pesticides to be mapped by PARA takes into consideration the assets commercialized in the country, the degree of toxicity of the active ingredient, the detection history of analyte, the availability of analytical standards for the identification of surveyed active substances, and whether the analyte in question can be detected by multiresidue methodology.

Since its inception in 2001, PARA has published annual reports, which showed increasing complexity of information over the years, reaching a higher level of detail in the report published in 2009 and 2010. There was an increase regarding the number of monitored crops as well as the amount of active surveyed in waste. However, there are gaps and questions regarding the PARA, due to changes in MRLs, information discrepancies between reports, such as the type of pesticide evaluated each year and the main vegetables selected for analysis. PARA's first annual report was published in 2001, but the reference lists containing the MRLs for the crops analyzed by PARA were made available for public access from 2007. The 2007 MRL list features a total of 104 substances used as pesticides in Brazilian nine crops—lettuce, banana, potato, carrot, orange, apple, papaya, strawberry, and tomato. In 2008, a new list of MRLs which added eight new crops was published, including pineapple, rice, onions, beans, mango, bell pepper, cabbage, and grapes, bringing the total to 17 crops. The 2008 MRL added to the list 63 more active analytes, bringing the total to 167 substances. Such report provided for the first time the unauthorized substances (US) notation, which indicates the presence of a substance banned for agricultural purposes in the country. The main assays employed by government labs for this purpose are based on the gas chromatography-based method.[28]

A comparative analysis of MRL lists shows that the number of analytes has been increasing over the years. Likewise, an increase in permissivity concerning previously prohibited substances is seen. Moreover, the MRLs increased for some crops from year to year. Over the reporting period, several crops were included for analysis, starting from nine crops in 2007 to 22 in 2012.

In general, it is estimated that one-third of the food consumed daily by the Brazilian population is contaminated by pesticide residues over the permitted limits. The compilation of PARA's data indicates the average level of food contamination samples observed throughout the analysis. Over 90% of bell pepper crops are contaminated by pesticides. Similarly, foods like strawberry, cucumber, and lettuce showed more than 50% of samples contaminated by pesticide residues above the permitted limits, or by the presence of US.

The program facilitated mapping of the use of various US for each type of crop, and the presence of active ingredients banned in the country. It also identifies various foods with pesticide residue levels above the allowed limit. Interestingly, the report shows that every year the same crops show residue levels above that permitted; however, it is not clear the extent to which punitive measures or sanctions are taken against those who insist on using products and amounts outside the established standards. There is also no monitoring report of the presence of glyphosate in food, although this product represents about 40% of agrochemical sales in the country. It is worth noting that among the 50 active ingredients mostly used in crops in Brazil, 22 are banned in the European Union.[27]

The details of the results reported by PARA in the period between 2001 and 2012 are presented in Section 13.3 of this chapter.

13.3.2   PESTICIDE RESIDUE EVALUATION IN WATER

The relationship between pesticides, environment, and human health is notorious, either because of their soil contamination potential; of water bodies, which may have their water used for human or animal consumption; direct contact with people who apply pesticides; or through the consumption of foods with residues of these substances.

However, in Brazil, since the beginning of the use of pesticides in the 1970s until the late 1980s, there seemed to be no major concerns about the possible presence of pesticides in soil, surface water and groundwater, even with the likelihood of contamination by these products.[48]

The MMA of Brazil also draws attention to the environmental contamination of air, soil, and water as a result of agricultural processes. The MMA highlights the complexity of the assessment of pesticide behavior due to physical displacement and chemical and biological transformation, which can lead to formation of by-products with properties different from the initial product that may further damage health or environment (www.mma.gov.br).

There are specific rules that have defined the potability parameters of Brazilian waters. The first of them—Ministerial Decree No. 56 of 1977, MH—allowed the presence of 12 types of pesticides in water for human consumption. In 1990 was established a new Ordinance (No. 36/1990), allowing the presence of 13 types of pesticides. The third standard was MH Ordinance No. 518/2004[22] which extended the amount of permitted pesticides to 22. The standard in force today is the MH Ordinance No. 2,914 of 2011,[15] which allows the presence of 27 types of pesticides;[109] 15 inorganic chemicals (heavy metals); 15 organic chemicals (solvents); 7 secondary chemicals from household disinfection, and the use of algaecides in water sources and treatment plants. On one hand, this development can be interpreted as positive, because as the use of pesticides is recent and growing in Brazil (last 50 years), the Brazilian government has sought to adapt to the new products which are used on the country's crops. On the other hand, data from Figure 13.3 are worrying because as a precaution, water used for human consumption should not allow the presence of any pesticide residue.[96]

Table 13.4 details the substances and the maximum allowable limits for water targeted for human consumption in the 1990 Ordinance: 2004 and 2011. Some pesticides were allowed in the 2004 Ordinance, but no longer in the current Ordinance since 2011. By 2011, the following substances were permitted: heptachlor and heptachlor epoxide (0.03 µg/L); hexachlorobenzene (1 µg/L); pentachlorophenol (9 µg/L); and propanil (20 µg/L).

Another relevant standard in terms of control of pesticide residues in water is the Resolution No. 357/2005[20] from the National Environmental Council. It classifies water bodies and establishes the conditions and effluent discharge standards. While MH Ordinance No. 2,914/2011[15] makes clear which substances are pesticides, CONAMA Resolution No. 357[20] does not. However, the fact that the two standards, both existing in the country, show a significant differences in the MRLs allowed is of great concern.

Table 13.2 makes a comparison of the standards parameters, indicating the MRLs of substances called pesticides (as Ordinance MH No. 2914/2011) for Class 1 waters (with good conditions for human consumption) and Class

3 (limit class for human consumption, requiring treatment) of CONAMA Resolution 357,[20] and the MRLs of Ordinance MH No. 2,914.[15] Blank items correspond to nonexisting substances in that standard.

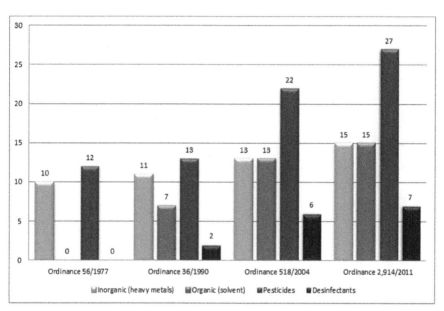

FIGURE 13.3   Amount of evaluated substances in water potability parameters. *Source*: Souza, L. C. A Critical Analysis of the Water Quality for Human Consumption in the Municipality of Francisco Beltrao—PR. Master Dissertation. UNIOESTE, Francisco Beltrão, PR, 2014.

While the Ordinance of the MH lists 27 pesticides for which the presence of residues in water for human consumption is allowed, the CONAMA resolution included only 14. This indicates increase of permissiveness for the presence of pesticides in water in Brazil. Nevertheless, there are substances for which MRLs considerably changed between the 2005 standard and the 2011 standard. The MRL for endosulfan changed from 0.22 to 20 µg/L; endrin, from 0.2 to 0.6 µg/L; glyphosate from 280 to 500 µg/L (glyphosate + AMPA); 2,4,5-trichlorophenoxyacetic acid, from 2 to 30 µg/L; and trifularina from 0.2 to 20 µg/L.

In addition to this increase in MRLs, there are criticisms of notions of MRL and *acceptable daily intake* (ADI).[62,99] Petersen[82] considers that both are derived from a Cartesian approach that is improperly applied to toxicology but used for transmitting confidence regarding supposed *tolerance limits* related to pesticide contamination in food and water. Another problem is the "establishment of MRLs acceptable for each of the active ingredients

TABLE 13.2   Comparison the Maximum Residue Limits (MRLs) for Pesticides Allowed in Brazilian Standards.

Substances	Ordinance MH No. 36/1990	Ordinance MH No. 528/2004	Ordinance MH No. 2,914/2011 (µg/L) Potability standard for Pesticides (Appendix VII)	CONAMA resolution No. 357/2005 (µg/L) Class 1 (human consumption with treatment)	CONAMA resolution No. 357/2005 (µg/L) Class 3 (human consumption with advanced treatment)
Alachlor	–	20	20	20	–
Aldrin + Dieldrin	0.03	0.03	0.03	0.005	0.03
Atrazine	–	2	2	2	2
Carbendazim + Benomyl	–	–	120	–	–
Carbofuran	–	–	7	–	–
Chlordane (cis + trans)	0.3	0.2	0.2	0.04	0.3
2,4-D	100	30	30	4	30
2,4,5-T	–	–	30	2	–
Diuron	–	–	90	–	–
Dichlorodiphenyltrichloroethane (DDT) [p,p'-DDT + p,p'-dichlorodiphenyldichloroethylene (DDE) (DDE + p,p'-Dichlorodiphenyldichloroethane (DDD))]	1	2	1	0.002	1
Endosulfan	–	20	20	0.056	0.22
Endrin	0.2	0.6	0.6	0.004	0.2
Glyphosate	–	500	–	65	280
Glyphosate + AMPA	–	–	500	–	–
Lindane	3	2	2	0.02	2

TABLE 13.2 (Continued)

Substances	Ordinance MH No. 36/1990	Ordinance MH No. 528/2004	Ordinance MH No. 2,914/2011 (µg/L) Potability standard for Pesticides (Appendix VII)	CONAMA resolution No. 357/2005 (µg/L) Class 1 (human consumption with treatment)	CONAMA resolution No. 357/2005 (µg/L) Class 3 (human consumption with advanced treatment)
Mancozeb	—	—	180	—	—
Methamidophos	—	—	12	—	—
Metolachlor	—	10	10	10	—
Molinate	—	6	6	—	—
Parathion methyl	—	—	9	—	—
Pendimethalin	—	20	20	—	—
Permethrin	—	20	20	—	—
Prophenophos	—	—	60	—	—
Simazine	—	2	2	2	—
Tebuconazole	—	—	180	—	—
Terbufos	—	—	1.2	—	—
Trifluralin	—	20	20	0.2	—
Chlorpyrifos + chlorpyrifos-oxon	—	—	30	—	—

without establishing a maximum number of ingredients per sample, the sum of their concentrations or their combined effects" (Neto[74] p. 68).

The Ordinance from MH No. 2,914/2011[15] was target of criticism from researchers working in the field of public health in relation to the issue of pesticides, mainly because of very low monitoring of these substances in water intended for human consumption. Considering the 430 active ingredients registered in Brazil, only 27 are required to undergo periodic reviews.[109] In addition, there is an illegal entry problem of banned pesticides in Brazil through neighboring countries, such as Paraguay and Bolivia, among others.

These 27 pesticide active ingredients need to be monitored every 6 months by all municipalities. This number represents <10% of current active ingredients approved for use in the country. From 2007 to 2010, approximately 169,000 pesticide analytical results were prepared and evaluated, although approximately 980,000 would be expected if all municipalities registered their analyses. This shows that only 9–17% of municipalities registered their data in Water Quality for Human Consumption System (SISAGUA), a Brazilian database for monitoring pesticides in drinking water. The results shows that the pesticide active ingredients most consumed in Brazil from 2009 to 2012 (more than 1000 t/year for each ingredient) were glyphosate, mineral oil, 2,4-D, atrazine, sulfur, methamidophos, vegetable oil, carbendazim, acephate, mancozeb, and diuron. It is important to emphasize that approximately 30% of the 27 pesticides in the current ordinance are no longer authorized for use in Brazil. Among those that have been canceled are aldrin/dieldrin, chlordane, dichlorodiphenyltrichloroethane (**DDT**), endrin, and lindane. Aldicarb, methamidophos, and endosulfan were canceled recently.[109]

13.3.3 THE ISSUE OF PESTICIDES REGISTRATION IN BRAZIL

When requesting the registration of a new pesticide, the manufacturer must present the three competent bodies studies to prove the efficacy and safety of the product. These studies are prepared by laboratories contracted by companies, not by government agencies. Thus, nothing prevents the manufacturer to manipulate the information or to present only the results that are of interest. It is the responsibility of government agencies only to evaluate the submitted studies, comparing them with other published studies in the scientific literature.

There are some federal Brazilian agencies with different roles in the assessment of claims for pesticides registration intended for agricultural use.

Each agency makes its analysis within its area of competence: MAPA evaluates the agronomic efficacy of the product; ANVISA evaluates the risks to health of the population; IBAMA assesses the risks to the environment. Federal agencies are responsible for monitoring the manufacture and formulation of pesticides, while the federal states are responsible for overseeing the transportation, storage, usage, and disposal of packaging.[77]

Thus, it is up to the registrant agencies (MAPA, ANVISA, and IBAMA) to prove product risks—otherwise, the product is released for consumption. Londres[62] points out that one of the obstacles that hinders the assessment of the Ministry is to establish direct relationships between exposure to a certain product and the development of chronic health problems.

Through the Decrees No. 4,074/2002[107] and 5,981/2006,[19] it was possible to expedite the analysis and marketing of products based on active ingredients with expired patents. This helped to increase the entry of new products in the domestic market. The regulatory infrastructure has been insufficient to meet the high demand of product registration for agrochemical companies. This has led to pressure from groups linked to agribusiness to streamline the pesticide registration process. On one hand, there are a few people who work in pesticide regulation in Brazil. On the other, there is strong pressure from politicians, usually linked to companies that produce pesticides for release of the production or marketing of a product.[77]

Another important issue concerns the registration costs of a new active ingredient and the validity of these records. While in the United States, the costs are about US$ 630,000 and the validity of the registration of a pesticide is 15 years, in Brazil the registration fee for a new a.i. is a maximum of US$ 1000 and the period of validity is undetermined.[78]

Although the registrations are granted for an indefinite period, a fact that facilitates the permanence of pesticides on the market, Brazilian standards require certain conditions for the control of these registrations. Registrant agencies should conduct a reassessment of environmental safety and health risks, as well as agronomic efficacy of pesticide products, the results of which can lead even to the cancellation of the registration,[62] like aldicarb, methamidophos, and endosulfan.[109]

Based on data from ANVISA (2014) on pesticide registration balloting, Pelaez et al.[77] highlight the following information:

- By June 2014, the registration requests list indicated about 1500 processes (products) awaiting the start of the evaluation process.
- Only 5% of registration claims correspond to products based on new a.i.

- 20% of the products in the registration queue are based on a.i. on the lists of substances potentially to be banned in the EU market.
- 61% of registration requests are made by sales representatives of companies that do not have factories in Brazil. Thus, there is no guarantee that they would bear the costs of potential environmental or human health damage, or compensation to farmers if the pesticide does not have the expected efficacy.
- The time of the process of pesticide registration requests has increased. Between January 2010 and June 2014, it is estimated that the average time from the date of the submission of the Protocol to the completion of the analysis, increased from 25 to 44 months.

13.4   OVERVIEW OF PESTICIDE RESIDUE MONITORING IN FOOD IN BRAZIL

In the period between 2001 and 2007, the Sanitary Surveillance (VISAs) of 16 of the 23 Brazilian states participated in the PARA. The collection sites were in retail outlets in large supermarket chains of the capitals of the states participating in the program. The value of marketed pesticides in Brazil in 2007 was US$ 5.4 billion (second in the ranking), reaching the top of the world rank 2008 onward.[7]

The crops selected for analysis by the program were defined taking into account: *per* capita annual consumption of their food in kg, provided by the minimum food consumption in a typical family used to calculate the ADI of pesticides; cropping systems and pest management of different cultures; and the availability of these foods in trade in different states involved in the program. Thus, in this period, ANVISA selected 92 a.i. through (1) the data collected on the use of pesticides in the country; (2) the information available in laboratories regarding pesticide residues in food usually detected; and (3) the availability in the laboratory and/or in the market of analytical standards necessary for the determination thereof.

The results indicate that the period between 2001 and 2007, 7321 samples were analyzed. The greater issue was not the pesticides use above the limit, but the use of prohibited substances (reported as US). The crop with the highest presence of pesticides in absolute terms and in average was the strawberry (minimum of 37.7% and a maximum of 54.6% of the samples). Lettuce and tomato had maximum values above 40%, while papaya reached 37.6% and carrots 19.5%. It is noteworthy that the data from 2001 to 2007 were restricted to samples collected in 16 of the 27 states. Ideally, the program coverage should be extended to other Brazilian states.

In the 2008 report, five more participating states were added, and the Federal District, and 17 crops and 167 substances were monitored. Due to the ban or restrictions that have been applied internationally to several a.i., Brazil made a toxicological re-evaluation of the a.i. in effect at the time, considering also those which had been banned or prohibited in the country. This step represented a breakthrough in terms of food security for the country. As a result, several companies went to court, questioning PARA reports; requesting the extension of the residue levels detected in food, or requiring the release of use of products not authorized in Brazil.

Analyzing PARA evolution between 2002 and 2008 (Table 13.3), it is observed that there was expansion of the program, through the inclusion of new crops, unmonitored between 2001 and 2007. Note that, except for bananas, lettuce, strawberries, and tomatoes, other crops analyzed showed an increase in the percentage of unsatisfactory results, for instance, reaching 300% increase over the previous year (carrot). These data demonstrate that despite the prohibitions and sanctions established in the country, much of the crop still had pesticide residue levels higher than allowed. It stands out among the new foods analyzed, bell pepper (64.4%) and grapes (32.7%) with a high rate of contamination by pesticides.

TABLE 13.3 Chronological Analysis of the Evolution of the Percentage of Unsatisfactory Results in Some Crops Analyzed by the Program.

Crop	2002	2003	2004	2005	2006	2007	2008
Lettuce	8.64	6.67	14	46.45	28.68	40.00	19.80
Banana	6.53	2.22	3.59	3.65	NA	4.32	1.03
Potato	22.20	8.65	1.79	0	0	1.36	2.00
Carrot	0	0	19.54	11.30	NA?	9.93	30.39
Orange	1.41	0	4.91	4.70	0	6.04	14.85
Papaya	19.50	37.56	2.50	0	NA	17.21	17.31
Apple	4.04	3.67	4.96	3.07	5.33	2.90	3.92
Strawberry	46.03	54.55	39.07	UA	37.68	43.62	36.05
Tomato	26.10	0	7.36	4.38	2.01	44.72	18.27

NA, Not analyzed. Data taken from the 2008 report—PARA.
Source: PARA report.[18]

Likewise, there are a large number of US present in such unsatisfactory samples (Table 13.4).

TABLE 13.4 Data from Unauthorized Active Ingredients Found in Some Unsatisfactory Samples Analyzed and Reported in PARA 2008.

Culture	% Unsatisfactory	Unauthorized ingredients found in samples
Pineapple	9.47	Acephate, cypermethrin, dithiocarbamates, dimethoate
Lettuce	19.80	Acephate, carbaryl, carbendazim, chlorpyrifos, deltamethrin, dimethoate, fenpropathrin, methamidophos, methomyl, tebuconazole
Rice	4.41	Cyproconazole, flutriafol, methamidophos, myclobutanil
Banana	1.03	Fenarimol
Potato	2.00	Endosulfan
Onion	2.91	Acephate
Carrot	30.39	Acephate, chlorpyrifos, dimethoate, methamidophos, prophenophos
Bean	2.92	Cyproconazole, diuron
Orange	14.85	Cypermethrin, endosulfan, esfenvalerate, parathion methyl, prochloraz, prophenophos
Apple	3.92	Dichlorvos, triazophos
Papaya	17.31	Acephate, acetamiprid, cyfluthrin, dimethoate, endosulfan, epoxiconazol, methamidophos, methidathion
Strawberry	36.05	Acephate, captan, chlorfenapyr, chlorothalonil, chlorpyrifos, deltamethrin, endosulfan, folpet, methamidophos, prochloraz, tetradifon
Pepper	64.36	Bifenthrin, bromopropilate, carbendazim, cypermethrin, chlorpyrifos, dicofol, endosulfan, esfenvalerate, fenpropathrin, fenarimol, lambda-cyhalothrin, methamidophos, permethrin, procymidone, prochloraz, prophenophos, tebuconazol, triazophos
Cabbage	8.82	Carbendazim epoxiconazole, phenthoate, methamidophos, procymidone, tebuconazole
Tomato	18.27	Aldicarb, allethrin, cyproconazole, chlorpyrifos, chlorpyrifos methyl, folpet, methamidophos
Grape	32.67	Acephate, cypermethrin, chlorfenapyr, chlorpyrifos, deltamethrin, dimethoate, endosulfan, fenpropathrin, methamidophos, tetradifon

Source: PARA report.[18]

All foods analyzed had at least one unauthorized a.i. Among the ingredients, acephate and methamidophos were predominant. In samples of lettuce, strawberries, peppers, and grapes, 10 or more unauthorized ingredients were found. Concerning bell pepper, there were 19 unauthorized ingredients, and 4 above the MRL.

The data show that the use of prohibited substances, as well as the presence of substances above the MRL, continued in 2008. This indicates that there was no effective punishment for those that ignored the standards posted by ANVISA. It is observed, for example, the presence of residues of a.i. such as methamidophos on crops for which its use is unauthorized, such as lettuce, rice, carrot, papaya, strawberries, peppers, cabbage, and grapes, or in crops in which its use is restricted by ANVISA, as in the tomato. This points to a serious problem, because this a.i. is authorized, for example, in culture of tomato, and its application is through aerial spraying, tractor or center pivot, implementing rules not employed in the production of the tomato. It is clear that during this period aerial spraying of pesticides was still permitted, which created a huge risk of occupational exposure and contamination of the population living around these farms.

Since 2009, there has been a major concern in detailing the results of food analysis. There have been improvements in analytical skills, as well as the integration of reference laboratories for such analysis, optimizing the standardization of findings and allowing safe comparisons. In 2009, 20 types of food and 234 analyzed a.i. were monitored. It has expanded considerably the number of samples (3130), including samples collected in all states. The amount of a.i. evaluated was also considerably expanded between 2007 and 2009, as shown in Table 13.5.

TABLE 13.5 Summaries of the Data Presented by PARA between 2007 and 2009.

Year	2007	2008	2009
Participant states	16	16	26
Analyzed samples	1198	1773	3130
Number of crops	9	17	20
Active ingredients analyzed	104	167	234

Source: PARA report.[17]

In 2009, Sample Management System of PARA (SISGAP) was installed and ANVISA also began to report the methodology applied in these analyses, which represents a breakthrough in terms of critical analysis of the reports, especially for the scientific community. According to the document, the collection method is similar to that employed in the United States and some European countries. Of the 3130 samples analyzed, 29% were found unsatisfactory, 2.8% above the MRL, and 23.8% of samples with US. In 2.4% of the samples from grape and peppers presented residues above the MRL and unauthorized concomitant were found in the same sample. Bell pepper continues as the most contaminated culture for residues (about 80% of samples) and also showed high percentage of samples contaminated with unauthorized products (almost 65%), and the concomitant presence of US and above the MRL in 12.1% of the samples. Overall, much of the crop showed a high rate of contamination by unauthorized products. Detailed data for MRL and culture are presented in Table 13.6.

TABLE 13.6 Number of Samples Tested by Crop and Unsatisfactory Results of PARA 2009, Subdivided into Results Showed that Unauthorized Substances (US), Authorized (>MRL), and the Previous Two Situations in the Same Sample for Some Crops Evaluated.

Product	% US	% Over MRL	% Over MRL + presence of US
Pineapple	28.3	10.3	5.5
Lettuce	37.7	0	0.7
Rice	26.5	0	0.6
Banana	1.8	1.8	0
Beet	32	0	0
Onion	16.3	0	0
Carrot	24.8	0	0
Cabbage	32.6	6.2	5.4
Papaya	21.2	12.9	4.7
Strawberry	38.3	8.6	3.98
Cucumber	51.4	2.1	1.4
Bell pepper	64.8	3	12.1
Tomato	31.3	0	1.4
Grape	35.2	8.5	12.7

US, Unauthorized substances; MRL, maximum residue limit.
Source: PARA report.[17]

According to ANVISA, the above permitted MRL indicate the incorrect use of certain substances, whether for the number of applications (higher than recommended in the product information) by increasing the amount of a.i. applied *per* hectare; or for failure to observe the withdrawal period between the pesticide application date and the food harvest time. It is also observed that there was uneven application of substances, or use of substances suitable for other crops, and also use of substances unauthorized for use in Brazil, predominantly *pyrethroids*, carbamates, organophosphates, and azole derivatives.

In 2010, samples of all Brazilian states were analyzed, except São Paulo, which created a monitoring program for itself (Paulista Food Program). However, as the state of São Paulo provides much of the country's agricultural production, it should also be evaluated by PARA, regardless of owning a residue assessment program itself.

In 2010, ANVISA monitored 18 kinds of food in PARA. A total of 2488 samples were analyzed and 28% of them presented residues above the allowed limit or presence of unauthorized products. In the 2012 report, about 90% of bell pepper samples showed presence of unauthorized agricultural inputs, followed by cucumber (56%), lettuce (52%), and carrots (49%). Overall, nearly 92% of bell pepper samples were outside of the MRL and/ or used agricultural inputs not allowed in the country. Of the total sample assessed in 2010, 27.9% showed irregularities concerning MRLs and/or accused presence of unauthorized products.

According to this report, 24.3% of the samples used unauthorized pesticides for that specific crop or even unauthorized for marketing in the country. The a.i. found in most of the samples in general was organophosphates, which also predominated as the main products in the samples shown to be above the MRL. Within the unsatisfactory samples evaluated in 2012, 30% had input residues that were in the process of toxicological re-evaluation process or were discontinued for sale in Brazil in 2010, a fact that draws attention to the severity of the potential risk of disease development which the Brazilian population has been exposed to in 2010. It is noteworthy that in 37% of samples, pesticide residues were detected.

The report also points out that:

1. About 57% of farms where pesticides were used did not receive agronomic guidance.
2. Almost 85% of laborers worked on small farms, and most of them were illiterate.

PARA submitted a joint report for 2011–2012. In this PARA the MRL calculation formula was given:

In the toxicological evaluation for pesticide registration purposes conducted by ANVISA, it is calculated the Maximum Daily Intake

Theory (MDIT). It is defined by the ratio: sum of the average of the daily consumption per capita of each food and its LMR divided by the body weight of each individual. In other words, MDIT estimates the maximum amount of pesticides in foods that theoretically a person can ingest daily (Brazil).[14]

Based on this formula, it is observed that the calculation of the MRL is subject to individual variations that are not taken into account by the program. According to ANVISA, based on PARA, several pesticides, especially organophosphates, have been banned or suffered use restriction in the country (e.g., carbendazim and methamidophos), but they still appeared in the 2011–2012 report.

Samples of all states of the federation were collected in this report. There was a 25% increase in the number of monitored food. In 2011, 1628 samples were analyzed, and there was no residue detected in 22% of them. The results were unsatisfactory in 36% of samples, with 2.3% above the MRL and 32% with unauthorized pesticides. It was observed that 90% of bell pepper samples were unsatisfactory, the same for 67% of carrot samples and 44% of cucumber samples. The data indicate the same pattern of 2010 nonconformities in relation to products (organophosphates). The presence of active tebufempirade and azaconazole in grape suggest smuggling because these substances have never been registered in Brazil. Another a.i. was detected, aldicarb (in Portuguese, *Chumbinho*) in a sample of rice. This a.i. is the pesticide with the highest acute toxicity described. It is not suitable for the cultivation of rice and had been deregistered in October 2012 by manufacturer's request. As the data presented in the report were partial, a discussion of the findings cannot be held here.

From 2013, there was no provision of new reports of analysis of pesticide residues in the program's web page. It is important to highlight that the main pesticide used in large scale in the country, glyphosate, was never included in multiresidue analysis adopted by PARA. The program was also limited to examine only products in natura, and there is no reference to the analysis of processed or manufactured products.

13.5 OVERVIEW OF PESTICIDE RESIDUE MONITORING WATER IN BRAZIL

Taking into consideration that the Brazilian legislation on pesticides is recent and that monitoring compliance with legal requirements has been insufficient, it is reasonable to state that lack monitoring and research on the social and environmental consequences of pesticide use in Brazil. However, there are efforts of public institutions researchers in this direction, such as federal and state universities, the Oswaldo Cruz Foundation (FIOCRUZ), the Brazilian Association of Collective Health (ABRASCO), and the National Cancer Institute (INCA). In terms of scientific production, it is worth mentioning the journal *Ciência & Saúde Coletiva* under the responsibility of ABRASCO. In addition to a special number of issues dedicated to publications related to the impacts of pesticides on health and the environment [*12* (1), 2007 [Online]. http://www.scielo.br/scielo.php?script=sci_issuetoc&pid=1413-812320070001&lng=pt&nrm=iso (accessed December 3, 2015)], this journal has more than 50 articles discussing aspects related to pesticides.

The lack of available official data for consultation is another problem. In 2011, only 24% of Brazilian municipalities conducted pesticide analysis in water according to the criteria of the MH, so that 76% of municipalities do not even have access to information on the contamination of the drinkable water. Only four states (São Paulo, Mato Grosso do Sul, Tocantins and Paraná) monitored more than 40% of the municipalities, fulfilling the sampling plan required.[108]

An investigation of the Federal Public Ministry of the State of Mato Grosso do Sul revealed that the laboratory responsible for the MH water monitoring tests have analyzed only 55% of the substances mandated by law (15 of 27 substances).[27] There is inefficiency in monitoring the a.i. required by law, due to the low amount of public laboratories that perform toxicological analysis of pesticides.[74]

Albuquerque et al.[4] present a critical review of pesticides in Brazilian freshwaters, with information of official agency records and peer-reviewed scientific literature. Studies about the occurrence of pesticides in freshwaters in Brazil are scarce and concentrated in a few sampling sites in 5 of the 27 states. Herbicides (21) accounted for the majority of the substances investigated, followed by fungicides (11), insecticides (10), and plant growth regulators (1).

Oliveira et al.[75] investigated levels of pesticides in the fish *Prochilodus costatus* caught in São Francisco River, one of most important rivers in Brazil. Thirty-six fish were captured in three different areas, and samples of the dorsal muscle and pooled viscera were collected for toxicological analysis. They evaluated the presence of 150 different classes of insecticides, fungicides, herbicides, and acaricides. Organophosphorus and carbamate pesticides were detected at the highest levels in the caught fish. Among the

41 organophosphorus pesticides surveyed, 9 types were detected (chlorpyrifos, diazinon, dichlorvos, disulfoton, ethion, etrimfos, phosalone, phosmet, and pyrazophos) in the muscle, viscera pool, or both in 22 (61.1%) fish. Sampled tissues of 20 (55.6%) fish exhibited at least one of the eight evaluated carbamate pesticides and their metabolites. In conclusion, *P. costatus* fish caught in the São Francisco River contained residues of 17 different pesticides, in both muscles and the viscera pool, indicating heavy environmental contamination by pesticides in the study area.

Da Silva et al.[35] analyzed residues of organochlorines pesticides (OCs) and polychlorinated biphenyl (PCBs) in mullets and croakers collected in marine water in south coast of Rio de Janeiro. The results points that 23 OC pollutants were detected in croakers at GB and 20 in mullets and all PCBs congeners investigated in the study were present in the two species.

For the survey research related to monitoring of water bodies in Brazil, two reference documents were used. One of them is the Dossier of Brazilian Association of Public Health.[27] This document comes from a compilation of several studies, reflections, and collective forums held by a group of professionals involved with the problem of pesticides and/or the defense of agroecology. Gomes and Barizon[48] presented a literature review on research related to environmental contamination by pesticides in Brazil between 1992 and 2011.

One of the most alarming cases of contamination by pesticides registered in Brazil occurred in the municipality of Lucas do Rio Verde, Mato Grosso. In 2006, residential areas, crops, and people were contaminated with paraquat, used in drying of soybeans. Surveys conducted between 2007 and 2010 in the municipality of Lucas do Rio Verde,[83] found that (1) in 2010, it was used an average of 136 L of pesticides *per* inhabitant;[71] (2) there were spraying aircraft within 10 m of drinking water sources and streams, livestock, and residences, disregarding federal law, which prohibits spraying within 500 m of these sites; (3) pesticide residues were identified in 83% of 12 drinking water wells, especially schools; 56% from wet samples and 25% from air, in schools monitored for 2 years; (4) residues were present in 88% of blood and urine samples from teachers, being the level in rural areas twice greater when compared to urban areas; (5) identification of dichlorodiphenyldichloroethylene (DDE) pesticides, endosulfan, deltamethrin, and DDT in 100% of breast milk samples from 62 mothers in 2010; (6) presence of pesticide residues in sediments from two lakes and from the blood of frogs in the area as well as congenital malformations in frogs; and (7) malformation and death of earthworms exposed to glyphosate and 2,4-D.[34]

In the state of Ceará (CE), agrochemical sales increased from 1649 t in 2005 to 3284 t in 2009.[85] Marinho[65] identified contamination of water intended for human consumption in communities in the municipalities of Limoeiro do Norte and Quixere. Amongst the 24 water samples collected in triplicate from channels used for irrigation, tanks of water supply companies of municipalities, and deep wells, all of them showed the presence of at least 3 a.i. of pesticides. There were samples with ten different active ingredients. The analysis technique used was liquid chromatography-mass spectrometry with electrospray ionization (LC/MS-ESI). Data from a report of the Water Resources Management Company (COGERH) of Ceará showed that 10 samples analyzed in Jandaíra Aquifer, the same location of Marinho's sampling,[65] 6 revealed the presence of pesticides [diazona (3 samples); flutriafol (2); cyromazine (1); propiconazole I and II (1); ametryn (1)].

In rice-growing regions of southern Brazil, pesticide residues were found in 21 samples of surface water collected in three periods in season 2007–2008: before cultivation, during cultivation, and just after water drainage for crops harvest.[93] There are contamination records of water by pesticides such as carbofuran, clomazone, quincloraque, bentazone, 2,4-D, fipronil, and propanil in rice-growing areas[49,64] and imidacloprid, atrazine, and clomazone in tobacco-growing areas[13] in the state of Rio Grande do Sul.

In Paraiba do Sul River waters in Minas Gerais, in the vicinity of sugar cane plantations, residues of di-Syston, fosdrin, and malathion were found.[26] Fumes et al.[44] developed a method for the determination of six pesticides (tebuthiuron, carbofuran, atrazine, metribuzine, ametryn, and bifenthrin) in sugarcane juice samples commercially available in markets from different cities from São Paulo state, Brazil.

In Pantanal, fluvial plain region with rich biodiversity, water was analyzed in 25 points of confluence of the Rivers Paraguay, Cuiabá, São Lourenço, Vermelho, Itiquira, Correntes, Taquari, Coxim, Negro, Aquidauana and Miranda, as well as in 25 points of wetlands plain, in backwaters and irrigation channels of rice fields of the Miranda river basin (Paraguay river tributary). Sampling occurred during 2005 in the dry phase (May) and at the beginning of the rainy season (December), when pesticides, nutrients, and soil are discharged into the rivers with the first floods. Overall, 32 a.i. were analyzed by gas chromatography and mass spectrometry. In Miranda river were detected dieldrin, p,p'-DDE (DDT metabolite) in 100% of samples (1.2 to 14.4 mg/kg); and p,p'-DDT (1.0 mg/kg) in a sample on the Miranda river and in a rice culture irrigation canal, exceeding the legal parameters. Pyrethroid residues were found in concentration above the limits permitted and DDT, which is banned in Brazil since 1985 was also identified.[36,69]

Gomes and Barizon[48] presented various studies related to the presence of pesticide residues in water and sediment. In the mountain region of Rio de Janeiro, OC compounds such as DDT, benzene hexachloride (BHC), and lindane are still used, especially in vegetable crops, contrary to current legislation.[80] Moreira et al.[72] and Alves and Oliveira-Silva[8] found anticholinesterase pesticide concentrations in values up to eight times above the permitted limit in two points of an important water course in the city of Nova Friburgo-RJ.

In São Paulo, Corbi et al.[33] analyzed sediment from 11 streams in the central region, in areas with sugar cane cultivation, grazing and riparian forest. It was observed the presence of 16 OCs such as aldrin, BHC, endrin, DDT, endosulfan I, endosulfan II, and sulfate. Aldrin was present in the sediment of all streams and in high concentrations, ranging from 7.14 µg/kg in a sample of water from a river located in pasture area to 1787 µg/kg in a stream in sugar cane area. Lichi[61] reported the presence of residues of 10 a.i. (BHC, DDT, lindane, chlordane, aldrin, DDE, endrin, dieldrin, DDD, and heptachlor) in water in natura, sludge in natura, decanted sludge in water treatment plant and treated water in 16 municipalities, with equitable geographical representation of Paraná, and found that OC insecticides have accumulated in the bottom sediments and suspended materials.

In the State Park Terras Alto Ribeira (PETAR) located in the Ribeira Valley/SP, it was analyzed samples of water, sediment, and fish during the rainy season in January 2000. The results indicated that the fauna of PETAR is exposed to 20 different types of pesticides, which were found in water samples and sediments, as reported by Tomita.[110]

In the state of Rio Grande do Sul, glyphosate was found in irrigated rice fields with water from the Mirim Lagoon at concentrations above 7 µg/L which is the maximum allowed by the US Environmental Protection Agency.[68] In irrigated rice production areas in the coastal plain and the western border of Rio Grande do Sul (end state), residues of 11 herbicide, 2 insecticides, and 7 fungicides were found in 2007 and 2008 collections.[67]

In all rice-growing regions of southern Brazil, there was the presence of at least one pesticide occurring in groundwater.[92]

Flores[42] and Flores et al.[43] found the BHC and DDT OCs in sediments of Ribeirão São Bartolomeu, in Viçosa-MG, as well as Heptachlor epoxide and Endrin. In Paty do Alferes county, mountain region of Rio de Janeiro, Veiga et al.[100] identified water contamination by organophosphates and carbamates in areas of tomato cultivation.

In areas under intensive cultivation of cotton, corn, and soybean, Dores et al.[37,38] indicated the presence of some herbicides in groundwater region, which is usually used for human consumption.

The examples presented above highlight the contamination of water by different types of pesticides in several Brazilian regions. Therefore, these contaminants are linked to different types of crops developed in these regions, such as rice, soybean, cotton, sugarcane, tomato, among others. From the contamination of water also occurs contamination of fish and other animals.

13.6 THE ISSUE OF HUMAN CONTAMINATION BY PESTICIDES

Labor accidents and intoxications due to pesticides have experience and increase in Brazil according to Information System of Compulsory Conditions (SINAN). In addition, Brazilian pesticide market also expanded in the last decade, which is associated with increased consumption of those substances.[86] Depending on the exposure levels, pesticides have the potential to be hazardous to human health. There is a huge body of evidence supporting an association between pesticide exposure and incidence of human disease such as several types of cancer, neurodegenerative disorders, birth defects, reproductive disorders, among others. Populations living near agricultural areas are more exposed to pesticides and are more susceptible to developing diseases. Pesticides affect the health of agricultural workers and pesticides sprayers experience higher morbidity compared with other occupations.[66]

There are several types of pesticides, including insecticides, herbicides, and fungicides, which may impact health of children and adults. The exposure mode to pesticides may differ. Briefly, subjects may be directly exposed to pesticides, as farmers, or indirectly exposed to pesticides, as consumers. The direct exposure to pesticides includes pesticides handlers and sprayers, causing harm mainly through dermal absorption and/or respiratory inhalation routes. Pesticides handlers and sprayers need to combine, carry, and apply pesticides. However, until the moment, there is a lack of a defined method to measure the occupational exposure to pesticides.[90] Indirect exposure to pesticides involves predominantly diet and household use. Harm caused by respiratory exposure to pesticides can vary from a simple airway irritation to asthma and lung cancer.[105]

The environment has a huge influence on children's health. The health effects of children exposure to environmental pollutants has been revised in Brazil and it shows that most effect is related to respiratory problems as wheezing, asthma, and pneumonia. In more severe cases, effects can be related to leukemia, adverse pregnancy outcomes, and congenital abnormalities. In most studies, the effect occurrence is due to parental exposure to pesticides before and during pregnancy, and during lactation.[46] Indeed, a prenatal and postnatal exposure to pesticides can impact children's respiratory health. Mothers' exposures to chemicals from domestic and occupational pesticides during pregnancy period have increased risk of asthma and wheezing development in offspring as compared to nonexposed mothers. Studies conducted in Spain showed a positive association between prenatal exposure to OC insecticide and development of childhood asthma. The detection of dichlorodiphenyldichloroethylene (DDE), with a median concentration of 1 ng/mL in the cord serum at the birth, was associated with childhood persistent wheezing and doctor-diagnosed asthma after 4 years old.[63] In Brazil, the presence of OC and its metabolites was found in breast milk, umbilical cord, and in maternal and children' blood as an indicative of continued exposure to those pesticides.[46]

Several pediatric respiratory diseases can be associated with nonoccupational early life exposure to organophosphate pesticide. Chemicals present in pesticides can act on children respiratory rout by decreasing pulmonary function in school-age children living in agricultural areas.[84]

Exposure to indoor home pesticide, such as insecticides and residential herbicides, is positively associated with a greater risk to develop childhood leukemia and hematopoietic malignancies.[29] In utero and postnatally exposure to pesticides are associated with leukemia in children younger than 2 years old.[46] Besides, a study conducted in India with 132 children with leukemia showed a significant increase in risk factor of developing childhood leukemia associated with maternal work in agricultural areas and outdoor exposure to pesticides during pregnancy as compared to mother's occupation as house wife, workers and in industry.[56]

In utero exposure to OC pesticides is associated with autism and impairment of neuropsychological development. Human researches corroborated by animal's assays shows that children's cognitive deficits is associated with DDE and DDT exposure.[87] Chemicals from pesticides are able to modify cellular energy metabolism causing nervous system disorders as neurodevelopmental disorders, neurodegenerative diseases, and neurobehavioral disorders. OC exposure induces neuronal alterations by several mechanisms including damaging mitochondria, increasing reactive oxygen species (ROS) production, altering hormonal response, and interfering in neuronal myelination.[88] In Brazilian rural microregions with intense agricultural activity, deaths among infants <1-year old caused by central nervous system and cardiovascular system malformations were positively correlated with *per capita* pesticides consumption.[111]

Pesticides are also able to induce neuronal apoptosis by increasing peroxinitrite levels leading to protein nitration, lipid peroxidation, and DNA fragmentation. In addition, pesticides influence on nitric oxide and peroxisome proliferator-activated receptor gamma coactivator 1 alpha (PGC-1α) disrupting mitochondrial electron transport chain, inducing increased oxidative stress, and culminating in neuronal apopotosis.[70]

Several lifestyle factors are associated with the development of neuronal diseases. For example, pesticide exposure is recognized as a potent risk factor for the Parkinson's disease development.[47] Parkinson's disease is the second most common neurodegenerative disorders in the elderly. Genetic and nongenetic factors, such as environmental aspects, contribute to Parkinson's disease outcome.[2] Herbicides and insecticides ingredients have an association with Parkinson's disease especially in agricultural areas. A study conducted at Nebraska State showed that exposure to pesticides ingredients as atrazine, broxomy, alachlor, metribuzin, and glyphosate is related to Parkinson's diseases incidence.[104]

Besides neuronal diseases, pesticides are related to the induction of several distinct disorders. For instance, the indirect exposure to pesticides may be related to end-stage renal disease. A study reporting this correlation showed that wives of pesticide applicators, having indirect contact with pesticides through their husbands' use, showed increased incidence of end-stage renal disease.[57] Nevertheless, the association of pesticides and renal disease still needs to be better evaluated. Increased risk of hematological and hepatic damage is also associated with OC metabolites. A Brazilian study showed that in a population of 415 male and 432 female, the exposure to OC metabolites increased levels of eosinophilia and lowed hemoglobin and erythrocyte count. Pesticides exposure induces hepatic alterations in this population, observed by higher levels of bilirubin, glutamic-oxaloacetic transaminase, and glutamic-pyruvic transaminase.[45]

It is well known that environmental factors are related to the development of several types of cancer by affecting gene regulatory mechanisms. Pesticides can be genotoxic, and they have been associated with human cancer development. Genotoxicity is one key characteristic of a carcinogen that includes DNA damage, intercalation, gene mutation, cytogenetic change, among others.[91] Moreover, the incidence of cancer is greater among direct contact with pesticides.[3] In fact, genotoxic effects of pesticides have been demonstrated by its induction of higher frequencies of micronuclei, kariorrhexis, karyolysis, and binucleated cells in oral mucosa found in subjects with contact to pesticides.[1]

Another characteristic of a carcinogen is the ability to induce epigenetic alteration, such as DNA methylation, histone modification, and altering microRNA expression.[91] The induction of oncogenes mutation by pesticides is a mechanism related to cancer development. The exposure to chemicals from pesticides can alter histone deacetylases, histone acetyltransferases, and DNA methyltransferase activity favoring cellular growth and proliferation and increase transcription expression of oncogenes culminating in cancer development.[98] Pesticides have the potential to change gene regulation by altering epigenetic markers. Epigenetic alteration induced by pesticides may be one of several mechanisms by which pesticides harm human health.[32] For breast cancer, researchers have shown an association of pesticide exposure and its incidence.[9,97] Initiation of breast cancer may be induced by pesticides that possess a chemical structure similar to aryl hydrocarbon receptor, as OC, and studies have focusing on highlighting the mechanisms involved in its induction. It seems that pesticides can induce breast cancer cell proliferation by long-term exposure through the induction of aryl hydrocarbon receptor and estrogen receptor α signaling. This signaling pathway stimulates an imbalance between CYP1A1 and CYP1B1 enzymes modifying 2-hydroxyestradiol/4-hydroxyestradiol ratio favoring breast cancer carcinogenesis.[60] Chemicals structures of pesticides are similar to natural hormones and may have an endocrine-like interference. As a consequence, it may be associated with development of hormone-related tumors, as breast, testis, thyroid, prostate and ovary.[54] However, there are studies that do not support an association with pesticide exposure and breast cancer risk.[39,51] Testicular cancer is not common; however, its incidence increased in the past few years in young adult male in Brazil. The development of testicular cancer is unclear, but it may be associated with environmental factors, such as early life exposure to pesticides. Although several pesticides have been banned, pesticides such as OCs can persist in the environment and cause adverse effects. Concerning OC metabolism pathways, a case-control study comprising 1500 men in France showed that these compounds can induce additional single nucleotide polymorphisms altering the risk to develop testicular cancer when subjects are exposed to those chemicals in early life.[10]

Association of prostate cancer and pesticide exposure in the general population is still inconclusive. A meta-analysis of epidemiological evidence in the United States and Europe shows that there was no association between OC pesticide and its metabolites (including alachlor, chlordane, oxychlordane, chlordecone, dieldrin, endosulfan, heptachlor, hexachlorobenzene, methoxychlor, hexachlorocyclohexane, congeners, mirex, nonachlor, and toxaphene) exposure and the development of prostate cancer.[58] However, a systematic review from the same group pooled results into low and high levels of exposure to pesticides and authors found that high exposure to pesticide may be associated with prostate cancer in the United States, Canada, Japan, Australia, and Malaysia.[59] Besides, *per capita* sales of pesticides correlates with mortality rates by prostate, larynx, esophagus, and pancreas cancer found in men in Brazilian states.[30]

The relationship between bladder cancer and pesticides is still unclear and data shown in the literature are controversial. The incidence of bladder cancer has also increased in the past few years in Brazil and several authors attribute this fact to the increase in pesticide use. Actually, in certain regions increased incidences of bladder cancer among ever use of pesticides have been noted. In a cohort of 802,905.7 men in the Agricultural Health Study, 321 male applicators were diagnosed with bladder cancer during the follow-up period of 13–17 years. Increased risks for bladder cancer were attributed to several herbicides and insecticides.[55] The exposure to low levels of contaminated drinking water in America is an indirect form of pesticide intake and it has been suggested to be linked to bladder cancer.[31] Bladder cancer can be modulated by genetic polymorphism induced by pesticides exposure. As some pesticides lead to ROS generation, researchers evaluated the polymorphisms associated with antioxidant enzymes. A study with 953 cases of bladder cancer among male agricultural worker in Egypt showed a positive relationship between time of pesticide exposure and diagnoses of bladder cancer; moreover, increased odds of bladder cancer was greater for those with low or intermediate activity of for the enzyme NQO1 (NAD(P)H: quinone oxidoreductase) and superoxide 2.[5] This association was not observed for OC pesticides in a study cohort of 140 bladder cancer in Canary Island population, for instance. In this study, OC was detected in the serum of most patients and control and it did not related to bladder cancer. However, this study confirmed the greater incidence of genetic polymorphism of genes encoding xenobiotic-metabolizing enzyme for risk associated with bladder cancer.[11]

A few data are available regarding the impact of pesticides on hepatocellular carcinoma. The exposure to OC pesticides is also associated with increased incidence of hepatocellular carcinoma among males living in agriculturally intensive areas in California evaluated between 2000 and 2009.[102] Stratifying the analysis by sex, hepatocellular carcinoma seems to be related to pesticide exposure independently of the sex; however, for males, it has a strong association.[81] Nevertheless, others authors suggest that there is no association between hepatocellular carcinoma and pesticide exposure.[106]

Recently, pesticide exposures have been associated with adult lymphoma as well. Insecticide exposure may be related to neoplasm development, as difuse large B-cell lymphoma and chronic lymphocytic leukemia/small lymphocyte lymphoma.[89]

Despite the large amount of published studies in the literature worldwide, it is still not completely clear by what mechanisms pesticide components impact human health. Additional studies are necessary. In this regards, more detailed data are needed for classes of pesticides in association with disease development.

In Brazil, there is not enough scientific documentation of the chronic population exposure to pesticides. What is known concerns cases of acute poisoning (although underreported), and there is also some information related to occupational exposure.

Waissmann[103] (pp. 20–21) lists international research that linked exposure to pesticides with noncommunicable chronic diseases such as cancers, endocrine diseases, neuropathies, and others. Among cancers, we highlight the lymphohematopoietic, hormone-dependent, and gastroesophageal. Estrogenic and/or anti-estrogenic and/or anti-androgenic effects, as well as other endocrine effects have been detected. Despite a predominance of animal research, the author points out potential associations of pesticide use with pituitary, thyroid, lipid metabolism, and glucose metabolism changes, the menstrual cycle, testicular and sperm alterations, and sexual impotence complaints. Organophosphates, carbamates, OCs, pyrethroids and others can lead to acute (exposure at high doses) or chronic (low dose) effects.

Silva et al.[94] warn of the risks to the health of rural workers using pesticides. By analyzing cases of cancer in the state of Rio Grande do Sul, Jobim et al.[52] did not discount the influence of pesticide exposure as a contributing factor to the higher cancer mortality rate observed in the micro region of Ijuí (rice production) in relation to Rio Grande do Sul and Brazil. The authors point to the need for further research on the subject in Brazil.

Although there is little documentation to the deleterious effects of chronic exposure to pesticides in Brazil, the published results are alarming. A survey conducted in the milk of 62 nursing mothers living in the city of São Lucas do Rio Verde, located in a heavily agricultural region of the country, revealed the presence of 10 pesticides. All samples showed the presence of at least one type of pesticide, suggesting that maternal exposure may have occurred due to occupation, environment and/or food.[76]

In Rio Verde, state of Goiás, in 2013, an episode was documented where dozens of children, teachers, and staff of a school were exposed to a rain of pesticides (thiamethoxam and lambda-cyhalothrin) because of an unsuccessful aerial spraying. Such exposure brought several acute consequences, but what is striking is the range of chronic sequelae documented after this episode of poisoning: girls with menstrual irregularities, with two episodes of bleeding per month; teachers who have developed potentially associated cancers to exposure to pesticides and allergic reactions (Carneiro et al.,[27] p. 438).

Tobacco grower families of the Rio Azul region, in the state of Paraná, were identified as potential carriers of chronic health problems resulting from occupational exposure to pesticides. Cases have been documented with clinical symptoms associated with neurotoxicity as neuropathies, psychological disorders, depressive disorders, and suicide attempts (Carneiro et al.,[27] p. 490).

A cross-sectional study in the farming community of Nova Friburgo, Rio de Janeiro, also documented deleterious effects of chronic occupational exposure to pesticides in Brazil. In addition to acute poisoning, the study identified symptoms of chronic poisoning involving traces of neuropathy in about 13% of the subjects and traces of neurobehavioral syndrome and psychiatric disorders in almost 30% of respondents.[72]

Many of the a.i. used in agricultural cultivation is classified based on their acute toxicity as moderately or slightly toxic. Thus, there is no way to classify them according to their toxic potential compared to chronic human exposure.

13.7 CONCLUDING REMARKS

The information presented in this chapter indicates that the problem of the pesticides consumption in Brazil is extremely relevant to human and animal health and the environment. The lack of control in the pesticide registration process; flexibility in relation to MRLs in food and water; inadequate monitoring systems which are also poorly structured; underreporting of cases of human poisoning associated with pesticides; and the lack of supervision and punishment of those who produce, sell, transport, and use pesticides are examples of serious problems that the country has in relation to a more effective control on the use of these products. The information presented here highlights that the environment is impaired by pesticides in several levels of complexity, from water until living organisms. This scenario has been aggravated by the lack of adequate governmental surveillance concerning the buy and sell of pesticides, in both industry and end consumers.

In addition to a debate on a scientific level, we must consider that there are economic and political intentions associated with this matter. On one hand, there is the state's failure regarding a more efficient control; on the other hand, there is a movement of question and resistance to widespread use of pesticides in Brazil, highlighted by the National Campaign Against Pesticides. The Dossier ABRASCO[27] is not only one of the main documents that expose several problems but also indicate possible new ways, focusing on the promotion of agroecology as a scientific movement, political, and ideological alternative.

Despite PARA's implementation to represent a breakthrough in the monitoring of pesticide residues in food in Brazil, it is expected that it evaluates less than 50% of pesticides used in the country. In addition, the program does not analyze the presence of the main a.i. currently used in the country, the glyphosate, which has been deemed as a potential carcinogen to humans by the International Agency for Research on Cancer (IARC). These program failures point to the lack of structure in the supervision and control of pesticide use in the country where there is the aggravating factor of political influence regarding the decision-making about limits and sanctions that are adopted. There is also the necessity to expand the analytical method employed, because according to ANVISA, the multiresidue method adopted as a reference does not apply to the detection of dithiocarbamates, paraquat, and glyphosate.

It stands out as aggravating that, in Brazil, aerial spraying of pesticides is still carried out, which facilitates the widespread contamination of the environment and population. It is also important to note that a.i. banned in several countries due to its high toxicity and health risks are still authorized in Brazil.

According to the dossier ABRASCO (Brazilian Association of Public Health), in the period analyzed by PARA, the consumption of pesticides has increased by 288% in the country. But the same evolution within the quantitative parameters for the program did not occur. It is noteworthy that the program does not perform the monitoring of processed/industrial products, which certainly present contamination by pesticide residues.

Even with scientific and technological advances, hard work and political will are still necessary to measure the actual effects of exposure to pesticides. In Brazil, we are far from being able to give an overview of the impact of this exposure, as studies are scarce and limited to monitor acutely the health of exposed rural workers. To date, no research line has received incentives to develop studies that investigate the impact of this exposure on the health of the population exposed directly/indirectly in the long run. In this sense, the INCA published a document positioning itself about the potential damage associated with the use of pesticides linked to cancer. This document also underscores the concern that there is no reduction in the consumption of fruits and vegetables—essential to the fight against cancer and other chronic diseases development—but to intensify the fight against the use of pesticides and environmental contamination.

Some discussion points should be encouraged in Brazilian society. It is not known the exact harmfulness of a.i. that are available for use in the market. And despite the social mobilization on the subject, there are no signs of the implementation of a risk identification policy and chronic monitoring of human exposure to pesticides. Considering that there is no effective method for the removal of pesticide residues in food, that there is little incentive for the production of organic food, and that the regulation of the use of these substances in the country is far off from protecting human health, the issue of food security in Brazil remains a serious public health problem, which remains neglected and far from being resolved.

KEYWORDS

- **Brazil**
- **pesticides**
- **regulation**
- **monitoring**
- **water**
- **food**

REFERENCES

1. Adad, L. M.; de Andrade, H. H.; Kvitko, K.; Lehmann, M.; Cavalcante, A. A.; Dihl, R. R. Occupational Exposure of Workers to Pesticides: Toxicogenetics and Susceptibility Gene Polymorphisms. *Genet Mol Biol.* **2015**, *38* (3), 308–315. DOI:10.1590/S1415-475738320140336.

2. Agim, Z. S.; Cannon, J. R. Dietary Factors in the Etiology of Parkinson's Disease. *Biomed. Res. Int.* **2015**, *2015*, 672838. DOI:10.1155/2015/672838.

3. Alavanja, M. C; Ross, M. K; Bonner, M. R. Increased Cancer Burden among Pesticide Applicators and Others Due to Pesticide Exposure. *CA Cancer J. Clin.* **2013**, *63* (2), 120–142. DOI:10.3322/caac.21170.

4. Albuquerque, A. F.; Ribeiro, J. S.; Kummrow, F.; Nogueira, A. J. A.; Montagner, C. C.; Umbuzeiro, G. A. Pesticides in Brazilian Freshwaters: A Critical Review. *Environ. Sci. Process Impacts* **2016**, *18*, 779–787.

5. Amr, S.; Dawson, R.; Saleh, D. A.; Magder, L. S.; St George, D. M.; El-Daly, M.; Squibb, K.; Mikhail, N. N.; Abdel-Hamid, M.; Khaled, H.; Loffredo, C. A. Pesticides,

Gene Polymorphisms, and Bladder Cancer among Egyptian Agricultural Workers. *Arch. Environ. Occup. Health* **2015,** *70* (1), 19–26. DOI:10.1080/19338244.2013.853646.

6. Ananias, P. Preface. In *GM Crops—Risks and Uncertainties: More than 750 Studies Despised by Regulators of GMOs*; Ferment, G., Malgarejo, L., Fernandes, G. B., Ferraz, J. M., Ed.; Ministry of Agrarian Development: Brasília, 2015.

7. ANVISA. Agência Nacional de Vigilância Sanitária. Programa de Análise de Resíduos de Agrotóxicos em Alimentos (PARA). Relatório de Atividades de 2011 e 2012. Brasília: Agência Nacional de Vigilância Sanitária; 2013. *Bol. Epidemiol.* **2013,** *44* (17) [Online]. http://bit.do/bol44 (accessed November 05, 2015).

8. Alves, S. R.; Oliveira-Silva, J. J. Evaluation of Environments Contaminated by Pesticides. In *Is It Poison or Medicine? Pesticides, Health and Environment*; Peres, F., Ed.; Publisher/Fiocruz: Rio de Janeiro, 2003; pp 137–156.

9. Arrebola, J. P.; Belhassen, H.; Artacho-Cordón, F.; Ghali, R.; Ghorbel, H.; Boussen, H.; Perez-Carrascosa, F. M.; Expósito, J.; Hedhili, A.; Olea, N. Risk of Female Breast Cancer and Serum Concentrations of Organochlorine Pesticides and Polychlorinated Biphenyls: A Case-Control Study in Tunisia. *Sci. Total Environ.* **2015,** *520*, 106–113. DOI:10.1016/j.scitotenv.2015.03.045.

10. Béranger, R.; Pérol, O.; Bujan, L.; Faure, E.; Blain, J.; Le Cornet, C.; Flechon, A.; Charbotel, B.; Philip, T.; Schüz, J.; Fervers, B. Studying the Impact of Early Life Exposures to Pesticides on the Risk of Testicular Germ Cell Tumors during Adulthood (TESTIS Project): Study Protocol. *BMC Cancer* **2014,** *14*, 563. DOI:10.1186/1471-2407-14-563.

11. Boada, L. D.; Henríquez-Hernández, L. A.; Zumbado, M.; Almeida-González, M.; Álvarez-León, E. E.; Navarro, P.; Luzardo, O. P. Organochlorine Pesticides Exposure and Bladder Cancer: Evaluation from a Gene-Environment Perspective in a Hospital-Based Case-Control Study in the Canary Islands (Spain). *J Agromed.* **2015,** 21, 34–42.

12. Bombardi, L. M. Pesticide Poisoning in Brazil and the Violation of Human Rights. In *Human Rights in Brazil 2011: Report*; Merlino, T., Mendonça, M. L., Eds.; Social Network for Justice and Human Rights: São Paulo, 2011.

13. Bortoluzzi, E. C.; Rheinheimer, D. S.; Gonçalves, C. S.; Pellegrini, J. B. R.; Zanella, R.; Copetti, A. C. C.; et al. Contamination of Surface Water by Pesticides on the Basis of Land Use in Agudo Watershed, RS. *Rev. Bras. Engenhar. Agríc. Ambiental.* **2006,** *10* (4), 881–887.

14. Brazil. Ministry of Health. National Health Surveillance Agency (ANVISA). *Report 2012 of Pesticide Residue Analysis Program in Food (PARA)*. Brasília, 2012 [Online]. http://portal.anvisa.gov.br/wps/content/Anvisa+Portal/Anvisa/Inicio/Agrotoxicos+e+ Toxicologia/Assuntos+de+Interesse/Programa+de+Analise+de+Residuos+de+Agrotox icos+em+Alimentos (accessed October 18, 2016).

15. Brazil. Ministry of Health. *Ordinance No. 2,914 of December 12, 2011* [Online]. http:// bvsms.saude.gov.br/bvs/saudelegis/gm/2011/prt2914_12_12_2011.html (accessed June 02, 2014).

16. Brazil. Ministry of Health. National Health Surveillance Agency (ANVISA). *Report 2010 of Pesticide Residue Analysis Program in Food (PARA)*. Brasília, 2010 [Online]. http://portal.anvisa.gov.br/wps/content/Anvisa+Portal/Anvisa/Inicio/Agrotoxicos+ e+Toxicologia/Assuntos+de+Interesse/Programa+de+Analise+de+Residuos+de+Agrot oxicos+em+Alimentos (accessed October 18, 2016).

17. Brazil. Ministry of Health. National Health Surveillance Agency (ANVISA). *Report 2009 of Pesticide Residue Analysis Program in Food (PARA)*. Brasília, 2009 [Online]. http://portal. anvisa.gov.br/wps/content/Anvisa+Portal/Anvisa/Inicio/Agrotoxicos+e+Toxicologia/

Assuntos+de+Interesse/Programa+de+Analise+de+Residuos+de+Agrotoxicos+em+Ali
mentos (accessed October 18, 2016).

18. Brazil. Ministry of Health. National Health Surveillance Agency (ANVISA). *Report 2008 of Pesticide Residue Analysis Program in Food (PARA)*. Brasília, 2008 [Online]. http://portal. anvisa.gov.br/wps/content/Anvisa+Portal/Anvisa/Inicio/Agrotoxicos+e+Toxicologia/ Assuntos+de+Interesse/Programa+de+Analise+de+Residuos+de+Agrotoxicos+em+Ali mentos (accessed October 18, 2016).

19. Brazil. *Decree No. 5,981 of December 06, 2006. Redrafts and Includes Provisions to Decree No. 4,074 of January 4, 2002*. Brasília, 2006.

20. Brazil. Ministry of Environment. *CONAMA Resolution No. 357 of March 17, 2005*. [Online] http://www.mma.gov.br/port/conama/res/res05/res35705.pdf (accessed June 02, 2014).

21. Brazil. Ministry of Health. *Decree No. 5,549 of September 22, 2005. Redrafts and repeals provisions of Decree No. 4074 of January 4, 2002*, Brasília, 2005.

22. Brazil. Ministry of Health. *Ordinance No. 518/2004*. Ministry of Health, Health Surveillance Agency, General Coordination of Environmental Health Surveillance Agency— Brasilia: Publisher of the Ministry of Health, 2005.

23. Brazil. *Decree No. 4,074 of January 4, 2002. Regulates Law 7802 of July 11, 1989* [Online]. http://www.planalto.gov.br/ccivil_03/decreto/2002/D4074.htm (accessed April 08, 2014).

24. Brazil. *Federal Law 9,974 of 6 June 2000. Changes Law No. 7,802 of July 11, 1989*. Brazilian Official Gazette: Brasilia, DF, June 7, 2000. Section 1 [Online] http://www. planalto.gov.br/ccivil_03/leis/L9974.htm (accessed November 18, 2013).

25. Brazil. *Federal Law 7,802 of 11 July 1989* [Online] http://www.planalto.gov.br/ ccivil_03/leis/l7802.htm (accessed April 05, 2014).

26. Capobiango, H. L. V.; Cardeal, Z. L. A Solid-Phase Microextraction Method for the Chromatographic Determination of Organophosphorus Pesticides in Fish, Water, Pota-toes, Guava and Coffee. *J. Braz. Chem. Soc.* **2005**, *16* (5), 907–914.

27. Carneiro, F. F.; Augusto, L. G. S.; Rigotto, R. M.; Friedrich, K.; Búrigo, A. C., Eds. *Dossier ABRASCO: a Warning about the Impacts of Pesticides on Health*. EPSJV: Rio de Janeiro; Expressão Popular: São Paulo, 2015.

28. Cesnik, H. B.; Gregorcic, A. Validation of the Method for Determination of Dithiocar-bamates and Thiuram Disulphide on Apple, Lettuce, Potato, Strawberry and Tomato Matrix. *Acta Chim. Sloven.* **2006**, *53*, 100–104.

29. Chen, M.; Chang, C. H.; Tao, L.; Lu, C. Residential Exposure to Pesticide during Child-hood and Childhood Cancers: A Meta-Analysis. *Pediatrics* **2015**, *136* (4), 719–729. DOI:10.1542/peds.2015-0006.

30. Chrisman, J.; Koifman, R. S.; Sarcinelli, P. N.; Moreira, J. C.; Koifman, R. J.; Meyer, C. Pesticide Sales and Adult Male Cancer Mortality in Brazil. *Int. J. Hyg. Environ. Health* **2009**, *212* (3), 310–321.

31. Colli, J. L.; Kolettis, P. N. Bladder Cancer Incidence and Mortality Rates Compared to Ecologic Factors among States in America. *Int. Urol. Nephrol.* **2010**, *42* (3), 659–665. DOI:10.1007/s11255-009-9655-5.

32. Collotta, M.; Bertazzi, P. A.; Bollati, V. Epigenetics and Pesticides. *Toxicology* **2013**, *307*, 35–41. DOI:10.1016/j.tox.2013.01.017.

33. Corbi, J. J.; Strixino, S. T.; Santos, A.; Del Grande, M. Environmental Diagnosis of Metals and Organochlorines in Adjacent Streams to Areas of Sugar Cane Cultivation (State of São Paulo, Brazil). *Química Nova* **2006**, *29* (1), 61–65.

34. Correia, F. V.; Moreira, J. C. Effects of Glyfhosate and 2.4-D on Earthworms (*Eisenia foetida*) in Laboratory Tests. *Bull. Environ. Contam. Toxicol.* **2010**, *85*, 264–268.
35. Da Silva, A. M. F.; Pavesi, T.; Rosa, A. C. S.; Santos, T. P.; Tabalipa, M. M.; Lemes, V. R.; Alves, S. R.; Sarcinelli, P. N. Organochlorines and Polychlorinated Biphenyl Environmental Pollution in South Coast of Rio de Janeiro State. *Mar. Pollut. Bull.* **2016**, *108* (1–2), 325–331.
36. Dores, E. F. G. C.; Calheiros, D. F. Pesticide Contamination in the Basin of the Miranda River, Pantanal (MS). *Rev. Bras. Agroecol.* **2008**, *3*, 202–205.
37. Dores, E. F. G. C.; Navickiene, S.; Cunha, M. L. F.; Carbo, L.; Ribeiro, M. L.; Freire, E. M. L. Multiresidue Determination of Herbicides in Environmental Waters from Primavera do Leste region (Middle West of Brazil) by SPE-GC-NPD. *J. Braz. Chem. Soc.* **2006**, *17* (5), 866–873.
38. Dores, E. F. G. C.; De Lamônica-Freire, E. M.; Ribeiro, M. L. Herbicides in Groundwater Used for Human Consumption in Primavera do Leste, Mato Grosso. *Braz. Congr. Ecotoxicol.* **2002**, *7*, Vitória, Es: Ecotox.
39. El-Zaemey, S.; Heyworth, J.; Glass, D. C.; Peters, S.; Fritschi, L. Household and Occupational Exposure to Pesticides and Risk of Breast Cancer. *Int. J. Environ. Health Res.* **2014**, 24 (2), 91–102. DOI:10.1080/09603123.2013.800958.
40. Facchini, L. A.; Souza, L. E. Presentation. In *Dossier ABRASCO: A Warning about the Impacts of Pesticides on Health*; Carneiro, F. F., et al., Eds.; EPSJV: Rio de Janeiro; Expressão Popular: São Paulo, 2015; pp 37–40.
41. Ferment, G.; Malgarejo, L.; Fernandes, G. B.; Ferraz, J. M., Eds.; *GM Crops—Risks and Uncertainties: More than 750 Studies Despised by Regulators of GMOs*. Ministry of Agrarian Development: Brasília, 2015.
42. Flores, A. V. Determination of Organochlorine Residues in Waters and Sediments. Master's Degree Dissertation: Federal University of Viçosa, MG, 2000.
43. Flores, A. V.; Queiroz, M. E. L. R.; Neves, A. A.; Goulart, S. M. Extraction and Analysis of Organochlorines in Sediments of the Ribeirão São Bartolomeu, Viçosa-MG. *Rev. Anal.* **2004**, *3* (13), 42–47.
44. Fumes, B. H.; F. N. Andrade, Neto, A. J. S.; Lanças, F. M. Determination of Pesticides in Sugarcane Juice Employing Microextraction by Packed Sorbent Followed by Gas Chromatography and Mass Spectrometry. *J. Separat. Sci.* **2016**, *39* (14), 2823–2830.
45. Freire, C.; Koifman, R. J.; Koifman, S. Hematological and Hepatic Alterations in Brazilian Population Heavily Exposed to Organochlorine Pesticides. *J. Toxicol. Environ. Health A* **2015**, *78* (8), 534–548.
46. Froes Asmus, C. I.; Camara, V. M.; Landrigan, P. J.; Claudio, L. A Systematic Review of Children's Environmental Health in Brazil. *Ann. Glob. Health* **2016**, *82* (1), 132–148.
47. Goldman, S. M. Environmental Toxins and Parkinson's Disease. *Annu. Rev. Pharmacol. Toxicol.* **2014**, *54*, 141–164. DOI:10.1146/annurev-pharmtox-011613-135937.
48. Gomes, M. A. F.; Barizon, R. R. M. Overview of Environmental Contamination by Pesticides and Agricultural Nitrate Origin in Brazil: Scenario 1992/2011. Embrapa Environment: Jaguariúna, 2014.
49. Grutzmacher, D. D.; Grutzmacher, A. D.; Agostinetto, D.; Loeck, A. E.; Roman, R.; Peixoto, S. C.; Zanella, R. Pesticide Monitoring in Two Water Sources in Southern Brazil. *Rev. Bras. Engenhar. Agríc. Ambien.* **2008**, *12* (6), 632–637.
50. Ianni, O. *The Era of Globalism*. Civilização Brasileira: Rio de Janeiro, 1996.

51. Ingber, S. Z.; Buser, M. C.; Pohl, H. R.; Abadin, H. G.; Murray, H. E.; Scinicariello, F. DDT/DDE and Breast Cancer: A Meta-Analysis. *Regul. Toxicol. Pharmacol.* **2013,** *67* (3), 421–433. DOI:10.1016/j.yrtph.2013.08.021.

52. Jobim, P. F. C.; Nunes, L. N.; Giugliani, R.; Cruz, I. B. M. Is There an Association between Cancer Mortality and Use of Pesticides?: A Contribution to the Debate. *Ciên. Saúde Colet.* **2010,** *15* (1), 277–288.

53. Kageyama, A. The New Brazilian Agricultural Pattern: From Rural Complex to Agro-industrial Complex. In *Agriculture and Public Policy*; Delgado, G., Ed.; IPEA: Brasilia, 1990; pp 48–65.

54. Koifman, S.; Koifman, R. J. Environment and Cancer in Brazil: An Overview from a Public Health Perspective. *Mutat. Res.* **2003,** 544 (2–3), 305–311.

55. Koutros, S.; Silverman, D. T.; Alavanja, M. C.; Andreotti, G.; Lerro, C. C.; Heltshe, S.; Lynch, C. F.; Sandler, D. P.; Blair, A.; Beane Freeman, L. E. Occupational Exposure to Pesticides and Bladder Cancer Risk. *Int. J. Epidemiol.* **2015,** pii: dyv195.

56. Kumar, A.; Vashist, M.; Rathee, R. Maternal Factors and Risk of Childhood Leukemia. *Asian Pac. J. Cancer Prev.* **2014,** *15* (2), 781–784.

57. Lebov, J. F.; Engel, L. S.; Richardson, D.; Hogan, S. L.; Sandler, D. P.; Hoppin, J. A. Pesticide Exposure and End-Stage Renal Disease Risk among Wives of Pesticide Applicators in the Agricultural Health Study. *Environ Res.* **2015,** *143* (Pt. A), 198–210. DOI:10.1016/j.envres.2015.10.002.

58. Lewis-Mikhael, A. M.; Olmedo-Requena, R.; Martínez-Ruiz, V.; Bueno-Cavanillas, A.; Jiménez-Moleón, J. J. Organochlorine Pesticides and Prostate Cancer, Is There an Association? A Meta-Analysis of Epidemiological Evidence. *Cancer Causes Control* **2015,** *26* (10), 1375–1392. DOI:10.1007/s10552-015-0643-z.

59. Lewis-Mikhael, A. M.; Bueno-Cavanillas, A.; Ofir Guiron, T.; Olmedo-Requena, R.; Delgado-Rodríguez, M.; Jiménez-Moleón, J. J. Occupational Exposure to Pesticides and Prostate Cancer: A Systematic Review and Meta-analysis. *Occup Environ Med.* **2015,** *7.* pii: oemed-2014-102692. DOI:10.1136/oemed-2014-102692.

60. L'Héritier, F.; Marques, M.; Fauteux, M.; Gaudreau, L. Defining Molecular Sensors to Assess Long-term Effects of Pesticides on Carcinogenesis. *Int. J. Mol. Sci.* **2014,** *15* (9), 17148–17161. DOI:10.3390/ijms150917148.

61. Lichi, O. A. B. Multielement Geochemistry in Environmental Management: Identification and Characterization of Natural Geochemical Provinces, Anthropogenic Changes in the Landscape, Favorable Areas for Mineral Prospecting and Risk Areas for Health in the State of Paraná, Brazil. PfD Dissertation, Federal University of Paraná, Curitiba, PR, 2001.

62. Londres, F. *Pesticides in Brazil: A Guide to Action in Defense of Life.* AS-PTA—Advisory Services for Alternative Agriculture Projects: Rio de Janeiro, 2011.

63. Mamane, A.; Raherison, C.; Tessier, J. F.; Baldi, I.; Bouvier, G. Environmental Exposure to Pesticides and Respiratory Health. *Eur. Respir. Rev.* **2015,** *24* (137), 462–473. DOI:10.1183/16000617.00006114.

64. Marchesan, E.; Sartori, G. M. S.; Avila, L. A.; Machado, S. L. O. M.; Zanella, R.; Primel, E. G.; Macedo, V. R. M.; Marchezan, M. G. Pesticide Residues in River Waters of the Central Depression of Rio Grande do Sul, Brazil. *Ciên. Rural* **2010,** *16* (2), 123–127.

65. Marinho, A. M. C. P. Contexts and Outlines of Agricultural Modernization in Municipalities in the Baixo Jaguaribe-EC: The Development and its Effects on Health, Labor and Environment. Ph.D. Dissertation, Department of Public Health Practice, School of Public Health, University of São Paulo, 2010.

66. Mathew, P.; Jose, A.; Alex, R. G.; Mohan, V. R. Chronic Pesticide Exposure: Health Effects among Pesticide Sprayers in Southern India. *Indian J. Occup. Environ. Med.* **2015,** *19* (2), 95–101. DOI:10.4103/0019-5278.165334.

67. Mattos, M. L. T.; Martins, J. F. S.; Nunes, C. D. M.; Moura Neto, F. P.; Magalhães Júnior, A.; Petrini, J. A.; Santos, I. B. Pesticide Monitoring in Pilot Areas of Integrated Rice Production in the Outer Coastal Plain and the Western Border of Rio Grande do Sul. Embrapa Temperate Climate: Pelotas, RS, 2011; pp 35–39.

68. Mattos, M. L. T.; Peralba, M. C. R.; Dias, S. L. P.; Prata, F.; Camargo, L. Environmental Monitoring of Glyphosate and Its Metabolite (Aminomethylphosphonic Acid) in Rice Farming Water. *Pesticides: Rev. Ecotoxicol. Meio Amb.* **2002,** *12,* 143–154.

69. Miranda, K.; Cunha, M. L. F.; Dores, E. F. G. C.; Calheiros, D. Pesticide Residues in River Sediments from the Pantanal Wetland, Brazil. *J. Environ. Sci. Health B* **2008,** *43* (8), 717–722. DOI:10.1080/03601230802388843.

70. Modgil, S.; Lahiri, D. K.; Sharma, V. L.; Anand, A. Role of Early Life Exposure and Environment on Neurodegeneration: Implications on Brain Disorders. *Transl. Neurodegen.* **2014,** *3,* 9. DOI:10.1186/2047-9158-3-9.

71. Moreira, J. C.; Peres, F.; Simões, A. C.; Pignati, W. A.; Dores, E. C.; Vieira, S. N.; Strüssmann, C.; Mott, T. Contamination of Surface Water and Rain by Pesticides in a State of Mato Grosso Region. *Ciên. Saúde Colet.* **2012,** *17* (6), 1557–1568.

72. Moreira, J. C.; Jacob, S. C.; Peres, F.; Lima, J. S.; Meyer, A.; Oliveira-Silva, J. J.; Sarcinelli, P. N.; Batista, D. F.; Egler, M.; Faria, M. V. C; de Araújo, A. J.; Kubota, A. H.; de Soares, M. O.; Alves, S. R.; Moura, C. M.; Curi, R. Integrated Evaluation of Impact Assessment of Pesticide Use on Human Health within an Agricultural Community of Nova Friburgo, RJ. *Ciên. Saúde Colet.* **2002,** *7* (2), 299–311.

73. National Health Surveillance Agency (ANVISA)/Federal University of Parana (UFPR). *Seminar "Mercado de agrotóxico e regulação".* ANVISA: Brasília, 2012.

74. Neto, M. L. F. Brazilian Standard of Water Potability: Analysis of Agrochemical Parameters of a Risk Assessment Approach. Ph.D. Dissertation. National School of Public Health Sergio Arouca, Oswaldo Cruz Foundation, Rio de Janeiro, 2010.

75. Oliveira, F. A.; Reis, L. P.; Soto-Blanco, B.; Melo, M. M. Pesticides Residues in the *Prochilodus costatus* (Valenciennes, 1850) Fish Caught in the Sao Francisco River, Brazil. *J. Environ. Sci. Health B* **2015,** *50* (6), 398–405.

76. Palma, D. C. A. Pesticides in Human Milk of Mothers Living in Lucas do Rio Verde, MT. Master Dissertation. Federal University of Mato Grosso: Cuiabá, 2011.

77. Pelaez, V. M.; Silva, L. R.; Guimarães, T. A.; Teodorovicz, F. D. T. (Dis)coordination Policy for Agrochemical Industry in Brazil. *Rev. Bras. Inov.* **2015,** *14,* 153–178.

78. Pelaez, V.; Silva, L.; Borges, E. Regulation of Pesticides: A Comparative Analysis. *Sci. Publ. Policy* **2013,** *40* (5), 644–656.

79. Pelaez, V; Terra, F.; Silva, L. R. The Regulation of Pesticides in Brazil: Between Market Power and the Protection of Health and the Environment. *Rev. Econ.* **2010,** *36* (1), 27–48.

80. Peres, F.; Moreira, J. C. Health and Environment in Relation to the Consumption of Pesticides in an Agricultural Hub of the State of Rio de Janeiro, Brazil. *Cadern. Saúde Públ.* **2007,** *23* (4), 612–621.

81. Persson, E. C.; Graubard, B. I.; Evans, A. A.; London, W. T.; Weber, J. P.; LeBlanc, A.; Chen, G.; Lin, W.; McGlynn, K. A. Dichlorodiphenyltrichloroethane and Risk of Hepatocellular Carcinoma. *Int. J. Cancer* **2012,** *131* (9), 2078–2084. DOI:10.1002/ijc.27459.

82. Petersen, P. Preface. In *Dossier ABRASCO: A Warning about the Impacts of Pesticides on Health*; Carneiro, F. F., Augusto, L. G. S., Rigotto, R. M., Friedrich, K., Búrigo, A. C., Eds.; EPSJV: Rio de Janeiro; Expressão Popular: São Paulo, 2015; pp 27–36.

83. Pignati, W.; Dores, E. F.; Moreira, J. C.; Perez, F. *Executive Summary of the Impacts of Pesticides on Health and the Environment in the Municipalities of the "Interior" of Mato Grosso, Brazil*, Cuiabá, 2014 (mimeo).

84. Raanan, R.; Balmes, J. R.; Harley, K. G.; Gunier, R. B.; Magzamen, S.; Bradman, A.; Eskenazi, B. Decreased Lung Function in 7-Year-Old Children with Early-Life Organophosphate Exposure. *Thorax* **2015**. pii: thoraxjnl-2014-206622. DOI:10.1136/thoraxjnl-2014-206622.

85. Rigotto, R. M. Agrochemicals, Work and Health: Vulnerabilities, Resistance in the Context of Agricultural Modernization in Baixo Jaguaribe/CE. UFC/Expressão Popular: Fortaleza, 2011.

86. Rigotto, R. M.; Vasconcelos, D. P.; Rocha, M. M. Pesticide Use in Brazil and Problems for Public Health. *Cad. Saude Publ.* **2014**, *30* (7), 1360–1362.

87. Rosas, L. G.; Eskenazi, B. Pesticides and Child Neurodevelopment. *Curr. Opin. Pediatr.* **2008**, *20* (2), 191–197.

88. Saeedi Saravi, S. S.; Dehpour, A. R. Potential Role of Organochlorine Pesticides in the Pathogenesis of Neurodevelopmental, Neurodegenerative, and Neurobehavioral Disorders: A Review. *Life Sci.* **2016**, *145*, 255–264.

89. Schinasi, L. H.; De Roos, A. J.; Ray, R. M.; Edlefsen, K. L.; Parks, C. G.; Howard, B. V.; Meliker, J. R.; Bonner, M. R.; Wallace, R. B.; LaCroix, A. Z. Insecticide Exposure and Farm History in Relation to Risk of Lymphomas and Leukemias in the Women's Health Initiative Observational Study Cohort. *Ann Epidemiol.* **2015**, *25* (11), 803–810. e4. DOI:10.1016/j.annepidem.2015.08.002.

90. Selmi, G. F. R.; Trape, A. Z. Health Protection for Rural Workers: The Need to Standardize Techniques for Quantifying Dermal Exposure to Pesticides. *Cad. Saude Publ.* **2014**, *30* (5), 952–960.

91. Smith, M. T.; Guyton, K. Z.; Gibbons, C. F.; Fritz, J. M.; Portier, C. J.; Rusyn, I.; DeMarini, D. M.; Caldwell, J. C.; Kavlock, R. J.; Lambert, P.; Hecht, S. S.; Bucher, J. R.; Stewart, B. W.; Baan, R.; Cogliano, V. J.; Straif, K. Key Characteristics of Carcinogens as a Basis for Organizing Data on Mechanisms of Carcinogenesis. *Environ. Health Perspect.* **2015**, *124*, 713–721.

92. Silva, D. R. O.; Avila, L. A.; Agostinetto, D.; Bundt, A. D. C.; Primel, E. G.; Caldas, S. S. Pesticide Occurrence in Groundwater Areas Adjacent to Rice Fields. *Quím. Nova* **2011**, *34* (5), 748–752.

93. Silva, D. R. O.; Avila, L. A.; Agostinetto, D.; Magro, T. D.; Oliveira, E.; Zanella, R.; Noldin, J. A. Pesticide Monitoring in Surface Water of Rice Growing Regions in Southern Brazil. *Ciên. Rural* **2009**, *39* (9), 2283–2389.

94. Silva, J. C.; Silva, E. N.; Faria, H. P.; Pinheiro, T. M. M. Pesticides and Work: A Dangerous Combination for the Health of Rural Workers. *Ciên. Saúde Colet.* **2005**, *10* (4), 891–903.

95. Spadotto, C. A.; Gomes, M. A. F. *Pesticides in Brazil*, 2015 [Online] http://www.agencia.cnptia.embrapa.br/gestor/agricultura_e_meio_ambiente/arvore/CONTAG01_40_210200792814.html (accessed November 13, 2015).

96. Souza, L. C. A Critical Analysis of the Water Quality for Human Consumption in the Municipality of Francisco Beltrao-PR. Master Dissertation. UNIOESTE: Francisco Beltrão, PR, 2014.

97. Tang, M.; Zhao, M.; Zhou, S.; Chen, K.; Zhang, C.; Liu, W. Assessing the Underlying Breast Cancer Risk of Chinese Females Contributed by Dietary Intake of Residual DDT from Agricultural Soils. *Environ. Int.* **2014**, *73*, 208–215. DOI:10.1016/j.envint.2014.08.001.

98. Vakonaki, E.; Androutsopoulos, V. P.; Liesivuori, J.; Tsatsakis, A. M.; Spandidos, D. A. Pesticides and Oncogenic Modulation. *Toxicology* **2013**, *307*, 42–45. DOI:10.1016/j.tox.2013.01.008.

99. Veiga, M. M. Contamination Analysis of Water Systems by Pesticides in a Small Rural Community in Southeast Brazil. *Cad. Saúde Públ.* **2006**, *22* (11), 2391–2399.

100. Veiga, M. M. Risk Analysis of Environmental Pesticide Contamination in Water Systems of Paty do Alferes County, RJ. In *National Conference on Industrial Engineering*; ENEGEP/ABEPRO: Porto Alegre; 2005, vol. 25, pp 4259–4264.

101. Vekic, A. The Pesticide Regulation in Brazil: Perceptions and Challenges. In *Lecture Slides Given at Seminar "Sector Dialogues on the Control and Regulation of Pesticides and Biocides."* Brasilia, 2014.

102. VoPham, T.; Brooks, M. M.; Yuan, J. M.; Talbott, E. O.; Ruddell, D.; Hart, J. E.; Chang, C. C.; Weissfeld, J. L. Pesticide Exposure and Hepatocellular Carcinoma Risk: A Case-Control Study Using a Geographic Information System (GIS) to Link SEER-Medicare and California Pesticide Data. *Environ Res.* **2015**, *143* (Pt. A), 68–82. DOI:10.1016/j.envres.2015.09.027.

103. Waissmann, W. Pesticides and Non-communicable Diseases. *Ciên. Saúde Colet.* **2007**, *12* (10), 20–21.

104. Wan, N.; Lin, G. Parkinson's Disease and Pesticides Exposure: New Findings from a Comprehensive Study in Nebraska, USA. *J. Rural Health* **2015**. DOI:10.1111/jrh.12154.

105. Ye, M.; Beach, J.; Martin, J. W.; Senthilselvan, A. Occupational Pesticide Exposures and Respiratory Health. *Int. J. Environ. Res. Public Health* **2013**, *10* (12), 6442–6471. DOI:10.3390/ijerph10126442.

106. Zhao, B.; Shen, H.; Liu, F.; Liu, S.; Niu, J.; Guo, F.; Sun, X. Exposure to Organochlorine Pesticides Is an Independent Risk Factor of Hepatocellular Carcinoma: A Case-Control Study. *J. Expo. Sci. Environ. Epidemiol.* **2012**, *22* (6), 541–548. DOI:10.1038/jes.2011.29.

107. Brazil. *Decree No. 4,074 of January 4, 2002. Regulates Law 7802 of July 11, 1989* [Online], 2002. http://www.planalto.gov.br/ccivil_03/decreto/2002/D4074.htm (accessed April 08, 2014).

108. Brazil, Ministry of Health, National Health Surveillance Agency (ANVISA). Pesticides Monitoring in Water for Human Consumption in Brazil, 2011/2012. *Epidemiol. Bull. Bras* **2013**, *44* (17), 1–10 [Online]. http://portalarquivos.saude.gov.br/images/pdf/2014/junho/11/BE-2013-44--17----Agrot--xicos.pdf (accessed April 15, 2014).

109. Barbosa, A. M. C.; Solano, M. L. M.; Umbuzeiro, G. A. Pesticides in Drinking Water—The Brazilian Monitoring Program. *Front. Public Health,* **2015**, *3*, 246. DOI:10.3389/fpubh.2015.00246

110. Tomita, R. Y.; Beyruth, Z. Toxicology of Pesticides in Aquatic Environment. *O Biológico* **2002**, *64* (2), 135–142.

111. Cremonesi, M.; Chiesa, C.; Strigari, L.; Ferrari, M.; Botta, F.; Guerriero, F.; De Cicco, C.; Bonomo, G.; Orsi, F.; Bodei, L.; Di Dia, A.; Grana, C. M.; Orecchia, R. Radioembolization of Hepatic Lesions from a Radiobiology and Dosimetric Perspective. *Front. Oncol.* **2014**, *4*, 210. DOI:10.3389/fonc.2014.00210.

CHAPTER 14

APTAMERS AS ADVANCED NANOSENSING TOOLS IN FOOD SAFETY

ASHISH SACHAN*

Toxam Inc., ON, Canada

*E-mail: asachan@toxam.ca

CONTENTS

ABSTRACT

Food-borne toxicants and food contamination is a major risk for human populations. Ensuring food safety for more than 7.5 billion people on this planet remains a major challenge for scientists around the world. There is a continual need and ever-increasing importance for the development of sensitive, specific, easy-to-use, and cost-effective biosensor platforms for food safety applications. The requirement for the development of such systems has gained even more importance due to the increased risks in the food-supply chains in this current age of global production and global food distribution systems. The technological advancements in molecular biology have enabled researchers to change the properties of biomolecules on demand, thereby generating customized assay systems with nanosensing capabilities. In this regard, aptamers have emerged as powerful nanosensing tools and as viable alternatives to antibody based immunoassays to detect a wide range of ligands in the areas of food safety. Such aptamer-based nanosensing platforms are continuously being developed and improved to fulfill the complex molecular recognition needs in overall food supply chain. The extraordinary ability of aptamers that can be custom designed and bind tightly and specifically to their ligands underlines their immense utility in such nanosensor technologies. The current chapter focuses on the potential use of aptamers in food safety by borrowing specific examples, as a proof of concept, from advanced aptasensing platforms used in the detection of mycotoxins in different food matrices.

14.1 INTRODUCTION

The detection and quantification of molecules has enormous importance in the field of nanosensor development.[1,2] In the era of stringent regulatory requirements and modern day challenges in global food safety, there is a constant analytical need for the development of sensitive, specific, fast, reliable, reusable, and cost-effective analytical methods, and in the context of this chapter, the need is there for such detection methods in the area of food safety involving all aspects of the complex global food-supply chain. The last few decades have witnessed a remarkable progress in in-vitro selection and amplification techniques[3,4] that have allowed the discovery of specific oligonucleotide sequences, single-stranded DNA (ssDNA) or single-stranded RNA, referred to as aptamers. The word aptamer is derived from the Latin word "aptus," meaning "to fit."[5] The techniques for optimizing, stabilizing, and delivering nucleic acid-based detection technologies, in its varied form, are already in commercial use today. In the past decade, aptamers have emerged as powerful in-vivo and in-vitro nanosensing tools. Aptamers with their high specificity and strong affinity for their target molecules are very powerful tools for analytical, diagnostic, and therapeutic applications at dissociation constants in the range of 100 pM.[6] Since the first reports that appeared in early 1990s[5] enthusiasm for aptamer activities has grown enormously with prospects of applications spanning every area of biomedical sciences including their expanded use in the area of food toxicology. Synthetic oligonucleotide aptamers display high affinity binding to recognize virtually any class of target molecules with high specificity[7-9] and can discriminate between closely related targets[10] that have small structural differences in their molecular configurations.[9,11] Tertiary structures of aptamers adopt a vast number of three dimensional shapes and bind tightly to specific toxicologically relevant targets and specific ligands such as cell surface proteins,[12,13] carbohydrates,[14] liposomes,[15] peptides,[16] small molecules,[17,18] metal ions,[19,20] complex structures such as viruses,[21,22] prion,[23,24] and whole living cells.[25] With the knowledge that aptamers are already used in recognition of various class of molecules, it is not difficult to envision that aptamers with their diverse, targeted, and with relevant chemical modifications have found their application to a number of ligands important in food safety.[26,27] Presently, most of the analytical platforms in biomedical research are based on antibodies. The application of aptamers for target detection offers several advantages over immunological assays.[28] One of the many advantages of aptamers is that the generation of aptamers is animal independent and they can be tested and operated in nonphysiological buffers at nonphysiological temperatures. In the backdrop of traditional biosensing techniques available in the market, aptamers provide an interesting alternative in terms of their specificity, sensitivity, simplicity, rapid and cost-effective approaches.

Current sensor approaches rely on assays which require several analytical steps with the requirement of various reagents for a single assay, all of which are difficult for development of analytical methods intended for onsite field applications. Therefore, considerable efforts have been made to overcome such limitations by use of aptamers as simple assay systems in analytical applications involving detection of food toxins and food contaminants. A very wide array of aptamer-based biosensor targets for food safety has evolved over the recent years. Some of the examples that demonstrate the versatility of such aptasensing platforms include the detection of antibiotics such as chloramphenicol,[29,30] tetracycline,[31,32] streptomycin,[33,34] tobramycin,[35] kanamycin,[36,38] and neomycin.[39] Also the agrochemical contamination of the food has been addressed thought the use of aptasensor platforms for the detection of pesticides[40–45] and fungicides such as malachite green and its residues in food.[46,47] There are details of reported use of aptasensing toward detection of heavy metals such as arsenic and mercury,[48–52] detection of bisphenol associated with food packaging,[53,54] and for the detection of melamine in food products.[55] Aptamers have been used in biomolecular screening of bacteria such as *Bacillus thuringiensis, Campylobacter jejuni, Escherichia coli, Listeria monocytogenes, Salmonella typhimurium, Staphylococcus aureus,*[56–64] and toxins such as botulinum[65] and contaminants such as sulforhodamine,[66] and allergens such as egg-white lysozyme,[67,68] and gluten.[69] Aptasensors have also been reported for toxins such as abrin[70,71] and ricin.[72,73] Aptasensor for endocrine disrupters such as 17 beta-estradiol[74] which is a hormone of immense importance in animal husbandry has also been reported. Aptamer-based nanosensing platforms have now been reported for food-surface contamination,[67] contamination by virus[75,76] and prions.[77] The fundamental principles governing these aptamer-based nanosensing platforms, because of the unique properties of aptamers, in food safety can be used for almost any facet-related food safety. Compared to the long history of antibodies, the aptamer science is relatively in the stage of infancy to protein antibodies; however, the extent to which the aptamer technology has progressed in a short amount of time is phenomenal nonetheless. As indicative from the examples above, the aptamers are being utilized in plethora of issues related to food safety in relation to antibiotics in food, pesticide contamination, allergens, heavy metals, viruses, prions, food surface contamination, and pathogens, etc. Apart from the above examples, the aptasensing platform has also been used for another area of major concern involving mycotoxins. The worldwide contamination of mycotoxins in foods and feeds is a significant and continuing global problem causing significant economic losses. Borrowing the examples from mycotoxin research the chapter demonstrates and reflects the robust and varied application of aptamers in the very diverse aptasensing platforms for sensitive, specific, rapid, and cost-effective nanosensing tools as being utilized in areas of food toxicology.

14.2　SELECTION OF APTAMERS AND ADVANTAGES OF APTAMER OVER ANTIBODIES

Traditionally, the aptamers are produced through an iterative evolutionary process called "Systematic Evolution of Ligands by EXponential enrichment" (SELEX).[4,5] Over the years, the SELEX combinatorial approach has developed as a scientific discipline in itself. This very important process of SELEX, the detailed mention of which is beyond the preview of this chapter, has evolved over the years through many innovative technologies, including the use of SELEX in food safety assessments.[13,25,26,78–85] The SELEX is a highly controllable process performed more conventionally in vitro. Several variants of SELEX have evolved over the years, since its first mention in 1990. The process of SELEX provides opportunities to choose conditions that completely match the final-selected aptamer application in different areas of food safety. The method is an iterative process based on selection and amplification of the anticipated tight-binding aptamer to its ligand thereby providing unpredictable and unimaginable molecular configurations of nucleic acids.[4] The start library for selection of aptamers contains ssDNA or RNA oligonucleotides with a central region of randomized sequences[86,87] of up to 10^{15} different sequences. The starting library of nucleic acids is incubated with the molecule of interest for subsequent transcription, reverse transcription and DNA amplification. The population of molecules that bound the target is eluted and amplified by polymerase chain reaction (PCR). Aptamer sequence is determined by sequencing of the resultant clones which then can be generated by chemical synthesis. The selection cycles in aptamer discovery are usually repeated 5–12 times to obtain the aptamers with highest affinity. Need-dependent functional group modifications can be attached to the aptamer postselection by synthetic approaches which enormously increasing the realm of aptamer functionality.[86]

In comparison to antibodies, aptamer have a number of advantages for their use in nanosensor applications.[88] The main advantage is overcoming of the use of animals for their production. The aptamers are isolated by in vitro methods that are independent of animals thereby eliminating the concern for the toxicity or low immunogenicity of the analyte. Also the process of aptamer discovery is relatively much faster than that compared to the development of antibodies. This gives aptamers a faster turnaround time for their commercial applications compared to antibodies. An in vitro combinatorial library can be generated against any target of choice. In addition, generation of antibodies is under physiological conditions which considerably limits their use when used outside the limits of physiological pH or temperature or under nonphysiological buffer conditions. Such a limitation can be overcome in the development of aptamers which can be selected for functionality in the milieu and conditions similar to their final use. Aptamer selection process can also be modified such that selected aptamers only recognize the part of the analyte. After selection, aptamer sequences can be produced by chemical synthesis with no batch-to-batch variation. The repeated chemical synthesis of the aptamer sequence thus generated has reduced cost of production compared to the antibodies. Further postselection modifications to the aptamer can be introduced thereby enhancing the stability, affinity, and specificity of these molecules. Aptamers can also be selected against difficult target haptens, such as toxins or prions and organic molecules with molecular weights of 100–10,000 Da.[10] The selected aptamers can bind to their targets with high affinity and discriminate between closely related targets, for example, the specificity of theophylline aptamer which does not bind to caffeine.[89]

14.3 APTAMER-BASED NANOSENSING PLATFORMS: EXAMPLES FROM MYCOTOXINS

Mycotoxin contamination grains and forages result in immense economic losses and health risk throughout the world.[90] Mycotoxins are toxic secondary metabolites[91,92] produced by several species of *Aspergillus*, *Fusarium*, *Penicillium*, *Claviceps*, and *Alternaria*. These toxic metabolites, which are nonvolatile and low molecular weight secondary metabolic products, when ingested by humans or farm animals result in market losses of millions of dollars per year and also cause human health risks.[93] Several factors affect the mold production such as the genetic factors, fungal strain specificity and variation, the humidity, temperature, concentration of oxygen in the environment and condition of the soil.[94] Mycotoxins are responsible for several central nervous system effects, carcinogenic, teratogenic, mutagenic, embryotoxic, oestrogenic, immunosuppression, and growth retardation and production losses in the livestock industry.[95] The economic impact of mycotoxins include increased healthcare costs, both animal and human fatality, production losses in animals, and food and feed wastage.[96] Mycotoxin contamination of forages, cereals, fruits,[97] spices, nuts, dairy products, animal feed, coffee, wine, beer, and cocoa frequently occurs in the field following infection of plants with specific pathogenic fungi or with symbiotic endophytes. The most common mycotoxins are aflatoxins, ochratoxin-A produced by *Aspergillus* and *Penicillium* species,[98] trichothecenes, zearalenone (ZEN), fumonisins, ergot, and T-2 toxin.[99] The mycotoxins of greatest significance in foods and feeds are aflatoxins which are produced mainly by *Aspergillus flavus*, *Aspergillus parasiticus*, *Aspergillus nominus*, and *Fusarium moniliforme*. The detection of mycotoxins is a complicated task. Several analytical methods[100–102] have been reported for the measurement of specific mycotoxins in food and feedstuffs. The detection of mycotoxin is made difficult because of the different metabolites generated by the fungi.[103] Owing to the huge economic and health losses by mycotoxins, there is significant global emphasis for the sensitive, specific, and rapid-detection procedures for their detection.

Aptasensing platforms often times use the properties of optical sensors. An optical sensor system also consists of a recognition element and a transduction method of translating the recognition event into an observable, quantifiable optical signal, and, ultimately, an electronic signal. In optical sensor systems, the recognition step is generally based on the absorption, emission or light-scattering interactions with the analytes. Several aptasensor platforms in addition to optical sensors have been developed against mycotoxins. For example, amongst others, Ochratoxin A DNA aptamer was first reported by Cruz-Aguado and Penner,[104] whereas aptamer for ZEN was reported by Chen et al.[105] OTA detection, which has maximum reported admissible limit of 2 mg/kg, has remained the primary focus in aptasensor platforms. Several advanced assays have been employed in such aptamer-based nanosensing. Fluorescence-based proximity-dependent DNA-scaffolded silver nanocluster (DNA/AgNC) probes are of low cost, have high quantum yield, and are photostable because of the properties of AGNCs. Chen et al. reported an inexpensive, sensitive and selective signal-on aptasensor based on DNA-scaffolded AGNCs and magnetic beads for OTA in wheat samples. Using such an approach, the authors could detect OTA, based on fluorescence intensity, in the range of 0.01–0.3 ng/ml with limit of detection (LOD) of 2 pg/ml.[106] The label-free aptamer-based detection of OTA using localized surface plasmon resonance (LSPR) was demonstrated by Park et al. The authors could detect OTA in concentrations of less than 1 nM, and the aptamer in this method as an added advantage could be regenerated thereby rendering the overall method more cost effective.[107] The structure switching properties of aptamer have been effectively used in nanosensing platforms. The aptamers can switch from duplex structures in the presence of complimentary sequence to target induced structural change in the aptamer bound to OTA. This structure switching property of the aptamer (5′-GAT CGG GTG TGG GTG GCG TAA AGG GAG CAT CGG ACA-FAM-3′) was used in a fluorescence dequenching assay for the detection of OTA in OTA-spiked corn samples.[108] The LOD was calculated to be 0.8 ng/mL and based on the fluorescence readings the OTA could be detected in the range of 1–100 mg/ml. Ochratoxin A has also been detected by surface plasmon resonance-based aptasensor in wine and peanut oil.[109] The aptamer was (5′-GATCGGGTGTGGGTGGCGTAAAGGGAGCATCGGACA-3′ Biotin) biotin tagged at the 3′ end. Electrochemical-based aptasensor based on the combination of hairpin–DNA probe and restriction endonuclease for site specific DNA cleavage, resulted in the development of highly sensitive and specific probe for the detection of OTA.[110] The method could detect OTA in the linear range of 1.0–20 pg/ml with the LOD of 0.4 pg/ml of OTA. Aptasensor platforms for detection of mycotoxins have also benefitted from the use of catalytic property of horseradish peroxidase (HRP)-mimicking DNAzyme.[111] Aptamer–DNAzyme hairpin was used in the aptasensing of OTA extracted from wine samples.[112] The method involved the development of a nucleotide sequence which included G-rich oligonucleotides which could form the hemin–G-quadruplex complex, the anti-OTA aptamer, and the blocking tail. The colorimetric signal readout of the activity of HRP-mimicking DNAzyme in this hairpin-based assay was dependent on the opening of the hairpin upon formation of the aptamer–OTA complex. The authors reported OTA detection in the linear range of up to 10 nM with the LOD value of 2.5 nM in wine samples. Magnetic nanoparticles (MNPs)[113] have also been increasingly used in aptamer-based nanosensing platforms. Biotinylated aptamer conjugated with NH_2 group functionalized Fe_3O_4 MNPs along with upconversion nanoparticles (UCNPs) were used for the detection of OTA in luminescence-based bioassay in maize samples.[114] The detection limit of 0.1 pg/ml could be achieved for the detection of OTA.

Recently, the ultrasensitive aptasensing platform using luminescence resonance energy transfer (LRET) between core/shell upconversion nanoparticles (CS-UCNPs) as luminophores and graphene oxide as a universal quencher was used for the detection of OTA.[115] The authors reported immense benefit of using UCNPs compared to fluorescence-based probes in terms of enhanced sensitivity of OTA detection in food because of low signal-to-noise ratio by use of UCNPs. The authors used the wavelength of 541 nm as the luminescence output of UCNPs and reported the detection of OTA in the linear range of 0.001–250 ng/ml with LOD of 0.001 ng/ml. The high specificity of the assay was confirmed by testing the aptamer for binding against OTB, AFB1, fumonisin B1 (FB1), ZEN, and T2 mycotoxins.

Mishra et al. reported an impedimetric aptasensor for OTA in cocoa beans which are widely used by the chocolate industry. The 36 nucleotide long DNA aptamer, 5′-GATCGGGTGTGGGTGGCGTAAAGGGAGCATCG-GACA-3′, used by the authors was amino modified at the 5′ end was used for OTA detection under optimized experimental conditions. Based on the impedimetric responses, the aptasensor had a detection range of 0.15–10 ng/ml and the linear detection range of 01.5–2.5 ng/ml. The detection limit reported by the authors was 01.15 ng/ml which the authors concluded met the standards set by European Union (EU) regulation of 2 mg/kg OTA in cocoa products.[116]

Aptamer for aflatoxin M1 was reported by Malhotra et al.[117] Guo et al. reported an aptasensor for mycotoxin aflatoxin M1 (AFM1) which is an important toxic component in dairy products.[118] The authors utilized the unique property, of the structural/conformational change upon target recognition, of the aptamer to selectively bind to AFM1 and the quantitative real-time polymerase chain reaction (RT-qPCR) technique to enhance the sensitivity of the assay. The authors reported the high selectivity of the aptasensor against OTA, ZEN, FB1, AFB1, and AFB2 with LOD for AFM1 at 0.03 ng/l. The authors mentioned the advantage of this aptasensing method in the detection of AFM1 in infant rice cereal. Electrochemical impedance spectroscopy (EIS) was used in an aptasensing platform for the detection of AFM1 in real samples of milk.[119] The 21 nucleotide long aptamer of the sequence 5'-ACT-GCT-AGA-GAT-TTT-CCA-CAT-3' was hexaethyleneglycol modified at the 5' end and linked covalently on to the screen-printed carbon electrodes (SPCEs) through carbodiimide immobilization. The aptamer-functionalized surface of SPCEs was used for the detection of extracted AFM1 in different varieties of milk available in the market which included the raw milk, microfiltered full fat milk, pasteurized full fat milk, and pasteurized skimmed milk. The signal output was recorded as a function of inhibition of Faradaic response and the increase in the electron transfer resistance in the presence and absence of AFM1. This aptasensor method resulted in the detection of AFM1 after extraction from the milk in the range of 20–1000 ng which also fulfilled the limit of AFM1 in milk set by the EU Legislation at 50 ng/kg.

A split DNA enzyme probe-based aptasensor was used for the detection of AFB1 corn samples.[120] The corn samples were spiked in the laboratory with 0, 0.1, 0.3, 0.5, and 1.0 ng/kg of AFB1. The detection of AFB1 based on this DNA enzyme probe was based on the loss of peroxidase activity of G-quadraplex upon conformational change of aptamer and split DNA complex resulting from aptamer binding to AFB1. This colorimetric detection method had LOD of 0.054 ng/ml using absorption spectroscopy. Another important advantage of this aptamer-based assay was that detection limit of 0.1 ng/ml could be recorded using naked eyes thereby making the method suitable of onsite detections without the use of complicated instrumentation. The aptamer-based indirect competitive dipstick assay[121] involving biotin modified aptamer and cy5 modified DNA probes was used for the detection of AFB1 with detection limits of 0.1 ng/ml AFB1 in buffer and 0.3 ng/g in the artificially spiked corn samples. The authors reported the portable advantage of this assay toward the development of onsite detection technologies. Gold nanoparticles (AuNPs) have been increasingly used in the aptasensing platforms due to their property of intradistance-based color change. The property of aptamers, in the absence of their target molecules, to adsorb to the AuNPs to prevent salt induced aggregation of AuNPs has been used in these assays involving AuNPs.[122,123] The sensitive and selective detection of AFB1 in the range of 80–270 nM with the LOD of 7 nM was reported in the artificially spiked peanut and rice samples, by the use of colorimetric and chemiluminescence detection method based on interaction of AuNPs and the anti-AFB1 aptamer.[124] This was another of the aptasensor method for food safety which could be used onsite for the detection of AFB1 by naked eye read out of the color change of AuNPs dependent upon their dispersed or aggregated state in the salt medium. The interaction of aptamer with the AuNPs (aptamer–AuNp complex) in the absence of AFB1 prevented the aggregation of AuNPs leaving the color of the test solution red. However, in the presence of AFB1 in the test solution the aptamers, due to their higher affinity bound to AFB1 (aptamer–AFB1 complex). The resultant aptamer binding with their specific target left the AuNPs to aggregate which in turn changed the color of the solution to purple blue. The novel method by the use of RT-qPCR-based nanosensing for detection of AFB1 in infant rice cereal and Chinese wildrye hay has been reported.[125] The method involved the biotin-labeled aptamer (5'-GTT GGGCACGTGTT-GTCTCTCTGTGTCTCGTGCCCTTCGCTAGGCCC-biotin-3') that was coupled to the streptavidin coated PCR tubes. The complimentary sequence (5'-ACACGTGCCCAACAATCTGGTTTAGCTACGCCTTCCCCGTG-GCGATGTTTCTTAGCGCCTTAC-3') was hybridized with the aptamer on the surface of the PCR tubes. The assay resulted in the AFBI concentration-dependent change in PCR amplification signal. The presence of AFB1 in the PCR reaction mix resulted in the formation of aptamer–AFB1 complex which resulted in the AFB1 concentration dependent release of the hybridized complimentary ssDNA thereby affecting the RT-qPCR signal. High sensitivity with LOD of 25 fg/ml was obtained in this method. The specificity of this method was tested against other mycotoxins which included OTA, ZEN, α-zearalenol, FB1, AFM1, AFB2, AFG1, and AFG. In the another study, the advantage of RT-qPCR in aptasensing platforms for determination of trace levels of OTA in red wine was demonstrated by use of fluorescence RT-qPCR and high sensitivity with LOD 1 fg/ml was obtained.[126] Fluorescence resonance energy transfer was used in aptamer-sensing platform using AuNPs and covalently linked aptamers to quantum dots (QD) in the detection of AFB1 in rice and peanut samples.[127] QDs which provide large Stokes shift were conjugated with aptamers which in the absence of AFB1 bound to AuNPs due to electrostatic interaction. This aptamer-induced proximity of QDs with AuNPs resulted in the quenching of fluorescence of QDs because of the quenching effect of AuNPs. However, in the presence of AFB1, due to higher binding affinity of the aptamer to its target AFB1 compared to electrostatic adsorption to AuNPs, the QD–aptamer–AFB1 conjugation was formed which resulted in the dissociation of QD–aptamer–AuNP complex. AFB1 concentration-dependent separation of QD from AuNP resulted in the proportional increase in signal output from the QDs at the excitation wavelength of 350 nm. The method resulted in the AFB1 detection in the concentration range of 10–400 nM with LOD value of 3.4 nM.

UCNPs aptasensing platforms have advantages over traditional downconversion fluorophores in terms of lack of autofluorescence, large Stokes shift, photochemical stability, and low toxicity.[128,129] High-throughput aptasensing platform for the detection of ZEN-spiked corn and beer samples was employed using fluorescent UCNPs and MNPs.[130] The aptamers were immobilized on the MNPs while the complimentary sequence for the aptamer (cDNA) was immobilized on the UCNPs. In the test solution, the cDNA hybridized with the aptamer resulting in the aptamer-MNPs/cDNA-UCNPs complex. However, in the presence of ZEN, the aptamer preferentially bound with ZEN resulting in the separation of the aptamer-MNPs/cDNA-UCNPs complex. The luminescence spectra taken in the presence and absence of ZEN for the quantitative measurement of ZEN showed a concentration-dependent decrease in fluorescent intensities and the LOD for beer and corn which were determined to be 0.007 µg/L and 0.126 µg/kg, respectively.

Several other fluorescence,[131,132] electrochemical,[126,133–136] electrochemiluminescent,[137] fluorescence polarization,[138] QDs,[139] EIS,[140] circular dichroism spectroscopy,[99] enzyme-linked oligonucleotide assay (ELONA)[141]-based enzyme-linked-based aptasensing formats had been applied for the detection of mycotoxins. Simultaneous detection of mycotoxins in a multiplex platform has been reported by different authors.[142,143]

14.4 CONCLUSION

The signaling aptamers hold great promise in nanosensor-based food safety applications. Aptasensors allow low cost, rapid, specific detection, and precise quantification of analytes to cater to the modern day challenges in food safety analysis. The advantages of aptamer-based nanosensors has been displayed on a many number of biosensing platforms. Progress in biosensors has mainly been made possible by harnessing the unique properties of aptamers which include their structure switching property and the implementation of advances in transduction technologies. The aptamer-based nanosensors provide a huge opportunity for replacing the conventional immunoassay employed in food safety applications. The high affinities, specificities, low cost, and rapid detection of aptamer-based biosensing platforms are poised to play an indispensable role in the complex global food safety net. The aptasensors for detection of mycotoxins have their own challenges because of the low concentrations of mycotoxins in the samples and also because of the matrix complexity which affects the sensitivity of the aptamer-based detection platform. However, such limitations are being overcome by designing aptamers with higher specificity and sensitivity. Technologies have been developed for aptasensing platforms which allow miniaturization and also allow such assays to be performed onsite in the absence of sophisticated laboratory equipment. The current chapter demonstrates the versatility and scope of aptamer-based nanosensing platforms in food safety by using mycotoxin detection as representative examples. Nanosensing platforms for mycotoxin detection provide reflection of the versatility of aptamer-based nanosensors that can cater to biosensing needs of the complex food safety systems. Aptasensing has effectively embraced in such assays, the custom aptamers, advances in colorimetric techniques and fluorescence-labeled probes, advances in molecular biology techniques, and the advances in transduction mechanisms used in such sensors. These advances, to name a few, have included the use of advanced nanoparticles and QDs, the use of electrochemical methods, LSPR, LRET, use of DNAzymes, advances in impedimetric biosensors, EIS, and ELONA. While aptamers have been used in a lot of nanosensing platforms, not all of the advances made in the laboratories have been able to make it to the actual market applications. There is a greater need to bridge the gap between academic advancements made in aptasensor platforms and the actual-market realization of such advancements made. The future of aptamer sensing in food safety will need advances in self-reporting nanosensing assays which can be employed at all stages of the food-supply chain. Aptasensors in conjugation with multiplexing, lab-in-the-chip, high throughput, will alter the nanosensing landscape in food safety applications in the near future with direct consequences that will alleviate the food safety concerns of the public, allow early detection of any contamination and prevent economic losses resulting from frequent recalls. Such aptasensing platforms will not only prove to be valuable tools in the animal husbandry industry but also to achieve the goal of food security and food safety for the everincreasing global population.

KEYWORDS

- **nanosensor**
- **aptamers**
- **food safety**
- **mycotoxins**

REFERENCES

1. Rasooly, A. Biosensor Technologies. *Methods (San Diego, Calif.)* **2005,** *37* (1), 1–3.
2. Jhaveri, S.; Rajendran, M.; Ellington, A. D. In Vitro Selection of Signaling Aptamers. *Nat. Biotech.* **2000,** *18* (12), 1293–1297.
3. Jayasena, S. D. Aptamers: An Emerging Class of Molecules that Rival Antibodies in Diagnostics. *Clin. Chem.* **1999,** *45* (9), 1628–1650.
4. Tuerk, C.; Gold, L. Systematic Evolution of Ligands by Exponential Enrichment: RNA Ligands to Bacteriophage T4 DNA Polymerase. *Science* **1990,** *249* (4968), 505–510.
5. Ellington, A. D.; Szostak, J. W. In Vitro Selection of RNA Molecules that Bind Specific Ligands. *Nature* **1990,** *346* (6287), 818–822.
6. Jellinek, D.; Lynott, C. K.; Rifkin, D. B.; Janjic, N. High-Affinity RNA Ligands to Basic Fibroblast Growth Factor Inhibit Receptor Binding. *Proc. Natl. Acad. Sci. U.S.A.* **1993,** *90* (23), 11227–11231.
7. Vicens, M. C.; Sen, A.; Vanderlaan, A.; Drake, T. J.; Tan, W. Investigation of Molecular Beacon Aptamer-Based Bioassay for Platelet-Derived Growth Factor Detection. *ChemBioChem* **2005,** *6* (5), 900–907.
8. Wochner, A.; Menger, M.; Orgel, D.; Cech, B.; Rimmele, M.; Erdmann, V. A.; Glokler, J. A DNA Aptamer with High Affinity and Specificity for Therapeutic Anthracyclines. *Anal. Biochem.* **2008,** *373* (1), 34–42.
9. Jenison, R. D.; Gill, S. C.; Pardi, A.; Polisky, B. High-Resolution Molecular Discrimination by RNA. *Science* **1994,** *263* (5152), 1425–1429.
10. Tombelli, S.; Minunni, M.; Mascini, M. Analytical Applications of Aptamers. *Biosens. Bioelectron.* **2005,** *20* (12), 2424–2434.
11. Sachan, A.; Ilgu, M.; Kempema, A. M.; Kraus, G. A.; Nilsen-Hamilton, M. Specificity and Ligand Affinities of the Cocaine Aptamer: Impact of Structural Features and Physiological NaCl. *Anal. Chem.* **2016,** *88* (15), 7715–7723.
12. Lupold, S. E.; Hicke, B. J.; Lin, Y.; Coffey, D. S. Identification and Characterization of Nuclease-Stabilized RNA Molecules that Bind Human Prostate Cancer Cells via the Prostate-Specific Membrane Antigen. *Cancer Res.* **2002,** *62* (14), 4029–4033.
13. Takahashi, M.; Sakota, E.; Nakamura, Y. The Efficient Cell-SELEX Strategy, Icell-SELEX, Using Isogenic Cell Lines for Selection and Counter-Selection to Generate RNA Aptamers to Cell Surface Proteins. *Biochimie* **2016,** *131*, 77–84.

14. Shao, Z.; Li, Y.; Yang, Q.; Wang, J.; Li, G. A Novel Electrochemical Method to Detect Cell Surface Carbohydrates and Target Cells. *Anal. Bioanal. Chem.* **2010**, *398* (7–8), 2963–2967.

15. Janas, T.; Janas, T. The Selection of Aptamers Specific for Membrane Molecular Targets. *Cell. Mol. Biol. Lett.* **2011**, *16* (1), 25–39.

16. Nieuwlandt, D.; Wecker, M.; Gold, L. In Vitro Selection of RNA Ligands to Substance P. *Biochemistry* **1995**, *34* (16), 5651–5659.

17. Mannironi, C.; Di Nardo, A.; Fruscoloni, P.; Tocchini-Valentini, G. P. In Vitro Selection of Dopamine RNA Ligands. *Biochemistry* **1997**, *36* (32), 9726–9734.

18. Kang, K.; Sachan, A.; Nilsen-Hamilton, M.; Shrotriya, P. Aptamer Functionalized Microcantilever Sensors for Cocaine Detection. *Langmuir* **2011**, *27* (23), 14696–14702.

19. Wrzesinski, J.; Ciesiolka, J. Characterization of Structure and Metal Ions Specificity of Co^{2+}-Binding RNA Aptamers. *Biochemistry* **2005**, *44* (16), 6257–6268.

20. Qu, H.; Csordas, A. T.; Wang, J.; Oh, S. S.; Eisenstein, M. S.; Soh, H. T. Rapid and Label-Free Strategy to Isolate Aptamers for Metal Ions. *ACS Nano* **2016**, *10* (8), 7558–7565.

21. Han, S. R.; Lee, S. W. Inhibition of Japanese Encephalitis Virus (JEV) Replication by Specific RNA Aptamer against JEV Methyltransferase. *Biochem. Biophys. Res. Commun.* **2017**, *483* (1), 687–693.

22. Kim, M. Y.; Jeong, S. Inhibition of the Functions of the Nucleocapsid Protein of Human Immunodeficiency Virus-1 by an RNA Aptamer. *Biochem. Biophys. Res. Commun.* **2004**, *320* (4), 1181–1186.

23. Gilch, S.; Schatzl, H. M. Aptamers against Prion Proteins and Prions. *Cell. Mol. Life Sci.* **2009**, *66* (15), 2445–2455.

24. Mashima, T.; Nishikawa, F.; Kamatari, Y. O.; Fujiwara, H.; Saimura, M.; Nagata, T.; Kodaki, T.; Nishikawa, S.; Kuwata, K.; Katahira, M. Anti-prion Activity of an RNA Aptamer and its Structural Basis. *Nucleic Acids Res.* **2013**, *41* (2), 1355–1362.

25. Quang, N. N.; Miodek, A.; Cibiel, A.; Duconge, F. Selection of Aptamers against Whole Living Cells: From Cell-SELEX to Identification of Biomarkers. *Methods Mol. Biol. (Clifton, N.J.)* **2017**, *1575*, 253–272.

26. Liu, X.; Zhang, X. Aptamer-Based Technology for Food Analysis. *Appl. Biochem. Biotechnol.* **2015**, *175* (1), 603–624.

27. Teng, J.; Yuan, F.; Ye, Y.; Zheng, L.; Yao, L.; Xue, F.; Chen, W.; Li, B. Aptamer-Based Technologies in Foodborne Pathogen Detection. *Front. Microbiol.* **2016**, *7*, 1426.

28. Nezlin, R. Use of Aptamers in Immunoassays. *Mol. Immunol.* **2016**, *70*, 149–154.

29. Mehta, J.; Van Dorst, B.; Rouah-Martin, E.; Herrebout, W.; Scippo, M. L.; Blust, R.; Robbens, J. In Vitro Selection and Characterization of DNA Aptamers Recognizing Chloramphenicol. *J. Biotechnol.* **2011**, *155* (4), 361–369.

30. Duan, Y.; Wang, L.; Gao, Z.; Wang, H.; Zhang, H.; Li, H. An Aptamer-Based Effective Method for Highly Sensitive Detection of Chloramphenicol Residues in Animal-Sourced Food Using Real-Time Fluorescent Quantitative PCR. *Talanta* **2017**, *165*, 671–676.

31. Berens, C.; Thain, A.; Schroeder, R. A Tetracycline-Binding RNA Aptamer. *Bioorg. Med. Chem.* **2001**, *9* (10), 2549–2556.

32. Wang, S.; Liu, J.; Yong, W.; Chen, Q.; Zhang, L.; Dong, Y.; Su, H.; Tan, T. A Direct Competitive Assay-Based Aptasensor for Sensitive Determination of Tetracycline Residue in Honey. *Talanta* **2015**, *131*, 562–569.

33. Taghdisi, S. M.; Danesh, N. M.; Nameghi, M. A.; Ramezani, M.; Abnous, K. A Label-Free Fluorescent Aptasensor for Selective and Sensitive Detection of Streptomycin in Milk and Blood Serum. *Food Chem.* **2016**, *203*, 145–149.

34. Zhou, N.; Wang, J.; Zhang, J.; Li, C.; Tian, Y. Selection and Identification of Strepto-mycin-Specific Single-Stranded DNA Aptamers and the Application in the Detection of Streptomycin in Honey. *Talanta* **2013**, *108*, 109–116.

35. Zhou, C. S.; Jiang, Y. X.; Wang, J.; Ma, B. C.; Li, M. L.; Fang, X. H. High Sensitive Tobramycin Detection Using a Novel Signaling Aptamer. *Chem. J. Chin. Univ.—Chin.* **2006**, *27* (5), 826–829.

36. Wang, C.; Wang, Q.; Chen, D. Resonance Light Scattering Method for Detecting Kanamycin in Milk with Enhanced Sensitivity. *Anal. Bioanal. Chem.* **2017**, *409* (11), 2839–2846.

37. Ha, N. R.; Jung, I. P.; La, I. J.; Jung, H. S.; Yoon, M. Y. Ultra-sensitive Detection of Kanamycin for Food Safety Using a Reduced Graphene Oxide-Based Fluorescent Apta-sensor. *Sci. Rep.* **2017**, *7*, 40305.

38. Song, K. M.; Cho, M.; Jo, H.; Min, K.; Jeon, S. H.; Kim, T.; Han, M. S.; Ku, J. K.; Ban, C. Gold Nanoparticle-Based Colorimetric Detection of Kanamycin Using a DNA Aptamer. *Anal. Biochem.* **2011**, *415* (2), 175–181.

39. Ling, K.; Jiang, H.; Zhang, L.; Li, Y.; Yang, L.; Qiu, C.; Li, F. R. A Self-assembling RNA Aptamer-Based Nanoparticle Sensor for Fluorometric Detection of Neomycin B in Milk. *Anal. Bioanal. Chem.* **2016**, *408* (13), 3593–3600.

40. Cao, F.; Lu, X.; Hu, X.; Zhang, Y.; Zeng, L.; Chen, L.; Sun, M. In vitro Selection of DNA Aptamers Binding Pesticide Fluoroacetamide. *Biosci. Biotechnol. Biochem.* **2016**, *80* (5), 823–832.

41. Bala, R.; Sharma, R. K.; Wangoo, N. Development of Gold Nanoparticles-Based Apta-sensor for the Colorimetric Detection of Organophosphorus Pesticide Phorate. *Anal. Bioanal. Chem.* **2016**, *408* (1), 333–338.

42. Wang, L.; Liu, X.; Zhang, Q.; Zhang, C.; Liu, Y.; Tu, K.; Tu, J. Selection of DNA Aptamers that Bind to Four Organophosphorus Pesticides. *Biotechnol. Lett.* **2012**, *34* (5), 869–874.

43. He, J.; Liu, Y.; Fan, M.; Liu, X. Isolation and Identification of the DNA Aptamer Target to Acetamiprid. *J. Agric. Food Chem.* **2011**, *59* (5), 1582–1586.

44. Tian, Y.; Wang, Y.; Sheng, Z.; Li, T.; Li, X. A Colorimetric Detection Method of Pesti-cide Acetamiprid by Fine-Tuning Aptamer Length. *Anal. Biochem.* **2016**, *513*, 87–92.

45. Bai, W.; Zhu, C.; Liu, J.; Yan, M.; Yang, S.; Chen, A. Gold Nanoparticle-Based Colori-metric Aptasensor for Rapid Detection of Six Organophosphorous Pesticides. *Environ. Toxicol. Chem.* **2015**, *34* (10), 2244–2249.

46. Feng, X.; Gan, N.; Zhang, H.; Yan, Q.; Li, T.; Cao, Y.; Hu, F.; Yu, H.; Jiang, Q. A Novel "Dual-Potential" Electrochemiluminescence Aptasensor Array Using CdS Quantum Dots and Luminol–Gold Nanoparticles as Labels for Simultaneous Detection of Mala-chite Green and Chloramphenicol. *Biosens. Bioelectron.* **2015**, *74*, 587–593.

47. Stead, S. L.; Ashwin, H.; Johnston, B.; Dallas, A.; Kazakov, S. A.; Tarbin, J. A.; Sharman, M.; Kay, J.; Keely, B. J. An RNA-Aptamer-Based Assay for the Detection and Analysis of Malachite Green and Leucomalachite Green Residues in Fish Tissue. *Anal. Chem.* **2010**, *82* (7), 2652–2660.

48. Ye, B. F.; Zhao, Y. J.; Cheng, Y.; Li, T. T.; Xie, Z. Y.; Zhao, X. W.; Gu, Z. Z. Colorimetric Photonic Hydrogel Aptasensor for the Screening of Heavy Metal Ions. *Nanoscale* **2012**, *4* (19), 5998–6003.

49. Liu, C. W.; Huang, C. C.; Chang, H. T. Highly Selective DNA-Based Sensor for Lead(II) and Mercury(II) Ions. *Anal. Chem.* **2009**, *81* (6), 2383–2387.

50. Wu, S.; Duan, N.; Shi, Z.; Fang, C.; Wang, Z. Dual Fluorescence Resonance Energy Transfer Assay between Tunable Upconversion Nanoparticles and Controlled Gold Nanoparticles for the Simultaneous Detection of Pb(2)(+) and Hg(2)(+). *Talanta* **2014,** *128*, 327–336.

51. Kaur, H.; Kumar, R.; Babu, J. N.; Mittal, S. Advances in Arsenic Biosensor Development—A Comprehensive Review. *Biosens. Bioelectron.* **2015,** *63*, 533–545.

52. Song, L.; Mao, K.; Zhou, X.; Hu, J. A Novel Biosensor Based on Au@Ag Core–Shell Nanoparticles for SERS Detection of Arsenic(III). *Talanta* **2016,** *146*, 285–290.

53. Mirzajani, H.; Cheng, C.; Wu, J.; Chen, J.; Eda, S.; Najafi Aghdam, E.; Badri Ghavifekr, H. A Highly Sensitive and Specific Capacitive Aptasensor for Rapid and Label-Free Trace Analysis of Bisphenol A (BPA) in Canned Foods. *Biosens. Bioelectron.* **2017,** *89* (Pt. 2), 1059–1067.

54. Chung, E.; Jeon, J.; Yu, J.; Lee, C.; Choo, J. Surface-Enhanced Raman Scattering Aptasensor for Ultrasensitive Trace Analysis of Bisphenol A. *Biosens. Bioelectron.* **2015,** *64*, 560–565.

55. Gu, C.; Xiang, Y.; Guo, H.; Shi, H. Label-Free Fluorescence Detection of Melamine with a Truncated Aptamer. *Analyst* **2016,** *141* (14), 4511–4517.

56. Duan, N.; Wu, S.; Zhu, C.; Ma, X.; Wang, Z.; Yu, Y.; Jiang, Y. Dual-Color Upconversion Fluorescence and Aptamer-Functionalized Magnetic Nanoparticles-Based Bioassay for the Simultaneous Detection of *Salmonella typhimurium* and *Staphylococcus aureus*. *Anal. Chim. Acta* **2012,** *723*, 1–6.

57. McMasters, S.; Stratis-Cullum, D. N. Evaluation of Aptamers as Molecular Recognition Elements for Pathogens Using Capillary Electrophoretic Analysis. In *Smart Medical and Biomedical Sensor Technology IV*; Cullum, B. M., Carter, J. C., Eds.; 2006; vol. 6380, pp U83–U90.

58. Dwivedi, H. P.; Smiley, R. D.; Jaykus, L. A. Selection and Characterization of DNA Aptamers with Binding Selectivity to *Campylobacter jejuni* Using Whole-Cell SELEX. *Appl. Microbiol. Biotechnol.* **2010,** *87* (6), 2323–2334.

59. Dua, P.; Ren, S.; Lee, S. W.; Kim, J. K.; Shin, H. S.; Jeong, O. C.; Kim, S.; Lee, D. K. Cell-SELEX Based Identification of an RNA Aptamer for *Escherichia coli* and Its Use in Various Detection Formats. *Mol. Cells* **2016,** *39* (11), 807–813.

60. Zhang, L.; Huang, R.; Liu, W.; Liu, H.; Zhou, X.; Xing, D. Rapid and Visual Detection of *Listeria monocytogenes* Based on Nanoparticle Cluster Catalyzed Signal Amplification. *Biosens. Bioelectron.* **2016,** *86*, 1–7.

61. Ma, X.; Song, L.; Zhou, N.; Xia, Y.; Wang, Z. A Novel Aptasensor for the Colorimetric Detection of *S. typhimurium* Based on Gold Nanoparticles. *Int. J. Food Microbiol.* **2017,** *245*, 1–5.

62. Sheikhzadeh, E.; Chamsaz, M.; Turner, A. P.; Jager, E. W.; Beni, V. Label-Free Impedimetric Biosensor for *Salmonella typhimurium* Detection Based on Poly[Pyrrole-*co*-3-Carboxyl-Pyrrole] Copolymer Supported Aptamer. *Biosens. Bioelectron.* **2016,** *80*, 194–200.

63. Wang, X.; Huang, Y.; Wu, S.; Duan, N.; Xu, B.; Wang, Z. Simultaneous Detection of *Staphylococcus aureus* and *Salmonella typhimurium* Using Multicolor Time-Resolved Fluorescence Nanoparticles as Labels. *Int. J. Food Microbiol.* **2016,** *237*, 172–179.

64. Baumstummler, A.; Lehmann, D.; Janjic, N.; Ochsner, U. A. Specific Capture and Detection of *Staphylococcus aureus* with High-Affinity Modified Aptamers to Cell Surface Components. *Lett. Appl. Microbiol.* **2014,** *59* (4), 422–431.

65. Fetter, L.; Richards, J.; Daniel, J.; Roon, L.; Rowland, T. J.; Bonham, A. J. Electrochemical Aptamer Scaffold Biosensors for Detection of Botulism and Ricin Toxins. *Chem. Commun. (Camb.)* **2015,** *51* (82), 15137–15140.

66. Holeman, L. A.; Robinson, S. L.; Szostak, J. W.; Wilson, C. Isolation and Characterization of Fluorophore-Binding RNA Aptamers. *Fold. Des.* **1998,** *3* (6), 423–431.

67. Boushell, V.; Pang, S.; He, L. Aptamer-Based SERS Detection of Lysozyme on a Food-Handling Surface. *J. Food Sci.* **2017,** *82* (1), 225–231.

68. Tran, D. T.; Janssen, K. P.; Pollet, J.; Lammertyn, E.; Anne, J.; Van Schepdael, A.; Lammertyn, J. Selection and Characterization of DNA Aptamers for Egg White Lysozyme. *Molecules* **2010,** *15* (3), 1127–1140.

69. Amaya-Gonzalez, S.; de-Los-Santos-Alvarez, N.; Miranda-Ordieres, A. J.; Lobo-Castanon, M. J. Sensitive Gluten Determination in Gluten-Free Foods by an Electrochemical Aptamer-Based Assay. *Anal. Bioanal. Chem.* **2015,** *407* (20), 6021–6029.

70. Hu, J.; Ni, P.; Dai, H.; Sun, Y.; Wang, Y.; Jiang, S.; Li, Z. Aptamer-Based Colorimetric Biosensing of Abrin Using Catalytic Gold Nanoparticles. *Analyst* **2015,** *140* (10), 3581–3586.

71. Tang, J. J.; Yu, T.; Guo, L.; Xie, J. W.; Shao, N. S.; He, Z. K. In Vitro Selection of DNA Aptamer against Abrin Toxin and Aptamer-Based Abrin Direct Detection. *Biosens. Bioelectron.* **2007,** *22* (11), 2456–2463.

72. Lamont, E. A.; He, L.; Warriner, K.; Labuza, T. P.; Sreevatsan, S. A Single DNA Aptamer Functions as a Biosensor for Ricin. *Analyst* **2011,** *136* (19), 3884–3895.

73. Ding, S.; Gao, C.; Gu, L. Q. Capturing Single Molecules of Immunoglobulin and Ricin with an Aptamer-Encoded Glass Nanopore. *Anal. Chem.* **2009,** *81* (16), 6649–6655.

74. Yildirim, N.; Long, F.; Gao, C.; He, M.; Shi, H. C.; Gu, A. Z. Aptamer-Based Optical Biosensor for Rapid and Sensitive Detection of 17Beta-estradiol in Water Samples. *Environ. Sci. Technol.* **2012,** *46* (6), 3288–3294.

75. Escudero-Abarca, B. I.; Suh, S. H.; Moore, M. D.; Dwivedi, H. P.; Jaykus, L. A. Selection, Characterization and Application of Nucleic Acid Aptamers for the Capture and Detection of Human Norovirus Strains. *PLoS ONE* **2014,** *9* (9), e106805.

76. Moore, M. D.; Escudero-Abarca, B. I.; Suh, S. H.; Jaykus, L. A. Generation and Characterization of Nucleic Acid Aptamers Targeting the Capsid P Domain of a Human Norovirus GII. Strain. *J. Biotechnol.* **2015,** *209*, 41–49.

77. Xiao, S. J.; Hu, P. P.; Wu, X. D.; Zou, Y. L.; Chen, L. Q.; Peng, L.; Ling, J.; Zhen, S. J.; Zhan, L.; Li, Y. F.; Huang, C. Z. Sensitive Discrimination and Detection of Prion Disease-Associated Isoform with a Dual-Aptamer Strategy by Developing a Sandwich Structure of Magnetic Microparticles and Quantum Dots. *Anal. Chem.* **2010,** *82* (23), 9736–9742.

78. Duan, Y.; Gao, Z.; Wang, L.; Wang, H.; Zhang, H.; Li, H. Selection and Identification of Chloramphenicol-Specific DNA Aptamers by Mag-SELEX. *Appl. Biochem. Biotechnol.* **2016,** *180* (8), 1644–1656.

79. Gotrik, M. R.; Feagin, T. A.; Csordas, A. T.; Nakamoto, M. A.; Soh, H. T. Advancements in Aptamer Discovery Technologies. *Acc. Chem. Res.* **2016,** *49* (9), 1903–1910.

80. Wu, Y. X.; Kwon, Y. J. Aptamers: The "Evolution" of SELEX. *Methods (San Diego, Calif.)* **2016,** *106*, 21–28.

81. Darmostuk, M.; Rimpelova, S.; Gbelcova, H.; Ruml, T. Current Approaches in SELEX: An Update to Aptamer Selection Technology. *Biotechnol. Adv.* **2015,** *33* (6 Pt. 2), 1141–1161.

82. Yan, A.; Levy, M. Cell Internalization SELEX: In Vitro Selection for Molecules that Internalize into Cells. *Methods Mol. Biol. (Clifton, N.J.)* **2014**, *1103*, 241–265.

83. Shamah, S. M.; Healy, J. M.; Cload, S. T. Complex Target SELEX. *Acc. Chem. Res.* **2008**, *41* (1), 130–138.

84. Djordjevic, M. SELEX Experiments: New Prospects, Applications and Data Analysis in Inferring Regulatory Pathways. *Biomol. Eng.* **2007**, *24* (2), 179–189.

85. Gopinath, S. C. Methods Developed for SELEX. *Anal. Bioanal. Chem.* **2007**, *387* (1), 171–182.

86. Gold, L.; Polisky, B.; Uhlenbeck, O.; Yarus, M. Diversity of Oligonucleotide Functions. *Annu. Rev. Biochem.* **1995**, *64*, 763–797.

87. Famulok, M.; Mayer, G. Aptamers as Tools in Molecular Biology and Immunology. In *Combinatorial Chemistry in Biology*; Famulok, M., Winnacker, E.-L., Wong, C.-H., Eds.; Springer Berlin Heidelberg: Berlin, Heidelberg, 1999; pp 123–136.

88. Singh, G.; Manohar, M.; Adegoke, A. A.; Stenström, T. A.; Shanker, R. Novel Aptamer-Linked Nanoconjugate Approach for Detection of Waterborne Bacterial Pathogens: An Update. *J. Nanopart. Res.* **2016**, *19* (1), 4.

89. Zimmermann, G. R.; Wick, C. L.; Shields, T. P.; Jenison, R. D.; Pardi, A. Molecular Interactions and Metal Binding in the Theophylline-Binding Core of an RNA Aptamer. *RNA* **2000**, *6* (5), 659–667.

90. Kuiper-Goodman, T. Risk Assessment to Humans of Mycotoxins in Animal-Derived Food Products. *Vet. Hum. Toxicol.* **1991**, *33* (4), 325–332; discussion 332–333.

91. Williams, P. P. Effects of T-2 Mycotoxin on Gastrointestinal Tissues: A Review of In Vivo and In Vitro Models. *Arch. Environ. Contam. Toxicol.* **1989**, *18* (3), 374–387.

92. Rotter, B. A.; Prelusky, D. B.; Pestka, J. J. Toxicology of Deoxynivalenol (Vomitoxin). *J. Toxicol. Environ. Health* **1996**, *48* (1), 1–34.

93. Meulenberg, E. P. Immunochemical Methods for Ochratoxin A Detection: A Review. *Toxins* **2012**, *4* (4), 244–266.

94. Klich, M. A. Soil Fungi of Some Low-Altitude Desert Cotton Fields and Ability of Their Extracts to Inhibit *Aspergillus flavus*. *Mycopathologia* **1998**, *142* (2), 97–100.

95. GR, O. B.; Fakhoury, A. M.; Payne, G. A. Identification of Genes Differentially Expressed during Aflatoxin Biosynthesis in *Aspergillus flavus* and *Aspergillus parasiticus*. *Fungal Genet. Biol.: FG&B* **2003**, *39* (2), 118–127.

96. Hussein, H. S.; Brasel, J. M. Toxicity, Metabolism, and Impact of Mycotoxins on Humans and Animals. *Toxicology* **2001**, *167* (2), 101–134.

97. Stinson, E. E.; Osman, S. F.; Heisler, E. G.; Siciliano, J.; Bills, D. D. Mycotoxin Production in Whole Tomatoes, Apples, Oranges, and Lemons. *J. Agric. Food Chem.* **1981**, *29* (4), 790–792.

98. Klaric, M. S.; Rasic, D.; Peraica, M. Deleterious Effects of Mycotoxin Combinations Involving Ochratoxin A. *Toxins* **2013**, *5* (11), 1965–1987.

99. Chen, X.; Huang, Y.; Duan, N.; Wu, S.; Xia, Y.; Ma, X.; Zhu, C.; Jiang, Y.; Wang, Z. Screening and Identification of DNA Aptamers against T-2 Toxin Assisted by Graphene Oxide. *J. Agric. Food. Chem.* **2014**, *62* (42), 10368–10374.

100. Boenke, A. Method Validation for Mycotoxin Determinations in Food and Feedstuffs. *TrAC: Trends Anal. Chem.* **1998**, *17* (1), 10–17.

101. Gilbert, J.; Anklam, E. Validation of Analytical Methods for Determining Mycotoxins in Foodstuffs. *TrAC: Trends Anal. Chem.* **2002**, *21* (6–7), 468–486.

102. Turner, N. W.; Subrahmanyam, S.; Piletsky, S. A. Analytical Methods for Determination of Mycotoxins: A Review. *Anal. Chim. Acta* **2009**, *632* (2), 168–180.

103. Nielsen, K. F. Mycotoxin Production by Indoor Molds. *Fung. Genet. Biol.* **2003,** *39* (2), 103–117.

104. Cruz-Aguado, J. A.; Penner, G. Determination of Ochratoxin A with a DNA Aptamer. *J. Agric. Food Chem.* **2008,** *56* (22), 10456–10461.

105. Chen, X.; Huang, Y.; Duan, N.; Wu, S.; Ma, X.; Xia, Y.; Zhu, C.; Jiang, Y.; Wang, Z. Selection and Identification of ssDNA Aptamers Recognizing Zearalenone. *Anal. Bioanal. Chem.* **2013,** *405* (20), 6573–6581.

106. Chen, J.; Zhang, X.; Cai, S.; Wu, D.; Chen, M.; Wang, S.; Zhang, J. A Fluorescent Aptasensor Based on DNA-Scaffolded Silver-Nanocluster for Ochratoxin A Detection. *Biosens. Bioelectron.* **2014,** *57,* 226–231.

107. Park, J. H.; Byun, J. Y.; Mun, H.; Shim, W. B.; Shin, Y. B.; Li, T.; Kim, M. G. A Regeneratable, Label-Free, Localized Surface Plasmon Resonance (LSPR) Aptasensor for the Detection of Ochratoxin A. *Biosens. Bioelectron.* **2014,** *59,* 321–327.

108. Chen, J.; Fang, Z.; Liu, J.; Zeng, L. A Simple and Rapid Biosensor for Ochratoxin A Based on a Structure-Switching Signaling Aptamer. *Food Control* **2012,** *25* (2), 555–560.

109. Zhu, Z.; Feng, M.; Zuo, L.; Zhu, Z.; Wang, F.; Chen, L.; Li, J.; Shan, G.; Luo, S. Z. An Aptamer Based Surface Plasmon Resonance Biosensor for the Detection of Ochratoxin A in Wine and Peanut Oil. *Biosens. Bioelectron.* **2015,** *65,* 320–326.

110. Zhang, J.; Chen, J.; Zhang, X.; Zeng, Z.; Chen, M.; Wang, S. An Electrochemical Biosensor Based on Hairpin–DNA Aptamer Probe and Restriction Endonuclease for Ochratoxin A Detection. *Electrochem. Commun.* **2012,** *25,* 5–7.

111. Wang, C.; Dong, X.; Liu, Q.; Wang, K. Label-Free Colorimetric Aptasensor for Sensitive Detection of Ochratoxin A Utilizing Hybridization Chain Reaction. *Anal. Chim. Acta* **2015,** *860,* 83–88.

112. Yang, C.; Lates, V.; Prieto-Simon, B.; Marty, J. L.; Yang, X. Aptamer-DNAzyme Hairpins for Biosensing of Ochratoxin A. *Biosens. Bioelectron.* **2012,** *32* (1), 208–212.

113. Wu, X.; Hu, J.; Zhu, B.; Lu, L.; Huang, X.; Pang, D. Aptamer-Targeted Magnetic Nanospheres as a Solid-Phase Extraction Sorbent for Determination of Ochratoxin A in Food Samples. *J. Chromatogr. A* **2011,** *1218* (41), 7341–7346.

114. Wu, S.; Duan, N.; Wang, Z.; Wang, H. Aptamer-Functionalized Magnetic Nanoparticle-Based Bioassay for the Detection of Ochratoxin A Using Upconversion Nanoparticles as Labels. *Analyst* **2011,** *136* (11), 2306–2314.

115. Dai, S.; Wu, S.; Duan, N.; Chen, J.; Zheng, Z.; Wang, Z. An ultrasensitive aptasensor for Ochratoxin A Using Hexagonal Core/Shell Upconversion Nanoparticles as Luminophores. *Biosens. Bioelectron.* **2017,** *91,* 538–544.

116. Mishra, R. K.; Hayat, A.; Catanante, G.; Ocana, C.; Marty, J. L. A Label-Free Aptasensor for Ochratoxin A Detection in Cocoa Beans: An Application to Chocolate Industries. *Anal. Chim. Acta* **2015,** *889,* 106–112.

117. Malhotra, S.; Pandey, A. K.; Rajput, Y. S.; Sharma, R. Selection of Aptamers for Aflatoxin M1 and their Characterization. *J. Mol. Recogn.* **2014,** *27* (8), 493–500.

118. Guo, X.; Wen, F.; Zheng, N.; Li, S.; Fauconnier, M. L.; Wang, J. A qPCR Aptasensor for Sensitive Detection of Aflatoxin M1. *Anal. Bioanal. Chem.* **2016,** *408* (20), 5577–5584.

119. Istamboulie, G.; Paniel, N.; Zara, L.; Reguillo Granados, L.; Barthelmebs, L.; Noguer, T. Development of an Impedimetric Aptasensor for the Determination of Aflatoxin M1 in Milk. *Talanta* **2016,** *146,* 464–469.

120. Seok, Y.; Byun, J. Y.; Shim, W. B.; Kim, M. G. A Structure-Switchable Aptasensor for Aflatoxin B1 Detection Based on Assembly of an Aptamer/Split DNAzyme. *Anal. Chim. Acta* **2015,** *886,* 182–187.

121. Shim, W. B.; Kim, M. J.; Mun, H.; Kim, M. G. An Aptamer-Based Dipstick Assay for the Rapid and Simple Detection of Aflatoxin B1. *Biosens. Bioelectron.* **2014,** *62,* 288–294.

122. Yang, C.; Wang, Y.; Marty, J. L.; Yang, X. Aptamer-Based Colorimetric Biosensing of Ochratoxin A Using Unmodified Gold Nanoparticles Indicator. *Biosens. Bioelectron.* **2011,** *26* (5), 2724–2727.

123. Rouah-Martin, E.; Mehta, J.; van Dorst, B.; de Saeger, S.; Dubruel, P.; Maes, B. U.; Lemiere, F.; Goormaghtigh, E.; Daems, D.; Herrebout, W.; van Hove, F.; Blust, R.; Robbens, J. Aptamer-Based Molecular Recognition of Lysergamine, Metergoline and Small Ergot Alkaloids. *Int. J. Mol. Sci.* **2012,** *13* (12), 17138–17159.

124. Hosseini, M.; Khabbaz, H.; Dadmehr, M.; Ganjali, M. R.; Mohamadnejad, J. Aptamer-based Colorimetric and Chemiluminescence Detection of Aflatoxin B1 in Foods Samples. *Acta Chim. Slov.* **2015,** *62* (3), 721–728.

125. Guo, X.; Wen, F.; Zheng, N.; Luo, Q.; Wang, H.; Wang, H.; Li, S.; Wang, J. Development of an Ultrasensitive Aptasensor for the Detection of Aflatoxin B1. *Biosens. Bioelectron.* **2014,** *56,* 340–344.

126. Kuang, H.; Chen, W.; Xu, D.; Xu, L.; Zhu, Y.; Liu, L.; Chu, H.; Peng, C.; Xu, C.; Zhu, S. Fabricated Aptamer-Based Electrochemical "signal-off" Sensor of Ochratoxin A. *Biosens. Bioelectron.* **2010,** *26* (2), 710–716.

127. Sabet, F. S.; Hosseini, M.; Khabbaz, H.; Dadmehr, M.; Ganjali, M. R. FRET-Based Aptamer Biosensor for Selective and Sensitive Detection of Aflatoxin B1 in Peanut and Rice. *Food Chem.* **2017,** *220,* 527–532.

128. Wu, S.; Duan, N.; Li, X.; Tan, G.; Ma, X.; Xia, Y.; Wang, Z.; Wang, H. Homogenous Detection of Fumonisin B(1) with a Molecular Beacon Based on Fluorescence Resonance Energy Transfer between NaYF4:Yb, Ho Upconversion Nanoparticles and Gold Nanoparticles. *Talanta* **2013,** *116,* 611–618.

129. Wu, S.; Duan, N.; Zhu, C.; Ma, X.; Wang, M.; Wang, Z. Magnetic Nanobead-Based Immunoassay for the Simultaneous Detection of Aflatoxin B1 and Ochratoxin A Using Upconversion Nanoparticles as Multicolor Labels. *Biosens. Bioelectron.* **2011,** *30* (1), 35–42.

130. Wu, Z.; Xu, E.; Chughtai, M. F.; Jin, Z.; Irudayaraj, J. Highly Sensitive Fluorescence Sensing of Zearalenone Using a Novel Aptasensor Based on Upconverting Nanoparticles. *Food Chem.* **2017,** *230,* 673–680.

131. Duan, N.; Wu, S.-J.; Wang, Z.-P. An Aptamer-based Fluorescence Assay for Ochratoxin A. *Chin. J. Anal. Chem.* **2011,** *39* (3), 300–304.

132. Wu, S.; Duan, N.; Zhang, W.; Zhao, S.; Wang, Z. Screening and Development of DNA Aptamers as Capture Probes for Colorimetric Detection of Patulin. *Anal. Biochem.* **2016,** *508,* 58–64.

133. Bonel, L.; Vidal, J. C.; Duato, P.; Castillo, J. R. An Electrochemical Competitive Biosensor for Ochratoxin A Based on a DNA Biotinylated Aptamer. *Biosens. Bioelectron.* **2011,** *26* (7), 3254–3259.

134. Zheng, W.; Teng, J.; Cheng, L.; Ye, Y.; Pan, D.; Wu, J.; Xue, F.; Liu, G.; Chen, W. Hetero-Enzyme-Based Two-Round Signal Amplification Strategy for Trace Detection of Aflatoxin B1 Using an Electrochemical Aptasensor. *Biosens. Bioelectron.* **2016,** *80,* 574–581.

135. Huang, L.; Wu, J.; Zheng, L.; Qian, H.; Xue, F.; Wu, Y.; Pan, D.; Adeloju, S. B.; Chen, W. Rolling Chain Amplification Based Signal-Enhanced Electrochemical Aptasensor for Ultrasensitive Detection of Ochratoxin A. *Anal. Chem.* **2013,** *85* (22), 10842–10849.

136. Wu, J.; Chu, H.; Mei, Z.; Deng, Y.; Xue, F.; Zheng, L.; Chen, W. Ultrasensitive One-Step Rapid Detection of Ochratoxin A by the Folding-Based Electrochemical Aptasensor. *Anal. Chim. Acta* **2012**, *753*, 27–31.

137. Wang, Z.; Duan, N.; Hun, X.; Wu, S. Electrochemiluminescent Aptamer Biosensor for the Determination of Ochratoxin A at a Gold-Nanoparticles-Modified Gold Electrode Using *N*-(aminobutyl)-*N*-Ethylisoluminol as a Luminescent Label. *Anal. Bioanal. Chem.* **2010**, *398* (5), 2125–2132.

138. Cruz-Aguado, J. A.; Penner, G. Fluorescence Polarization Based Displacement Assay for the Determination of Small Molecules with Aptamers. *Anal. Chem.* **2008**, *80* (22), 8853–8855.

139. Wang, L.; Chen, W.; Ma, W.; Liu, L.; Zhao, Y.; Zhu, Y.; Xu, L.; Kuang, H.; Xu, C. Fluorescent Strip Sensor for Rapid Determination of Toxins. *Chem. Commun. (Camb.)* **2011**, *47* (5), 1574–1576.

140. Rivas, L.; Mayorga-Martinez, C. C.; Quesada-Gonzalez, D.; Zamora-Galvez, A.; de la Escosura-Muniz, A.; Merkoci, A. Label-Free Impedimetric Aptasensor for Ochratoxin-A Detection Using Iridium Oxide Nanoparticles. *Anal. Chem.* **2015**, *87* (10), 5167–5172.

141. Wang, Y. K.; Zou, Q.; Sun, J. H.; Wang, H. A.; Sun, X.; Chen, Z. F.; Yan, Y. X. Screening of Single-Stranded DNA (ssDNA) Aptamers against a Zearalenone Monoclonal Antibody and Development of a ssDNA-Based Enzyme-Linked Oligonucleotide Assay for Determination of Zearalenone in Corn. *J. Agric. Food. Chem.* **2015**, *63* (1), 136–141.

142. Wang, C.; Qian, J.; An, K.; Huang, X.; Zhao, L.; Liu, Q.; Hao, N.; Wang, K. Magneto-Controlled Aptasensor for Simultaneous Electrochemical Detection of Dual Mycotoxins in Maize Using Metal Sulfide Quantum Dots Coated Silica as Labels. *Biosens. Bioelectron.* **2017**, *89* (Pt 2), 802–809.

143. Wu, S.; Duan, N.; Ma, X.; Xia, Y.; Wang, H.; Wang, Z.; Zhang, Q. Multiplexed Fluorescence Resonance Energy Transfer Aptasensor between Upconversion Nanoparticles and Graphene Oxide for the Simultaneous Determination of Mycotoxins. *Anal. Chem.* **2012**, *84* (14), 6263–6270.

INDEX

Printed and bound by CPI Group (UK) Ltd, Croydon, CR0 4YY

23/10/2024

01777705-0011